Lecture Notes in Computer Science

Lecture Notes in Computer Science

Vol. 296: R. Janßen (Ed.), Trends in Computer Algebra. Proceedings, 1987. V, 197 pages. 1988.

Vol. 297: E.N. Houstis, T.S. Papatheodorou, C.D. Polychronopoulos (Eds.), Supercomputing. Proceedings, 1987. X, 1093 pages. 1988.

Vol. 298: M. Main, A. Melton, M. Mislove, D. Schmidt (Eds.), Mathematical Foundations of Programming Language Semantics. Proceedings, 1987. VIII, 637 pages. 1988.

Vol. 299: M. Dauchet, M. Nivat (Eds.), CAAP '88. Proceedings, 1988. VI, 304 pages. 1988.

Vol. 300: H. Ganzinger (Ed.), ESOP '88. Proceedings, 1988. VI, 381 pages. 1988.

Vol. 301: J. Kittler (Ed.), Pattern Recognition. Proceedings, 1988. VII, 668 pages. 1988.

Vol. 302: D.M. Yellin, Attribute Grammar Inversion and Source-to-source Translation. VIII, 176 pages. 1988.

Vol. 303: J.W. Schmidt, S. Ceri, M. Missikoff (Eds.), Advances in Database Technology – EDBT '88. X, 620 pages. 1988.

Vol. 304: W.L. Price, D. Chaum (Eds.), Advances in Cryptology – EUROCRYPT '87. Proceedings, 1987. VII, 314 pages. 1988.

Vol. 305: J. Biskup, J. Demetrovics, J. Paredaens, B. Thalheim (Eds.), MFDBS 87. Proceedings, 1987. V, 247 pages. 1988.

Vol. 306: M. Boscarol, L. Carlucci Aiello, G. Levi (Eds.), Foundations of Logic and Functional Programming. Proceedings, 1986. V, 218 pages. 1988.

Vol. 307: Th. Beth, M. Clausen (Eds.), Applicable Algebra, Error-Correcting Codes, Combinatorics and Computer Algebra. Proceedings, 1986. VI, 215 pages. 1988.

Vol. 308: S. Kaplan, J.-P. Jouannaud (Eds.), Conditional Term Rewriting Systems. Proceedings, 1987. VI, 278 pages. 1988.

Vol. 309: J. Nehmer (Ed.), Experiences with Distributed Systems. Proceedings, 1987. VI, 292 pages. 1988.

Vol. 310: E. Lusk, R. Overbeek (Eds.), 9th International Conference on Automated Deduction. Proceedings, 1988. X, 775 pages. 1988.

Vol. 311: G. Cohen, P. Godlewski (Eds.), Coding Theory and Applications 1986. Proceedings, 1986. XIV, 196 pages. 1988.

Vol. 312: J. van Leeuwen (Ed.), Distributed Algorithms 1987. Proceedings, 1987. VII, 430 pages. 1988.

Vol. 313: B. Bouchon, L. Saitta, R.R. Yager (Eds.), Uncertainty and Intelligent Systems. IPMU '88. Proceedings, 1988. VIII, 408 pages. 1988.

Vol. 314: H. Göttler, H.J. Schneider (Eds.), Graph-Theoretic Concepts in Computer Science. Proceedings, 1987. VI, 254 pages. 1988.

Vol. 315: K. Furukawa, H. Tanaka, T. Fujisaki (Eds.), Logic Programming '87. Proceedings, 1987. VI, 327 pages. 1988.

Vol. 316: C. Choffrut (Ed.), Automata Networks. Proceedings, 1986. VII, 125 pages. 1988.

Vol. 317: T. Lepistö, A. Salomaa (Eds.), Automata, Languages and Programming. Proceedings, 1988. XI, 741 pages. 1988.

Vol. 318: R. Karlsson, A. Lingas (Eds.), SWAT 88. Proceedings, 1988. VI, 262 pages. 1988.

Vol. 319: J.H. Reif (Ed.), VLSI Algorithms and Architectures – AWOC 88. Proceedings, 1988. X, 476 pages. 1988.

Vol. 320: A. Blaser (Ed.), Natural Language at the Computer. Proceedings, 1988. III, 176 pages. 1988.

Vol. 321: J. Zwiers, Compositionality, Concurrency and Partial Correctness. VI, 272 pages. 1989.

Vol. 322: S. Gjessing, K. Nygaard (Eds.), ECOOP '88. European Conference on Object-Oriented Programming. Proceedings, 1988. VI, 410 pages. 1988.

Vol. 323: P. Deransart, M. Jourdan, B. Lorho, Attribute Grammars. IX, 232 pages. 1988.

Vol. 324: M.P. Chytil, L. Janiga, V. Koubek (Eds.), Mathematical Foundations of Computer Science 1988. Proceedings. IX, 562 pages. 1988.

Vol. 325: G. Brassard, Modern Cryptology. VI, 107 pages. 1988.

Vol. 326: M. Gyssens, J. Paredaens, D. Van Gucht (Eds.), ICDT '88. 2nd International Conference on Database Theory. Proceedings, 1988. VI, 409 pages. 1988.

Vol. 327: G.A. Ford (Ed.), Software Engineering Education. Proceedings, 1988. V, 207 pages. 1988.

Vol. 328: R. Bloomfield, L. Marshall, R. Jones (Eds.), VDM '88. VDM – The Way Ahead. Proceedings, 1988. IX, 499 pages. 1988.

Vol. 329: E. Börger, H. Kleine Büning, M.M. Richter (Eds.), CSL '87. 1st Workshop on Computer Science Logic. Proceedings, 1987. VI, 346 pages. 1988.

Vol. 330: C.G. Günther (Ed.), Advances in Cryptology – EUROCRYPT '88. Proceedings, 1988. XI, 473 pages. 1988.

Vol. 331: M. Joseph (Ed.), Formal Techniques in Real-Time and Fault-Tolerant Systems. Proceedings, 1988. VI, 229 pages. 1988.

Vol. 332: D. Sannella, A. Tarlecki (Eds.), Recent Trends in Data Type Specification. V, 259 pages. 1988.

Vol. 333: H. Noltemeier (Ed.), Computational Geometry and its Applications. Proceedings, 1988. VI, 252 pages. 1988.

Vol. 334: K.R. Dittrich (Ed.), Advances in Object-Oriented Database Systems. Proceedings, 1988. VII, 373 pages. 1988.

Vol. 335: F.A. Vogt (Ed.), CONCURRENCY 88. Proceedings, 1988. VI, 401 pages. 1988.

Vol. 336: B.R. Donald, Error Detection and Recovery in Robotics. XXIV, 314 pages. 1989.

Vol. 337: O. Günther, Efficient Structures for Geometric Data Management. XI, 135 pages. 1988.

Vol. 338: K.V. Nori, S. Kumar (Eds.), Foundations of Software Technology and Theoretical Computer Science. Proceedings, 1988. IX, 520 pages. 1988.

Vol. 339: M. Rafanelli, J.C. Klensin, P. Svensson (Eds.), Statistical and Scientific Database Management. Proceedings, 1988. IX, 454 pages. 1989.

Vol. 340: G. Rozenberg (Ed.), Advances in Petri Nets 1988. VI, 439 pages. 1988.

Vol. 341: S. Bittanti (Ed.), Software Reliability Modelling and Identification. VII, 209 pages. 1988.

Vol. 342: G. Wolf, T. Legendi, U. Schendel (Eds.), Parcella '88. Proceedings, 1988. 380 pages. 1989.

Vol. 343: J. Grabowski, P. Lescanne, W. Wechler (Eds.), Algebraic and Logic Programming. Proceedings, 1988. 278 pages. 1988.

Vol. 344: J. van Leeuwen, Graph-Theoretic Concepts in Computer Science. Proceedings, 1988. VII, 459 pages. 1989.

Vol. 345: R.T. Nossum (Ed.), Advanced Topics in Artificial Intelligence. VII, 233 pages. 1988 (Subseries LNAI).

Vol. 346: M. Reinfrank, J. de Kleer, M.L. Ginsberg, E. Sandewall (Eds.), Non-Monotonic Reasoning. Proceedings, 1988. XIV, 237 pages. 1989 (Subseries LNAI).

Vol. 347: K. Morik (Ed.), Knowledge Representation and Organization in Machine Learning. XV, 319 pages. 1989 (Subseries LNAI).

Vol. 348: P. Deransart, B. Lorho, J. Maluszyński (Eds.), Programming Languages Implementation and Logic Programming. Proceedings, 1988. VI, 299 pages. 1989.

Vol. 349: B. Monien, R. Cori (Eds.), STACS 89. Proceedings, 1989. VIII, 544 pages. 1989.

Vol. 350: A. Törn, A. Žilinskas, Global Optimization. X, 255 pages. 1989.

Vol. 351: J. Díaz, F. Orejas (Eds.), TAPSOFT '89. Volume 1. Proceedings, 1989. X, 383 pages. 1989.

Lecture Notes in Computer Science

Edited by G. Goos and J. Hartmanis

389

D.H. Pitt D.E. Rydeheard
P. Dybjer A.M. Pitts A. Poigné (Eds.)

Category Theory and Computer Science

Manchester, UK, September 5–8, 1989
Proceedings

Springer-Verlag
Berlin Heidelberg New York London Paris Tokyo Hong Kong

Editors

David H. Pitt
Department of Mathematics, University of Surrey
Guildford, Surrey, GU2 5XH, UK

David E. Rydeheard
Department of Computer Science, The University
Manchester, M13 9PL, UK

Peter Dybjer
Chalmers University of Technology, Department of Computer Sciences
S-41296 Göteborg, Sweden

Andrew M. Pitts
University of Cambridge Computer Laboratory, New Museum Site
Pembroke Street, Cambridge, CB2 3QG, UK

Axel Poigné
GMD, F2 G2, Schloss Birlinghoven, Postfach 1240
D-5205 Sankt Augustin 1, FRG

CR Subject Classification (1987): D.2.1, D.3.1, D.3.3, F.3, F.4.1

ISBN 3-540-51662-X Springer-Verlag Berlin Heidelberg New York
ISBN 0-387-51662-X Springer-Verlag New York Berlin Heidelberg

Printing and binding: Druckhaus Beltz, Hemsbach/Bergstr.
2145/3140-543210 – Printed on acid-free paper

Preface

This collection of papers records some applications of category theory to computer science. The papers were presented at the third in a series of biennial conferences on Category Theory and Computer Science. The proceedings of the previous two conferences (Guildford 1985 and Edinburgh 1987) also appear in the Springer Lecture Notes in Computer Science, Numbers 240 and 283.

Traditionally, category theory has been used in mathematics as an abstract language for the formulation of definitions and the organisation of concepts in areas such as algebra and topology. Applications in computer science draw upon this and other aspects of the theory. It may help the reader if we briefly outline several of the more important links that have been established between category theory and computing.

One of the key ideas is the representation of programming languages as categories. This is particularly appropriate for languages based upon typed lambda calculi where the types become objects in a category and lambda terms (programs) become arrows. Conversions between programs are treated as equality, or alternatively, making the conversions explicit, as 2-cells. Composition is substitution of programs for free variables. Multiple variables are handled by admitting categories with finite products. This treatment enforces a stratification based upon the types of variables and expressions. For example, languages with type variables lead to indexed (or fibred) categories. Constructs in programming languages correspond to structure within categories, and categories with sufficient structure delimit the semantics of a language.

Correctness issues in programming require the introduction of logics for reasoning about program behaviour. Logics arising in category theory, for example internal logics in toposes, are, in general, intuitionistic, and in this sense appropriate for reasoning about constructions and programs. Propositional logics can be modelled directly as categories whose objects are propositional formulae and whose arrows are formal proofs. Predicate logics then become indexed categories of propositional logics. Logical constants are introduced as adjoints. Here we see a general algebraic framework for the analysis of programs, programming languages and their logics.

Other applications of category theory may be found in the development of universal algebra (theories for both program specification and computation) and of domain theory (for application in denotational semantics). Altogether, category theory provides a rich setting for the investigation of some foundational issues in computer science. The papers collected here illustrate recent progress in this area.

The organisers would like to thank the Science and Engineering Research Council for their support under the Logic for IT Initiative and the London Mathematical Society for their support. The Department of Computer Science at the University of Manchester kindly hosted the meeting. Local arrangements were ably handled by Jane Gray.

D.H.Pitt
D.E.Rydeheard
P.Dybjer
A.Pitts
A.Poigné

Organising and Program Committee

S. Abramsky, P. Dybjer, P.L. Curien, D.H. Pitt, A.M. Pitts, A. Poigné, D.E. Rydeheard, D.T. Sannella, E. Wagner.

Referees

S. Abramsky, S. Ambler, A. Asperti, I. Bethke, J. Bradfield, D.P. Carlisle, T. Coquand, P.L. Curien, M. Dam, P. Dybjer, H. Ehrig, F. Fages, M. Fourman, D. Gurr, T. Hagino, W. Harwood, M. Hedberg, B. Hilken, H. Huwig, J.M.E. Hyland, P.T. Johnstone, K.Karlsson, Y. Lafont, F. Lamarche, P. Mathieu, M. Mauny, J.Mitchell, W.P.R.Mitchell, E. Moggi, K. Moody, L.Moss, F. Nielson, D.H. Pitt, A.M.Pitts, A.Poigné, J.C.Raoult, G. Rosolini, D.E. Rydeheard, H.Sander, D.T.Sannella, B.Steffen, T.Streicher, P.Taylor, E. Wagner, G. Winskel.

Contents

Coherence and Valid Isomorphism in Closed Categories Applications of Proof Theory to Category Theory in a Computer Scientist Perspective

(Notes for an Invited Lecture)

Guiseppe Longo
(in collaboration with Andrea Asperti and Roberto Di Cosmo)
Dipartimento di Informatica, Università di Pisa

Many applications of Category Theory to Computer Science do not derive directly from the pure Theory of Categories but from the interplay of this Theory with Proof Theory. Two relevant recent cases may be quoted to support this claim.

The recent understanding of modularity in functional programming as polymorphism, in the sense of higher order λ-calculi, is endebted to a wise blend of the formal approach derived from Proof Theory (Girard's system F) and its categorical semantics (by indexed or internal categories). In particular, ideas from Proof Theory have been used in developing core type-systems, whose extensions were often suggested by (or checked against) categorical models.

In Linear Logic, the formal system has been inspired by categorical structures. Moreover, progress in a complete categorical description of this system has been made jointly with the use of these very same descriptions for the purposes of applications to parallel computations.

One should observe that in these examples, as well as in most other cases, Category Theory has been applied to (the understanding of) Proof Theory. This has been the role of Categorical Logic or, more specifically, of the Topos Theoretic semantics of Intuitionistic Logic. I believe, though, that fruitful interactions in scientific knowledge always go "two ways": a "reflective equilibrium" of theories is the only approach to reliable foundations and unified understanding in Mathematics and in Computer Science as well.

I will then hint two examples where Proof Theoretic methods have been applied to obtain results in Category Theory, the converse of the prevailing interaction. The first, and main, application is essentially due to Lambek and Mints and studies coherence problems suggested by Mac Lane and Kelly. The other is a recent charaterization theorem for (natural) isomorphisms in various closed categories due to several people, including the author. As we shall see, there is some duality between the two (classes of) results. Moreover, both use similar methods, based on λ-calculi, as calculi of proofs, and in particular, on the relation among composition of morphisms, cut-elimination and β-reduction.

In Kelly&MacLane[1971] a series of results are given which guarantee that, under certain circumstances, diagrams in monoidal and monoidal closed categories commute. In other words, they prove *equations* between morphisms obtained from basic ones (identity, unit and counit of the adjunction between tensor and exponent...) and closed under composition and image under the tensor and exponent functors. The observation, due to Lambek, is that a crucial (and hard) step in their work, the proof of closure under composition of a certain class of morphisms, corresponds exactly to what proof-theorists call "cut-elimination". Mints and others greatly simplified the original proof and extended the result to other categories. In particular, Mints proved that, in suitable assumptions, any deduction is interpreted by a unique morphism, in Cartesian Closed Categories (CCC's).

In Bruce&Longo[1985] we started from a problem suggested by the denotational semantics of recursive definitions. The meaning of recursively defined data types, e.g. $A = C\times(A\rightarrow A)$, is given by finding *specific* categories, basically Scott domains or others as CCC's, where A (and C) could be interpreted as objects, $\times$ and $\rightarrow$ as the intended functors and $=$ as (natural) isomorphism. Which equations among types are then valid in *all* CCC's?

This may be expressed in proof-theoretic terms as follows. Consider intuitionistic positive calculus, IPC, i.e. Intuitionistic Logic with only $\rightarrow$ and $\times$ (i.e. conjunction). The question then is: which sequents $A \vdash B$ and $B \vdash A$ are such that

(id) the composition of "$A \vdash B$ and $B \vdash A$" and of "$B \vdash A$ and $A \vdash B$" reduce, by cut-elimination, to the one step deductions $A \vdash A$ and $B \vdash B$, respectively ?

Indeed, the investigation of proofs as λ-terms (and, thus, of cut-elimination as β-reduction) answers the problem. In short, let $T_=$ be the following theory of equality

<u>AXIOMS</u>:

	(ι)	$A = A$
	(ρ)	$T\times A = A$
	(γ)	$A \times B = B \times A$
	(α)	$(A \times B) \times C = A \times (B \times C)$
	(λ)	$A \times B \rightarrow C = A \rightarrow (B \rightarrow C)$
	(τ)	$T \rightarrow A = A$
	(τε)	$A \rightarrow T = T$
	(δσ)	$A \rightarrow B \times C = (A \rightarrow B) \times (A \rightarrow C)$

<u>RULES</u>:

(functor) $$\frac{A = A' \qquad B = B'}{A \circledcirc B = A' \circledcirc B'}$$ where © is $\rightarrow$ or $\times$.

$$\text{(inverse)} \qquad \frac{A = B}{B = A}$$

$$\text{(compose)} \qquad \frac{A = B \quad B = C}{A = C}$$

Write $T_= \vdash A = B$ when $A = B$ is derivable in the above theory.

When interpreting connectives as functors and $=$ as (natural) isomorphism, this theory has plenty of models (take say a monoidal category and interpret $\rightarrow$ as the constant functor in the second argument....). However, a lot of hacking on λ-terms, based on work in Dezani[1976] and DiCosmo[in progress], answers the proof theoretic question:

A and B satisfy **(id)** iff $T_= \vdash A = B$.

Categorically, this characterizes the natural isomorphisms between "$\rightarrow - \times$–functors" which hold in all CCC's. Namely, view at formulae A and B as at "functor schemata" and define "transformation schemata" $\eta: A \rightarrow B$ in the 'natural' way (e.g. given a categorical model **C** of IPC, $A,B : \mathbf{C} \rightarrow \mathbf{C}$ define functors and η a natural transformation, when symbols are suitably instantiated). Then one has that

(iso) there exist transf. schemata $\eta: A \rightarrow B$, $\delta: B \rightarrow A$ such that, in all CCC's, $\delta \circ \eta = i$ and $\eta \circ \delta = i$ iff $T_= \vdash A = B$.

The result was given for the $\rightarrow$-fragment of IPC (i.e. with no $\times$) and for Girard's system *F*, and thus for (higher order) type-theories and their categorical models, in Bruce&Longo[1985]. DiCosmo recently made the non-trivial extension to IPC and related systems. We hope to have soon a full account of this work (the basic ideas will be presented in Asperti&Longo[1989]).

Note finally that (iso) may be also understood as a "reverse" implication of Mac Lane's coherence results: these essentially say that (iso)morphisms constructed out of some basic ones are unique. Conversely, we proved that each valid isomorphism can be given by the basic isomorphisms (ι), (ρ)... (δσ) in $T_=$ above.

There are still no computer science applications of either result, that I know, even though our work was somewhat inspired by denotational semantics. However, coherence results and their "dual" may be essential in discussions, say, on the unicity of interpretation or, even, in checking that a given categorical meaning of programs is well defined. Finally, as a wild guess, one may consider the possibility of type checking "up to valid isomorphisms" and extend by this the running typed systems.

References

Asperti A., Longo G. [1989] **Applied category theory: an introduction to categories, types and structures for the working computer scientist,** M.I.T. Press, to appear.

Bruce K., Longo G. [1985] "Provable isomorphisms and domain equations in models of typed languages," (Preliminary version) **1985 A.C.M. Symposium on Theory of Computing (STOC 85)**, Providence (R.I.) , May, (263-272).

Dezani M. [1976] "Characterization of normal forms possessing an inverse in the λβη-calculus," **Theor. Comp. Sci.**, 2 (323-337).

Kelly G.M., Mac Lane S.,[1971] "Coherence in closed categories", **J. Pure Appl. Algebra,** n° 1, (97-140); erratum, ibid. n° 2, 219.

Lambek J., [1968] "Deductive systems and categories I, Syntactic calculus and residuated categories", **Math. Systems Theory 2**, (287-318).

Lambek J., [1969] "Deductive systems and categories II, Standard constructions and closed categories", **Lecture Notes in Math., Vol. 86**, Springe r-Verlag, Berlin (76-122).

Mac Lane S. [1963] "Natural associativity and commutativity", Rice University Studies 49 (28-46).

Mac Lane S. [1976] "Topology and logic as a source of algebra" (retiring Presidential address), Bull Amer. **Math. Soc. 82, n°. 1**, (1-40).

Minc G.E. [1972] "A cut elimination theorem for relevant logics" (Russian English summary), **Investigations in constructive mathematics and mathematical logic V. Zap. Naucn Sem. Leningrad Otdel mat. Inst. Steklov (LOMI) 32**, (90-97); 156.

Minc G.E., [1977] "Closed categories and the theory of proofs (Russian, English Summary), **Zap Naucn. Sem. Leningrad Otdel Mat. Inst. Steklov (LOMI) 68**, (83-114); 145. (Trans circulated as preprint).

Minc G.E. [197?] "A simple proof of the coherence theorem for cartesian closed categories" **Bibliopolis** (to appear in translation from russian).

Minc G.E. [197?] "Proof theory and category theory" **Bibliopolis** (to appear in translation from russian).

An Algebraic View of Interleaving and Distributed Operational Semantics for CCS

Ugo Montanari[1]
Dipartimento di Informatica, Università di Pisa
Corso Italia 40, I-56100 Pisa, Italy. E_mail: ugo@di.unipi.it.uucp

Daniel N. Yankelevich
Escuela Superior Latino Americana de Informatica
Parque Pereira Iraola, Buenos Aires, Argentina. E_mail: dani@eslai.edu.ar

Abstract

In this paper we describe CCS models in terms of categories of structured transition systems: we define two categories CCS and CMonRCCS for representing interleaving and "truly concurrent", distributed aspects of CCS (without recursion). Among the objects of CCS and CMonRCCS we choose two *standard* models, called M and $\mathcal{M}$ respectively. We show that our interleaving model M essentially coincides with the classical transition system of CCS, while the distributed model $\mathcal{M}$ faithfully expresses the issues about decentralized control and multiple representations of agents discussed in a recent paper by the first author in collaboration with P. Degano and R. De Nicola [4]. Consistency of distributed and interleaving semantics is also proved.

The advantage of defining categories of models instead of simply models is that within the same category we can distinguish *initial* transition systems, whose transitions are proofs, and *interpreted* transition systems, which embody synchronization algebras. Another advantage of categories is the use of free adjoints of forgetful functors for constructing the models and for comparing transition systems having similar, but different, structures.

1. Introduction

The operational semantics of concurrent languages has often been defined in terms of labelled transition systems. In particular, the Structural Operational Semantics (SOS) approach developed by Plotkin [13] derives the transitions by using inference rules driven by the syntactic structure of the states, which are terms of the language. The semantics of the Calculus for Communicating Systems (CCS) is defined in this style [11].

A drawback of this approach is that it describes transitions between global states only, and does not offer a full account of the causal dependencies between the actions (possibly due to independent/parallel subsystems) which are performed when passing from one state to another. As a result, transition systems provide concurrent languages with a so-called *interleaving operational semantics*.

[1]Research performed in part while visiting ESLAI with the support of the Italian Foreign Ministry, Programma di Cooperazione e Sviluppo.

Several authors have studied how to equip concurrent languages with operational definitions which are also able to describe their causal and distributed behaviour. In particular, some recent work by the first author in collaboration with P. Degano and R. De Nicola [3,4] has shown how the centralized assumption on transition systems can be relaxed by letting Petri Nets [14] play the rôle of "distributed" transition systems. In [4], an Augmented Condition/Event (A-C/E) System[2] called Σ_{CCS} is defined, whose events are derived by means of inference rules, following the SOS style.

An interesting aspect of the semantics defined by Σ_{CCS} is that truly concurrent behavior is exhibited also in the presence of nondeterminism or recursion. For instance, the part of Σ_{CCS} corresponding to agent E = αNIL|βNIL + γNIL is depicted in Fig. 1. The initial case (i.e. the global state) corresponding to E contains two conditions; three events, labelled by α, β and γ, are enabled by them. The two conditions, which are obtained by decomposing E, are αNIL|id + γNIL and id|βNIL + γNIL. Conditions are constructed as special terms representing sequential processes and are called *grapes*, where the special constructs _|id and id|_ express the fact that a process is put in (left or right) parallel with another process. Events α and β have disjoint preconditions, and thus can fire in parallel. Conversely, both pairs <α,γ> and <β,γ> have intersecting preconditions and thus cannot fire in parallel. According to the interleaving transition relation we have αNIL|βNIL + γNIL —α→ NIL|βNIL. Here when event α fires, the case {NIL|id, id|βNIL + γNIL} is reached, which contains a *non–updated* sequential process, i.e. a process where the γNIL choice is still present. This is a necessary consequence of the *distributed control* in Petri Nets: a condition *not* involved in the transition *cannot* be modified. However, we must still make sure that the set {NIL|id, id|βNIL + γNIL} has the same behaviour as the sets of grapes obtained by decomposing agent NIL|βNIL, i.e.: {NIL|id,id|βNIL}. Thus, since we need to represent global choices in a completely distributed way, we have to give up a one-to-one correspondence between cases and agents.

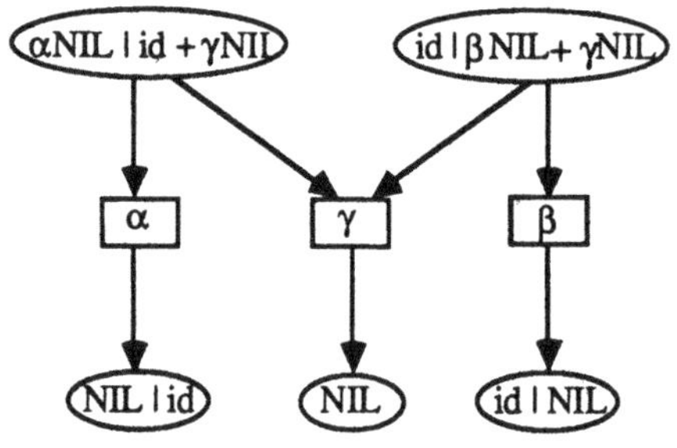

Fig. 1. The part of Σ_{CCS} corresponding to CCS agent αNIL|βNIL + γNIL. The initial case consists of the two topmost conditions.

This many-to-one relation is reflected in the result, proved in [4], which establishes the relation between Σ_{CCS} and the transition system Ψ which defines the interleaving semantics of CCS: a (non

2 A-C/E Systems are a minor extension of C/E Systems, where the enabling condition is modified for an event with intersecting preset and postset and where the simplicity requirement is lifted.

injective) homomorphism is defined between the case graph[3] of Σ_{CCS} and Ψ. The homomorphism guarantees the isomorphism of the synchronization trees obtained by unfolding the graphs starting from two corresponding nodes, and thus this result proves the consistency of the distributed and the interleaving semantics of CCS.

The adequacy of the distributed semantics of CCS defined in this way has been confirmed by later results. In fact, Olderog introduces in [12] a Place/Transition System similar to Σ_{CCS}. More importantly, it is shown in [6] that the labelled event structure obtained by unfolding Σ_{CCS} from a case corresponding to an agent E, essentially[4] coincides with the *denotational* semantics of E according to Winskel's definition [16].

Specifying a suitable transition system is only the first step for defining the interleaving semantics of CCS. In fact, transition systems provide a very intensional semantics, and some quotient must be defined to identify equivalent programs. Unfolding the transition system and identifying isomorphic trees provides a first equivalence (which actually coincides with the *strong observational equivalence*). At this level the *expansion* theorem starts to hold, which derives an equivalent agent without parallel composition from every given agent. Several observation mechanisms can then be defined (weak observational equivalence, testing equivalence, various axiomatizations, etc.), which achieve the desired levels of abstraction. The same method can be followed for the distributed semantics of CCS (where, however, the expansion theorem in its simpler form does not hold). Actually, a general approach based on node-labeled trees called NMS [7] works for both the interleaving and the distributed case, providing a simple way for defining and comparing different equivalences. In this paper, as in [4], we are not interested in defining any extensional semantics for CCS. On the contrary, we intend to be as intensional as possible, in order to permit the definition of the widest possible range of equivalences in a successive step.

Let us now move to an apparently different subject. In recent work by the first author in collaboration with J. Meseguer [9,10], a new algebraic definition for Place/Transition Petri nets has been proposed. Some formal account of it is given in Section 2.2.; here we shortly describe aims and results. A Petri net is seen as a graph where the nodes form a free commutative monoid. The generators of the monoid are the places and the monoidal operation means union of finite multisets of places. A category <u>Petri</u> of nets is defined.

A closure operation C[_], seen as a free adjoint in categorical terms, allows us to extend the monoidal structure to the arcs, thus deriving from a net N a commutative monoid structure C[N] on a reflexive[5] graph, which turns out to be the case graph[6] of N. Here the monoidal operation has the semantical meaning of firing transitions in parallel, while identity transitions associated to nodes represent idle tokens. The category of such structures, where for the sake of generality the monoids on the nodes

[3] The case graph of an A-C/E System is the labelled graph whose nodes are cases, i.e. sets of conditions, and whose arcs are (possibly multiple) event firings, labelled by such events. Actually, to make the comparison possible, the above result applies to a modified version of the case graph of ΣCCS, qualified as *abstract interleaving*, where arcs are labelled by actions rather than events, and arcs corresponding to multiple event firings are deleted.

[4] Actually, the result applies to a slightly different A-C/E System. It can be obtained from ΣCCS by allowing multiple copies of an event whenever it can be derived with many different proofs. The construction of the "free" version of the transition system for CCS shown in this paper provides ground for a more direct result.

[5] I.e. with an "identity" loop on every node.

[6] C[N] could rather be called the *marking graph* of N, since N is a (not necessarily 1-safe) P/T net rather than a C/E system.

and on the arcs are not necessarily free, is called CMonRGraph.

A further closure operation introduces a sequentialization operation on the arcs. Thus now the arcs represent computations, or, more generally, equivalence classes of computations. Such abstract computations are often able to express in a natural way the semantics of concurrent systems: for instance Petri nonsequential processes [14] can be represented in this way [5]. Several categories are defined, where categorical product and coproduct have the semantical meaning of parallel and nondeterministic composition of nets. It is important to notice that morphisms in these categories represent some sort of simulation, and thus they express a specification-implementation relation. In particular, a category ImplPetri is defined, whose objects are Petri nets and whose morphisms make single transitions in the specification correspond to whole computations in the implementation. These constructions[7] (but not the definition of category Petri given in Section 2.2) extend and generalize previous results by Winskel [15].

In this paper, our aim is defining a category of graphs, structured by means of CCS operations, and then adding the monoidal operation studied in the Petri net context. The models in the latter category will be similar to Σ_{CCS}, but hopefully with a more satisfactory algebraic structure.

More precisely, the content of the paper is as follows. In Section 2 we present a short introduction to both CCS and Petri nets as monoids. In Section 3 we describe a category CCS of graphs with the CCS operations on both nodes and arcs. For the sake of simplicity, recursion is not considered (but see the comments in the Final Remarks). Category CCS has an obvious forgetful functor to Graph (the category of graphs), which has a free adjoint. Applying this free adjoint to the empty graph we generate the initial object in CCS, which we call T_{CCS}. T_{CCS} is essentially the transition system of CCS with a free synchronization algebra, where the arcs are (labelled by) their proofs. Similar models have been considered for instance in [2,3]. The synchronization algebra of CCS defines a quotient on T_{CCS} and thus another object in CCS, which we call M. The only difference between M and the transition system of CCS is that M has absorbent error transitions, since our arc operations are total.

In Section 4 we introduce another category, modelling "truly concurrent" CCS. We call it CMonRCCS, since it has the structure of both CMonRGraph and CCS. The operations on nodes and arcs are: i) CCS operations; ii) the monoidal operation $\oplus$ of CMonRGraph; and iii) the _|id and id|_ operations expressing the "spatial" structure of the system. However, the operations are not free: the following *decomposition* axiom

$u|v = u|id \oplus id|v$

interprets composition of parallel agents as disjoint union. Furthermore, $\oplus$ distributes over most of the other operations. The effect of these axioms is that of introducing agent decompositions in the style of [3,4]. However, the axioms *equate* CCS agents to their decompositions. For instance, going back to the example in Fig. 1, we can write

$\alpha NIL|\beta NIL + \gamma NIL \;=\; \alpha NIL|id + \gamma NIL \;\;\oplus\;\; id|\beta NIL + \gamma NIL$

i.e., differently than in [3,4], both structures are present in the model at the same time and are related by the axioms.

When considering the corresponding axioms on transitions, an interesting fact arises: as a

[7] The definition of ImplPetri is not reported in this paper, since the discussion about morphisms as abstractions, while potentially very interesting for CCS (see for instance [8] and the comments in Section 5) is here only partially relevant.

consequence of the distributivity axiom between ⊕ and the nondeterministic composition operation +, we can equate transitions with different sources and targets, which thus are identified as well. Interestingly enough, the states equated in this way are CCS agents and all their, updated or not, distributed versions. For our example, we can write

$$\mathrm{NIL}|\mathrm{id} \oplus \mathrm{id}|\beta\mathrm{NIL} = \mathrm{NIL}|\mathrm{id} \oplus \mathrm{id}|\beta\mathrm{NIL} + \gamma\mathrm{NIL} = \mathrm{NIL}|\beta\mathrm{NIL}.$$

Therefore, the fact that an agent has many distributed representations - a necessary consequence of decentralized control according to our previous discussion about Σ_{CCS} - is here the consequence of a distributivity axiom. In turn, the distributivity axioms between ⊕ and all the other operations express the semantically interesting fact that an agent can be seen as a collection of sequential processes, and that its transitions can be considered as collections of transitions of its parts.

As in the case of CCS, category CMonRCCS has an initial object D_{CCS} and a forgetful functor U[_] to CCS with a free adjoint, which we call $\mathcal{A}$[_]. The model we propose for distributed CCS is $\mathcal{M} = \mathcal{A}[M]$, i.e. the element of CMonRCCS which embodies the synchronization algebra of CCS. D_{CCS} and $\mathcal{M}$ have, at the same time, the structure of CCS transition systems and of case graphs. However, $\mathcal{M}$ cannot be considered the case graph of any net, since its monoid of nodes is not free.

The main advantage of $\mathcal{M}$ with respect to Σ_{CCS} is that the latter is constructed in a two–step fashion: first certain inference rules related to CCS operations are used to build a Petri net, and then the operational semantics of nets is used to define global transitions and computations. Instead, $\mathcal{M}$ has at the same time the operations of both CCS and nets. Furthermore, the identifications among non-updated states - expressed in [4] by an auxiliary homomorphism defined between the case graph of Σ_{CCS} and the interleaving transition system of CCS - are in $\mathcal{M}$ a direct consequence of the axioms.

Consistency of truly concurrent and interleaving semantics of CCS is proved by showing that the morphism between M and U[$\mathcal{M}$] is injective, i.e. M can be completely recovered from $\mathcal{M}$. The relation between $\mathcal{M}$ and Σ_{CCS}, while clearly quite close, requires further, careful analysis. In fact, it is first necessary to transform Σ_{CCS} in a P/T net Σ°_{CCS} and then to take its case graph $C[\Sigma^{\circ}_{CCS}]$. On the other side, the forgetful functor V[_] from CMonRCCS to CMonRGraph should be applied to $\mathcal{M}$ yielding V[$\mathcal{M}$]. There should be a "folding" morphism in CMonRGraph from $C[\Sigma^{\circ}_{CCS}]$ to V[$\mathcal{M}$], which should be related to the tp-homomorphism defined in [4]. It is also necessary to take care of the lack of recursion construct in $\mathcal{M}$.

2. CCS and Petri Nets as Monoids

2.1. Calculus of Communicating Systems (CCS)

Here we briefly introduce the relevant definitions for Milner's Calculus of Communicating Systems (CCS) [11]. First we recall the operators of CCS, and then we present the traditional interleaving operational semantics. Neither recursion nor observational equivalence are considered.

Definition 2.1.1. *(CCS agents)*

Let $\Delta = \{\alpha, \beta, \gamma \dots\}$ be a fixed set and $\Delta^- = \{\alpha^- \mid \alpha \in \Delta\}$; then $\Lambda = \Delta \cup \Delta^-$ (ranged over by λ) will be used to denote the set of *visible actions*. Moreover, symbol τ will be used to denote a distinguished *invisible action* not in Λ; $\Lambda \cup \{\tau\}$ will be ranged over by μ.

Here, CCS **agents** are the terms generated by the following BNF-like grammar:

$$E ::= NIL \mid \mu E \mid E\backslash\alpha \mid E[\phi] \mid E + E \mid E|E$$

where ϕ is a bijection from $\Lambda \cup \{\tau\}$ to itself which preserves τ and the operation $^-$ of complementation.

Example 2.1.2.

$E = \alpha\beta|\gamma + \theta$ and $E' = \beta|\gamma$ are CCS agents.

Precedence between operators is $\mu > [\phi] > \backslash\alpha > | > +$. NIL is often omitted.

The CCS labelled transition system has CCS agents as nodes and the triples $E_1 \text{—}\mu\text{→} E_2$ defined below as transitions from E_1 to E_2.

Definition 2.1.3. *(Derivation relation)*

The **derivation** relation $E_1 \text{—}\mu\text{→} E_2$ is defined as the least relation satisfying the following axiom and inference rules:

Act)	$\mu E \text{—}\mu\text{→} E$				
Res)	$E_1 \text{—}\mu\text{→} E_2$	**and**	$\mu \notin \{\alpha, \alpha^-\}$	**implies**	$E_1\backslash\alpha \text{—}\mu\text{→} E_2\backslash\alpha$,
Rel)	$E_1 \text{—}\mu\text{→} E_2$	**implies**	$E_1[\phi] \text{—}\phi(\mu)\text{→} E_2[\phi]$		
Sum)	$E_1 \text{—}\mu\text{→} E_2$	**implies**	$E_1+E \text{—}\mu\text{→} E_2$	**and**	$E+E_1 \text{—}\mu\text{→} E_2$
Com)	$E_1 \text{—}\mu\text{→} E_2$	**implies**	$E_1|E \text{—}\mu\text{→} E_2|E$	**and**	$E|E_1 \text{—}\mu\text{→} E|E_2$
	$E_1 \text{—}\lambda\text{→} E_2$	**and**	$E'_1 \text{—}\lambda^-\text{→} E'_2$	**implies**	$E_1|E'_1 \text{—}\tau\text{→} E_2|E'_2$

This relation completely specifies the CCS operational semantics which, given an agent, determines the actions (and sequences of actions) it may perform, and the new agents which are obtained as results.

Example 2.1.4.

The triple $\alpha|\beta + \gamma \text{—}\alpha\text{→} NIL|\beta$ belongs to the derivation relation.

2.2. Petri Nets are monoids

This is a brief account of the definition of Petri Nets proposed in [9,10]. Standard Petri Nets theory can be found in [14].

Definition 2.2.1. *(Petri)*

A **(Place/Transition) Petri Net** is a graph where the nodes form a free commutative monoid. Namely, a net is a quadruple $N=(S^{\oplus},T,\partial_0,\partial_1)$, where $S^{\oplus}$ is the free commutative monoid of **nodes** over a set of **places** S; T is a set of **transitions**; and ∂_0,∂_1: $T \rightarrow S^{\oplus}$ are the functions associating to every transition a source and a target node respectively.

A net morphism h from N to N' is a pair of functions <f,g> , f: $T \rightarrow T'$ and g: $S^{\oplus} \rightarrow S'^{\oplus}$, where g is a monoid morphism, such that $g \circ \partial_0 = \partial_0' \circ f$ and $g \circ \partial_1 = \partial_1' \circ f$, i.e. such that the diagram commutes. This defines a category <u>Petri</u>.

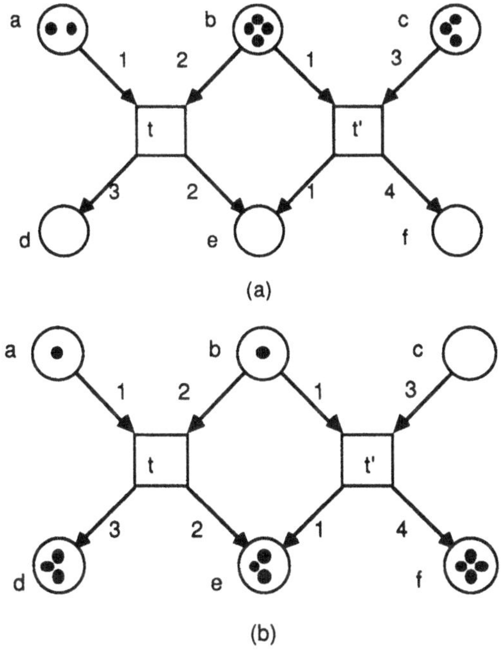

Fig. 2. The concurrent firing of two transitions in a Place/Transition Petri net.

Example 2.2.2.

Consider the net in Fig. 2. It has S={a,b,c,d,e,f} and T={t,t'}. The numbers on the arrows specify how many tokens are consumed or generated as a consequence of firing the transition. Picture (a) describes a state before the concurrent firing of t and t', picture (b) describes the state reached after the firing.

The state of picture (a) is described by the element $2a \oplus 4b \oplus 3c$ of $S^{\oplus}$, and the transition t is an arrow from $a \oplus 2b$ to $3d \oplus 2e$.

Definition 2.2.3. *(Reflexive commutative monoids)*

A reflexive commutative monoid structure consists of a reflexive graph $(V,T,\partial_0,\partial_1,id)$, where both the set of nodes V and the set of arcs T are commutative monoids and where ∂_0 and ∂_1 are monoid homomorphisms. There is a reflexive graph structure determined by function *id*: $V \rightarrow T$, which is a monoid homomorphism. Node u is source and target of the arc *id* (u), which is called the **identity** of u and denoted, by coercion, u as well. A morphism is a pair <f,g> of monoid morphisms such that the diagram commutes. This defines a category CMonRGraph.

There exist a forgetful functor from CMonRGraph to Petri that has a left adjoint C[_]. It associates to each Petri Net N its *case graph* C[N]. Both node and arc monoids in C[N] are free, and have as generators the places of N, and the transitions plus the identities of N, respectively.

Example 2.2.4.

In the case graph C[N] of the net N in Fig. 2, there is an arrow $t \oplus t' : a \oplus 3b \oplus 3c \to 3d \oplus 3e \oplus 4f$ corresponding to the concurrent firing of both transitions, and also an arrow $a \oplus b \oplus t \oplus t' : 2a \oplus 4b \oplus 3c \to a \oplus b \oplus 3d \oplus 3e \oplus 4f$ corresponding to the transition displayed in Fig. 2, which has two idle tokens.

3. A category for CCS

We now define a category of transition systems structured with CCS operations.

Definition 3.1. *(CCS)*

A CCS **transition system** is a graph $\partial_0, \partial_1 : T \to V$ with operations on the nodes V and on the arcs T. V and T are called **states** and **transitions**. The operations which yield states are those of CCS, namely

NIL, μu, $u \backslash \alpha$, $u[\phi]$, $u+v$ and $u|v$.

Free terms E built with these operations are called *agents*. Agents are sometime used for denoting the state [E] they evaluate to, when the operations are interpreted.

The operations which yield transitions are:

$[\mu,E>$, $t \backslash \alpha$, $t[\phi]$, $t<+E$, $E+>t$, $t \rfloor E$, $E \lfloor t$ and $t_1|t_2$

The operations must satisfy a set of axioms establishing that corresponding operations on transitions and states commute with ∂_0 and ∂_1. If t is a transition from u to v, i.e. if $t: u \to v$, and if $t': u' \to v'$, we have:

$[\mu,E>$: $\mu E \to E$

$t\backslash\alpha$: $u\backslash\alpha \to v\backslash\alpha$

$t[\phi]$: $u[\phi] \to v[\phi]$

$t<+E$: $u+E \to v$

$E+>t$: $E+u \to v$

$t\rfloor E$: $u|E \to v|E$

$E\lfloor t$: $E|u \to E|v$

$t|t'$: $u|u' \to v|v'$

Notice that neither the algebra of states nor the algebra of transitions is required to be free. **CCS morphisms** are graph morphisms <f,g>, where f is a morphism for the algebra of transitions and g is a morphism for the algebra of states. We call this category CCS.

Notice that operations yielding transitions are sometimes parametrized with actions and with *agents*, but of course the resulting transitions have *states* as sources and targets. For instance, we might write, more explicitly, $[\mu,E\rangle: \mu[E] \rightarrow [E]$.

Category CCS has an initial transition system. Its transitions are proofs of derivations.

Property 3.2. *(Constructing the initial object T_{CCS})*

CCS has an obvious forgetful functor to Graph. Applying its left adjoint to the empty graph yields the initial object of CCS, which we call T_{CCS}. The set of nodes of T_{CCS} is the carrier of the word algebra on the node operations of CCS, namely agents, and the transitions T are freely generated by the following inference rules:

		$[\mu,E\rangle$:	$\mu E \rightarrow E \in T$
$t{:}E_1 \rightarrow E_2 \in T$	implies	$t\backslash\alpha$:	$E_1\backslash\alpha \rightarrow E_2\backslash\alpha \in T$
$t{:}E_1 \rightarrow E_2 \in T$	implies	$t[\Phi]$:	$E_1[\Phi] \rightarrow E_2[\Phi] \in T$
$t{:}E_1 \rightarrow E_2 \in T$	implies	$t\langle +E$:	$E_1+E \rightarrow E_2 \in T$
	and	$E+\rangle t$:	$E+E_1 \rightarrow E_2 \in T$
	and	$t\rfloor E$:	$E_1 \mid E \rightarrow E_2 \mid E \in T$
	and	$E\lfloor t$:	$E \mid E_1 \rightarrow E \mid E_2 \in T$
$t{:}E_1 \rightarrow E_2,\ t'{:}E'_1 \rightarrow E'_2 \in T$	implies	$t \mid t'$:	$E_1 \mid E'_1 \rightarrow E_2 \mid E'_2 \in T$

Example 3.3.

The transition $[\alpha,NIL\rangle\rfloor\beta \langle + \gamma$ is in T. It corresponds to the proof of $\alpha\mid\beta + \gamma \xrightarrow{\alpha} NIL\mid\beta$.

Standard CCS calculus embodies a synchronization algebra: we want to express it in our framework. Furthermore, there are transitions of T_{CCS} which do not correspond to derivations, since they are proofs of erroneous transitions. For example, the transition $[\alpha,NIL\rangle\backslash\alpha$ does not make sense. We solve both problems by defining another object M, called standard model, in CCS.

Definition 3.4. *(Standard model of CCS)*

The **standard model M** is the object of CCS obtained by applying to the transitions of T_{CCS} the quotient defined by the congruence relation $\equiv$

$$t \equiv t' \quad \text{iff} \quad l(t) = l(t') \text{ and } \partial_i(t) = \partial_i(t'),\ i = 0,1$$

where

$$l: T \rightarrow \Lambda\cup\{\tau\}\cup\{*\}$$

is defined by structural induction as follows

$$l([\mu,E\rangle) = \mu$$

$$l(t\backslash\alpha) = \text{if } l(t) \notin \{\alpha,\bar{\alpha}\} \text{ then } l(t) \text{ else } *$$

$$l(t[\Phi]) = \text{if } l(t)=* \text{ then } * \text{ else } \Phi(l(t))$$

$$l(t\langle+E) = l(E+\rangle t) = l(t\rfloor E) = l(E\lfloor t) = l(t)$$

$$l(t\mid t') = \text{if } l(t)=\overline{l(t)} \text{ then } \tau \text{ else } *.$$

Property 3.5. *(The transitions of M are labelled)*

The labelling function l can be defined on the transitions of M:

$l([t]_{/\equiv}) = l(t)$.

Example 3.6.

The transition $[\alpha,NIL>\backslash\alpha$ has label $*$.

Example 3.7.

The two transitions

$[\alpha,NIL> <+ \alpha NIL$ and $\alpha NIL +> [\alpha,NIL>$

of T_{CCS} are mapped into the same transition of M. Thus the (unique) morphism from T_{CCS} to M is not an isomorphism. Another example is as follows:

$[\alpha,NIL>\backslash\beta <+ \alpha(NIL\backslash\beta)$ and $(\alpha NIL)\backslash\beta +> [\alpha,NIL\backslash\beta>$

are mapped into the same transition of M.

We can now relate our standard model with the traditional operational semantics of CCS.

Theorem 3.8.

Let $E_1 \text{—}\mu\rightarrow E_2$ be the derivation relation. We have

$E_1 \text{—}\mu\rightarrow E_2$ iff $t: E_1 \rightarrow E_2$ is a transition of M, with $l(t) = \mu$.

Proof. Given a transition t of T_{CCS}, it is easy to prove by induction that if $t: E_1 \rightarrow E_2$, with $l(t) = \mu$, then there is a proof, isomorphic to t, of derivation $E_1 \text{—}\mu\rightarrow E_2$, and viceversa.

4. A category for "truly concurrent" CCS

In this section, we will define a category of truly concurrent CCS models following the idea, described in [3,4], of decomposing agents into sequential processes. This category, which we will call CMonRCCS, has as objects transition systems having, at the same time, the algebraic structure of CCS and the monoidal structure of CMonRGraph.

Definition 4.1. *(CMonRCCS)*

A **concurrent CCS transition system** is a graph $\partial_0,\partial_1: T \rightarrow V$ with operations on the states V and on the transitions T. The operations which yield states are those of CCS

NIL, μu, $u\backslash\alpha$, $u[\phi]$, $u+v$ and $u|v$

the commutative monoid operation, together with its unity

$u \oplus v$ and $\emptyset$

and the two unary operations

$u|id$ and $id|u$.

For those operations, we have a decomposition axiom

$u|v = u|id \oplus id|v$

and the distributivity axioms of $\oplus$ with respect to all the other operations, with the exception of μu and $u|v$

$(u \oplus v)\backslash\alpha = u\backslash\alpha \oplus v\backslash\alpha$

$(u \oplus v)[\phi] = u[\phi] \oplus v[\phi]$

$(u \oplus v)+(u' \oplus v') = u+u' \oplus u+v' \oplus v+u' \oplus v+v'$

$(u \oplus v)|id = u|id \oplus v|id$

$id|(u \oplus v) = id|u \oplus id|v$.

The operations which yield transitions are those of CCS

$[\mu,E>$, $t\backslash\alpha$, $t[\phi]$, $t<+E$, $E+>t$, $t \rfloor E$, $E \lfloor t$ and $t_1|t_2$

the commutative monoid operation, together with its unity

$t_1 \oplus t_2$ and $\emptyset$

the identity transitions

id (u)

usually coerced to u, and two unary operations

$t|id$ and $id|t$.

For those operations, we have again decomposition axioms

$t \rfloor E = t|id \oplus id|E$

$E \lfloor t = E|id \oplus id|t$

distributivity axioms

$(t_1 \oplus t_2)\backslash\alpha = t_1\backslash\alpha \oplus t_2\backslash\alpha$

$(t_1 \oplus t_2)[\phi] = t_1[\phi] \oplus t_2[\phi]$

$(t_1 \oplus t_2)<+E = t_1<+E \oplus t_2<+E$

$E+>(t_1 \oplus t_2) = E+>t_1 \oplus E+>t_2$

$(t_1 \oplus t_2)|id = t_1|id \oplus t_2|id$

$id|(t_1 \oplus t_2) = id|t_1 \oplus id|t_2$

$(u \oplus t_1)|t_2 = u|id \oplus t_1|t_2$

$t_1|(u \oplus t_2) = id|u \oplus t_1|t_2$

and, as for CCS, commutativity of corresponding operations on transitions and states with ∂_0 and ∂_1. If $t: u \to v$, and $t': u' \to v'$, we have:

$[\mu,E>$:	$\mu E \to E$
$t\backslash\alpha$:	$u\backslash\alpha \to v\backslash\alpha$
$t[\phi]$:	$u[\phi] \to v[\phi]$
$t<+E$:	if $t \neq id(u)$ then $u+E \to v$ else $u+E \to u+E$
$E+>t$:	if $t \neq id(u)$ then $E+u \to v$ else $E+u \to E+u$
$t|t'$:	$u|u' \to v|v'$
$t \oplus t'$:	$u \oplus u' \to v \oplus v'$
$\emptyset$:	$\emptyset \to \emptyset$

u: u → u

t|id: u|id → v|id

id|t: id|u → id|v

Notice that properties t⌋E: u|E → v|E and E⌊t: E|u → E|v are consequences of the decomposition axioms. Finally, we have commutativity of corresponding operations on states and transitions with *id*

$id(u)\backslash\alpha = id(u\backslash\alpha)$

$id(u)[\phi] = id(u[\phi])$

$id(u)\text{<+}E = id(u+E)$

$E\text{+>}id(u) = id(E+u)$

$id(u)|id(v) = id(u|v)$

$id(u)\oplus id(v) = id(u\oplus v)$

$id(\emptyset) = \emptyset$

$id(u)|\mathrm{id} = id(u|\mathrm{id})$

$\mathrm{id}|id(u) = id(\mathrm{id}|u)$

Notice that properties $id(u)\rfloor E = id(u|E)$ and $E\lfloor id(u) = id(E|u)$ are, as before, consequences of the decomposition axioms.

Concurrent CCS morphisms are reflexive graph morphisms <f,g>, where f is a morphism for the algebra of transitions and g is a morphism for the algebra of states. We call this category CMonRCCS.

Category CMonRCCS has an initial transition system DCCS. As usual, it can be generated by a free construction.

Property 4.2. *(Constructing the initial object DCCS)*

CMonRCCS has an obvious forgetful functor to Graph. Applying its left adjoint to the empty graph yields the initial object of CMonRCCS, which we call DCCS. The states and the transitions of DCCS are constructed by taking the word algebras for the state and transition operations of CMonRCCS respectively and by taking the quotient with respect to the axioms.

We have the following important property.

Property 4.3. *(Identifying distributed representations)*

In DCCS, the following equalities hold

ups) $\partial_1(t) \oplus w = \partial_1(t) \oplus w{+}E$

$\partial_1(t) \oplus w = \partial_1(t) \oplus E{+}w$

where t is any transition which is not an identity.

Proof. Let

$t_1 = (t \oplus w) \text{<+} E : ((u \oplus w) + E \rightarrow v \oplus w$ and

$t_2 = t \text{<+} E \oplus w \text{<+} E : u{+}E \oplus w{+}E \rightarrow v \oplus w{+}E$

where t: u → v is not an identity. We have $t_1 = t_2$ due to the distributivity axiom for <+ above. Thus $v \oplus w = v \oplus w{+}E$. Similarly for the second equation.

Example 4.4.

Let us consider our running example $E = \alpha NIL|\beta NIL + \gamma NIL$. We have:

$$([\alpha,NIL>|id \oplus id|\beta NIL) <+ \gamma NIL : \ \alpha NIL|id + \gamma NIL \ \oplus \ id|\beta NIL+ \gamma NIL \rightarrow NIL|id \ \oplus id|\beta NIL.$$

For the distributivity of <+ we have:

$$([\alpha,NIL>|id \oplus id|\beta NIL) <+ \gamma NIL \ = \ [\alpha,NIL>|id <+ \gamma NIL \ \oplus \ id|\beta NIL <+ \gamma NIL.$$

On the other hand

$$[\alpha,NIL>|id <+ \gamma NIL \ \oplus \ id|\beta NIL <+ \gamma NIL :$$

$$\alpha NIL|id + \gamma NIL \ \oplus \ id|\beta NIL + \gamma NIL \rightarrow NIL|id \ \oplus id|\beta NIL + \gamma NIL$$

since in $id|\beta NIL <+ \gamma NIL$ the transition $id|\beta NIL$ is an identity. Thus

$$NIL|id \ \oplus id|\beta NIL \ = \ NIL|id \ \oplus id|\beta NIL + \gamma NIL.$$

The same result could be obtained directly from the first ups) equation by letting in it $t = [\alpha,NIL>|id$, $w = id|\beta NIL$ and $E = \gamma NIL$.

The fact that an agent has many, but equivalent, distributed representations is thus the consequence of a distributivity axiom. In turn, distributivity axioms express the semantically interesting fact that an agent can be seen as a collection of sequential processes, and that its transitions can be considered as collections of transitions of its parts.

There are obvious forgetful functors from CMonRCCS to both CCS and CMonRGraph. We call them U[_] and V[_]. U[_] has a left adjoint that we call $\mathcal{A}$[_]. We can now introduce our standard concurrent transition system for CCS.

Definition 4.5. *(The standard concurrent transition system $\mathcal{M}$)*

Let $\mathcal{A}$[_] be the left adjoint to the forgetful functor U[_] from CMonRCCS to CCS and let M be the standard transition system for CCS. The object $\mathcal{M} = \mathcal{A}[M]$ of CMonRCCS is the **standard concurrent transition system** for CCS.

We can now prove our main result, i.e. the consistency, in our framework, of the interleaving and the distributed operational semantics of CCS. Of course the comparison must take place in the category with less structure, i.e. CCS.

Theorem 4.6. *(The distributed semantics extends the interleaving semantics)*

The unique morphism in CCS between M and U[$\mathcal{M}$] is injective on both states and transitions.

Outline of the proof. The proof consists tentatively of five steps, which we describe here in short.

1) To prove that the morphism is injective on the states, it is enough to prove in DCCS that $E_1 \neq E_2$ for every pair E_1 and E_2 of different agents. Similarly, for the transitions we can prove $t_1 \neq t_2$ for every pair t_1 and t_2 of different proofs built up without ⊕, |id, id| and identities. Once the state part is proved, the transition part follows easily.

2) To prove in DCCS that $E_1 \neq E_2$ for every pair E_1 and E_2 of different agents, it is enough to consider the algebra of states, with the corresponding decomposition and distributivity axioms, adding property ups) as an axiom.

3) The remaining steps are *ad absurdum*. If there is a demonstration of $E_1=E_2$ in the formal system of 2), then there exists a demonstration of $E'_1=E'_2$ in the formal system consisting of terms in normal form[8] and with the single axiom ups), where both left and right members are considered normalized. Agents E'_1 and E'_2 will not, in general, be equal to E_1 and E_2, but will be subagents of them.

4) In the formal system described in 3), two terms that are equal can differ only for summations at the outermost level. The normalizations of two such agents are nonintersecting sets.

5) For every proof $\underline{E}_1 \equiv u_1 = u_2 = ... = u_n \equiv \underline{E}_2$, we have $\cap\, u_i \neq \emptyset$ and thus the theorem is proved. Actually, to show this last step we have to take advantage of the auxiliary result that it is not possible to prove $\underline{E}' = \underline{E}'' \oplus u$, u nonempty. Now suppose that $u_1 = t_1 \oplus t_2 \oplus ... \oplus t_i \oplus ... \oplus t_n$. Then, without loss of generality, we assume that, in applying ups) the first time, $\partial_1(t) = t_1 \oplus ... \oplus t_i$. Thus we have that $u_2 = t_1 \oplus ... \oplus t_i \oplus (\underline{t_{i+1} + E}) \oplus ... \oplus (\underline{t_n + E})$. In the next application of ups), $\partial_1(t')$ must be included in $t_1 \oplus ... \oplus t_i$. In fact if $\partial_1(t)$ contains a grape of the form u+v, either $\partial_1(t) = \underline{E' + E}''$ or it contains a grape not of the form u+v. On the other hand, by the auxiliary result the first alternative is impossible. Therefore t_1 does belong to u_n.

Example 4.7.

In the unique morphism in <u>CCS</u> between M and $U[\mathcal{M}]$, the transition

$[\alpha, NIL> \rfloor \beta <+ \gamma$

of M is mapped in the same transition of $U[\mathcal{M}]$, which however, in $\mathcal{M}$, coincides with the transition

$[\alpha, NIL>|id <+ \gamma NIL \oplus id|\beta NIL <+ \gamma NIL$.

Instead, the transition

$[\alpha, NIL>|id <+ \gamma NIL \oplus id|[\beta NIL> <+ \gamma NIL$

of $\mathcal{M}$ is a generator of $U[\mathcal{M}]$, with no corresponding transition in M.

5. <u>Final Remarks</u>

In the paper, we have defined two categories of structured graphs for modelling interleaving and truly concurrent aspects of CCS. For the sake of simplicity, we did not consider recursion. However, it should be easy to handle the recursive construct in our algebraic framework: we could add variables, the recursion operator and the axiom

rec x.E = E[rec x.E/x].

[8] A term t representing a state is in normal form iff t = t1⊕t2⊕...⊕tn, where ti contains neither ⊕ nor occurrences of | outside the outermost μE context. Terms in normal form can be considered as sets of terms without ⊕. Such terms are called *grapes* in [4]. Given an agent E, there is a unique term $\underline{E}$ in normal form, called the *normalization* of E, which is obtained applying the decomposition and distributivity axioms to the occurrences of | external to the outermost μE context, as to bring all ⊕ operators to the outermost level. The normalization $\underline{E}$ of E corresponds to the complete set of grapes for E as defined in [4].

Probably it would be necessary to change the construction of $\mathcal{M}$, since in the presence of recursion it would contain too many identifications. E.g. there would be only one transition in $\mathcal{M}$, besides the identity, from (rec x.a) | (rec x.a) to itself, while one would like instead to detect which of the two processes actually moved.

The advantage of defining categories instead of simply models is that within the same category we can distinguish *initial* transition systems, whose transitions are proofs, and *interpreted* transition systems, which embody synchronization algebras. Another advantage of categories was the use of forgetful functors and their free adjoints for constructing the models and for comparing transition systems having different structures.

Following the ideas of [9,10], we plan to extend the approach of this paper by defining a category CatCCS, whose objects are small categories. They are obtained from the concurrent transition systems of CMonRCCS by adding a sequentialization operation _;_ and some axioms. As in the case of Petri Nets, this extension introduces abstract computations, seen as equivalence classes of transitions, which may correspond to the operational notion of observation [7]. Furthermore, morphisms in CatCCS define a specification-implementation relation on models, able to handle also atomic actions and transactions [8].

To give an idea of the possible axioms and of the identifications caused by them between computations, we consider a "functoriality" axiom for $\oplus$:

$$(t_1 \oplus t'_1) ; (t_2 \oplus t'_2) = (t_1 ; t_2) \oplus (t'_1 ; t'_2)$$

and similar axioms for the other operators.

We apply these axioms to our running example.

We have

$$[\alpha,\mathrm{NIL}> \rfloor \beta <+ \gamma \, ; \, \mathrm{NIL} \lfloor [\beta,\mathrm{NIL}> =$$

$$([\alpha,\mathrm{NIL}>|\mathrm{id} \oplus \mathrm{id}|\beta\mathrm{NIL}) <+ \gamma \, ; \; \mathrm{NIL}|\mathrm{id} \oplus \mathrm{id}|[\beta,\mathrm{NIL}> =$$

$$([\alpha,\mathrm{NIL}>|\mathrm{id} \oplus \mathrm{id}|\beta\mathrm{NIL} \; ; \; \mathrm{NIL}|\mathrm{id} \oplus \mathrm{id}|[\beta,\mathrm{NIL}>) <+ \gamma =$$

$$(([\alpha,\mathrm{NIL}>|\mathrm{id};\mathrm{NIL}|\mathrm{id}) \oplus (\mathrm{id}|\beta\mathrm{NIL};\mathrm{id}|[\beta,\mathrm{NIL}>)) <+ \gamma =$$

$$([\alpha,\mathrm{NIL}>|\mathrm{id} \oplus \mathrm{id}|[\beta,\mathrm{NIL}>) <+ \gamma =$$

$$(\mathrm{id}|[\beta,\mathrm{NIL}> \oplus [\alpha,\mathrm{NIL}>|\mathrm{id}) <+ \gamma =$$

$$((\mathrm{id}|[\beta,\mathrm{NIL}>;\mathrm{id}|\mathrm{NIL}) \oplus (\alpha\mathrm{NIL}|\mathrm{id};[\alpha,\mathrm{NIL}>|\mathrm{id})) <+ \gamma =$$

$$(\mathrm{id}|[\beta,\mathrm{NIL}> \oplus \alpha\mathrm{NIL}|\mathrm{id} \, ; \, \mathrm{id}|\mathrm{NIL} \oplus [\alpha,\mathrm{NIL}>|\mathrm{id}) <+ \gamma =$$

$$(\mathrm{id}|[\beta,\mathrm{NIL}> \oplus \alpha\mathrm{NIL}|\mathrm{id}) <+ \gamma \, ; \, \mathrm{id}|\mathrm{NIL} \oplus [\alpha,\mathrm{NIL}>|\mathrm{id} =$$

$$\alpha\mathrm{NIL} \lfloor [\beta,\mathrm{NIL}> <+ \gamma \, ; \, [\alpha,\mathrm{NIL}> \rfloor \mathrm{NIL}.$$

In short:

$$[\alpha,\mathrm{NIL}> \rfloor \beta <+ \gamma \, ; \, \mathrm{NIL} \lfloor [\beta,\mathrm{NIL}> = \alpha\mathrm{NIL} \lfloor [\beta,\mathrm{NIL}> <+ \gamma \, ; \, [\alpha,\mathrm{NIL}> \rfloor \mathrm{NIL}$$

$$= ([\alpha,\mathrm{NIL}>|\mathrm{id} \oplus \mathrm{id}|[\beta,\mathrm{NIL}>) <+ \gamma$$

i.e. two concurrent events can be executed in both orders and in parallel. Identifications of this type give ground for proving characterizations of the transitions of the models of CatCCS in terms of partial orderings. This follows the line of [5], where transitions of T[N], the small category obtained closing a Petri net N with respect to parallel and sequential composition, have been shown isomorphic to the nonsequential processes of N.

To take full advantage of the algebraic approach outlined in this paper, it should be possible to "unfold" the transition systems in either CMonRCCS or CatCCS, obtaining tree-like models (similar to synchronization trees, or, better, to labeled event structures [16] and NMS [7]), where a notion of observational equivalence could be defined. Alternatively, it could be possible to define a notion of abstraction homomorphism [17] directly within CatCCS, thus recovering observational equivalence in terms of a final universal property.

References

1. Badouel, E., Une Construction Systématique de Modèles à Partir de Spécifications Opérationnelles Structurelles, *Report nº 764, INRIA*, 1988.
2. Boudol, G. and Castellani, I., Permutation of Transitions: an Event Structure Semantics for CCS and SCCS, in: *J. W. de Bakker, W. T. de Roever and G. Rozenberg (Eds.), Linear Time, Branching Time and Partial Order in Logics and Models for Concurrency, Springer LNCS 354*, pp. 411-427, 1988.
3. Degano, P., De Nicola, R. and Montanari, U., Partial Ordering Derivations for CCS, *Proc. 5th Int. Conf. on Fundamentals of Computation Theory (L. Budach, ed.), LNCS 199*, pp. 520-523, Springer-Verlag, 1985.
4. Degano, P., De Nicola, R. and Montanari, U., A Distributed Operational Semantics for CCS Based on Condition/Event Systems, *Acta Informatica 26*, pp. 59-91 (1988).
5. Degano, P., Meseguer, J., Montanari, U., Axiomatizing Net Computations and Processes, *Proc. 4th Symp. on Logics in Computer Science*, IEEE 1989.
6. Degano, P., De Nicola, R., Montanari, U., On the Consistency of "Truly Concurrent" Operational and Denotational Semantics, *Proc. 3rd Symp. on Logics in Computer Science*, IEEE 1988, pp. 133-141.
7. Degano, P. and Montanari, U., Concurrent Histories: A Basis for Observing Distributed Systems, *Journal of Computer and System Sciences, Vol. 34*, April/June 1987, No.2/3, pp. 422-461.
8. Gorrieri, R., Marchetti, S. and Montanari, U., A^2CCS: A Simple Extension Of CCS For Handling Atomic Actions, *Proc. CAAP 1988, Springer LNCS 299*, pp. 258-270.
9. Meseguer, J. and Montanari, U., Petri Nets are Monoids, *Technical Report SRI-CSL-88-3, CS Lab., SRI International*, January 1988, also *Information and Computation*, to appear.
10. Meseguer, J. and Montanari, U., Petri Nets are Monoids: A New Algebraic Foundation for Net Theory, *Proc. 3rd Symp. on Logics in Computer Science*, IEEE 1988, pp. 155-164.
11. Milner, R., Notes on a Calculus for Communicating Systems, in: *Control Flow and Data Flow: Concepts of Distributed Programming (M. Broy, ed.), NATO ASI Series F, Vol. 14*, pp. 205-228, Springer-Verlag, 1984.
12. Olderog, E.-R., Operational Petri Net Semantics for CCSP, in: *Advances in Petri Nets 1987, (G. Rozenberg, ed.) LNCS 266*, pp. 196-223, Springer-Verlag, 1987.
13. Plotkin, G., A Structural Approach to Operational Semantics, *Technical Report DAIMI FN-19, Aarhus University, Department of Computer Science*, Aarhus, 1981.
14. Reisig, W., *Petri Nets: An Introduction*, EACTS Monographs on Theoretical Computer Science, Springer-Verlag, 1985.
15. Winskel, G., Petri Nets, Algebras, Morphisms and Compositionality, *Info. and Co., 72*, 197-238 (1987).
16. Winskel, G., Event Structures for CCS and Related Languages, *PROC. 9th ICALP, LNCS 140*, Springer-Verlag, 1982, pp. 561-576.
17. Montanari, U. and Sgamma, M., Canonical Representatives for Observational Equivalence Classes, *Proc. Colloquium On The Resolution Of Equations In Algebraic Structures*, Lakeway, Texas, May 4-6, 1987, North-Holland, 1989, to appear.

Temporal Structures

Ross Casley† Roger F. Crew† José Meseguer‡ Vaughan Pratt†

June 30, 1989

† Dept. of Computer Science, Stanford University, Stanford, CA 94305
‡ SRI International, Menlo Park, CA 94025 and
Center for the Study of Language and Information, Stanford University, Stanford, CA 94305
‡ Supported by the Office of Naval Research under contracts N00014-86-C-0450 and N00014-88-C-0618, and by the National Science Foundation under grant CCR-8707155.

Abstract

We have been developing a process specification language PSL based on an algebra of labeled partial orders. The order encodes temporal precedence of events, and the event labels represent the actions performed. In this paper we extend this basis to encompass other temporal metrics by generalizing partial orders to generalized metric spaces, an interpretation of enriched categories due to Lawvere.

Two needs then arise: a means of specifying kinds of spaces, and a well-defined semantics for PSL relative to a given kind. We define kinds to be semiconcrete symmetric monoidal (ssm) categories, forming the category **SSM**. We find in **SSM** not only kinds of spaces, with and without labels, but their underlying metrics and kinds of labeling alphabets, including certain basic bicomplete kinds $\mathbf{1}, \mathbf{2}, \mathbf{3}, \ldots$, $\bar{\mathbf{R}}_+$, etc. We equip **SSM** with functors ! and $\triangleright$, where $\mathcal{D}!$ denotes the category of spaces on a metric $\mathcal{D}$ and $\mathcal{D} \triangleright \mathcal{E}$ that of $\mathcal{D}$-structured $\mathcal{E}$-labeled spaces. Finally we establish the continuity of these operators.

We define the kind language KL whose terms are formed via the operators ! and $\triangleright$ from constants for the basic kinds. A KL kind is the denotation of a KL term, by induction on which we obtain that all KL kinds are bicomplete. We give a uniform semantics for PSL relative to *any* KL kind, whether a metric, a metric space, an alphabet, or a labeled metric space. That this semantics is well-defined follows from its adherence to universal constructions and bicompleteness of KL kinds.

KL kinds include $\mathbf{1}!$ = sets, $\mathbf{1} \triangleright \mathbf{1}!$ = pointed sets, $\mathbf{2}!$ = preordered sets, $\mathbf{2}! \triangleright \mathbf{1}!$ = labeled preordered sets, $\mathbf{1}!!$ = categories, $\mathbf{2}!!$ = order-enriched categories, $\mathbf{1}!!!$ = 2-categories, $\mathbf{3}!$ = causal spaces, $\mathbf{3}'!$ = prossets, and $\bar{\mathbf{R}}_+!$ = premetric spaces.

1 Introduction

There is much interest these days in giving a precise meaning to the term *concurrent process*. The general approach we start from defines a concurrent process to be a set of behaviors, where a behavior is a set of events. The idea is that each such set of events is a possible behavior of that process, just as each string of a language is a possible utterance in that language.

There are two levels of structure in this approach. The upper level is disjunctive: a process exhibits *one* of its behaviors. The lower level is conjunctive: a behavior performs *all* of its events. This is the *disjunctive normal form* interpretation of the set-of-sets model of concurrent computation.

These two levels are the respective homes of *nondeterminism* and *concurrency*. A nondeterministic choice takes the form of a *necessarily* mutually exclusive selection of a behavior. Concurrent execution is the *possibly* simultaneous execution of the events of a behavior.

This basic framework is now rounded out by imposing suitable structure on each layer. For example we may partially order either layer. An ordering $b < c$ of behaviors "attracts" by weakening the mutual exclusion of b and c to a temporal ordering: both may now happen but b must happen first. This is the nature of an (acyclic) automaton—comparability of states is a necessary condition for both states to appear in the same computation, and the states must appear in the specified order. Dually a comparison $u < v$ of events "repels" by imposing a sequentiality constraint on u and v, a temporal ordering. This is the nature of a schedule. Each layer thus assigns the *same* meaning to comparability and *opposite* meanings to incomparability.

We like to think of those who study suitable structures for each layer as respectively "nondeterminists" and "concurrencists." The strategy of both is to simplify the study of their space by imposing the appropriate extremal structure on the other space. Nondeterminists linearly order the lower layer to make behaviors merely traces (strings). Concurrencists discretely order the upper layer to make processes merely sets of behaviors. Making both simplifications leads to the study of formal languages.

Dual to a formal language as a set of (labeled) chains is the notion of a chain of (labeled) sets. The duality here is in the sense of conjunctive normal form as dual to disjunctive normal form. Relatively little has been done on modeling computation from this perspective, but it is worth pursuing for the additional insights this view has to offer. The main question here is whether there exists a DNF-CNF duality for processes that is as natural as the duality for Boolean algebras defined by negation. The structural evidence from lattice theory would seem to indicate not.

Some concurrencists have adopted the slogan "true concurrency" for their perspective. Robin

Milner has with tongue in cheek described his fellow nondeterminists as working on "false concurrency." However the results of the two camps are not so much contradictory as complementary, and a more informative term would be "true nondeterminism."

There is also ongoing work to reconcile the two layers. A central concern of such work is how the structures of the two layers must interact; we mention here only the work of Winskel [Win80, Win88], Rabinovich and Trakhtenbrot [RT88], and Crew [Cre89]. A recurring theme at this boundary is the Dedekind-Birkhoff-Stone duality between posets and distributive lattices typical of the lower and upper structures respectively.

This paper concerns the concurrency layer, without regard for the nondeterminism layer. Our main question is as follows. *What structures suit true concurrency?* That is, of the various structures that one might impose on the conjunctive level, which are suitable as a semantic foundation for existing or anticipated constructs of process languages?

Our basic premise has been that the best language constructs are those of a universal character, namely categorical limits and colimits. We arrived at this premise by the following steps. First we settled on what we felt was a reasonable notion of behavior, namely labeled partial orders, which can also be thought of as partially ordered multisets or *pomsets* [Gra81, Pra82, Gis84, Pra84]. Next we developed a process language having pomsets (and sets of pomsets) as its domain of discourse [Pra85a, Pra86]. Then we took the part of the language dealing just with individual pomsets and observed that its constructs were all expressible as categorical universals in a succinctly described category of pomsets. From this we leapt to the conclusion that the desirable operations were those expressible as such universals. This last step was taken some two years ago, and in the intervening period we have had no occasion to regret it.

What seems to have made the categorical approach work for us is the use of partial as opposed to total orders. The strings that arise in formal languages as well as in trace-theoretic accounts of concurrent computation [BHR84] can be thought of as totally ordered multisets (the string *abab* is a multiset with two a's and two b's, totally ordered). A pomset is the same notion without the requirement of totality. Whereas pomsets are easily persuaded to band together as categories, strings are much less gregarious in that regard, and we have seen no category of strings offering the advantages of the categories of pomsets that we have been considering. The most glaring deficiency is the shortage of limits and colimits in any plausible category of strings.

While pomsets are attractively categorically, they do present some limitations from a computer science perspective. One limitation is that they treat time as two-valued. The meaning of $u \leq v$ in a pomset is that event u must precede event v.

Real-world schedules typically convey more information than mere temporal precedence. We may for example wish to distinguish strict and nonstrict precedence, giving rise to a three-valued metric logic. A similar distinction arises between necessary and accidental precedence, corresponding to the other possible three-valued metric logic of delays. Or we may require v to follow u by at least three, or at most five, microseconds; such requirements pervade the timing logic of computer hardware. In the context of network queuing problems we may have a Poisson distribution of possible delays between u and v, indicating the need for a suitable stochastic metric logic.

Common to binary relations and metrics on a set X is that they are representable as a square matrix indexed in each dimension by elements of X, that is, a function $\delta : X^2 \to D$ for a suitable choice of D. For binary relations D is $\{0,1\}$, for metric spaces it is the (nonnegative) reals, and so on. But they also have in common a "triangle inequality." The expression of this inequality for partial orders is via the transitivity law $\delta(u,v) \wedge \delta(v,w) \to \delta(u,w)$. For metric spaces we have $\delta(u,v) + \delta(v,w) \geq \delta(u,w)$.

It can be seen that δ equips the space of events with a *metric logic*. The distances of a metric logic are schizophrenic: they combine additively, via $\wedge$ or $+$ as spatial measurements, and they compare logically, via $\to$ or $\geq$, by strength of constraint.

The natural place for these notions to come together is monoidal categories. And the natural construction of generalized metric spaces from a monoidal category $\mathcal{D}$ of such distances is that of enrichment: a metric space is a $\mathcal{D}$-category.[1]

Besides a metric δ assigning distances to pairs of points we wish also to assign labels to individual points. Thus in addition to changing the metric logic we may also wish to change the kind of alphabets supplying such labels.

So where do we look for kinds of behaviors entailing other notions of time such as those just contemplated, each capable of coexisting with various kinds of labeling? Most importantly such kinds, organized as categories, should have "enough" (e.g. all) limits and colimits, allowing us to rely on their existence in giving a semantics for PSL. An easy way of finding and describing such kinds would also be nice.

The answer we offer here is to work in a single category having enough structure to support both metrics and labeling. We define a category **SSM** each of whose objects is a *semiconcrete* category, defined as a category $\mathcal{D}$ equipped with a functor $U_{\mathcal{D}} : \mathcal{D} \to$ **Set**. We define a kind to be an object of **SSM**.

We shall use kinds found in **SSM** in various ways: as metrics, as kinds of alphabets, as kinds of metric spaces, and as kinds of labeled structures. Any kind may be used in any of these ways. Even if a kind does not appear to have been constituted as a category of metric spaces or labeled structures this will not stop us using it as though it were.

[1]In [Pra84] one of us proposed adapting semirings, as used in the Floyd-Warshall algorithm [AHU74], for formalising concurrent real-time processes. As pointed out in [Pra89], monoidal categories are to semirings as categories are to posets.

To obtain spaces with metrics we define a functor ! on **SSM**. Any kind $\mathcal{D}$ may be used as a metric to yield the kind $\mathcal{D}!$ of $\mathcal{D}$-metric spaces. For example $\mathcal{D}$ may be the kind **2** of truth values, which as a metric gives rise to the kind **2**! of preordered sets. The metric $\bar{\mathbf{R}}_+$ gives rise to the kind $\bar{\mathbf{R}}_+!$ of premetric spaces, spaces with a nonnegative real-valued metric δ satisfying the triangle inequality $\delta(u,v)+\delta(v,w)\geq\delta(u,w)$ and $\delta(u,u)=0$.

To obtain spaces with labeled points we define a functor $\triangleright$ on **SSM**. Any kind $\mathcal{D}$ of spaces may be equipped with labels from objects of any other kind $\mathcal{E}$ to yield the kind $\mathcal{D} \triangleright \mathcal{E}$ of $\mathcal{E}$-labeled $\mathcal{D}$-spaces. For example if $\mathcal{D}$ is the kind **Pos** of posets and $\mathcal{E}$ is the kind **Set** of sets then **Pos** $\triangleright$ **Set** is the kind of set-labeled posets, in which each poset is equipped with a set constituting an alphabet of point labels for that poset, together with a function assigning labels to points. The kind **1** $\triangleright$ **Set** consists of one-point spaces each with an alphabet; the assignment of a symbol of Σ to the point has the effect of making this kind the category of pointed sets.

We take $\mathcal{D}!$ to be the category of $\mathcal{D}$-categories or categories *enriched in* a monoidal category $\mathcal{D}$ [EK66, Kel82]. The idea of viewing enriched categories as metric spaces is due to Lawvere [Law73]. And we take $\mathcal{D} \triangleright \mathcal{E}$ to be the comma category $U_{\mathcal{D}} \downarrow U_{\mathcal{E}}$, where $U_{\mathcal{D}}, U_{\mathcal{E}}$ are "forgetful" functors to **Set** that we associate with every kind.

With ! and $\triangleright$ in mind as a semantics, we formulate a *kind language* KL whose terms are formed via the operators ! and $\triangleright$ from constants for the basic kinds. A KL kind is the denotation of a KL term. By structural induction on terms of KL, appealing to completeness of the basic KL kinds and continuity of ! and $\triangleright$, we obtain that all KL kinds are bicomplete.

We give a uniform semantics for PSL relative to *any* KL kind, whether a metric, a metric space, an alphabet, or a labeled metric space. That this semantics is well-defined follows from its adherence to universal constructions and bicompleteness of KL kinds.

There is a strong element of typelessness throughout. As far as KL is concerned there is only one category of kinds, namely **SSM**. And as far as PSL is concerned there is only one category of behaviors, namely a fixed KL kind.

The remainder of this paper is organized as follows. Section 2 presents our process specification language PSL, both informally and formally. Section 3 defines the category **SSM** and gives examples of its objects. Sections 4 and 5 treat ! and $\triangleright$ respectively, defining them as functors on **SSM** and establishing their continuity properties. Section 6 summarizes with conclusions and directions for further work.

2 The Process Specification Language PSL

The language we shall describe here amounts to that part of the process language of [Pra86] defined for single behaviors. It therefore omits those process operations that are defined only for sets of behaviors, notably linearization, union, intersection, star, augment closure, and prefix closure.

The behavior operations that we shall treat here are concurrence, orthocurrence, concatenation, local product, and local concatenation. These suffice to give insight into our approach to defining the semantics of the process language. For more background, rationale, and examples than we can provide here the reader should consult [Pra86]. To give some flavor of how the language works we will give the opening example from that paper, and then supplement it with an example not in that paper.

The following was the first of ten process specification problems posed for participant solution at an SERC/STL workshop on concurrency held in Cambridge, England in 1983.

> The "channel" between endpoints "a" and "b" can pass messages in both directions simultaneously, until it receives a "disconnect" message from one end, after which it neither delivers nor accepts messages at that end. It continues to deliver and accept messages at the other end until the "disconnect" message arrives, after which it can do nothing. The order of messages sent in a given direction is preserved.

Some participants submitted two-page solutions. The best at the time was seven temporal logic axioms supplied by Koymans-de Roever—the solution of one of us [Pra85b] was a dozen lines of predicate calculus. With considerable subsequent effort we eventually obtained the following "one-liner" logically equivalent to our original predicate calculus "12-liner."

$$\pi\underline{\lambda}(((M^* \otimes A) \otimes AB) \| ((M^* \otimes B) \otimes BA)) \wedge \neg\underline{\Diamond}^{\pm}\delta$$

The disconnect property is expressed to the right of the $\wedge$; we will not bother with it here. The two directions of the channel are the two arguments to $\|$. Each direction consists of a set M^* of possible strings from a message alphabet M, i.e. a set of behaviors. The next few operations act pointwise on this set, so it suffices to say what happens to a single message sequence s of M^*. The product with A "stamps" each message m in s with its origin to yield a sequence of (m, A)'s. The subsequent product with AB blows up each message in this sequence to two copies in their own sequence, stamping the first copy with a second A and the second with a B. The order on the result is the direct product of the constituent orders. Thus a message originating at A has the form $((m, A), A)$ while the same message as subsequently received at B has the form $((m, A), B)$. The product $2718 \otimes AB$ for example is:

$$
\begin{array}{ccccccccc}
((2,A),A) & \longrightarrow & ((7,A),A) & \longrightarrow & ((1,A),A) & \longrightarrow & ((8,A),A) & \longrightarrow & \\
\downarrow & & \downarrow & & \downarrow & & \downarrow & & \\
((2,A),B) & \longrightarrow & ((7,A),B) & \longrightarrow & ((1,A),B) & \longrightarrow & ((8,A),B) & \longrightarrow &
\end{array}
$$

In the other direction of the two-way channel the roles of A and B are interchanged, giving two more kinds of message: $(m, B), B$ transmitted from B and $(m, B), A$ received at A.

The two channels are then combined via $||$, whose pointwise effect on pairs of grids is simply to juxtapose the grids (coproduct of posets). The π and $\underline{\lambda}$ then form the prefix closure (which we may ignore) and *local* linearization of their respective arguments. Local linearization augments the partial order to a linear order locally, meaning for events at a single location. We followed the convention that the location was given by the second element of a pair; thus the receipt $((m, A), B)$ and the transmission $((n, B), B)$ are colocated messages, albeit travelling in opposite directions. Colocation meant that $\underline{\lambda}$ forced them to be comparable in the temporal order, forming the set of all such locally linear augments. The net effect is a pair of linear orders, one consisting of receipts and transmissions at A and the other similarly at B, together with the AB transmissions, which preserve order among themselves (messages are received in the order transmitted), and the BA transmissions, which are pretty much independent of the AB transmissions.

An example not in [Pra86] is that of a bank of memory cells. Typically in other process languages notions such as memory cells and reading and writing are taken as primitive. The primitives of our language are more primitive than that, and memory cells along with all other computational structures are built from quite trivial structures.

The syntax of a write-once memory is to write and then read repeatedly, expressible as the regular expression $W; R*$ where W and R are atomic behaviors, i.e. the set of all behaviors of the form $W; R^i$. The semantics of such a memory is $M = (W; R*) \otimes V$ where V is a domain of values to be written to and read from the memory, represented as a set of atomic behaviors. A typical behavior of M is determined by choosing a value $v \in V$ and an integer i, yielding the behavior $(W, v); (R, v)^i$. This is a sequence of $i + 1$ events, the action of the first of which is to write v to memory, and of the remainder to read v back i times. Memory is thus achieved by the choice of v *outside* the scope of the expression where i is chosen.

M may now be extended to a write-many memory $N = M^* = \epsilon \cup M \cup M;M \cup \ldots$, allowing it to write any number of times. The reads between writing u and writing the next v all return the value u. A variant of this is to take $N = M^{\underline{*}} = \epsilon \cup M \cup M\underline{;}M \cup \ldots$, *local* star, whereby consecutive runs of M are *locally* concatenated, analogously to local linearization: no temporal precedence is forced between input and output of the memory, allowing the possibility that old values may be read back for a while after a new value has been written. This models a memory where the effect of the write may be internally delayed.

N may be turned into a bank $A \cdot N$ (copower of A with N) of memory cells where A is a set of memory locations, i.e. forming a coproduct of as many copies of N as there are locations.

It may be thought that these operations were chosen to favor a category theory semantics at the possible expense of the programmer. In fact this language was designed prior to giving any thought to its categorical formulation. The principal criteria exercised in the choice of operations were utility, mathematical naturality, and enough expressive power to beat down our 12-line predicate-calculus axiomatization of the SERC problem to one or two lines. The semantics given in [Pra86] for the language made no reference to category theory.

Let us now enumerate those operations of our language that are definable for single behaviors. This will suffice to illustrate the ways in which we use universal constructions to obtain a diversity of useful operations. In particular it will show how operations can be designed to take advantage of metric and labeling information when it exists, yet be defined independently of the existence of such information so as to be meaningful even when it does not exist.

It should be understood that all operations are defined at a single kind, namely an object $\mathcal{D}$ of **SSM**. The definitions may refer to $\mathcal{D}$, but will not assume that $\mathcal{D}$ has any a priori structure such as $\mathcal{D} = \mathcal{D}'! \rhd \mathcal{E}$. The definitions may construct categories *from* $\mathcal{D}$ however.

Although these operations are defined independently of any particular structure on $\mathcal{D}$, they are best understood in particular structures such as **Pom**. We will therefore briefly describe their effect in **Pom**.

Disjoint Concurrence. The *disjoint concurrence* $\mathtt{dconcur}(p,q)$, or $p+q$, of behaviors p and q is their coproduct $p+q$ in $\mathcal{D}$. In **Pom** this juxtaposes pomsets and relabels them to make their alphabets disjoint. One application is in bringing modules such as integrated circuits into a common arena before applying a wiring operation, where one needs to keep the pins separate. Using ordinary concurrence would result in all pin 13's bearing the same label; thus subsequently connecting a wire to one pin 13 would result in connecting it to all pin 13's.

Concurrence. In **Pom** the concurrence of two pomsets just means their juxtaposition. For example whereas the disjoint concurrence of two copies of action a is a two-event behavior whose actions are a_0, a_1 respectively, the concurrence of these two copies of a is a two-event behavior both of whose actions are a.

Thus if p and q have a common alphabet Σ we want the concurrence $p||q$ to be the coproduct

$p+q$ in the category of pomsets with a fixed alphabet Σ, i.e. in $\mathbf{Pom}_\Sigma = \mathbf{Pos} \triangleright \Sigma$, $\mathbf{Pom}_\Sigma$ being the *fiber* of $\mathbf{Pom} = \mathbf{Pos} \triangleright \mathbf{Set}$ at the set Σ. Here Σ is a one object category with underlying set $U_\Sigma(\cdot) = \Sigma$.

Two issues arise here. First, we need a definition that will apply to any $\mathcal{D}$, not just one of the form $\mathcal{D} \triangleright \mathcal{E}$. Second, $\mathbf{Pom}_\Sigma$ is not in **SSM** unless Σ is a singleton because the Σ we have defined has a forgetful functor that is only strong monoidal when Σ is a singleton.

We address both issues by requiring the label coordination implicit in the information that p and q have the same alphabet to be made explicit. This information takes the form of two morphisms to a coordinating object k, with $k = \Sigma$ in the motivating application but with other forms of coordination also allowed. (Thus we obtain a more general operation.)

The concurrence $\mathtt{concur}(p \to k \leftarrow q) = r \to k$ of two morphisms to k is their coproduct in $\mathcal{D}/k$[2], also a morphism to k. We refer to r as $p||q$. When p, q and context determine k and the morphisms, or when operations, e.g. alphabet(p), are provided for so determining them, $\mathtt{concur}(p \to k \leftarrow q)$ may be written $p||q$ to so indicate.

Disjoint concurrence can be described as a special case of concurrence, namely where $\Sigma = 1$, the final object of $\mathcal{D}$. In this case the object p conveys no more or less information than the morphism $p \to 1$.

Orthocurrence. The *orthocurrence* $p \otimes q$ of two behaviors (which we previously wrote $p \times q$ [Pra86]) is defined as their tensor product. It describes any flow situation in which p and q flow past each other. For example in **Pom** a message sequence 010 may flow along a channel TR (Transmit-Receive); then $010 \otimes TR$ is

$$\begin{array}{ccccccc} 0 & & & & & & \\ \downarrow & & T & & (0,T)\to & (1,T)\to & (0,T) \\ 1 & \otimes & \downarrow & = & \downarrow & \downarrow & \downarrow \\ \downarrow & & R & & (0,R)\to & (1,R)\to & (0,R) \\ 0 & & & & & & \end{array}$$

This would be direct product of posets except that we need to reckon with the possibility of repeated labels, as in the example. Thus it would seem that we want something like product of functor categories, or of comma categories. However this is not always the case. When taking the product of metric spaces, if in the above example we have a delay of one second from 0 to 1 and two seconds from T to R then the delay from $(0,T)$ to $(1,R)$ should not be

[2] $\mathcal{D}/k$ is the comma or slice category of "objects over" k, i.e. morphisms to k

the *maximum* of these two delays, as would happen if we insisted on the product, but rather their *sum*, namely the tensor product of the underlying metric. As will be seen in section 4, this tensor product lifts, by construction of $\mathcal{D}!$, to a tensor product on $\mathcal{D}!$ which has the intended effect of tensoring the delays that arise at diagonals in this fashion, rather than using their product, here max. Thus in this case tensor product gives the expected outcome.

Fortunately $\mathcal{D}$ always has a tensor product $\otimes$, allowing us to define orthocurrence simply as that product without reference to metrics on spaces, or distances between points in spaces.

In **Pom** orthocurrence happens to be ordinary product. When the metric is numeric however, as we just saw, this need no longer be the case. Indeed orthocurrence is ordinary product just when $\mathcal{D}$ is cartesian closed, this being the definition of cartesian closed.

Concatenation. There are two forms of concatenation, disjoint and ordinary, paralleling the situation for concurrence. We treat here only disjoint concatenation, where in the case of say **Pom** the label sets of the two arguments are made disjoint.

In **Pom** the disjoint concatenation $p;q$ is as for the disjoint concurrence $p+q$ with the additional constraint that every event of p precede every event of q. Now we cannot refer explicitly to order. Moreover we would really prefer a more general notion of concatenation if possible, where we could specify say a real-valued delay between p and q when $\mathcal{D}$ has a quantitative metric.

We address both these needs by supplying an additional parameter, namely the delay. In the case of concatenation in **Pom** the object can be the canonical two-event concatenation, namely a two element chain with identity as its labeling function. For numerical delays the language may have constants for say 0 seconds or π fortnights. Common to all these objects must be that it has a two-element underlying set, the "canonical behavior" with that delay. Thus the operation is $\mathtt{concat}(p,d,q)$ where p,d,q are objects of $\mathcal{D}$.

We shall depend in two places on the counit ϵ of the adjunction formed by the forgetful functor U from the model to **Set** together with its left adjoint D.

Now given p and q, the behaviors to be concatenated, and a behavior d with two elements in its underlying set, we construct arrows $U(d)\times U(p\otimes q)\to 2\times U(p)\times U(q)\to U(p)+U(q)$; the first arrow uses $U(d)\cong 2$ by hypothesis, and U strong, while the second is the standard if-then-else arrow $2\times a\times b\to a+b$. Now apply D to the composite and extend on the left and right to obtain $DU(d)\otimes DU(p\otimes q)\to D(U(d)\times U(p\otimes q))\to D(U(p)+U(q))\cong DU(p)+DU(q)\to p+q$. The first arrow is from D strong, the second is from before, the third is from D a left adjoint, and the last is $\epsilon_p+\epsilon_q$. Take this as the top edge of a pushout.

Now form the left edge of the pushout as $\epsilon_d\otimes 1$ where 1 is the identity on $DU(p\otimes q)$. The pushout then has the desired concatenation $\mathtt{concat}(p,d,q)$ in the bottom right hand corner.

This construction generalizes straightforwardly to obtain the "pomset definable" operations described by [Gis84] and [Pra85a].

Local product. Given arrows $p \to L \leftarrow q$, their limit is their *local product* $p \times_L q$. When L is a set we think of its elements as locations. In that case $p \times_L q$ forms a coproduct over the locations l in L of $p_l \times q_l$ where p_l, q_l denote the inverse images of l. We use local product in the next definition.

Local concatenation. As for local product the data are arrows $p \to L \leftarrow q$. The meaning of local concatenation of p and q over L is that p and q run concurrently, subject to the constraint at each location that every event of p at that location precede every event of q at that location. We form the standard quotient $p \times_L q \to p \times q$ then apply U to obtain $U(p \times_L q) \to U(p \times q) \cong U(p) \times U(q)$ using U a right adjoint. This now can be matched to the start of the construction for concatenation, substituting $\times_L$ for $\otimes$, and the rest follows as for concatenation.

3 The Category SSM

Our goal here is to define the category **SSM** of semiconcrete symmetric monoidal categories, the source of all kinds of behavior. Unlike cartesian closed categories, monoidal categories have had almost no exposure in computer science. We shall therefore aim this account at an audience comfortable with Mac Lane chapters I-IV (IV = adjunctions). We first define a monoidal category, referring the reader to chapter VII of Mac Lane [Mac71] for a more comprehensive account. We take as examples some of the monoidal categories we will be using later in various applications. We then define the notions of monoidal functor, strong monoidal functor, and semiconcrete category, the terminology "strong" for functors being taken from [KS74].

3.1 Monoidal Categories

Informally, a monoidal category amounts to a structure that is both a monoid and a category. Formally a *strict monoidal category* $\mathcal{D} = (D, \otimes, I)$[3] is a category D together with a functor $\otimes : D^2 \to D$ called the *tensor product* and an object I of D called the *unit*, such that the object part and the morphism part of D each form a monoid under $\otimes$ with respective identities I and 1_I.

The structure $(\mathbf{Set}, \times, \{0\})$ with $\times$ cartesian product is not strict monoidal because $X \times (Y \times Z)$ is not equal to $(X \times Y) \times Z$ unless one of X, Y, or Z is empty. Yet they *are* isomorphic.

[3]Mac Lane uses $\in$ for $\otimes$.

Moreover there is a particular isomorphism we would like to be able to *assume* is the one meant when we say that they are isomorphic, namely $\alpha_{XYZ} : (X \times Y) \times Z \to X \times (Y \times Z)$ defined as $\alpha(\langle\langle x, y\rangle, z\rangle) = \langle x, \langle y, z\rangle\rangle$. Similarly particular isomorphisms take the place of the two identity laws for the unit $I = \{0\}$.

We therefore define a more general notion. A *monoidal category* $\mathcal{D} = (D, \otimes, I, \alpha, \lambda, \rho)$ is as for strict monoidal categories but with the three equations replaced by specified natural isomorphisms $\alpha : (c \otimes d) \otimes e \cong c \otimes (d \otimes e)$, $\lambda : I \otimes d \cong d$ and $\rho : d \otimes I \cong d$. Certain *coherence conditions* must be met, whose effect is that *every* identity of the theory of monoids, not just the three axioms, is unambiguously and coherently associated with a specific natural isomorphism. A strict monoidal category is a monoidal category whose specified natural isomorphisms are all identity morphisms, making it unnecessary to specify them, identities being unique.

A *symmetric* monoidal category has an additional natural isomorphism $\gamma : d \otimes e \cong e \otimes d$, with the coherence conditions being expanded correspondingly to maintain a coherent correspondence between equations of the theory of commutative monoids and the specified natural isomorphisms of the given symmetric monoidal category.

When $\otimes$ has a right adjoint in one or both arguments we call the category *closed* or *biclosed* respectively, abbreviating "closed monoidal category" to "closed category." Since left adjoints preserve colimits, $\otimes$ closed in say the right argument implies $A \otimes (B + C) \cong A \otimes B + A \otimes C$ for the coproduct $B + C$ and $A \otimes 0 \cong 0$ for the initial object 0 (and symmetrically for a category closed on the left), in practice a very useful semicondition for testing closedness. A symmetric closed category is automatically biclosed.

When a category has either finite coproducts or finite products they determine an obvious symmetric monoidal structure. In the latter case, if that structure is closed (hence biclosed), we call the category *cartesian closed.* By this definition cartesian closed is a property of categories rather than monoidal categories, and can be conveniently defined without going into the details of monoidal categories.

3.2 Examples

(1) Every successor ordinal n as an n-object category with morphisms $i \leq j$ forms a cartesian closed category, taking I to be the final object (whence successor) and $\otimes$ to be min. Other choices for $\otimes$ and I lead to monoidal categories that will not be cartesian closed. There are 2^{n-1} symmetric categories on n for which $\otimes$ is idempotent, all strict since n has no nonidentity isomorphisms. Exactly half of these are closed, namely those satisfying $n \times 0 = 0$. This half exhibits a very pretty structure: $\otimes$ as a sup defines an order such that if we enumerate the elements of n in their numerical order, then they descend in $\otimes$ order until I is reached, then they ascend again. Furthermore the descent interleaves with the ascent to make the $\otimes$ order

linear, imparting structure to the enumeration 2^{n-2}.

The most important case is **2**, defined as the unique closed structure on the ordinal 2, for which $\otimes$ is $\wedge$ and $I = 1$. (The other idempotent monoidal structure for **2** takes $\otimes$ to be $\vee$ and $I = 0$.)

In modelling concurrency Gaifman has implicitly used each of the two idempotent symmetric closed structures on 3, one that we call $\mathbf{3}'$ in connection with work with one of us on strict vs. nonstrict temporal precedence [GP87] and the other, **3**, the cartesian closed one, in more recent work on causal vs. accidental precedence [Gai89].

The appropriate interpretations of the elements of $\mathbf{3}'$, in increasing order of strength, are as follows. 0 denotes no constraint, 1 denotes nonstrict order $\leq$, and 2 denotes strict order or nonzero delay $<$. That $1 \otimes 2 = 2$ in $\mathbf{3}'$ corresponds to the implication $u \leq v < w \rightarrow u < w$. This structure was used to define "prossets" (preordered specification sets) [GP87], which were used to prove Kahn's principle relative to a pomset-based semantics of nets [Pra86].

In the cartesian closed **3**, the interpretation of 0 is no constraint as before. However $1 \in \mathbf{3}$ denotes accidental order, and 2 denotes causal order. The identity $1 \otimes 2 = 1$ in **3** corresponds to the notion that if u accidentally precedes v whereas v *causes* (causally precedes) w, then one can infer from this that u accidentally precedes w, but one cannot infer that u causally precedes w.

It is noteworthy that the two idempotent symmetric closed structures on 3 should not only both have such natural interpretations but should both be showing up in the concurrency literature [GP87, Gai89], even if not described as such.

(2) We have already discussed **Set** with $\otimes$ taken to be cartesian product, as a basic example of a nonstrict category. This monoidal structure for **Set** is cartesian closed, and will always be the one we have in mind when referring to **Set** as a monoidal category.

A variation on **Set** that is of some interest is **Seti**, sets and their inclusions.[4] Coproduct and product in **Seti** are respectively union and intersection. Unfortunately **Seti** is not cartesian closed, there being no final object to serve as the set of all sets. Yet **Seti** does admit a monoidal structure, namely $\otimes = +$ and $I = 0$. However with this monoidal structure the inclusion of **Seti** into **Set** is not strong monoidal. There is however a trivial strong monoidal functor to **Set**, taking everything to 1, and a seemingly more useful one taking each set to its power set in **Set**, either of which could be taken as the "forgetful" functor of **Seti**.

(3) The reverse-ordered nonnegative reals including ∞ form a bicomplete cartesian closed category, taking $\otimes$ to be max and I to be 0. This monoidal category forms the basis for *ultrametric* spaces. The more useful monoidal structure takes $\otimes$ to be addition, a symmetric

[4]For purists, **Seti** is any maximal subcategory of **Set** which is a poset and consists only of monics of **Set**.

closed *but not cartesian closed* structure. This structure forms the basis for ordinary metric spaces. Note the distinction drawn here between the additive structure as a nonclosed monoidal structure vs. the underlying category as cartesian closed by virtue of the product supplying a closed monoidal structure. This is Lawvere's original example [Law73] of a single category admitting two useful closed monoidal structures; the ordinal **3** above provides another pair of useful examples. In general a single category may admit many monoidal structures, only one of which, up to isomorphism, can be cartesian closed.

(4) Take **SR** to be the category of sets of reals with morphisms reverse inclusions between sets (there is an arrow from X to Y just when Y is a subset of X), with $X \otimes Y$ the set $\{x + y | x \in X, y \in Y\}$ and $I = \{0\}$. This is a natural generalization of (2); the latter can then be seen as the fragment of **SR** consisting of the lower sets (if x is in the set and $y \leq x$ then y is in the set) that contain 0. The dual consisting of upper sets is equally useful. Combining these leads to convex sets or intervals where the two endpoints of the interval must be specified, with addition being describable as addition of endpoints. Another fragment is all singletons. **SR** works just as well as these various fragments and is to be preferred unless representability of sets is an issue.

(5) Any monoid automatically forms a monoidal category having a discrete underlying category (all morphisms identities). Among these we will find groups particularly useful as metrics for "rigid" spaces. This is the one category in these examples that is not bicomplete.

3.3 Functors

Just as strict monoidal categories motivate monoidal categories, so do strict monoidal functors motivate monoidal functors. A *strict monoidal functor* $F : \mathcal{D} \rightarrow \mathcal{D}'$ is a functor $F : D \rightarrow D'$ between their underlying categories satisfying $Fx \otimes' Fy = F(x \otimes y)$ and $I' = FI$ for objects and similarly for morphisms. (The unit morphism is often written 1 instead of I or 1_I.) We now give the general (nonstrict) definition.

Definition A *monoidal functor* $(F, \tau, t) : (\mathcal{D}, \otimes, I, \alpha, \lambda, \rho) \rightarrow (\mathcal{D}', \otimes', I', \alpha', \lambda', \rho')$ consists of a functor $F : D \rightarrow D'$, a natural transformation $\tau_{xy} : Fx \otimes' Fy \rightarrow F(x \otimes y)$, and a morphism $t : I' \rightarrow FI$ of D'.

We again have coherence conditions, which in this case require natural transformations constructed from τ and t to commute with the corresponding transformations constructed from α, λ, ρ (see [EK66]).

If $\mathcal{D}$ and $\mathcal{D}'$ are symmetric monoidal then a *symmetric* monoidal functor $F : \mathcal{D} \rightarrow \mathcal{D}'$ must satisfy an additional coherence condition to do with the symmetry γ.

Following [KS74], we call (F, τ, t) *strong monoidal* when τ and t are isomorphisms. Thus strong monoidal lies between strict monoidal, where τ and t are identities, and monoidal. The morphisms of **SSM** will be strong monoidal.

A strong monoidal functor need not be faithful. To see this take the three nonidentity arrows of the ordinal **3** to be $(0,0,1)$, $(1,0,2)$, and their composition $(0,0,2)$. Form T from **3** by adding an arrow $(1,1,2)$. Define $I = 0$ and $\otimes$ = coordinatewise max. The functor $T(0,-) : T \to$ **Set** is then strong but not faithful since it identifies the two parallel arrows. Conversely a faithful monoidal functor need not be strong, e.g. the forgetful functor from $\mathbf{Vct}_K \to$ **Set**.

Definition. The category **SSM** is the slice (comma) category **SSMonCat/Set**. Its objects are pairs $(\mathcal{D}, U)$ called *ssm*'s (for semiconcrete symmetric monoidal categories), where $U : \mathcal{D} \to$ **Set** is strong monoidal, and its morphisms $F : (\mathcal{D}, U) \to (\mathcal{D}', U')$ satisfy $U = U'F$.

The condition $U = U'F$ is strong, but reasonably so given that its main applications are to the underlying sets of labeled spaces, where the action of F is intended to influence the metric without influencing the points themselves. For example $F!$ may be a morphism from metric spaces to posets induced by a morphism F from reals to truth values; $F!$ then replaces numeric edge labels by truth values.

All the examples of monoidal categories in our list above belong to **SSM**. Only the last example is not bicomplete. The ordinal **1** is the null object of **SSM**.

4 Enriched Categories and the ! Operator

We now introduce the basic enriched category notions. Our hope is that by having the definitions not only close at hand but presented from the same perspective as the applications, enriched categories will not seem too inaccessible. Enriched categories or V-categories are treated briefly by Arbib and Manes [AM75] and exhaustively and precisely by Kelly [Kel82]. Section I-8 of Mac Lane also touches on them, without calling them such. As far as we are aware ours is the first proposed engineering application of Lawvere's vision [Law73] of enriched categories as a combined generalized metric and logic.

4.1 $\mathcal{D}$-categories

The essence of an enriched category may be found in an ordinary category C. Write $\delta(u,v)$ for the homset of morphisms from u to v. Associated with each triple u, v, w of objects of C is a function $m_{uvw} : \delta(u,v) \times \delta(v,w) \to \delta(u,w)$, namely composition, satisfying a composition law. And to each object u there is a function $j_u : \{\cdot\} \to \delta(u,u)$ picking out the identity element $1_u \in \delta(u,u)$.

We pass from this account of an ordinary category C to the notion of a category C *enriched in* $\mathcal{D}$, or $\mathcal{D}$-category, by the familiar passage from sets to objects *at the homset level.* That is, homsets become objects of $\mathcal{D}$ while functions between homsets and other sets become morphisms of $\mathcal{D}$. The set or class of objects of C do not participate in this passage: the objects of C remain a set or class. A functor $f : A \to B$ participates partially in this passage: when we view it as the $\mathcal{D}$-functor $f : A \to B$, its object part remains unchanged but its action on morphisms, viewed for each pair uv of objects of A as a function $\tau(u,v) : \delta^A(u,v) \to \delta^B(f(u),f(v))$, becomes a morphism of $\mathcal{D}$ between homobjects $\delta^A(u,v)$ and $\delta^B(f(u),f(v))$ of A, B respectively.

From this viewpoint, ordinary categories and functors between them are **Set**-categories and **Set**-functors, categories and functors enriched in **Set**.

Expressing this formally, a small[5] $\mathcal{D}$*-category* $A = (V, m, j)$, or *category enriched in* $\mathcal{D}$, consists of a set V together with a function $\delta : V^2 \to \mathrm{ob}(\mathcal{D})$, families of morphisms of $\mathcal{D}$, namely $m_{uvw} : \delta(u,v) \otimes \delta(v,w) \to \delta(u,w)$ (for composition) and $j_u : I \to \delta(u,u)$ (for the identities), such that for all objects u, v, w the following diagrams commute. These diagrams express associativity of composition, and the left and right identity laws, explained below.

$$
\begin{array}{ccccc}
(\delta(u,v) \otimes \delta(v,w)) \otimes \delta(w,x) & & \xrightarrow{\alpha_{\delta(u,v)\delta(v,w)\delta(w,x)}} & & \delta(u,v) \otimes (\delta(v,w) \otimes \delta(w,x)) \\
\downarrow{\scriptstyle m_{uvw} \otimes 1} & & & & \downarrow{\scriptstyle 1 \otimes m_{vwx}} \\
\delta(u,w) \otimes \delta(w,x) & \xrightarrow{m_{uwx}} & \delta(u,x) & \xleftarrow{m_{uvx}} & \delta(u,v) \otimes \delta(v,x)
\end{array}
$$

$$
\begin{array}{ccccc}
I \otimes \delta(u,v) & \xrightarrow{\lambda_{\delta(u,v)}} & \delta(u,v) & \xleftarrow{\rho_{\delta(u,v)}} & \delta(u,v) \otimes I \\
\downarrow{\scriptstyle j_u \otimes 1} & & \| & & \downarrow{\scriptstyle 1 \otimes j_v} \\
\delta(u,u) \otimes \delta(u,v) & \xrightarrow{m_{uuv}} & \delta(u,v) & \xleftarrow{m_{uvv}} & \delta(u,v) \otimes \delta(v,v)
\end{array}
$$

Readers more accustomed to working with graphs than categories may view a $\mathcal{D}$-category as an edge-labeled graph, with labels drawn from $\mathrm{ob}(\mathcal{D})$. If $\mathcal{D}$ has an initial object 0 we may follow the convention of omitting edges labeled with 0 (making labels unnecessary in the case of preorders).

For $\mathcal{D} = \mathbf{2}$ the diagrams expressing the associativity and identity laws hold vacuously since

---

[5] We will not be needing the concept of enrichment for any large categories in this paper.

$\mathcal{D}$ is a preorder. Thus composition and identity become

$$\begin{aligned} m_{uvw}: &\quad u \leq v \wedge v \leq w \quad \rightarrow \quad u \leq w \\ j_u: &\quad u \leq u \end{aligned}$$

expressing respectively transitivity and reflexivity. Thus **2**-categories (in the sense of categories enriched in **2**, as opposed to ordinary 2-categories which in this framework would be called **Cat**-categories) are just preorders. This example is closely related to our pomset model, with the preorder (in the original treatment a partial order) indicating event precedence.

For $\mathcal{D} = \bar{\mathbf{R}}_+$, also a preorder, we again may ignore the associativity and identity laws. Here we get

$$\begin{aligned} m_{uvw}: &\quad \delta(u,v) + \delta(v,w) \geq \delta(u,w) \\ j_u: &\quad 0 \geq \delta(u,u) \end{aligned}$$

In this case m expresses the triangle inequality while j forces $\delta(u,u) = 0$, as there are no negative distances. Thus $\bar{\mathbf{R}}_+$-categories are what we shall call premetric spaces (Lawvere calls them generalized metric spaces), lacking the other two Fréchet conditions $\delta(u,v) = \delta(v,u)$ and $\delta(u,v) = 0$ implies $u = v$. This example supports a more informative notion of concurrent behavior than pomsets, since now the *delay* between two events can be indicated in addition to their precedence relation.

For $\mathcal{D} = \mathbf{Cat}$ we obtain ordinary 2-categories, along the same lines as **Set** but with the addition of 2-cells between morphisms of the same homset.

4.2 $\mathcal{D}$-functors

Definition. A *$\mathcal{D}$-functor* $F : A \rightarrow B$ where A and B are $\mathcal{D}$-categories consists of a function $f : V_A \rightarrow V_B$ between object sets together with a family $\tau_{uv} : \delta_A(u,v) \rightarrow \delta_B(fu, fv)$ of morphisms between homobjects satisfying the following preservation of composition and identities conditions:

$$\begin{array}{ccc} \delta_A(u,v) \otimes \delta_A(v,w) & \xrightarrow{m_{uvw}} & \delta_A(u,w) \\ {\scriptstyle \tau_{uv} \otimes \tau_{vw}} \downarrow & & \downarrow {\scriptstyle \tau_{uw}} \\ \delta_B(fu,fv) \otimes \delta_B(fv,fw) & \xrightarrow{m_{fu,fv,fw}} & \delta_B(fu,fw) \end{array} \qquad \begin{array}{ccc} I & \xrightarrow{j_u} & \delta_A(u,u) \\ \Vert & & \downarrow {\scriptstyle \tau_{uu}} \\ I & \xrightarrow{j_{fu}} & \delta_B(fu,fu) \end{array}$$

The elements of a $\mathcal{D}$-functor are depicted on the left of the following figure, and compose as on the right.

$$
\begin{array}{ccc}
u & \xRightarrow{\delta_A(u,v)} & v \\
f \Big\updownarrow & \sigma_{uv} \Big\downarrow & \Big\updownarrow f \\
fu & \xRightarrow[\;]{\delta_B(fu,fv)} & fv
\end{array}
\qquad\qquad
\begin{array}{ccc}
u & \xRightarrow{\delta_A(u,v)} & v \\
f \Big\updownarrow & \sigma_{uv} \Big\downarrow & \Big\updownarrow f \\
fu & \xRightarrow{\delta_B(fu,fv)} & fv \\
g \Big\updownarrow & \tau_{uv} \Big\downarrow & \Big\updownarrow g \\
gfu & \xRightarrow{\delta_C(gfu,gfv)} & gfv
\end{array}
$$

It can be seen that **2**-functors are monotone functions, $\bar{\mathbf{R}}_+$-functors are contracting maps, and **Set**-functors are ordinary functors. In our computational application a $\mathcal{D}$-functor is an event map which maintains all temporal constraints, though the result may have more constraints than the source.

$\mathcal{D}$-functors compose pointwise, by composing their respective object function and homobject morphisms in the obvious way. This composition is easily verified to be associative and to have a left and right identity at every $\mathcal{D}$-category consisting of the appropriate identity morphisms. Hence the class of all $\mathcal{D}$-categories and $\mathcal{D}$-functors between them forms a (large) category, denoted $\mathcal{D}!$. The more usual term for $\mathcal{D}!$ is $\mathcal{D}$-**Cat**.

Lemma 1 *If* $\mathcal{D}$ *is in* **SSM** *so is* $\mathcal{D}!$.

The symmetric monoidal structure is immediate. Given $\mathcal{D}$-categories A and B, we define $A \otimes B$ as a $\mathcal{D}$-category with objects $V_{A\otimes B} = V_A \times V_B$ and with morphism objects of the form $\delta_{A\otimes B}((u,u'),(v,v')) = \delta_A(u,v) \otimes \delta_B(u',v')$. The unit object of $\mathcal{D}!$ is a category I with only one object 1 and with $\delta_I(1,1) = I$. It can then be seen that, up to isomorphism, the forgetful functor $\mathcal{D}!(I,\_)$ sends each $\mathcal{D}$-category A to the set V_A; therefore, $\mathcal{D}!$ is strong monoidal.

A notion of $\mathcal{D}$-natural transformation can also be defined, making $\mathcal{D}!$ a 2-category. This 2-category structure permits more succinct definitions of some of the limits we want for the operations of our programming language, e.g. concatenation as cocomma. However we have only begun to take advantage of this useful additional structure and do not treat it further here.

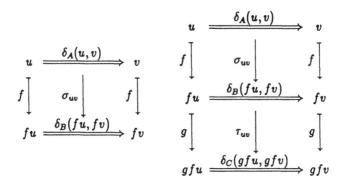

4.3 Enrichment as a Functor

We now consider the question of how relating different domains $\mathcal{D}$ of delay constraints by means of monoidal functors lifts to $\mathcal{D}!$. In other words, we wish to view ! as a functor on the category **SSM**. We discuss the lifting for the more general case of a monoidal functor and obtain our desired functor as a corollary.

A monoidal functor $F : \mathcal{D} \to \mathcal{D}'$ induces a functor $F! : \mathcal{D}! \to \mathcal{D}'!$ as follows.

Definition Given a monoidal functor $F : \mathcal{D} \to \mathcal{D}'$, we define the functor $F! : \mathcal{D}! \to \mathcal{D}'!$ *induced by F* as follows:

For any $\mathcal{D}$-category A, we get the $\mathcal{D}'$-category $F!A$ from

$$\begin{aligned} V_{F!A} &= V_A \\ \delta_{F!A}(u,v) &= F(\delta_A(u,v)) \\ m_{F!A,uvw} &= Fm_{A,uvw} \circ n_{\delta_A(u,v)\delta_A(v,w)} \\ j_{F!A,u} &= Fj_{A,u} \circ k \end{aligned}$$

For any $\mathcal{D}$-functor $f : A \to B$, we get a $\mathcal{D}'$-functor $F!f$ by using the same object function and by taking $(F!f)_{uv} = F(\tau_{uv})$

Proposition 2 *$F! : \mathcal{D}! \to \mathcal{D}'!$ is a functor.*

The proof is a diagram chase verifying that $F!A$ and $F!f$ really do satisfy the axioms for $\mathcal{D}'$-categories and $\mathcal{D}'$-functors.

For the case when F is a *strong* symmetric monoidal functor, the above proof shows in fact that $F!$ is so too, and this establishes our desired

Corollary 3 *The construction $\mathcal{D} \mapsto \mathcal{D}!$ is a functor $!_{-}$:* **SSM** $\to$ **SSM**.

4.4 Examples

Consider the case where $\mathcal{D}$ is discrete, having no morphisms other than identities. In our interpretation this would correspond to requiring definite delays rather than mere delay constraints.

We noted earlier that every monoid is automatically a discrete monoidal category. Furthermore when forming a $\mathcal{D}$ category A we have that the constituent composition m_{uvw} and identity j_u morphisms must be identities in $\mathcal{D}$.

$$\begin{aligned} m_{uvw} : \delta_A(u,v) \otimes \delta_A(v,w) &= \delta_A(u,w) \\ j_u : I &= \delta_A(u,u) \end{aligned}$$

for all objects $u, v, w \in A$. The remaining axioms for a $\mathcal{D}$-category are all trivially satisfied. (In a discrete category, every diagram commutes). However, we get as an immediate consequence that

$$m_{uvu} \circ j_u^{-1} : \delta_A(u,v) \otimes \delta_A(v,u) = I$$

namely, that $\delta_A(u,v)$ is invertible for all $u, v \in A$. Objects of $\mathcal{D}$ which have no inverses are never used in defining a $\mathcal{D}$-category. Thus, there is no loss of generality in assuming that $(\mathcal{D}, \otimes, I)$ is a group, abelian in the case that $(\mathcal{D}, \otimes, I)$ is symmetric.

A concrete example is given by the additive group on the reals viewed as a discrete category. We denote this category by $\mathbf{R}^{\mathbf{d}}$. In any $\mathbf{R}^{\mathbf{d}}$-category, A, if we choose a single object u (to be thought of as an origin), then all the homobjects of A can be reconstructed from the homobjects $\delta_A(u,v)$, using the identity $\delta_A(v,w) = \delta_A(u,w) - \delta_A(u,v)$. Thus $\mathbf{R}^{\mathbf{d}}$-categories are simply sets of "timestamped" events. In general we can think of $\mathcal{D}$-categories in the case of discrete $\mathcal{D}$ as *observations* (or *models* or *traces*) of the behaviors specified by more general enriched categories (it should be pointed out that an observation, in this sense, is nothing more than a rigid specification)

We now provide an example of a monoidal functor, we consider a functor $F : \mathbf{R}^{\mathbf{d}} \to \mathbf{2}$. Whereas $\mathbf{R}^{\mathbf{d}}$-categories are a form of observation, $\mathbf{2}$-categories, namely preorders, are a form of specification.

Since $\mathbf{R}^{\mathbf{d}}$ is discrete, F is completely determined by what it does to objects. The conditions that F be a monoidal functor reduce to the implications

$$\begin{aligned} n_{rs} &: F(r) \wedge F(s) \to F(r+s) \\ k &: 1 \to F(0) \end{aligned}$$

Thus if we stipulate $F(r) = 1$ iff $r \geq 0$, then it can be seen that we get a monoidal functor. (Notice that in this case F does not necessarily preserve tensor products, showing the inadequacy of strict monoidal functors.)

This functor $F : \mathbf{R}^{\mathbf{d}} \to \mathbf{2}$ induces a corresponding functor $F! : \mathbf{R}^{\mathbf{d}}! \to \mathbf{2}!$. For any particular timestamped set of events in $\mathbf{R}^{\mathbf{d}}!$, $F!$ will, as one might expect, produce the corresponding preorder on those events, namely the most restrictive preorder satisfied by this trace.

More generally, given a symmetric monoidal $F : \mathcal{D} \to \mathcal{D}'$, we say that an observation ($\mathcal{D}$-category) A is *F-allowed* by (alternatively, *F-satisfies*) a specification ($\mathcal{D}'$-category) B iff there exists a morphism $B \to F!(A)$.

This relates observations/traces best expressed in the language of $\mathcal{D}!$ to specifications best described in the language of $\mathcal{D}'!$, formalizing the notion of the maximally precise specification satisfied by a given trace or observation.

4.5 Metrics and the Process Language

We have given the meaning of the process language solely in terms of a monoidal category, namely a member of **SSM**, with no reference to the notion of metric. Let us now see what our language means when the monoidal structure is that of $\mathcal{D}!$. The following propositions tell the story.

Proposition 4 *If $\mathcal{D}$ has an initial object* 0 *and* $0 \otimes A \cong 0 \cong A \otimes 0$ *(automatic when $\mathcal{D}$ is closed) then $\mathcal{D}!$ has coproducts.*

A discrete category is a copower in **Cat** of 1, that is, the coproduct of a set of one-object categories, yielding the following similar result:

Proposition 5 *If $\mathcal{D}$ has an initial object* 0 *and* $0 \otimes A \cong 0 \cong A \otimes 0$*, then the forgetful functor* $U_{\mathcal{D}!} : \mathcal{D}! \to \mathbf{Set}$ *has a left adjoint* $D : \mathbf{Set} \to \mathcal{D}!$ *taking a set S to the corresponding* discrete *$\mathcal{D}$-category ($\delta_{DS}(u,v) = I$ if $u = v$, 0 otherwise).*

As usual, there is a dual result.

Proposition 6 *If $\mathcal{D}$ has a final object* 1, $U_{\mathcal{D}!} : \mathcal{D}! \to \mathbf{Set}$ *has a right adjoint* $E : \mathbf{Set} \to \mathcal{D}!$ *taking the set X to the corresponding* indiscrete *$\mathcal{D}$-category ($\delta_{DS}(u,v) = 1$ for all u, v).*

Meanwhile, for the case of general colimits we refer to a result of [BCSW83].

Theorem 7 *If $\mathcal{D}$ is cocomplete and $\mathcal{D}$ is closed then $\mathcal{D}!$ is cocomplete.*

There is also a similar but much easier result about limits (we will be interested in limits in $\mathcal{D}!$ when describing restriction, local orthocurrence, and local concatenation, all of which refer to the labels on events).

Theorem 8 *If $\mathcal{D}$ has limits of (small) type J, then so does $\mathcal{D}!$.*

Using these observations it may be verified that the operations of our process language behave for any $\mathcal{D}!$ essentially as in the motivating example **2**!. Orthocurrence for example acts analogously to direct product of posets, with the two edges from (u, v) to (u', v') formed by the component edges from u to u' and from v to v' being tensored together. Other operations behave similarly.

5 Comma Categories and the $\triangleright$ Operator

Our space crisis has been met with the ! operator, creating categories of spaces with *distances* between pairs of points. We turn now to the identity crisis for the points themselves: what are they?

From the point of view of sets as determined only up to isomorphism, an element of a set is pure identity: it is itself but it has no attributes. This is how we perceive the vertices of a graph or the states of an automaton. We follow standard practice in equipping points with attributes via *labels*. Given a labeling alphabet, a set Σ, we label the underlying set $U(p)$ of points of a space p with a function $\mu : U(p) \to \Sigma$. For p a poset this is the notion of pomset or partially ordered multiset as a labeled partial order.

Our framework can be simplified by not assuming that Σ is a pure set but may have (unwanted) structure, which we must strip off ourselves with the forgetful functor V for the category supplying Σ. Our labeling function then becomes $\mu : U(p) \to V(\Sigma)$.

But this is what it means to be an object of the comma category $U \downarrow V$. Moreover the morphisms of this comma category from $\mu : U(p) \to V(\Sigma)$ to $\mu' : U(p') \to V(\Sigma')$, namely pairs of maps $f : p \to p'$ and $t : \Sigma \to \Sigma'$ such that $V(t)\mu = \mu' U(f)$, turn out to be just what we need.

We therefore define the functor $- \triangleright - : \mathbf{SSM}^2 \to \mathbf{SSM}$ to take $\mathcal{D}$ and $\mathcal{E}$ to the comma category $\mathcal{D} \triangleright \mathcal{E} = (U_{\mathcal{D}} \downarrow V_{\mathcal{E}})$ equipped with a monoidal structure we describe at the end of this section.

A particular comma construction central to our interests is the one where C is **Set**, and U and V are the forgetful functors $U_{\mathcal{D}}$ and $U_{\mathcal{E}}$ associated with two semiconcrete monoidal categories $\mathcal{D}$ and $\mathcal{E}$.

5.1 Functors to Comma Categories

We wish to lift tensor products, limits and colimits that exist in the component categories to comma categories defined from them. To do this we must deal with functors into comma categories.

The following lemma can be applied whenever functors into a comma category are to be defined. It states that defining a functor from a category $\mathcal{I}$ into $U \downarrow V$, where $U : A \to C$ and $V : B \to C$, is essentially equivalent to defining functors $F : \mathcal{I} \to A$ and $G : \mathcal{I} \to B$ and a natural transformation between from UF to VG. Furthermore, natural transformations between such functors are essentially equivalent to a pair of natural transformations satisfying an obvious commutativity condition.

For any functor $F : \mathcal{X} \to \mathcal{Y}$ denote by $F^{\mathcal{I}}$ the functor from the functor category $\mathcal{X}^{\mathcal{I}}$ to $\mathcal{Y}^{\mathcal{I}}$ defined by "left-composition with F". Then we have:

Lemma 9 $(U \downarrow V)^{\mathcal{I}} \cong (U^{\mathcal{I}}) \downarrow (V^{\mathcal{I}})$.

Outline of Proof: The comma category $U \downarrow V$ has projections $\pi_0 : U \downarrow V \to \mathcal{A}$ and $\pi_1 : U \downarrow V \to \mathcal{B}$, and defines a natural transformation $\nu : U\pi_0 \to V\pi_1$. (In fact this is an alternative definition of the concept of a comma category.)

Hence, given a functor, $F : \mathcal{I} \to U \downarrow V$, we can define functors $\pi_0 F : \mathcal{I} \to \mathcal{A}$ and $\pi_1 F : \mathcal{I} \to \mathcal{B}$, and a natural transformation $\nu \circ F : U\pi_0 F \to V\pi_1 F$. This defines an object of the comma category $(U^{\mathcal{I}}) \downarrow (V^{\mathcal{I}})$. It is routine to extend this mapping to a functor $(U \downarrow V)^{\mathcal{I}} \to (U^{\mathcal{I}}) \downarrow (V^{\mathcal{I}})$ and show that it is an isomorphism of categories. ∎

We can use this lemma to prove a basic result about limits in comma categories. (Note: We are unaware of this theorem having appeared in this form. Mac Lane proves a special case as part of his proof of Freyd's Adjoint Functor Theorem [Mac71, page 123].)

Theorem 10 *Let $U : A \to C$ and $V : B \to C$ be functors, and $\mathcal{I}$ be a category. If $\mathcal{A}$ and $\mathcal{B}$ have all $\mathcal{I}$-limits and V preserves $\mathcal{I}$-limits then the comma category $U \downarrow V$ has all $\mathcal{I}$-limits*

Proof: Let $F : \mathcal{I} \to U \downarrow V$ be an $\mathcal{I}$-diagram in $U \downarrow V$. Then by lemma 9, F can be considered as a triple, $\langle F_1, \mu, F_2 \rangle$ where

$$F_1 : \mathcal{I} \to \mathcal{A}$$
$$F_2 : \mathcal{I} \to \mathcal{B}$$
$$\mu : UF_1 \to VF_2$$

Since $\mathcal{A}$ and $\mathcal{B}$ have $\mathcal{I}$-limits, both F_1 and F_2 have limits. Denote their limiting cones by $\lambda^1 : \lim F_1 \to F_1$ and $\lambda^2 : \lim F_2 \to F_2$ respectively.

Since V preserves $\mathcal{I}$-limits, $V(\lim F_2)$ is a limit of VF_2, with limiting cone $V\lambda^2$.

Now the composite natural transformation

$$U(\lim F_1) \to U\lambda^1 UF_1 \to \mu VF_2$$

is a cone from $U(\lim F_1)$ to VF_2, so by the universal property of limits there exists a unique arrow p such that the following diagram of functors and natural transformations commutes.

(Note: in this and following diagrams an object of C represents the constant functor to that object and an arrow of C represents the "constant" natural tranformation between such constant functors.)

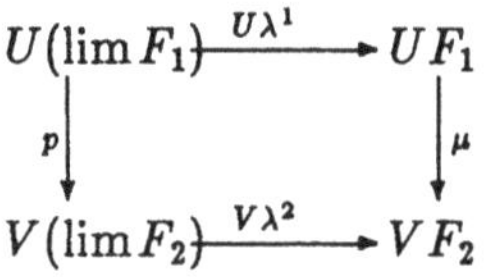

Now we claim that $\langle \lim F_1, p, \lim F_2 \rangle$ is the limit of F, and that the limiting cone is defined by the pair of natural transformations $\langle \lambda^1, \lambda^2 \rangle$.

For suppose we have an object, $\langle A, f, B \rangle$ and a cone $\kappa : \langle A, f, B \rangle \to F$. κ can be considered as a pair of natural transformations, $\kappa^1 : A \to F_1$ and $\kappa^2 : B \to F_2$. Since F_1 and F_2 have limits there must exist unique maps $a : A \to \lim F_1$ and $b : B \to \lim F_2$ which factor κ^1 and κ^2 through λ^1 and λ^2 respectively.

Consider the following diagram of natural transformations.

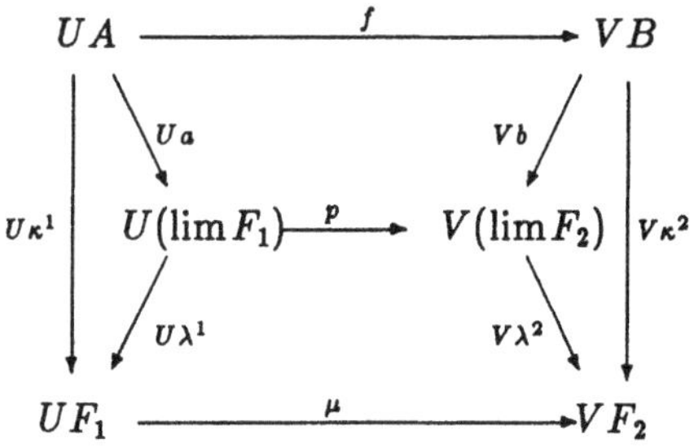

The outer square of this diagram commutes because κ is an arrow in the comma category. The two inner triangles commute by definition of a and b. The lower inner square commutes by definition of p. Now we wish to show that the upper inner square commutes. We do this by showing that the cone $(V\lambda^2).(Vb).f$ is identical to the cone $(V\lambda^2).p.(Ua)$. Since there must be a *unique* arrow which factors this cone through the limiting cone $V\lambda^2$, we can conclude that the arrows $(Vb).f$ and $p.(Ua)$ must be identical. This is the required commutativity condition. But we have

$$\begin{array}{rcll} (V\lambda^2).(Vb).f & = & (V\kappa^2).f & \text{definition of } b \\ & = & \mu.(U\kappa^1) & \text{outer square commutes} \\ & = & \mu.(U\lambda^1).(Ua) & \text{definition of } a \\ & = & (V\lambda^2).p.(Ua) & \text{lower square commutes} \end{array}$$

This proves the required commutativity, which shows that $\langle a, b\rangle$ is an arrow in $U \downarrow V$ and that it factors κ. That $\langle a, b\rangle$ is the unique such arrow follows from the fact that a must factor κ^1 through the limiting cone λ^1, so a is the unique such arrow. Similarly, b must factor κ^2 through λ^2, and so is unique. ∎

Note that the proof gives an explicit construction of limits in a comma category when the conditions of the theorem are satisfied.

The same technique allows us to construct colimits:

Corollary 11 *If A and B have all $\mathcal{I}$-colimits and U preserves $\mathcal{I}$-colimits then $U \downarrow V$ has all $\mathcal{I}$-colimits.*

Proof: For any functor $U : A \to C$, let U^{op} be the corresponding functor $U^{\mathrm{op}} : A^{\mathrm{op}} \to C^{\mathrm{op}}$. Observe that

$$(U \downarrow V)^{\mathrm{op}} = (V^{\mathrm{op}} \downarrow U^{\mathrm{op}}).$$

∎

We now return to the problem stated at the beginning of this section, which was to define a tensor product on $\mathcal{D} \rhd \mathcal{E}$.

To define $\otimes_{\mathcal{D},\mathcal{E}}$, observe that we wish to define a functor from $(U \downarrow V) \times (U \downarrow V)$ to the comma category $U \downarrow V$. By lemma 9 this can be done by defining three elements: a functor, $F_{\mathcal{D}}$ from $(U \downarrow V) \times (U \downarrow V)$ into $\mathcal{D}$, a similar functor, $F_{\mathcal{E}}$ into $\mathcal{E}$, and a natural transformation from $UF_{\mathcal{D}}$ to $VF_{\mathcal{E}}$.

We define the functor to $\mathcal{D}$ to be the composite

$$(U \downarrow V) \times (U \downarrow V) \overset{\pi_0 \times \pi_0}{\to} \mathcal{D} \times \mathcal{D} \overset{\otimes}{\to} \mathcal{D}$$

where π_0 is the projection of $(U \downarrow V) \to \mathcal{D}$.

A corresponding functor can be defined to $\mathcal{E}$:

$$U \downarrow V \times U \downarrow V \overset{\pi_1 \times \pi_1}{\to} \mathcal{D} \times \mathcal{D} \overset{\otimes}{\to} \mathcal{D}.$$

To define the required natural transformation, we paste together natural transformations as follows. Let $\nu : U\pi \to V\pi'$ be the standard natural transformation defined by the comma category. Let $\mathbf{d} : U \circ \otimes \to \times \circ (U \times U)$ be the natural isomorphism which makes U strong monoidal. Let $\mathbf{e}$ be the corresponding natural isomorphism for V. Then we can define the natural transformation we require to be

$$\mathbf{d} \circ (\mathbf{e}^{-1}.\nu^2)$$

Alternatively, by the interchange law, we could write this transformation as

$$(1_\times \circ \mathbf{e}^{-1}).(\mathbf{d} \circ \nu^2)$$

Under this definition we then obtain $\langle d, f, e\rangle \otimes \langle d', f', e'\rangle = \langle d \otimes d', f \otimes f', e \otimes e'\rangle$, where $f \otimes f'$ is the composite $\mathbf{e}^{-1}(f \times f')\mathbf{d}$.

This completes the definition of $\otimes$ on objects of the comma category. The unit object is $I_{\mathcal{D} \rtimes \mathcal{E}} = \langle I_{\mathcal{D}}, h, I_{\mathcal{E}}\rangle$, where h, being by the strong monoidal assumption a function between two one-point sets, is uniquely determined. Next we must show that this product is associative, symmetric, and $I_{\mathcal{D} \rtimes \mathcal{E}}$ is in fact an identity. We outline briefly the definition of the associativity natural transformation α. (Other required natural transformations are defined similarly.)

Again using lemma 9 defining such a natural transformation is equivalent to defining two natural transformations, one in each of the categories $\mathcal{D}$ and $\mathcal{E}$ (such that a certain commutativity condition is satisfied). The obvious definition is to take the two natural transformations to be the associativity transformations of $\mathcal{D}$ and $\mathcal{E}$.

In this case the commutativity condition amounts to the requirement that $V(\alpha).(f \otimes (f' \otimes f''))$ be equal to $((f \otimes f') \otimes f'').U(\alpha)$. This requirement follows directly from properties of monoidal functors and natural transformations. Having established that the transformation is well-defined, the coherence conditions follow trivially.

A fine example related to this construction is provided by the partially ordered multisets motivating this work. A pomset $p = \langle P, \mu, \Sigma\rangle$ is just a poset P, labeled by a function $\mu : V(P) \to \Sigma$ for some set Σ. This makes it an object of **Pom** = **Pos** $\triangleright$ **Set**. A morphism of pomsets $f : \langle P, \mu, \Sigma\rangle \to \langle P', \mu', \Sigma'\rangle$ consists of a monotone event map $f : P \to P'$ on the underlying posets together with an alphabet map or *translation* $t : \Sigma \to \Sigma'$ from Σ to Σ', such that $t\mu = \mu' f$ (i.e., the event map and translation are consistent with respect to the labelings). We can view **Pom** as the full subcategory of **Prom** = 2! $\triangleright$ **Set** that generalizes pomsets by allowing P to be a preorder.

In this paper, since the construction that we emphasize is $\mathcal{D} \triangleright \mathcal{E}$, we are interested only in labels drawn from sets. When more elaborately structured label sources are needed, as for the category **Pos** $\triangleright$ **2** of order ideals, the comma must be taken over the appropriate common denominator, here **Pos** (so that an order ideal is a triple $\langle P, f, \mathbf{2}\rangle$ for $f : P \to \mathbf{2}$ monotone). Many of the principles developed in this paper transcend this choice of denominator; fixing it at **Set** helps fix ideas.

To complete the description of the functor $\triangleright$ we must describe its action on functors F, G of **SSM**. This is much more straightforward than the object part. This is facilitated by the strong condition $U = U'F$ in the definition of the morphisms F of **SSM**. The functor

$F \triangleright G : \mathbf{SSM}^2 \to \mathbf{SSM}$ acts thus:

$$
\begin{array}{ccccc}
\mathcal{D} & \xrightarrow{U} & \mathbf{Set} & \xleftarrow{V} & \mathcal{E} \\
\downarrow{\scriptstyle F} & & \| & & \downarrow{\scriptstyle G} \\
\mathcal{D}' & \xrightarrow{U'} & \mathbf{Set} & \xleftarrow{V'} & \mathcal{E}'
\end{array}
\quad \xmapsto{\mathbf{SSM}^2 \to \mathbf{SSM}} \quad
\begin{array}{ccc}
\mathcal{D} \triangleright \mathcal{E} & \xrightarrow{UP} & \mathbf{Set} \\
\downarrow{\scriptstyle F \triangleright G} & & \| \\
\mathcal{D}' \triangleright \mathcal{E}' & \xrightarrow{U'P'} & \mathbf{Set}
\end{array}
$$

where P, P' are the respective left projections of comma [Mac71] (p.47 (5)) and $F \triangleright G$ takes each morphism $\langle f, s, g\rangle$ of $\mathcal{D} \triangleright \mathcal{E}$ to $\langle Ff, s, Gg\rangle$ as shown, where s denotes both the inner and outer square.

$$
\begin{array}{cccccccccc}
Fd & & U'Fd & = & Ud & \longrightarrow & Ve & = & V'Ge & & Ge \\
{\scriptstyle Ff}\downarrow & & {\scriptstyle U'Ff}\downarrow & & {\scriptstyle Uf}\downarrow & s & \downarrow{\scriptstyle Vg} & & \downarrow{\scriptstyle V'Gg} & & \downarrow{\scriptstyle Gg} \\
Fd' & & U'Fd' & = & Ud' & \longrightarrow & Ve' & = & V'Ge' & & Ge'
\end{array}
$$

6 The Kind Language KL

It is now straightforward to define a language KL for naming kinds. The constants of KL include the successor ordinals $\mathbf{1}, \mathbf{2}, \mathbf{3}, \ldots$, and **SR**, all of which are bicomplete cartesian closed categories. We shall not be terribly specific about the other constants since we are mainly concerned here with principles.

Terms of KL are formed from these constants via the unary operation ! and the binary operation $\triangleright$, with the obvious meanings.

We define a KL kind to be the denotation in **SSM** of a KL term.

Theorem 12 *All KL kinds are bicomplete.*

Proof: This follows from Theorem 10 and Corollary 11, and the fact that the forgetful functors of the basic kinds and those constructed by ! and $\triangleright$ have the necessary continuity properties. ∎

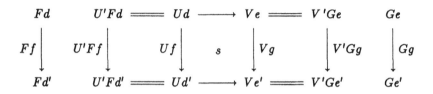

Theorem 13 *The interpretation of PSL in any KL kind $\mathcal{D}$ is well-defined.*

Proof: PSL is defined via limits and colimits in categories constructed from $\mathcal{D}$ via ! and $\triangleright$. Since $\mathcal{D}$ is a KL kind so are these others, whence all such limits and colimits exist. ∎

At the end of the introduction we drew attention to the typelessness of this enterprise. Both KL and PSL are typeless. KL's operations apply to any object $\mathcal{D}$ of **SSM**. PSL is only slightly more typed: each operation expects a J-diagram of a fixed shape J. Thus both concurrence and local product expect two arrows with a common codomain while orthocurrence just expects two objects.

It seems to us that typelessness of this kind makes for a simple language. If all the control structures of the language can be defined in such a typeless way it can do much to simplify both learning and implementing the language.

7 Coda

7.1 Directions For Further Work

We mention here a few directions for further work that are presently on our mind.

Working over **Cat**. Our techniques would appear to extend straightforwardly to monoidal 2-categories, where the forgetful functors are not to **Set** but to **Cat**. This turns the labeling function from the underlying set of a behavior to (the underlying set of) an alphabet into a labeling *functor*. Its functoriality can be put to good use, as in conveniently naming the monoidal category of order ideals.

SKL, a SubKind Language. Currently, methods for specifying subkinds of dynamic kinds are somewhat ad hoc, e.g., posets are acyclic **2**-categories. It would be useful to have a general scheme that produces subkinds for a wide variety of $\mathcal{D}$, such as a single approach to producing partial orders (as opposed to preorders) and say symmetric metric spaces.

Stochastic delay. Another possible choice for $\mathcal{D}$ is that of a category of probability distributions. In other words for each pair of events u, v we specify a probability density function $\rho_{uv} : \mathbf{R} \to [0,1]$ for the associated delay. Consecutive delays would be combined using convolution

$$\rho_{uv} \otimes \rho_{vw}(y) = \int_{-\infty}^{\infty} \rho_{uv}(x)\rho_{vw}(y-x)dxdy$$

The main problems include finding a suitable ordering on distributions and dealing with the fact that the various distributions on delays are not independent. There is also the question of the proper treatment of distributions which are not well behaved and continuous (e.g., the Dirac delta function).

7.2 Conclusions

We have shown how the partial order approach to modelling concurrency can be both extended and unified by restating it in category-theoretic terms.

The restatement is inspired by three observations. First, is that labeled structures (such as labeled sets,i.e. multisets, or labeled graphs or labeled partial orders) can be explicated using comma categories. Second, not only is a poset a very special case of an enriched category, but other enriched categories also arise naturally in the mathematics of concurrency. Third, we observed that the operations of pomset algebra corresponded to various limit and colimit constructions, thus giving a formal basis to our intuition that the algebra of processes is more or less independent of the notion of time.

Since we define operations using limits and colimits we must consider whether they exist. We have shown how to construct the required limits and colimits uniformly, dependent only on completeness properties of certain basic categories and continuity of the operators ! and $\triangleright$.

Finally, we have shown how both enrichment and labeling can be studied in the context of a single category, **SSM**, and furthermore that we can also use enrichment and labeling operators to construct many naturally occurring categories from simple ones. In other words, the study of all the models of concurrency we have considered here, and much else besides, can be construed as the study of the properties of **SSM**.

That the semantics for PSL should be well-defined for *any* KL kind is somewhat surprising. Despite PSL containing operations such as concatenation whose set-theoretic definition refers explicitly to a partial order, we have been able to define PSL without any assumption about the underlying model being of the form $\mathcal{D}$!.

One might presume from this that every concrete monoidal category must have an implicit metric somehow built into it that our definitions must be referring to indirectly. If so we have not yet found it. Certainly there is no "hidden inverse" for ! allowing us to find for any $\mathcal{D}$ a $\mathcal{D}'$ for which $\mathcal{D} = \mathcal{D}'!$, since $\mathcal{D}'!$ is always large and $\mathcal{D}$ may be small. In fact the underlying sets of the objects of many useful $\mathcal{D}$'s are all singletons, severely limiting any sense in which they could be construed as forming interesting metric spaces.

In conclusion, we see this work as a foundation, and much remains to be done in clarifying inherent properties of this model, in drawing connections between this model and others, and in applying the model to practice.

Acknowledgments

We are indebted to W. Lawvere both for his paper [Law73] clarifying the topic of enriched categories and for much helpful advice. We also appreciate the help we have received from the Sydney Category Seminar, especially A.J. Power and R.F.C. Walters.

References

[AHU74] A.V. Aho, J. E. Hopcroft, and J.D. Ullman. *The Design and Analysis of Computer Algorithms*. Addison-Wesley, Reading, Mass, 1974.

[AM75] M. Arbib and E. Manes. *Arrows, Structures, and Functors: The Categorical Imperative*. Academic Press, 1975.

[BCSW83] R. Betti, A. Carboni, R. Street, and R. Walters. Variation through enrichment. *Journal of Pure and Applied Algebra*, 29:109–127, 1983.

[BHR84] S.D. Brookes, C.A.R. Hoare, and A.D. Roscoe. A theory of communicating sequential processes. *Journal of the ACM*, 31:560–599, 1984.

[Cre89] R. Crew. Parametrized process categories. In *Proc. First International Conference on Algebraic Methods and Specification Techniques*, pages 39–42, Iowa City, May 1989.

[EK66] Samuel Eilenberg and G. Max Kelly. Closed categories. In S. Eilenberg, D. K. Harrison, S. MacLane, and H. Röhrl, editors, *Proceedings of the Conference on Categorical Algebra, La Jolla, 1965*, pages 421–562, Springer-Verlag, 1966.

[Gai89] H. Gaifman. Modeling concurrency by partial orders and nonlinear transition systems. In *Proc. REX School/Workshop on Linear Time, Branching Time and Partial Order in Logics and Models for Concurrency*, Springer-Verlag, Noordwijkerhout, The Netherlands, 1989.

[Gis84] J. Gischer. *Partial Orders and the Axiomatic Theory of Shuffle*. PhD thesis, Computer Science Dept., Stanford University, December 1984.

[GP87] H. Gaifman and V.R. Pratt. Partial order models of concurrency and the computation of functions. In *Proc. IEEE Symp. on Logic in Computer Science*, pages 72–85, Ithaca, NY, June 1987.

[Gra81] J. Grabowski. On partial languages. *Fundamenta Informaticae*, IV.2:427–498, 1981.

[Kel82] G.M. Kelly. *Basic Concepts of Enriched Category Theory: London Math. Soc. Lecture Notes. 64*, Cambridge University Press, 1982.

[KS74] G.M. Kelly and R. Street. Review of the elements of 2-categories. In *LNM 420*, Springer-Verlag, 1974.

[Law73] W. Lawvere. Metric spaces, generalized logic, and closed categories. In *Rendiconti del Seminario Matematico e Fisico di Milano, XLIII*, Tipografia Fusi, Pavia, 1973.

[Mac71] S. Mac Lane. *Categories for the Working Mathematician*. Springer-Verlag, 1971.

[Pra82] V.R. Pratt. On the composition of processes. In *Proceedings of the Ninth Annual ACM Symposium on Principles of Programming Languages*, January 1982.

[Pra84] V.R. Pratt. The pomset model of parallel processes: unifying the temporal and the spatial. In *Proc. CMU/SERC Workshop on Analysis of Concurrency, LNCS 197*, Springer-Verlag, Pittsburgh, 1984.

[Pra85a] V.R. Pratt. Some constructions for order-theoretic models of concurrency. In *Proc. Conf. on Logics of Programs, LNCS 193*, Springer-Verlag, Brooklyn, 1985.

[Pra85b] V.R. Pratt. Two-way channel with disconnect. In *The Analysis of Concurrent Systems: Proceedings of a Tutorial and Workshop, LNCS 207*, Springer-Verlag, 1985.

[Pra86] V.R. Pratt. Modeling concurrency with partial orders. *International Journal of Parallel Programming*, 15(1):33–71, February 1986.

[Pra89] V.R. Pratt. Enriched categories and the floyd-warshall connection. In *Proc. First International Conference on Algebraic Methods and Specification Techniques*, pages 177–180, Iowa City, May 1989.

[RT88] A. Rabinovich and B.A. Trakhtenbrot. Behavior structures and nets. *Fundamenta Informatica*, 11(4), 1988.

[Win80] G. Winskel. *Events in Computation*. PhD thesis, Dept. of Computer Science, University of Edinburgh, 1980.

[Win88] G. Winskel. A category of labelled petri nets and compositional proof system. In *Proc. Third Annual Symposium on Logic in Computer Science*, Computer Society Press, Edinburgh, 1988.

Compositional Relational Semantics for Indeterminate Dataflow Networks

Eugene W. Stark*
Department of Computer Science
State University of New York at Stony Brook
Stony Brook, NY 11794 USA

Abstract

Given suitable categories $\mathbf{T}, \mathbf{C}$ and functor $F : \mathbf{T} \to \mathbf{C}$, if X, Y are objects of $\mathbf{T}$, then we define an (X,Y)-*relation in* $\mathbf{C}$ to be a triple $(R, \underline{r}, \bar{r})$, where R is an object of $\mathbf{C}$ and $\underline{r} : R \to FX$ and $\bar{r} : R \to FY$ are morphisms of $\mathbf{C}$. We define an algebra of relations in $\mathbf{C}$, including operations of "relabeling," "sequential composition," "parallel composition," and "feedback," which correspond intuitively to ways in which processes can be composed into networks. Each of these operations is defined in terms of composition and limits in $\mathbf{C}$, and we observe that any operations defined in this way are preserved under the mapping from relations in $\mathbf{C}$ to relations in $\mathbf{C}'$ induced by a continuous functor $G : \mathbf{C} \to \mathbf{C}'$.

To apply the theory, we define a category **Auto** of concurrent automata, and we give an operational semantics of dataflow-like networks of processes with indeterminate behaviors, in which a network is modeled as a relation in **Auto**. We then define a category **EvDom** of "event domains," a (non-full) subcategory of the category of Scott domains and continuous maps, and we obtain a coreflection between **Auto** and **EvDom**. It follows, by the limit-preserving properties of coreflectors, that the denotational semantics in which dataflow networks are represented by relations in **EvDom**, is "compositional" in the sense that the mapping from operational to denotational semantics preserves the operations on relations. Our results are in contrast to examples of Brock and Ackerman, which imply that no compositional semantics is possible in terms of set-theoretic relations.

1 Introduction

Dataflow networks (see, *e.g.* [4,3,5,7,9,10]) consist of a collection of concurrently and asynchronously executing sequential processes that communicate by transmitting sequences or "streams" of "value tokens" over FIFO communication channels. Typically, a network is described as a directed graph, whose nodes are processes and whose arcs are communication channels. Each channel serves to connect an "output port" of one process to an "input port" of another process. "Determinate" (or functional) networks were first studied by Kahn [9], who gave an elegant fixed-point principle for determining the function computed by a network from the functions computed by the components.

"Indeterminate" (or non-functional) networks remain less well understood, despite extensive study. An interesting class of indeterminate processes are the "merge" processes, which shuffle together sequences of values from two input channels onto a single output channel. Brock and Ackerman have shown [3,4] that no naive generalization of Kahn's theory to indeterminate networks, obtained by replacing input/output functions with input/output relations, can be "compositional" in the sense that the mapping from operational to denotational semantics preserves

*Research supported in part by NSF Grant CCR-8702247.

the operations by which networks are built from component processes. Their examples use only functional processes and a weak form of merge process; thus, their results apply to essentially any interesting class of indeterminate networks.

The title of this paper advertises "compositional relational semantics for indeterminate dataflow networks." This reason this is not a contradiction with Brock and Ackerman's results is that their results apply only to the usual *set-theoretic* notion of a relation, whereas here we use instead a category-theoretic generalization. Our notion of relation involves two suitably complete categories **T** and **C**, and a suitable functor $F : \mathbf{T} \rightarrow \mathbf{C}$. If X and Y are objects of **T**, then by an "(X, Y)-relation in **C**" we mean a triple $(R, \underline{r}, \bar{r})$ where R is an object of **C**, and $\underline{r} : R \rightarrow FX$ and $\bar{r} : R \rightarrow FY$ are morphisms of **C**. We think of the objects of **T** as "types," and in our semantics of dataflow networks the types X and Y are simply the sets of input ports and output ports over which a process may communicate. An algebra of relations in **C** can be defined, including operations of "relabeling," "sequential composition," "parallel composition," and "feedback" that correspond intuitively to ways in which networks can be built from component processes. These operations have categorical definitions in terms of composition and limits in **C**. It is a simple observation (Theorem 1) that any operations definable in this way are preserved by the mapping from relations in **C** to relations in **C**$'$ induced by a continuous functor $G : \mathbf{C} \rightarrow \mathbf{C}'$.

To obtain a correct semantics of process networks, we must choose properly the category **C**, and this is a bit tricky. In this paper, we give two examples for **C**: (1) a category **Auto** of concurrent automata, and (2) a certain (non-full) subcategory **EvDom** of the category of Scott domains and continuous maps. The semantics based on **Auto** is "operational," because it involves automata and computation sequences, and it is not difficult to become convinced that it can serve as accurate model of the usual informal "token-pushing" operational semantics usually given for dataflow networks (see, *e.g.* [3]). The semantics based on **EvDom** is more "denotational," because it is defined using order-theoretic notions that do not necessarily have to do with computation. The objects of **EvDom** are the "event domains," which have been studied previously [6,20], but the morphisms we use are apparently new, and are crucial to our results. We obtain a coreflection between **Auto** and **EvDom**, and from the limit-preserving properties of coreflectors it follows immediately that the denotational semantics based on **EvDom** is "compositional" in the sense that the map from operational to denotational semantics preserves the operations of our relational algebra. Moreover, the denotational semantics is not "too abstract," in the sense that networks having distinct input/output relations in the ordinary set-theoretic sense, also determine distinct relations in **EvDom**. The semantics is not fully abstract, though.

In this paper, FX or $F(X)$ denotes the application of functor (or function) F to argument X. If $F : X \rightarrow Y$ and $G : Y \rightarrow Z$, then GF or $G \circ F$ denotes the composition of F and G. We use parentheses freely to increase readability. For basic definitions and terminology of category theory, we refer the reader to [8,11]. The text [15] and the unpublished [14] contain material on domain theory.

2 An Algebra of Relations

We begin by defining a category-theoretic generalization of the notion of a relation, and we show how to obtain various operations on relations that correspond intuitively to ways in which processes can be composed into networks.

2.1 Relations

Let **T** be a category equipped with a specified terminal object and binary products, and let **C** be a category with a specified terminal object, binary products, and equalizers of parallel pairs of arrows. Let $F : \mathbf{T} \rightarrow \mathbf{C}$ be a functor that preserves the specified terminal object and products.

Intuitively, we think of $\mathbf{T}$ as a category whose objects are "types" and whose morphisms are "relabeling maps." Pairs of objects of $\mathbf{T}$ will be used to index classes of relations, which are built in the category $\mathbf{C}$.

Formally, if X and Y are objects of $\mathbf{T}$, then an (X,Y)-*relation in* $\mathbf{C}$ (with respect to the functor F) is a triple $(R, \underline{r}, \bar{r})$, where R is an object of $\mathbf{C}$ and $\underline{r} : R \to FX$, $\bar{r} : R \to FY$ are morphisms of $\mathbf{C}$:

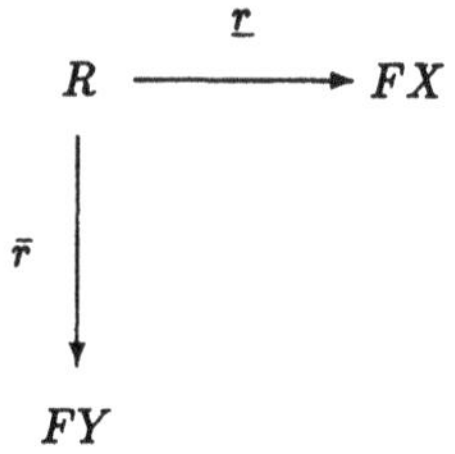

When there can be no confusion, we usually say "let R be an (X,Y)-relation in $\mathbf{C}$," and use corresponding lower-case letters $\underline{r}$, $\bar{r}$ to denote the morphisms.

If R and S are (X,Y)-relations in $\mathbf{C}$, then a *morphism* from R to S is a morphism $f : R \to S$ in $\mathbf{C}$ such that the obvious diagram commutes:

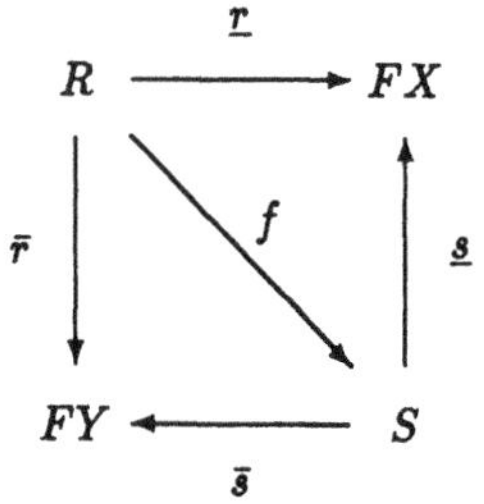

Let $\mathbf{Reln}_{\mathbf{C}}(X,Y)$ denote the category of (X,Y)-relations in $\mathbf{C}$ and their morphisms.

For example, if $\mathbf{T} = \mathbf{C} = \mathbf{Set}$, the category of sets and functions, and $F : \mathbf{T} \to \mathbf{C}$ is the identity functor, then an ordinary set-theoretic relation $R \subseteq X \times Y$ may be represented as an (X,Y)-relation in $\mathbf{C}$ by taking $\underline{r} : R \to X$ and $\bar{r} : R \to Y$ to be the projections on the first and second components, respectively.

2.2 Operations on Relations

We may define various operations on relations.

2.2.1 Relabeling

Suppose R is an (X,Y)-relation in $\mathbf{C}$. If $\phi : Y \to Y'$ is a morphism in $\mathbf{T}$, then the *output relabeling* of R by ϕ is the (X,Y')-relation $R;\phi = (R, \underline{r}, (F\phi)\bar{r})$. Similarly, if $\psi : X \to X'$ is a morphism in T, then the *input relabeling* of R by ψ is the (X',Y)-relation $\psi; R = (R, (F\psi)\underline{r}, \bar{r})$. Output relabeling by ϕ extends to a functor

$$(\text{-});\phi : \mathbf{Reln}_{\mathbf{C}}(X,Y) \to \mathbf{Reln}_{\mathbf{C}}(X,Y').$$

and input relabeling by ψ extends to a functor

$$\psi;(\text{-}) : \mathbf{Reln}_{\mathbf{C}}(X,Y) \to \mathbf{Reln}_{\mathbf{C}}(X',Y).$$

2.2.2 Sequential Composition

Relations may be composed in sequence as suggested by the picture:

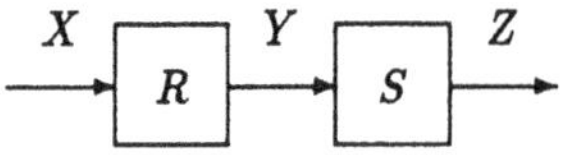

Formally, if R is an (X, Y)-relation in $\mathbf{C}$ and S is a (Y, Z)-relation in $\mathbf{C}$, then their *sequential composition* is the (X, Z)-relation $R; S = (R; S, \underline{r}s', \bar{s}r')$, where r' and s' are defined to make the square in the following diagram a pullback:

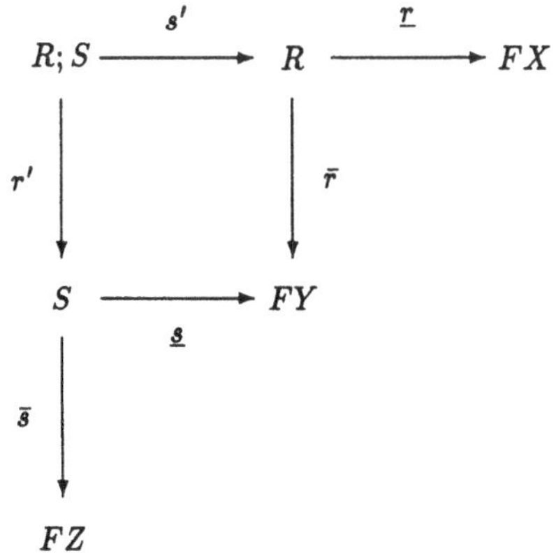

It should be noted that in the case $\mathbf{T} = \mathbf{C} = \mathbf{Set}$, with $F : \mathbf{T} \rightarrow \mathbf{C}$ the identity functor, the above definition of sequential composition of relations does not agree exactly with the standard set-theoretic definition. Specifically, if $R \subseteq X \times Y$ and $S \subseteq Y \times Z$, then

$$R; S \simeq \{(x, y, z) : (x, y) \in R, (y, z) \in S\}.$$

2.2.3 Parallel Composition

Relations may also be composed in parallel, as suggested by the picture:

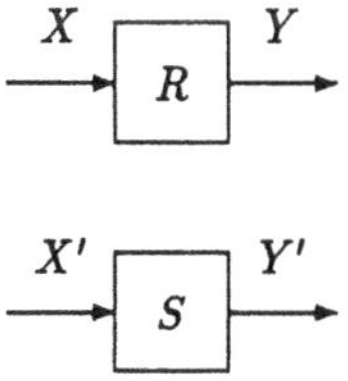

Formally, suppose R is an (X, Y)-relation in $\mathbf{C}$ and S is an (X', Y')-relation in $\mathbf{C}$. Then their *parallel composition* is the $(X \times X', Y \times Y')$-relation $R \| S = (R \times S, \underline{r} \times \underline{s}, \bar{r} \times \bar{s})$, where we have used the assumption that $F(X \times X') = FX \times FX'$ and $F(Y \times Y') = FY \times FY'$.

The mapping that takes the pair (R, S) to $R \| S$ extends to a functor

$$\| : \mathbf{Reln_C}(X, Y) \times \mathbf{Reln_C}(X', Y') \rightarrow \mathbf{Reln_C}(X \times X', Y \times Y').$$

2.2.4 Feedback

We can also define an operation of "feedback," corresponding to the picture:

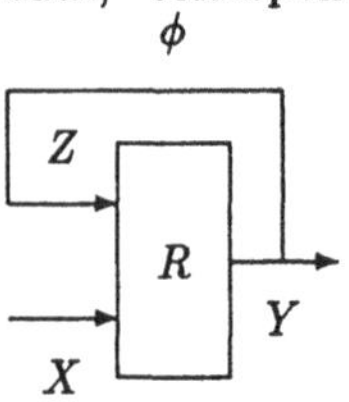

Formally, suppose R is an $(X \times Z, Y)$-relation in $\mathbf{C}$, and $\phi : Y \to Z$ is a morphism in $\mathbf{T}$. The *feedback* of R by ϕ is the (X, Y)-relation $(R_{\circlearrowleft\phi}, \pi_{FX}\underline{r}e, \bar{r}e)$, where e is the equalizer of $\pi_{FZ}\underline{r}$ and $(F\phi)\bar{r}$ in C:

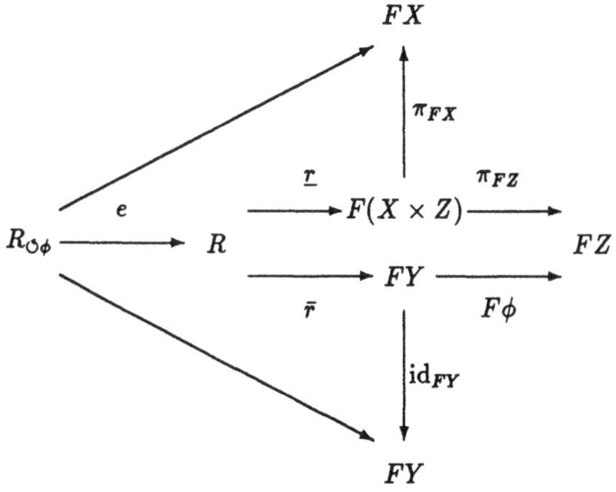

The map $(\text{ - })_{\circlearrowleft\phi}$ extends to a functor

$$(\text{ - })_{\circlearrowleft\phi} : \mathbf{Reln_C}(X \times Z, Y) \to \mathbf{Reln_C}(X, Y).$$

The operations we have defined are not independent: in particular, sequential composition is definable (up to isomorphism) in terms of parallel composition, output relabeling, and feedback. Formally, suppose R is an (X, Y)-relation in $\mathbf{C}$ and S is a (Y, Z)-relation in $\mathbf{C}$. Let $\pi_Y : Y \times Z \to Y$ and $\pi_Z : Y \times Z \to Z$ be the projections associated with the product $Y \times Z$ in $\mathbf{T}$. Then

$$R; S \simeq ((R \| S)_{\circlearrowleft\pi_Y}); \pi_Z.$$

2.3 Preservation of Operations by Functors

Suppose now we have two categories, $\mathbf{C}$ and $\mathbf{C}'$, equipped with specified terminal object, binary products, and equalizers, and two functors $F : \mathbf{T} \to \mathbf{C}$ and $F' : \mathbf{T} \to \mathbf{C}'$ that preserve the specified terminal object and products. If $G : \mathbf{C} \to \mathbf{C}'$ is a functor such that $GF = F'$, then each (X, Y)-relation $R = (R, \underline{r}, \bar{r})$ in $\mathbf{C}$ determines an (X, Y)-relation $GR = (GR, G\underline{r}, G\bar{r})$ in $\mathbf{C}'$. The map taking R to GR extends, for each (X, Y), to a functor

$$G_{X,Y} : \mathbf{Reln_C}(X, Y) \to \mathbf{Reln_{C'}}(X, Y).$$

Moreover, if G preserves the specified limits, then the ensemble of functors $\{G_{X,Y} : X, Y \in \mathbf{T}\}$ preserves the operations of relabeling, sequential composition, parallel composition, and feedback on relations in a way made precise by the following result:

Theorem 1 *Suppose* $\mathbf{C}, \mathbf{C}', F, F'$ *and* G *are as above. Then*

1. *If* $R \in \mathbf{Relnc}(X,Y)$, $\phi : Y \to Y'$, *and* $\psi : X \to X'$, *then*
$$G_{X,Y'}(R;\phi) = G_{X,Y}(R);\phi, \qquad G_{X',Y}(\psi;S) = \psi;G_{X,Y}(S).$$
2. *If* $R \in \mathbf{Relnc}(X,Y)$ *and* $S \in \mathbf{Relnc}(Y,Z)$, *then* $G_{X,Z}(R;S) = G_{X,Y}(R);G_{Y,Z}(S)$.
3. *If* $R \in \mathbf{Relnc}(X,Y)$ *and* $R' \in \mathbf{Relnc}(X',Y')$, *then*
$$G_{X\times X',Y\times Y'}(R\|R') = G_{X,Y}(R)\|G_{X',Y'}(R').$$
4. *If* $R \in \mathbf{Relnc}(X \times Z, Y)$ *and* $\phi : Y \to Z$, *then* $G_{X,Z}(R_{\circlearrowleft\phi}) = G_{X\times Z,Y}(R)_{\circlearrowleft\phi}$.

Proof – All of these operations have categorical definitions in terms of composition and limits, and these are preserved by the functor G. ∎

Obviously, the same reasoning applies to show that the functors $G_{X,Y}$ preserve any other operations on relations in **C** we might wish to consider, provided those operations can be defined in terms of composition and limits in **C**.

3 Concurrent Automata

In this section, we define a category **Auto** of concurrent automata. This category serves as our operational semantics for dataflow networks, in that we shall model networks as relations in **Auto**. The kind of automata we consider incorporate concurrency in the form of a binary *concurrency relation* $\|$ on a set E of *events*. The pair $(E, \|)$ is called a "concurrent alphabet." Intuitively, events represent primitive occurrences during computation. If events e and e' are both possible occurrences when in state q, and if e and e' are related by the concurrency relation, then e and e' can be executed in either order, with equivalent effect. Similar automata have been studied by Bednarczyk [2], and by the author [18,19].

3.1 Concurrent Alphabets

A *concurrent alphabet* is a set E, equipped with a symmetric, irreflexive binary relation $\|_E$, called the *concurrency relation.* Elements $e, e' \in E$ are said to *commute* if $e\|_E e'$, and a subset U of E is called *commuting* if every pair of its elements commutes. Let $\mathrm{Com}(E)$ denote the set of all finite commuting subsets of E. Suppose $U, V \in \mathrm{Com}(E)$. Then U and V are called *orthogonal,* and we write $U \perp_E V$, if $U \cup V \in \mathrm{Com}(E)$ and $U \cap V = \emptyset$.

A *morphism* from a concurrent alphabet E to a concurrent alphabet F is a function $\mu : \mathrm{Com}(E) \to \mathrm{Com}(F)$ such that

1. $\mu(\emptyset) = \emptyset$.
2. If $U \cup V \in \mathrm{Com}(E)$, then $\mu(U) \cup \mu(V) \in \mathrm{Com}(F)$, and $\mu(U \setminus V) = \mu(U) \setminus \mu(V)$.

Here the symbol $\setminus$ denotes set difference. Let **Alph** denote the category of concurrent alphabets and their morphisms.

Concurrent alphabets constitute the starting point for "trace theory" [1,12]. The above definition of morphism of concurrent alphabets was motivated by the author's study of "concurrent transition systems" [16,17,18]. The category **Alph** has a number of pleasant properties, although the author is not aware of it having been studied previously.

Lemma 3.1 *Suppose* $\mu : E \to F$ *is a morphism of concurrent alphabets. If* $U \cup V \in \mathrm{Com}(E)$, *then (1)* $\mu(U \cup V) = \mu(U) \cup \mu(V)$ *and (2)* $\mu(U \cap V) = \mu(U) \cap \mu(V)$.

Proof – (1) Observe that for arbitrary sets A, B, C, we have $C = A \cup B$ iff $A \setminus C = \emptyset = B \setminus C$, $C \setminus A = B \setminus A$, and $C \setminus B = A \setminus B$. Clearly, $\emptyset = \mu(U \setminus (U \cup V)) = \mu(U) \setminus \mu(U \cup V)$, and similarly $\mu(V) \setminus \mu(U \cup V) = \emptyset$. Since $\mu(U \cup V) \setminus \mu(V) = \mu((U \cup V) \setminus V) = \mu(U \setminus V) = \mu(U) \setminus \mu(V)$, and similarly, $\mu(V \cup U) \setminus \mu(U) = \mu(V) \setminus \mu(U)$, it follows that $\mu(U \cup V) = \mu(U) \cup \mu(V)$.

(2) $\mu(U \cap V) = \mu(U \setminus (U \setminus V)) = \mu(U) \setminus (\mu(U) \setminus \mu(V)) = \mu(U) \cap \mu(V)$. ∎

Lemma 3.2 *Suppose $\mu : E \to F$ is a morphism of concurrent alphabets. Then $\mu(e) \perp_F \mu(e')$ whenever $e \|_E e'$. Conversely, any function $\mu : E \to \mathrm{Com}(F)$ having this property extends uniquely to a morphism $\mu : E \to F$.*

Proof – If $\mu : E \to F$ is a morphism, and $e\|_E e'$, then $\mu(\{e\}) = \mu(\{e\} \setminus \{e'\}) = \mu(\{e\}) \setminus \mu(\{e'\})$, and similarly, $\mu(\{e'\}) = \mu(\{e'\}) \setminus \mu(\{e\})$, so $\mu(\{e\}) \cap \mu(\{e'\}) = \emptyset$.

Conversely, if $\mu : E \to \mathrm{Com}(F)$ has the stated property, then it extends to a morphism by defining $\mu(U) = \bigcup\{e \in U : \mu(e)\}$. Moreover, by Lemma 3.1, any extension of μ to a morphism must satisfy this relation, so the extension of μ is uniquely determined. ∎

Theorem 2 **Alph** *has finite limits.*

Proof – We show that **Alph** has: (1) a terminal object, (2) binary products, and (3) equalizers of parallel pairs.

(1) The empty concurrent alphabet is easily seen to be a terminal object (in fact a zero object) in **Alph**.

(2) Suppose E and F are concurrent alphabets. Let $E \otimes F$ denote the concurrent alphabet with elements $E + F$ (disjoint union), and with $\|_{E\otimes F} = \|_E \cup \|_F \cup (E \times F) \cup (F \times E)$. Note that the sets in $\mathrm{Com}(E \otimes F)$ are precisely those of the form $U + V$ with $U \in \mathrm{Com}(E)$ and $V \in \mathrm{Com}(F)$. Define projections $\pi_E : E \otimes F \to E$ and $\pi_F : E \otimes F \to F$ by $\pi_E(U) = U \cap E$ and $\pi_F(U) = U \cap F$. It is easy to check that π_E and π_F are morphisms, and that $E \otimes F$, equipped with π_E and π_F, has the universal property required of a categorical product.

(3) Suppose $\sigma, \tau : E \to F$ are morphisms in **Alph**. Define a nonempty set $U \in \mathrm{Com}(E)$ to be *equalizing* if $\sigma(U) = \tau(U)$. Call U *minimal equalizing* if it is equalizing and it has no proper equalizing subsets. We observe the following fact: if U and V are distinct minimal equalizing subsets of E, and $U \cup V \in \mathrm{Com}(E)$, then U and V are disjoint. For, if $U \cap V = W \neq \emptyset$, then $\sigma(W) = \sigma(U) \cap \sigma(V) = \tau(U) \cap \tau(V) = \tau(W)$, so W is equalizing, a contradiction with the assumed minimality of U, V. It follows from this observation that every equalizing $U \in \mathrm{Com}(E)$ can be written uniquely as a finite union of minimal equalizing subsets.

Now, let D be the set of all minimal equalizing subsets of E, and define $U\|_D V$ iff $U \cup V \in \mathrm{Com}(E)$. Define $\mu : D \to E$ to be the morphism that satisfies $\mu(\{U_1, \ldots, U_n\}) = \bigcup_k U_k$ whenever $\{U_1, \ldots, U_n\} \in \mathrm{Com}(D)$. One may now check that μ is an equalizer of σ and τ. ∎

3.2 Automata

An *automaton* is a tuple $A = (E, Q, q^\circ, T)$, where

- E is a concurrent alphabet of *events*, not containing the special symbol ϵ, called the *identity event*.

- Q is a set of *states*.

- $q^\circ \in Q$ is a distinguished *start state*.

- $T \subseteq Q \times (E \cup \{\epsilon\}) \times Q$ is a set of *transitions*. We write $t : q \xrightarrow{e} r$, or just $q \xrightarrow{e} r$, to denote a transition $t = (q, e, r) \in T$ or to assert the existence of such a transition in T.

These data are required to satisfy the following conditions:

(Identity) $q \xrightarrow{\epsilon} r$ iff $q = r$.

(Disambiguation) If $q \xrightarrow{e} r$ and $q \xrightarrow{e} r'$, then $r = r'$.

(Commutativity) For all states q and events $e, e' \in E$, if $e \|_E e'$, $q \xrightarrow{e} r$, and $q \xrightarrow{e'} r'$, then for some state s there exist transitions $r \xrightarrow{e'} s$ and $r' \xrightarrow{e} s$.

If $t : q \xrightarrow{e} r$, then q is called the *domain* $\mathrm{dom}(t)$ of t and r is called the *codomain* $\mathrm{cod}(t)$ of t. Transitions t and u are called *coinitial* if $\mathrm{dom}(t) = \mathrm{dom}(u)$. We say that event $e \in E$ is *enabled* in state q if there exists a transition $q \xrightarrow{e} r$ in T.

Intuitively, if $e \in E$, then a transition $q \xrightarrow{e} r$ represents a potential computation step of A in which event e occurs and the state changes from q to r. *Identity* transitions $q \xrightarrow{\epsilon} q$ do not represent computation steps of A; they serve merely to "pad" computations. The (Identity) condition ensures that this is all they can do. The (Disambiguation) condition ensures that the new state r in a transition $q \xrightarrow{e} r$ is uniquely determined by q and e. The (Commutativity) condition says that if two commuting events are enabled in the same state, then they can occur in either order with the same effect.

A *finite computation sequence* for an automaton A is a finite sequence γ of transitions of the form:

$$q_0 \xrightarrow{e_1} q_1 \xrightarrow{e_2} \ldots \xrightarrow{e_n} q_n.$$

The number n is called the *length* $|\gamma|$ of F. (By convention, if $n = 0$ then the computation sequence consists of the single state q_0, and no transitions.) Similarly, an *infinite computation sequence* for A is an infinite sequence of transitions:

$$q_0 \xrightarrow{e_1} q_1 \xrightarrow{e_2} \ldots.$$

We extend notation and terminology for transitions to computation sequences, so that if γ is a computation sequence, then the *domain* $\mathrm{dom}(\gamma)$ of γ is the state q_0, and if γ is finite, then the *codomain* $\mathrm{cod}(\gamma)$ of γ is the state q_n. We write $\gamma : q \to r$ to assert that γ is a finite computation sequence with domain q and codomain r. A computation sequence γ is *initial* if $\mathrm{dom}(\gamma)$ is the distinguished start state q°. If $\gamma : q \to r$ and $\delta : q' \to r'$ are finite computation sequences, then γ and δ are called *composable* if $q' = r$, in which case we define their *composition* to be the finite computation sequence $\gamma\delta : q \to r'$, obtained by concatenating γ and δ and identifying $\mathrm{cod}(\gamma)$ with $\mathrm{dom}(\delta)$.

If $A = (E, Q, q^\circ, T)$ and $A' = (E', Q', (q^\circ)', T')$ are automata, then a *morphism* from A to A' is a pair $\rho = (\rho_e, \rho_s)$, where $\rho_e : E \to E'$ is a morphism of concurrent alphabets, and $\rho_s : Q \to Q'$ is a function, such that

1. $\rho_s(q^\circ) = (q^\circ)'$.
2. Suppose $q \xrightarrow{e} r \in T$, with $e \neq \epsilon$. Then for every enumeration $\{e'_1, \ldots, e'_n\}$ of $\rho_e(\{e\})$, there exists a (necessarily unique) finite computation sequence

$$\rho_s(q) = r'_0 \xrightarrow{e'_1} r'_1 \xrightarrow{e'_2} \ldots \xrightarrow{e'_n} r'_n = \rho_s(r)$$

of A'.

We usually drop the subscripts on ρ_e and ρ_s, writing ρ for both.

Let **Auto** denote the category of automata and their morphisms. There is an obvious forgetful functor AuAl : **Auto** $\to$ **Alph**, which takes each automaton $A = (E, Q, q^\circ, T)$ to the concurrent alphabet E, and each morphism (ρ_e, ρ_s) of automata to the morphism ρ_e of concurrent alphabets.

Theorem 3 *The forgetful functor* AuAl : **Auto** $\rightarrow$ **Alph** *has a right adjoint* AlAu : **Alph** $\rightarrow$ **Auto**.

Proof – Given a concurrent alphabet E, let AlAu(E) be the one-state automaton $(E, \{*\}, *, T)$, where $T = \{*\} \times (E \cup \{\epsilon\}) \times \{*\}$. We claim that this construction defines the object map of a functor AlAu : **Alph** $\rightarrow$ **Auto**, which is right-adjoint to the forgetful functor AuAl : **Auto** $\rightarrow$ **Alph**. To prove this, it suffices to show that for each $E \in$ **Alph**, there exists an "evaluation map" ε_E : AuAl(AlAu$(E)) \rightarrow E$, universal from AuAl to E. We may simply take $\varepsilon_E = \mathrm{id}_E$. ∎

Theorem 4 **Auto** *has finite limits.*

Proof – We show that **Auto** has (1) a terminal object, (2) binary products, and (3) equalizers.

(1) It is easy to see that the one-state, one-transition automaton with the empty alphabet of events is a terminal object (in fact a zero object) in **Auto**.

(2) Suppose $A_1 = (E_1, Q_1, q_1^\circ, T_1)$ and $A_2 = (E_1, Q_2, q_2^\circ, T_2)$ are automata. Let $A_1 \times A_2$ be the automaton

$$A_1 \times A_2 = (E_1 \otimes E_2, Q_1 \times Q_2, (q_1^\circ, q_2^\circ), T),$$

where T is the set of all $((q_1, q_2), e, (r_1, r_2))$ such that one of the following conditions holds:

1. $e = \epsilon$, $q_1 = r_1$, and $q_2 = r_2$.
2. $e \in E_1$, $(q_1, e, r_1) \in T_1$, and $q_2 = r_2$.
3. $e \in E_2$, $(q_2, e, r_2) \in T_2$, and $q_1 = r_1$.

Define projections $\pi_i : A_1 \times A_2 \rightarrow A_i$ $(i \in \{1,2\})$ by letting $(\pi_i)_e : E_1 \otimes E_2 \rightarrow E_i$ be the projection in **Alph** and $(\pi_i)_s : Q_1 \times Q_2 \rightarrow Q_i$ be the projection in **Set**. It is straightforward to verify that $A_1 \times A_2$, equipped with projections π_1 and π_2, is a product in **Auto**.

(3) Suppose $\sigma, \tau : A \rightarrow A'$ are morphisms in **Auto**, where $A = (E, Q, q^\circ, T)$ and $A' = (E', Q', (q^\circ)', T')$. Let $\mu_e : D \rightarrow E$ be an equalizer of σ_e and τ_e in **Alph**, let $\mu_s : R \rightarrow Q$ be an equalizer of σ_s and τ_s in **Set**. Recall that D is the set of all minimal equalizing sets $U \in \mathrm{Com}(E)$. Define $B = (D, R, q^\circ, T)$, where T consists of all triples (q, U, r) such that to each enumeration $\{e_1, \ldots, e_n\}$ of U there corresponds a computation sequence

$$q = q_0 \xrightarrow{e_1} q_1 \xrightarrow{e_2} \ldots \xrightarrow{e_n} q_n = r$$

of A. Let $\mu = (\mu_e, \mu_s)$, then it is straightforward to verify that μ is an equalizer of σ and τ in **Auto**. ∎

4 Operational Semantics of Dataflow Networks

In this section, we give a formal operational semantics for dataflow networks as relations in **Auto**.

4.1 Port Sets

Let **Port** be the category whose objects are finite or countably infinite sets (of *ports*), and whose morphisms are the opposites of functions; thus, a morphism $f : P \rightarrowtail P'$ is a function from P' to P. The empty set $\emptyset$ is clearly a terminal object in **Port**, and if $+$ denotes disjoint union, then the set $P + P'$ is obviously a product of P and P' in **Port**. Let V be a fixed universe of *data values*, which we assume contains the set of natural numbers as a subset. If P is an object of **Port**, then define a *P-event* to be an element of the concurrent alphabet Events$(P) = P \times V$, where $(p, v) \| (p', v')$ iff

$p \neq p'$. Intuitively, we think of a P-event $e = (p, v)$ as representing the transmission of data value v over port p. We write port(e) to denote the port component p and value(e) to denote the value component v, of e.

Define the functor PoAl : **Port** $\rightarrow$ **Alph** to take each object P of **Port** to the concurrent alphabet Events(P), and each morphism $f : P \rightarrowtail P'$ of **Port** to the morphism ϕ : PoAl(P) $\rightarrow$ PoAl(P'), defined by

$$\phi(U) = \{e \in \mathrm{PoAl}(P') : (f(\mathrm{port}(e)), \mathrm{value}(e)) \in U\}$$

for all $U \in$ Com(Events(P)). Intuitively, we think of a morphism $f : P \rightarrowtail P'$ in **Port** as a relabeling map that labels each port in P' by a port in P. The corresponding morphism ϕ : PoAl(P) $\rightarrow$ PoAl(P') in **Alph** has the dual effects: (1) of deleting events for ports in P that are not the labels of ports in P', and (2) of duplicating events for ports in P that happen to be the labels of more than one port in P'.

Lemma 4.1 *The functor* PoAl *preserves finite products.*

Proof – Straightforward. ▮

Let PoAu : **Port** $\rightarrow$ **Auto** be the composite functor AlAu $\circ$ PoAl. Then since PoAl preserves finite products and AlAu is a right adjoint, hence preserves limits, it follows that PoAu also preserves finite products.

4.2 Port Automata

Define a *port signature* to be a pair (X, Y) of objects of **Port**. The elements of X are called *input ports* and those of Y, *output ports*. If (X, Y) is a port signature, then an (X, Y)-*port automaton* is an automaton $A = (E, Q, q^\circ, T)$, such that $E = \mathrm{Events}(X + Y + Z) = \mathrm{Events}(X) \otimes \mathrm{Events}(Y) \otimes \mathrm{Events}(Z)$ and the following condition holds:

(Receptivity) For all $q \in Q$ and all $a \in$ Events(X), there exists a transition $q \xrightarrow{a} r$ in T.

Elements of Events(X), Events(Y), and Events(Z) are called *input events*, *output events*, and *internal events*, respectively. Intuitively, the receptivity condition states that a port automaton is always prepared to receive arbitrary input.

A port automaton is *determinate* if the following additional condition holds:

(Determinacy) For all $q \in Q$ and all $b, b' \in$ Events($Y + Z$), if both b and b' are enabled in state q, then $b \| b'$.

Intuitively, determinate port automata make no internal choices between conflicting events.

As an example of how we can model a dataflow process as a port automaton, consider the case of a "merge" process, whose function is to shuffle together sequences of values arriving on two input ports into a single output sequence. Let i_1 and i_2 denote the two input ports and let o denote the single output port. We may represent the merge process as a port automaton

$$A_{\mathrm{mrg}} = (\mathrm{Events}(\{i_1, i_2\} + \{o\} + \{n\}), V^* \times V^* \times V^*, (\perp, \perp, \perp), T),$$

where $\perp$ denotes the empty string, and the set of transitions T contains a transition

$$((x_1, x_2, y), e, (x_1', x_2', y'))$$

iff one of the following conditions holds:

1. $e = \epsilon$, $x_1' = x_1$, $x_2' = x_2$, and $y' = y$.

2. $e = (i_1, v)$, $x_1' = x_1 v$, $x_2' = x_2$, and $y' = y$.

3. $e = (i_2, v)$, $x_1' = x_1$, $x_2' = x_2 v$, and $y' = y$.

4. $e = (n, 1)$, $v x_1' = x_1$, $x_2' = x_2$, and $y' = yv$.

5. $e = (n, 2)$, $x_1' = x_1$, $v x_2' = x_2$, and $y' = yv$.

6. $e = (o, v)$, $x_1' = x_1$, $x_2' = x_2$, and $vy' = y$.

It is straightforward to check that A_{mrg} satisfies the conditions for a port automaton, and that it is not determinate.

Intuitively, the state of A_{mrg} contains two "input buffers" and one "output buffer." Transitions of type (2) and (3) correspond to arriving input values being placed at the end of the appropriate input buffer. Transitions of type (4) and (5) are internal transitions that correspond to the indeterminate selection of input in one input buffer or the other to be moved to the output buffer. Transitions of type (6) correspond to the transmission of output from the output buffer. Similar constructions can be used to model many other kinds of dataflow processes.

4.3 Port Automata as Relations

An (X, Y)-port automaton $A = (E, Q, q^\circ, T)$ may be identified with the (X, Y)-relation $(A, \lambda_X, \lambda_Y)$ in **Auto** (with respect to the functor AlAu ∘ PoAl : **Port** → **Auto**), where

$$\lambda_X : A \to \mathrm{AlAu}(\mathrm{PoAl}(X)), \qquad \lambda_Y : A \to \mathrm{AlAu}(\mathrm{PoAl}(Y))$$

are the adjoint transforms of the projections

$$\pi_X : \mathrm{AuAl}(A) \to \mathrm{PoAl}(X), \qquad \pi_Y : \mathrm{AuAl}(A) \to \mathrm{PoAl}(Y).$$

(Recall that $\mathrm{AuAl}(A) = E = \mathrm{PoAl}(X) \otimes \mathrm{PoAl}(Y) \otimes \mathrm{PoAl}(Z)$.)

By making the above identification of port automata as relations in **Auto**, the definitions of Section 2.2 immediately become applicable, yielding operations of relabeling, sequential composition, parallel composition, and feedback on port automata. However, since not every (X, Y)-relation in **Auto** is an (X, Y)-port automaton, we do not know *a priori* that the class of port automata is closed under these operations.

Theorem 5 *The classes of port automata and of determinate port automata are closed under the following operations:*

1. *Input relabeling by bijections, and output relabeling by injections.*
2. *Sequential composition.*
3. *Parallel composition.*
4. *Feedback by injections.*

Proof – The proof simply requires substituting the characterizations of Theorem 4 into the definitions of the operations on relations, and checking that the conditions for port automata are satisfied in each case. We omit the details. ∎

By working through the details of the previous proof, one may convince oneself that the port automaton model is a reasonable operational semantics for dataflow networks. For example, suppose $A = (E, Q, q^\circ, T)$ is an $(X \times Z, Y)$-port automaton, and let ϕ be an injection from Z to Y, viewed as a morphism $\phi : Y \rightarrowtail Z$ in **Port**. Intuitively, we think of ϕ as specifying, for each input port in Z, a corresponding port in Y to which it is to be "connected" by a "feedback loop."

Applying the feedback operation $\circlearrowleft_\phi$ to A results (up to isomorphism) in an (X,Y)-port automaton $A' = (E \setminus \mathrm{Events}(Z), Q, q^\circ, T')$. Each input transition $q \xrightarrow{e} r$ of A with $\mathrm{port}(e) \in X$ is also a transition of A', and if $q \xrightarrow{e} r$ is an output transition of A with $\mathrm{port}(e) \in Y \setminus \phi(Z)$, then A' has that same output transition. However, if $q \xrightarrow{e} r$ is an output transition of A with $e = (\phi(z), v)$ for some $z \in Z$, then A' has instead a transition $q \xrightarrow{e} r'$, where r' is the unique state such that $r \xrightarrow{(z,v)} r'$ is a transition of A. Intuitively, A' behaves like A, except that in A' outputs of values on ports in $\phi(Z)$ occur simultaneously with inputs of the same values on the corresponding ports in Z.

4.4 Set-Theoretic Input/Output Relations

We now complete our operational semantics of dataflow networks by showing how the usual set-theoretic input/output relation between domains of "port histories" can be extracted from port automata.

Formally, if P is a set of ports, then a *P-history* is a mapping from P to the CPO V^∞ of all finite and infinite sequences of values in V, equipped with the prefix ordering $\sqsubseteq$. If H is a P-history, and $P' \subseteq P$, then we write $H|P'$ to denote the P'-history H' such that $H'(p) = H(p)$ for all $p \in P'$.

Suppose $A = (\mathrm{Events}(P), Q, q^\circ, T)$ is an (X,Y)-port automaton, where $P = X + Y + Z$. Then each finite or infinite computation sequence

$$\gamma = q_0 \xrightarrow{e_1} q_1 \xrightarrow{e_2} \ldots$$

of A determines a P-history H_γ as follows: for each $p \in P$, the sequence $H_\gamma(p)$ is the sequence of values $\mathrm{value}(e'_1), \mathrm{value}(e'_2), \ldots$, where $e'_1, e'_2, \ldots$, is the subsequence of $e_1, e_2, \ldots$ consisting of precisely those $e_k \neq \epsilon$ with $\mathrm{port}(e_k) = p$.

A computation sequence γ of A is called *completed* if there exists no computation sequence δ, such that $\mathrm{dom}(\gamma) = \mathrm{dom}(\delta)$, $H_\gamma|X = H_\delta|X$, and $H_\gamma \sqsubset H_\delta$. That is, completed computation sequences are those whose histories are maximal among all computation sequences having the same domain and input port history. This is actually a kind of "fairness" property (see [13]), which intuitively is true when "every enabled output or internal event, not in conflict with some other enabled output or internal event, eventually occurs." The *set-theoretic input/output relation* $\mathrm{Rel}(A)$ of A is the set of all pairs $(H_\gamma|X, H_\gamma|Y)$ such that γ is a completed initial computation sequence of A.

For example, the input/output relation of the automaton A_{mrg} defined in the previous section is the set of all $(H^{\mathrm{in}}, H^{\mathrm{out}})$, with H^{in} an $\{i_1, i_2\}$-history and H^{out} an $\{o\}$-history, such that $H^{\mathrm{out}}(o)$ is a shuffle of a prefix x_1 of $H^{\mathrm{in}}(i_1)$ and a prefix x_2 of $H^{\mathrm{in}}(i_2)$, subject to the following conditions:

1. If $H^{\mathrm{in}}(i_1)$ is finite, then $x_2 = H^{\mathrm{in}}(i_2)$.

2. If $H^{\mathrm{in}}(i_2)$ is finite, then $x_1 = H^{\mathrm{in}}(i_1)$.

3. If both $H^{\mathrm{in}}(i_1)$ and $H^{\mathrm{in}}(i_2)$ are infinite, then either $x_1 = H^{\mathrm{in}}(i_1)$ or else $x_2 = H^{\mathrm{in}}(i_2)$.

Thus, the input/output relation of A_{mrg} is the *angelic merge* relation [13]. It is also possible to construct a port automaton having a slightly different merging relation, called *infinity-fair merge*, as its input/output relation. However, it is shown in [13] that no port automaton can have as its input/output relation the *fair merge* relation, which is the set of all $(H^{\mathrm{in}}, H^{\mathrm{out}})$ such that $H^{\mathrm{out}}(o)$ is a shuffle of *all* of $H^{\mathrm{in}}(i_1)$ and *all* of $H^{\mathrm{out}}(i_2)$.

We may regard the set-theoretic input/output relation $\mathrm{Rel}(A)$ of an (X,Y)-port automaton A as an (X,Y)-relation in the category **Set** of sets and functions. To do this, let F be the functor $\mathrm{Hist} : \mathbf{Port} \to \mathbf{Set}$ that takes a set P of ports to the set $\mathrm{Hist}(P)$ of all port histories over P, and that takes a morphism $f : P \to P'$ to the corresponding duplication/restriction map

$\mathrm{Hist}(f) : \mathrm{Hist}(P) \to \mathrm{Hist}(P')$, given by $\mathrm{Hist}(f)(H)(p) = H(f(p))$ for all $p \in P'$. This functor is easily seen to preserve finite products. The set-theoretic input/output relation Rel(A) of an (X, Y)-port automaton A may now be regarded as the (X, Y)-relation $(\mathrm{Rel}(A), \pi_X, \pi_Y)$ in **Set**, where $\pi_X : \mathrm{Rel}(A) \to \mathrm{Hist}(X)$ and $\pi_Y : \mathrm{Rel}(A) \to \mathrm{Hist}(Y)$ are the obvious projections.

It is important to observe that, although we may regard the set-theoretic input/output relation of an (X, Y)-port automaton as an (X, Y)-relation in **Set**, and thereby obtain formal operations of relabeling, composition, product, and feedback on such relations, these operations are incompatible with the corresponding operations on port automata. Specifically, the feedback operation is not preserved by the mapping from port automata to input/output relations. Brock and Ackerman have given specific examples of this failure of commutativity, and it has come to be called the "Brock-Ackerman anomaly." The situation may be described succinctly as the failure of the set-theoretic relational semantics to be "compositional."

5 Denotational Semantics of Dataflow Networks

The goal of this section of the paper is to replace the category **Set**, which does not yield a compositional denotational semantics for dataflow networks, with a more highly structured category **EvDom**, for which a compositional semantics is obtained. We do this in two steps: first we define a category **TrDom** of "trace domains," which are domains that are embedded as certain normal subdomains of the domains of "traces" generated by a concurrent alphabet, then we throw away the embedding domains and obtain a category **EvDom**, whose objects are "event domains" and whose morphisms are certain continuous maps. We construct coreflections between **Auto** and **TrDom** and between **TrDom** and **EvDom**, with the right adjoints (coreflectors) going from **Auto** to **TrDom** and from **TrDom** to **EvDom**. It follows by Theorem 1 and the fact that right adjoints preserve limits, that the induced ensemble of functors, from (X, Y)-relations in **Auto** to (X, Y)-relations in **EvDom**, preserves the operations of relabeling, sequential and parallel composition, and feedback.

5.1 Domains

A (Scott) *domain* is an ω-algebraic, consistently complete CPO. A domain D is *finitary* if for all finite (=isolated=compact) elements $d \in D$ the set $\{d' \in D : d' \sqsubseteq d\}$ is finite. If D and E are domains, then a monotone map $f : D \to E$ is *continuous* if it preserves directed lubs, *strict* if $f(\perp_D) = \perp_E$, and *additive* if whenever d, d' are consistent elements of d, then $f(d), f(d')$ are consistent elements of E, and $f(e \sqcup e') = f(e) \sqcup f(e')$. Let **Dom** denote the category of domains and continuous maps.

A *subdomain* of D is a subset U of D, which is a domain under the restriction of the ordering on D, and is such that the inclusion of U in D is strict and continuous. A subdomain U of D is *normal* if for all $d \in D$, the set $\{e \in U : e \sqsubseteq d\}$ is directed.

Lemma 5.1 *A subdomain U of D is normal iff the inclusion of U in D is additive and reflects consistent pairs.*

Proof – Suppose U is a normal subdomain of D. If $e, e' \in U$ are consistent in D, then the set $V = \{e'' \in U : e'' \sqsubseteq e \sqcup_D e'\}$ is directed and contains e, e', so e and e' are consistent in U. Let $v = \bigsqcup_D V$, then $v \in U$ because $V \subseteq U$ and U is a subdomain of D. Clearly, $v \sqsubseteq e \sqcup_D e'$. Also, $e \sqcup_D e' \sqsubseteq v$, because v is an upper bound of e and e'. Hence $v = e \sqcup_D e' = e \sqcup_U e'$ is an upper bound for e and e' in U.

Conversely, suppose that whenever e and e' are consistent in D, then e, e' are consistent in U, and $e \sqcup_U e' = e \sqcup_D e'$. We claim that for each $d \in D$, the set $U_d = \{e \in U : e \sqsubseteq d\}$ is directed. If

$e, e' \in U_d$, then $e \sqsubseteq d$ and $e' \sqsubseteq d$, so e, e' are consistent in D, hence e, e' are consistent in U, and $e \sqcup_U e' = e \sqcup_D e' \sqsubseteq d$. Thus $e \sqcup_U e' \in U_d$. ∎

An *interval* of a domain D is a pair $I = (\mathrm{dom}(I), \mathrm{cod}(I)) \in D \times D$, with $\mathrm{dom}(I) \sqsubseteq \mathrm{cod}(I)$. Intervals I and J are *coinitial* if $\mathrm{dom}(I) = \mathrm{dom}(J)$. The interval I is an *identity* if $\mathrm{dom}(I) = \mathrm{cod}(I)$, and a nonidentity interval I is *prime* if there exists no $d \in D$ with $\mathrm{dom}(I) \sqsubset d \sqsubset \mathrm{cod}(I)$. It is $\sqcup$-*prime* if there exists a finite, nonempty set U of prime and identity intervals, such that $\mathrm{dom}(J) = \mathrm{dom}(I)$ for all $J \in U$, and $\mathrm{cod}(I) = \bigsqcup\{\mathrm{cod}(J) : J \in U\}$. Coinitial intervals I and J are *consistent* if $\mathrm{cod}(I)$ and $\mathrm{cod}(J)$ are consistent elements of D. If I and J are consistent, then the *residual of I after J* is the interval $I \setminus J = (\mathrm{cod}(J), \mathrm{cod}(I) \sqcup \mathrm{cod}(J))$.

In the sequel, we shall use the term "interval" exclusively to mean "interval with finite endpoints."

5.2 Trace Domains

Concurrent alphabets generate domains. Formally, suppose E is a concurrent alphabet. Let E^* denote the free monoid generated by E, then there is a least congruence $\sim$ on E^* such that $e \|_E e'$ implies $ee' \sim e'e$ for all $e, e' \in E$. The quotient $E^* /\!\sim$ is the *free partially commutative monoid* generated by E, and its elements are called *traces*. Define the relation $\sqsubseteq$ on $E^* /\!\sim$ by $[x] \sqsubseteq [y]$ iff there exists $z \in E^*$ with $[xz] = [y]$. This relation is a partial order with respect to which every consistent pair of elements has a least upper bound (this is not completely trivial to prove). We call $\sqsubseteq$ the *prefix* relation. By forming the ideal completion of the poset $(E^* /\!\sim, \sqsubseteq)$, we obtain a domain $\widehat{E}$. The map taking E to $\widehat{E}$ extends to a functor $(\widehat{-}) : \mathbf{Alph} \to \mathbf{Dom}$.

We may think of the elements of $\widehat{E}$ as equivalence classes of finite and infinite strings. The finite elements of E are the equivalence classes of finite strings. All the finite strings that are representatives of a given finite element of $\widehat{E}$ are permutations of each other, hence have the same length and the same number of occurrences of each element of E. The prime intervals of $\widehat{E}$ are those of the form $([x], [xe])$, where $x \in E^*$ and $e \in E$, and the $\sqcup$-prime intervals are those of the form $([x], [xe_1 \ldots e_n])$, where $x \in E^*$ and $\{e_1, \ldots, e_n\} \in \mathrm{Com}(E)$. Distinct coinitial prime intervals $I = ([x], [xe])$ and $I' = ([x], [xe'])$ are consistent iff $e \|_E e'$, in which case $I \setminus I' = ([xe'], [xe'e])$.

Lemma 5.2 *If E is a concurrent alphabet, then the domain $\widehat{E}$ is finitary.*

Proof – The finite elements of E are permutation equivalence classes of finite strings, and for finite $[x], [y]$ we have $[y] \sqsubseteq [x]$ iff there exists z with $[yz] = [x]$. Thus, the cardinality of $\{[y] : [y] \sqsubseteq [x]\}$ is bounded by the number of prefixes of permutations of x, which is finite. ∎

A *trace domain* is a pair (E, D), where E is a concurrent alphabet and D is a normal subdomain of $\widehat{E}$, such that every prime interval of D is also a prime interval of $\widehat{E}$. If (E, D) and (E', D') are trace domains, then a *morphism* from (E, D) to (E', D') is a morphism $\mu : E \to E'$ of concurrent alphabets such that the following two conditions hold:

1. $\widehat{\mu}(D) \subseteq D'$.

2. Whenever (d, d') is a prime interval of D, then $(\widehat{\mu}(d), \widehat{\mu}(d'))$ is a $\sqcup$-prime interval of D'.

Let **TrDom** denote the category of trace domains and their morphisms.

We say that an element d of a domain D is *secured* if there exists a finite chain

$$\bot = d_0 \sqsubseteq d_1 \sqsubseteq \ldots \sqsubseteq d_n = d,$$

such that each interval (d_k, d_{k+1}) is prime. We call such a chain a *securing chain* for d. We say that a domain D is secured if each of its finite elements is secured. It is easy to see that finitary domains are secured. We say that a subdomain U of D is *secured in D* if each finite element of U is secured as an element of D.

Lemma 5.3 *Suppose* (E, D) *is a trace domain. Then* D *is finitary and secured in* $\widehat{E}$.

Proof – Since D is a normal subdomain of $\widehat{E}$, every finite element of D is also a finite element of $\widehat{E}$. Since $\widehat{E}$ is finitary, so is D. Now, since D is finitary we know that D is secured (in itself). Since every prime interval in D is also a prime interval of E, it follows that every securing chain for $d \in D$ is also a securing chain for $d \in \widehat{E}$. Thus D is secured in E. ▮

Automata "unwind" to trace domains, and the unwinding map gives a coreflection between **Auto** and **TrDom**. Formally, suppose $A = (E, Q, q^\circ, T)$ is an automaton. Each finite or infinite computation sequence

$$\gamma = q_0 \xrightarrow{e_1} q_1 \xrightarrow{e_2} \cdots$$

of A, determines an element $\mathrm{tr}(\gamma)$ of the domain $\widehat{E}$, according to the definition

$$\mathrm{tr}(\gamma) = \bigsqcup_k e_1 e_2 \ldots e_k,$$

where concatenation denotes multiplication in the monoid of finite traces $E^* /\!\sim$, and we identify the identity event ϵ with the monoid identity, so that ϵ does not appear in the trace $e_1 \ldots e_k$. We call $\mathrm{tr}(\gamma)$ the *trace of* the computation sequence γ. Let $\mathrm{Traces}(A)$ denote the set of all traces of initial computation sequences of A.

Theorem 6 *Suppose* $A = (E, Q, q^\circ, T)$ *is an automaton. Then* $(E, \mathrm{Traces}(A))$ *is a trace domain. Moreover, the map taking* A *to* $(E, \mathrm{Traces}(A))$ *is the object map of a functor* AuTr : **Auto** → **TrDom**, *which is right-adjoint to a full embedding* TrAu : **TrDom** → **Auto**.

Proof – See the Appendix. ▮

Corollary 5.1 **TrDom** *has finite limits.*

Proof – This follows immediately from the previous result, but as an aid to intuition we give the explicit constructions.

(*Terminal Object*) The trace domain $(\emptyset, \{\bot\})$ is a terminal object (in fact a zero object) in **TrDom**.

(*Products*) Suppose (E_1, D_1), (E_2, D_2), are trace domains. Let $E = E_1 \otimes E_2$ be the product of E_1 and E_2 in **Alph**, and let $\pi_i : E_1 \otimes E_2 \to E_i$ $(i \in \{1,2\})$ be the associated projections. Let D be the set of all $x \in \widehat{E}$ such that $\widehat{\pi}_1(x) \in D_1$ and $\widehat{\pi}_2(x) \in D_2$. Then (E, D), equipped with the morphisms π_1 and π_2, is a product of (E_1, D_1) and (E_2, D_2) in **TrDom**.

(*Equalizers*) Suppose $\sigma, \tau : (E_1, D_1) \to (E_2, D_2)$ is a parallel pair of morphisms in **TrDom**. Let $\mu : E \to E_1$ be an equalizer of σ and τ in **Alph**. Define an element $d \in \widehat{E}$ to be *reachable* if there exists a chain

$$\bot = d_0 \sqsubseteq d_1 \sqsubseteq \ldots$$

of finite elements of $\widehat{E}$, such that $d = \bigsqcup_k d_k$, $\widehat{\mu}(d_k) \in D_1$ for all $k \geq 0$, and for all $k \geq 0$ the interval $(\widehat{\mu}(d_k), \widehat{\mu}(d_{k+1}))$ is $\sqcup$-prime. Let D be the set of all reachable elements of E. Then $\mu : (E, D) \to (E_1, D_1)$ is an equalizer of σ and τ in **TrDom**. ▮

5.3 Event Domains

We now wish to view a trace domain (E, D) as a domain in its own right, rather than as a normal subdomain of the domain $\widehat{E}$ of traces. In fact, this can be done. By discarding the concurrent alphabet component of trace domains we obtain a class of domains called "event domains," which have been studied previously. It is known that event domains are exactly those domains that are isomorphic to the "domains of configurations" of a certain kind of "event structure" [6,20]. A byproduct of our investigation is the following representation theorem, which the author has not seen explicitly stated before.

- Event domains are precisely those domains D for which there exists a concurrent alphabet E and an embedding $f : D \to \widehat{E}$ of D as a normal subdomain of $\widehat{E}$, such that if (d, d') is a prime interval of D, then $(f(d), f(d'))$ is a prime interval of $\widehat{E}$.

The formal definition of event domains requires a few preliminaries. Suppose D is a domain with the following property:

1. $I \setminus J$ is a prime interval whenever I and J are distinct, consistent prime intervals.

Then it is not difficult to see that the same property holds if "prime" is replaced by "$\sqcup$-prime."

If $I = (d, d')$ is a $\sqcup$-prime interval of D, then define

$$\mathrm{pr}(I) = \{(d, d'') : d \sqsubset d'' \sqsubseteq d' \text{ and } (d, d'') \text{ is prime}\}.$$

Call coinitial $\sqcup$-prime intervals I and J *orthogonal* if they are consistent and $\mathrm{pr}(I) \cap \mathrm{pr}(J) = \emptyset$. Note that coinitial prime intervals are orthogonal iff they are distinct and consistent. Let $\equiv$ be the least equivalence relation on $\sqcup$-prime intervals of D such that $I \equiv I \setminus J$ whenever I and J are orthogonal $\sqcup$-prime intervals. A straightforward induction using property (1) shows that if I is prime and $I \equiv I'$, then I' is also prime.

An *event domain* is a finitary domain D that satisfies property (1) above, and in addition satisfies:

2. $I \equiv J$ implies $I = J$, whenever I, J are coinitial prime intervals.

3. If I, I', J, J' are prime intervals such that $I \equiv I'$, $J \equiv J'$, I and J are coinitial, and I' and J' are coinitial, then I and I' are consistent iff J and J' are consistent.

A *morphism* from an event domain D to an event domain D' is a strict, additive, continuous function $f : D \to D'$ with the following two additional properties:

1. Whenever I is a prime interval of D, then $f(I)$ is a $\sqcup$-prime interval of D'.

2. Whenever I, J are distinct consistent prime intervals of D, then $f(I)$ and $f(J)$ are orthogonal.

Let **EvDom** denote the category of event domains and their morphisms, then **EvDom** is a (non-full) subcategory of **Dom**.

Lemma 5.4 *If (E, D) is a trace domain, then D is an event domain. Moreover, the map taking (E, D) to D extends to a (forgetful) functor* $\mathrm{TrEv} : \mathbf{TrDom} \to \mathbf{EvDom}$.

Proof – We first show that if E is a concurrent alphabet, then $\widehat{E}$ is an event domain. We have already seen (Lemma 5.2) that $\widehat{E}$ is finitary. It remains to verify axioms (1)-(3) for event domains.

(1) An interval I of $\widehat{E}$ is prime iff it is of the form $([x], [xe])$, with $x \in E^*$ and $e \in E$. Distinct, coinitial prime intervals $I = ([x], [xe])$ and $I' = ([x], [xe'])$ are consistent iff $e \|_E e'$, in which case $I \setminus I'$ is the interval $([xe], [xee'])$, which is prime.

(2) A straightforward induction shows that if $I \equiv I'$, where $I = ([x], [xe])$ and $I' = ([x'], [x'e'])$ are prime intervals, then $e = e'$. Hence if $I \equiv I'$ and I, I' are coinitial, then $I = I'$.

(3) If distinct prime intervals $I = ([x], [xe])$ and $J = ([x], [xe'])$ are consistent, then $e \|_E e'$. If $I' = ([x'], [x'e]) \equiv I$ and $J' = ([x'], [x'e']) \equiv J$, then clearly I' and J' must also be consistent.

Next, we show that if D is is a normal subdomain of $\widehat{E}$, such that an interval of D is prime in D iff it is also prime in $\widehat{E}$, then D is also an event domain. Suppose D is such a domain. Clearly, D has property (1) of an event domain. By an induction we see that if I and J are prime intervals of D, then $I \equiv J$ in D iff $I \equiv J$ in $\widehat{E}$. Properties (2)-(3) of an event domain follow easily from this fact.

Finally, to see that TrEv is a functor, note that given a morphism $\mu : (E,D) \to (E',D')$ in **TrDom**, the map $\widehat{\mu} : \widehat{E} \to \widehat{E}'$ restricts to a map $\mathrm{TrEv}(\mu) : D \to D'$. The prime intervals of D are those of the form $I = ([x],[xe])$ where $x \in E^*$ and $e \in E$. Then $\widehat{\mu}(I) = (\widehat{\mu}([x]), \widehat{\mu}([x])\widehat{\mu}(e))$, which is a $\sqcup$-prime interval of D'. Also, if I and J are distinct, consistent prime intervals of D, then $I = ([x],[xe])$ and $J = ([x],[xe'])$, where $e \|_E e'$. Since $\mu(e) \cap \mu(e') = \emptyset$, it follows that $\widehat{\mu}(I)$ and $\widehat{\mu}(J)$ are orthogonal. Hence $\mathrm{TrEv}(\mu)$ is a morphism of event domains. ∎

Theorem 7 *The forgetful functor* TrEv : **TrDom** $\to$ **EvDom** *has a left adjoint* EvTr, *such that* TrEv $\circ$ EvTr $\simeq 1$.

Proof – See the Appendix. ∎

Corollary 5.2 **EvDom** *has finite limits.*

5.4 Summary

We have now reached our goal. Dataflow networks have an operational semantics as port automata, which we identify with the corresponding relations in **Auto**. The functor $G = \mathrm{TrEv} \circ \mathrm{AuTr}$: **Auto** $\to$ **EvDom** preserves limits, hence induces an ensemble of functors

$$G_{X,Y} : \mathbf{Reln}_{\mathbf{Auto}}(X,Y) \to \mathbf{Reln}_{\mathbf{EvDom}}(X,Y)$$

which preserve not only the operations of relabeling, sequential composition, parallel composition, and feedback, but any operations that have categorical definitions in terms of composition and limits in **Auto**. Thus, the denotational semantics, in which a dataflow network denotes an (X,Y)-relation in **EvDom**, agrees with the operational semantics given by port automata. Moreover, the semantics is not "too abstract," in the sense that networks with distinct set-theoretic input/output relations receive distinct denotations. This can be verified by observing that the set-theoretic input/output relation of a port automaton can still be extracted from the corresponding relation in **EvDom**.

6 Port Automata and Causal Relations

Not every (X,Y)-relation in **EvDom** is the relation corresponding to an (X,Y)-port automaton. There are a number of reasons why this is so, but the easiest to see is that although the receptivity condition in the definition of port automata introduces an asymmetry between input and output, there is no such asymmetry in the definition of relations in **EvDom**. We would like to have a theorem that characterizes exactly, in terms of categorical properties of **EvDom**, those (X,Y)-relations that are the relations corresponding to (X,Y)-port automata and those that are the relations corresponding to determinate (X,Y)-automata. Note that the category **EvDom** has a substantial amount of structure with which to express such properties; for example, its hom-sets are partially ordered by virtue of its being a subcategory of the category **Dom**. At the moment, we have no exact characterization theorem, but we do have some partial results. In particular, we can show that the relation in **EvDom** corresponding to an (X,Y)-port automaton is *causal*, in a sense to be defined formally below, and that the class of causal relations in **EvDom** is closed under those relational operations that make sense for port automata. The notion of causality captures some of the input/output asymmetry exhibited by port automata.

Formally, let categories **T** and **C**, and functor $F : \mathbf{T} \to \mathbf{C}$ be as in Section 2. Suppose further that the category **C** is *pointed*; that is, it has a unique zero morphism $0_{A,B} : A \to B$ for every pair of objects A, B. Then an (X,Y)-relation R in **C** is *causal* if there exists a morphism $m : FX \to R$ such that $\underline{r}m = \mathrm{id}_{FX}$ and $\bar{r}m = 0$. Then $\underline{r}$ is a retraction and m is a section.

Theorem 8 *The class of causal relations in* **C** *is closed under the following operations:*

1. *Output relabeling by arbitrary morphisms and input relabeling by retractions.*
2. *Sequential composition on the output with arbitrary relations (hence also sequential composition on the input with causal relations).*
3. *Parallel composition.*
4. *Feedback.*

Proof – Omitted. ∎

Lemma 6.1 *The category* **EvDom** *is pointed.*

Proof – The one-point domain is a zero object. ∎

Theorem 9 *If R is the (X,Y)-relation in* **EvDom** *corresponding to an (X,Y)-port automaton, then R is causal.*

Proof – The receptivity property of an (X,Y)-port automaton A implies that for each input history $H \in \mathrm{Hist}(X)$, there is a corresponding initial computation sequence γ such that γ consists entirely of input or identity transitions, and $H_\gamma|X = H$. The map $m_0 : \mathrm{Hist}(X) \to R$ that takes each such H to the trace $\mathrm{tr}(\gamma) \in R$ is the m required to show R causal. ∎

It appears that we can go quite a bit further than this. By taking into account the order relation $\leq$ on the hom-sets of **EvDom**, one can see that if R is an (X,Y)-relation in **EvDom** corresponding to an (X,Y)-port automaton, then the section $m_0 : \mathrm{Hist}(X) \to R$ produced by the construction in the proof of the previous theorem is characterized by the properties $\underline{r}m_0 = \mathrm{id}_{\mathrm{Hist}(X)}$ and $m_0\underline{r} \leq \mathrm{id}_R$. That is, $(m_0, \underline{r})$ is an embedding-projection pair. One may also observe that for relations R corresponding to determinate automata, the class of all sections m, such that $\underline{r}m = \mathrm{id}_{\mathrm{Hist}(X)}$, is directed by $\leq$ and has m_0 as a least element. By taking the least upper bound of this collection (working now in the full subcategory of **Dom** whose objects are the event domains), we obtain a continuous map $\mu : \mathrm{Hist}(X) \to R$. By composing with $\bar{r}$, we obtain a continuous map $\bar{r}\mu : \mathrm{Hist}(X) \to \mathrm{Hist}(Y)$. Thus we see that determinate automata determine continuous maps from input to output. We expect that this observation can form the basis for a nice connection between the category-theoretic semantics of feedback we gave here in terms of equalizers, and the order-theoretic version given by Kahn in terms of least fixed points. We are currently attempting to work out the details of this connection.

7 Conclusion

By showing that the algebra of causal relations in **EvDom** constitutes a correct semantics for dataflow networks, we obtain a substantial amount of algebraic machinery for reasoning about such networks. Further clarification of the structure of those relations in **EvDom** that correspond to port automata should yield even more machinery. The study of relations in **EvDom** may also yield useful characterizations of "observational equivalence" of networks, and information about the structure of the fully abstract semantics with respect to this equivalence.

A Appendix: Proofs of Theorems 6 and 7

Theorem 6 *Suppose* $A = (E, Q, q^\circ, T)$ *is an automaton. Then* $(E, \mathrm{Traces}(A))$ *is a trace domain. Moreover, the map taking* A *to* $(E, \mathrm{Traces}(A))$ *is the object map of a functor* AuTr : **Auto** $\to$ **TrDom**, *which is right-adjoint to a full embedding* TrAu : **TrDom** $\to$ **Auto**.

Proof – We first show that Traces(A) is a normal subdomain of $\hat{E}$. To do this requires a detailed analysis of the structure of the set of initial computation sequences of A, and we perform this analysis using the notion of the "residual" $\gamma \setminus \delta$ of one finite computation sequence γ "after" another finite computation sequence δ. Intuitively, $\gamma \setminus \delta$ is obtained from γ by cancelling out transitions that, "up to permutation equivalence," also appear in δ.

Formally, the residual operation is a partial binary operation on coinitial pairs of computation sequences. We first define $\setminus$ for single transitions, and then extend to arbitrary finite computation sequences by induction on their length. For single transitions, suppose $t : q \xrightarrow{a} r$ and $u : q \xrightarrow{b} s$. If $a = \epsilon$, then $t \setminus u = (s \xrightarrow{\epsilon} s)$ and $u \setminus t = u$. If $a = b \neq \epsilon$, then we define $t \setminus u = (s \xrightarrow{\epsilon} s) = (r \xrightarrow{\epsilon} r) = u \setminus t$. If $\epsilon \neq a \neq b \neq \epsilon$, then $t \setminus u$ is defined iff $a \|_E b$, in which case the commutativity property of A implies there must exist (necessarily unique) transitions $s \xrightarrow{a} p$ and $r \xrightarrow{b} p$, which we take as $t \setminus u$ and $u \setminus t$, respectively.

Next, we extend to arbitrary finite computation sequences. Suppose $\gamma : q \to r$ and $\delta : q \to s$ are coinitial finite computation sequences. Then

1. If $|\gamma| = 0$ then $\gamma \setminus \delta = \mathrm{id}_s$, where id_s denotes the length-0 computation sequence from state s.

2. If $|\gamma| > 0$ and $|\delta| = 0$, then $\gamma \setminus \delta = \gamma$.

3. If $\gamma = t\gamma'$, δ is the single transition u, $t \setminus u$ is defined, and $\gamma' \setminus (u \setminus t)$ is defined, then
$$\gamma \setminus \delta = (t \setminus u)(\gamma' \setminus (u \setminus t)).$$

4. If $|\gamma| > 0$, $\delta = u\delta'$ with $|\delta'| > 0$, $\gamma \setminus u$ is defined, and $(\gamma \setminus u) \setminus \delta'$ is defined, then
$$\gamma \setminus \delta = (\gamma \setminus u) \setminus \delta'.$$

For a more detailed explanation of this operation and its properties, the reader is referred to [13,18].

We now use $\setminus$ to define a relation $\precsim$ on initial computation sequences as follows: For finite sequences γ and δ, define $\gamma \precsim \delta$ to hold precisely when $\gamma \setminus \delta$ is a sequence of identity transitions. Extend this definition to infinite sequences by defining $\gamma' \precsim \delta'$ iff for every finite prefix γ of γ', there exists a finite prefix δ of δ', such that $\gamma \precsim \delta$.

We observe the following facts about $\precsim$:

1. The relation $\precsim$ is a preorder, and the set of $\precsim$-equivalence classes, equipped with the induced partial ordering, is a domain whose finite elements are precisely the equivalence classes of finite initial computation sequences.

2. The map taking each $\precsim$-equivalence class to its trace is strict, additive, and continuous.

3. Initial computation sequences γ and δ have a $\precsim$-upper bound (given by $\gamma(\delta \setminus \gamma)$ or $\delta(\gamma \setminus \delta)$, which are $\precsim$-equivalent) iff the traces $\mathrm{tr}(\gamma)$ and $\mathrm{tr}(\delta)$ are consistent.

4. $\gamma \precsim \delta$ iff $\mathrm{tr}(\gamma) \sqsubseteq \mathrm{tr}(\delta)$.

Facts (1) and (2) are shown in [13] using the properties of the residual operation developed there. Facts (3) and (4) can be verified by straightforward inductive arguments from the definition of the residual operation. It follows from these facts and Lemma 5.1 that the map taking each $\precsim$-equivalence class to its trace is an isomorphism of the domain of equivalence classes of initial computation sequences of A to a normal subdomain of $\widehat{E}$. Since this subdomain is Traces(A), we conclude that Traces(A) is a normal subdomain of $\widehat{E}$.

To complete the proof that (E, Traces(A)) is a trace domain, we observe that the prime intervals $([\gamma],[\delta])$ in the domain of equivalence classes of initial computation sequences of A are those for which $\delta \setminus \gamma$ contains precisely one nonidentity transition. Since this implies that $\mathrm{tr}(\delta)$ is longer than $\mathrm{tr}(\gamma)$ by one symbol, it follows that prime intervals in Traces(A) are also prime intervals in $\widehat{E}$.

It remains to be shown the map AuTr, taking A to (E, Traces(A)), is the object map of a functor that is right-adjoint to a full embedding TrAu : **TrDom** $\rightarrow$ **Auto**. Given a trace domain (E, D), define the automaton $\mathrm{TrAu}(E,D) = (E, Q, q^\circ, T)$ as follows:

- Let Q be the set of finite elements of D, with $q^\circ = \bot$. Then Q is a set of equivalence classes of finite strings.
- Let T contain all transitions $[x]\xrightarrow{\epsilon}[x]$ with $[x] \in Q$ and all transitions $[x]\xrightarrow{e}[xe]$ with both $[x]$ and $[xe]$ in Q.

It is straightforward to verify that $\mathrm{TrAu}(E,D)$ is an automaton. Moreover, each morphism $\phi : (E,D) \rightarrow (E',D')$ in **TrDom** determines a morphism $\mathrm{TrAu}(\phi) : \mathrm{TrAu}(E,D) \rightarrow \mathrm{TrAu}(E',D')$, where $\mathrm{TrAu}(\phi)_e = \phi$, and $\mathrm{TrAu}(\phi)_s$ is the restriction of $\widehat{\phi}$ to a map from finite elements of D to finite elements of D'. To see that $\mathrm{TrAu}(\phi)$ is in fact a morphism of automata, note that the nonidentity transitions t of $\mathrm{TrAu}(E,D)$ correspond bijectively to the prime intervals $([x],[xe])$ in D. For each such interval, $(\widehat{\phi}([x]), \widehat{\phi}([xe]))$ is a $\sqcup$-prime interval of D'. hence is of the form

$$([x'], [x'e'_1] \sqcup \ldots \sqcup [x'e'_n]),$$

where $\{e'_1, \ldots, e'_n\}$ is an arbitrary enumeration of $\phi(\{e\})$, and $[x'e'_1], \ldots, [x'e'_n]$ are all in D'. It follows from this that to each enumeration $\{e'_1, \ldots, e'_n\}$ of $\phi(\{e\})$ there corresponds a computation sequence

$$\mathrm{TrAu}(\phi)_s([x]) = [x']\xrightarrow{e'_1}[x'e'_1]\xrightarrow{e'_2}([x'e'_1] \sqcup [x'e'_2])\xrightarrow{e'_3}\ldots\xrightarrow{e'_n}([x'e'_1] \sqcup \ldots \sqcup [x'e'_n]) = \mathrm{TrAu}(\phi)_s([xe])$$

of $\mathrm{TrAu}(E',D')$.

Clearly, the functor TrAu is faithful and injective on objects. It is also full, because if $(\psi_e, \psi_s) : \mathrm{TrAu}(E,D) \rightarrow \mathrm{TrAu}(E',D')$, then (1) an induction on the length of securing chains shows that $\psi_s([x]) = \widehat{\psi}_e([x])$ for each finite trace $[x] \in D$, and (2) it follows from the defining properties of automata that if $([x],[xe])$ is a prime interval of $\mathrm{Traces}(\mathrm{TrAu}(E,D))$, then $(\widehat{\psi}_e([x]), \widehat{\psi}_e([xe]))$ is a $\sqcup$-prime interval of $\mathrm{Traces}(\mathrm{TrAu}(E',D'))$. Hence if $\phi : (E,D) \rightarrow (E',D')$ is the morphism in **TrDom** with ϕ and ψ_e the same morphism of concurrent alphabets, then $\mathrm{TrAu}(\phi) = \psi$.

Given an automaton A, define the "evaluation map" $\varepsilon_A : \mathrm{TrAu}(\mathrm{AuTr}(A)) \rightarrow A$ to be the identity on events and to take each state $[x]$ of $\mathrm{TrAu}(\mathrm{AuTr}(A))$ to the state $\mathrm{cod}(\gamma)$ of A, where γ is an initial computation sequence of A having trace $[x]$. Since all such computation sequences γ determine the same state $\mathrm{cod}(\gamma)$ by commutativity, we know that ε_A is well-defined. To see that ε_A is in fact a morphism of automata, suppose that $[x]\xrightarrow{e}[xe]$ is a transition of $\mathrm{TrAu}(\mathrm{AuTr}(A))$. Then there exist initial computation sequences γ and δ of A, such that γ has trace $[x]$ and δ has trace $[xe]$. Then $\delta \setminus \gamma : \mathrm{cod}(\gamma) \rightarrow \mathrm{cod}(\delta)$ is a computation sequence of A that contains precisely one nonidentity transition $t : \mathrm{cod}(\gamma) \rightarrow \mathrm{cod}(\delta)$. Hence, to every enumeration of $\{e\}$ of $\varepsilon_A(\{e\})$ there corresponds a computation sequence $\mathrm{cod}(\gamma)\xrightarrow{e}\mathrm{cod}(\delta)$ of A, as required to show that ε_A is a morphism.

We claim that ε_A is universal from TrAu to A. Suppose $\psi : \mathrm{TrAu}(E', D') \to A$ is a morphism in **Auto**. We claim that there is a unique morphism $\phi : (E', D') \to \mathrm{AuTr}(A)$ in **TrDom** such that $\varepsilon_A \circ \mathrm{TrAu}(\phi) = \psi$. Since ε_A is the identity on events, ϕ and ψ must be the same morphism of concurrent alphabets, thus there is only one possible definition for ϕ. To show that this definition actually yields a morphism $\phi : (E', D') \to \mathrm{AuTr}(A)$ of trace domains, we must show: (1) that $\widehat{\phi}(D') \subseteq \mathrm{Traces}(A)$, and (2) if $([x'], [x'e'])$ is a prime interval in D', then $(\widehat{\phi}([x']), \widehat{\phi}([x'e']))$ is a $\sqcup$-prime interval in Traces(A).

To show (1), observe that $\widehat{\phi}(D') = \widehat{\psi}(D')$. An induction on the length of securing chains in D' shows that each finite $[x'] \in D'$ determines an initial computation sequence γ of A with $\mathrm{tr}(\gamma) = \widehat{\psi}([x'])$ and $\mathrm{cod}(\gamma) = \psi_s([x'])$. Thus, $\widehat{\psi}([x']) \in \mathrm{Traces}(A)$ for all finite $[x'] \in D'$. By continuity, $\widehat{\psi}(D') \subseteq \mathrm{Traces}(A)$.

To show (2), it suffices to show that if $[x] = \widehat{\phi}([x'])$ then $[xe] \in \mathrm{Traces}(A)$ for each $e \in \phi(\{e'\})$. Given $[x']$, obtain an initial computation sequence γ' of $\mathrm{TrAu}(E', D')$ with $\mathrm{cod}(\gamma') = [x']$, and a transition $t' : [x'] \xrightarrow{e'} [x'e']$ of $\mathrm{TrAu}(E', D')$. Using the fact that ψ is a morphism of automata, we may construct an initial computation sequence γ of A, with $\mathrm{tr}(\gamma) = [x]$ and $\mathrm{cod}(\gamma) = \psi_s([x'])$. Now, e' is enabled in state $[x']$ of A', so each element e of $\phi(\{e'\})$ is enabled in state $\mathrm{cod}(\gamma) = \psi_s([x'])$ of A. It follows that $[xe] \in \mathrm{Traces}(A)$ for each $e \in \phi(\{e'\})$. ∎

Theorem 7 *The forgetful functor* TrEv : **TrDom** $\to$ **EvDom** *has a left adjoint* EvTr, *such that* $\mathrm{TrEv} \circ \mathrm{EvTr} \simeq 1$.

Proof – We first define a mapping from event domains to trace domains. Given an event domain D, define the *events* of D to be the $\equiv$-classes of prime intervals of D. Let E_D be the set of all events of D. Say that events $e, e' \in E_D$ *commute*, and write $e \| e'$, when $e \neq e'$ and there exist representatives $I \in e$ and $I' \in e'$, such that I and I' are consistent. (By the properties of event domains, this implies that whenever $I \in e$ and $I' \in e'$ are coinitial, then they are consistent.) Then E_D, equipped with the relation $\|$, is a concurrent alphabet. Let A be the automaton $(E_D, Q, \perp, T)$, where Q is the set of finite elements of D and T contains all transitions $q \xrightarrow{\epsilon} q$ and all transitions $q \xrightarrow{[I]} r$ such that $I = (q, r)$ is a prime interval of D. Let $U_D = \mathrm{Traces}(A) \subseteq \widehat{E}_D$. It follows from Theorem 6 that (E_D, U_D) is a trace domain.

Next, we claim that D and U_D are isomorphic. To see this, note that the finite initial computation sequences γ of A are in bijective correspondence with the securing chains for finite elements d of D. Moreover, by definition of the commutativity relation $\|$ on E_D, any two securing chains for the same element d correspond to computation sequences having the same trace $[x] \in \widehat{E}_D$. Thus the mapping that takes d to the trace $[x]$ determined by a securing chain for d is a bijection between the finite elements of D and the finite elements of $U_D = \mathrm{Traces}(A)$. It can be shown that this mapping is monotone, hence by continuous extension we obtain an isomorphism $\mu_D : D \to U_D$.

Finally, we show that the forgetful functor TrEv has a left adjoint whose object map takes D to (E_D, U_D). Given D, let the "inclusion of generators" be the isomorphism $\mu_D : D \to U_D$ constructed above. We claim that μ_D is universal from D to TrEv. Suppose $(E', D') \in$ **TrDom** and $\phi : D \to D'$ in **EvDom** are given. Note that each prime interval I of D determines a $\sqcup$-prime interval $\phi(I)$ of D', which in turn determines a set $V_I \in \mathrm{Com}(E')$. If I and J are distinct consistent intervals, then $\phi(I)$ and $\phi(J)$ are orthogonal by the properties of the **EvDom**-morphism ϕ, hence $V_I \cap V_J = \emptyset$. Then an induction using the definition of $\equiv$ shows that if $I \equiv I'$, then $\phi(I) \equiv \phi(I')$, hence $V_I = V_{I'}$. Thus, the map taking each equivalence class $[I]$ to the corresponding V_I determines an **Alph**-morphism $\rho : E_D \to E'$.

Now, an induction on the length of securing chains in U_D shows that $\widehat{\rho}([x]) = \phi(\mu_D^{-1}([x]))$ for all finite $[x] \in U_D$, so $\widehat{\rho}(\mu_D(d)) = \phi(d) \in D'$ for all finite $d \in D$, hence for all $d \in D$ by continuity. Also, if $I = ([x], [xe])$ is a prime interval of U_D, then $e = [(d, d')]$, where $d = \mu_D^{-1}([x])$ and $d' = \mu_D^{-1}([xe])$. Since then $\widehat{\rho}(I) = (\phi(d), \phi(d'))$, which is a $\sqcup$-prime interval of D', it follows that $\widehat{\rho}$ maps prime intervals of U_D to $\sqcup$-prime intervals of D'. Thus, we have shown that $\rho : (E_D, U_D) \to (E', D')$ is a **TrDom**-morphism, and $\mathrm{TrEv}(\rho) \circ \mu_D = \phi$.

Finally, we note that any **TrDom**-morphism $\rho : (E_D, U_D) \to (E', D')$ satisfying $\mathrm{TrEv}(\rho) \circ \mu_D = \phi$ must satisfy $\widehat{\rho}([xe]) = \phi(\mu_D^{-1}([xe]))$ for all finite prime intervals $I = ([x], [xe]) \in U_D$. Let $d = \mu_D^{-1}([x])$ and $d' = \mu_D^{-1}([xe])$, then since $\widehat{\rho}([xe]) = \widehat{\rho}([x])\widehat{\rho}(e) = \phi(d)\widehat{\rho}(e)$ and $\phi(d') = \phi(d)[e'_1 \dots e'_n]$ for some $\{e'_1, \dots, e'_n\} \in \mathrm{Com}(E')$, it must be the case that $\widehat{\rho}(e) = [e'_1 \dots e'_n]$ and $\rho(\{e\}) = \{e'_1, \dots, e'_n\}$. Since every $e \in E_D$ is $[(d, d')]$ for some prime interval $(d, d') \in D$, for each $e \in E_D$ we can find a corresponding prime interval $([x], [xe])$ in U_D by taking $[x] = \mu_D(d)$ and $[xe] = \mu_D(d')$. It follows that the condition $\mathrm{TrEv}(\rho) \circ \mu_D = \phi$ uniquely determines ρ. ∎

References

[1] I. J. Aalbersberg and G. Rozenberg. Theory of traces. *Theoretical Computer Science*, 60(1):1–82, 1988.

[2] M. Bednarczyk. *Categories of Asynchronous Systems.* PhD thesis, University of Sussex, October 1987.

[3] J. D. Brock. *A Formal Model of Non-Determinate Dataflow Computation.* PhD thesis, Massachusetts Institute of Technology, 1983. Available as MIT/LCS/TR-309.

[4] J. D. Brock and W. B. Ackerman. Scenarios: a model of non-determinate computation. In *Formalization of Programming Concepts*, pages 252–259, Springer-Verlag. Volume 107 of *Lecture Notes in Computer Science*, 1981.

[5] M. Broy. Nondeterministic data-flow programs: how to avoid the merge anomaly. *Science of Computer Programming*, 10:65–85, 1988.

[6] P.-L. Curien. *Categorical Combinators, Sequential Algorithms, and Functional Programming. Research Notes in Theoretical Computer Science*, Pitman, London, 1986.

[7] A. A. Faustini. An operational semantics for pure dataflow. In *Automata, Languages, and Programming, 9th Colloquium*, pages 212–224, Springer-Verlag. Volume 140 of *Lecture Notes in Computer Science*, 1982.

[8] H. Herrlich and G. E. Strecker. *Category Theory. Sigma Series in Pure Mathematics*, Heldermann Verlag, 1979.

[9] G. Kahn. The semantics of a simple language for parallel programming. In J. L. Rosenfeld, editor, *Information Processing 74*, pages 471–475, North-Holland, 1974.

[10] G. Kahn and D. B. MacQueen. Coroutines and networks of parallel processes. In B. Gilchrist, editor, *Information Processing 77*, pages 993–998, North-Holland, 1977.

[11] S. Mac Lane. *Categories for the Working Mathematician.* Volume 5 of *Graduate Texts in Mathematics*, Springer Verlag, 1971.

[12] A. Mazurkiewicz. Trace theory. In *Advanced Course on Petri Nets*, GMD, Bad Honnef, September 1986.

[13] P. Panangaden and E. W. Stark. Computations, residuals, and the power of indeterminacy. In *Automata, Languages, and Programming*, pages 439–454, Springer-Verlag. Volume 317 of *Lecture Notes in Computer Science*, 1988.

[14] G. D. Plotkin. Domains: lecture notes. 1979. (unpublished manuscript).

[15] D. A. Schmidt. *Denotational Semantics: A Methodology for Language Development.* Allyn and Bacon, 1986.

[16] E. W. Stark. Concurrent transition system semantics of process networks. In *Fourteenth ACM Symposium on Principles of Programming Languages*, pages 199–210, January 1987.

[17] E. W. Stark. Concurrent transition systems. *Theoretical Computer Science*, 1989. (to appear).

[18] E. W. Stark. Connections between a concrete and abstract model of concurrent systems. In *Fifth Conference on the Mathematical Foundations of Programming Semantics*, Springer-Verlag. *Lecture Notes in Computer Science*, New Orleans, LA, 1989 (to appear).

[19] E. W. Stark. *On the Relations Computed by a Class of Concurrent Automata.* Technical Report 88-09, SUNY at Stony Brook Computer Science Dept., 1988.

[20] G. Winskel. *Events in Computation.* PhD thesis, University of Edinburgh, 1980.

Operations on Records
(Extended Abstract)*

Luca Cardelli
Digital Equipment Corporation
Systems Research Center

John C. Mitchell
Department of Computer Science
Stanford University

Abstract

We define a simple collection of operations for creating and manipulating record structures, where records are intended as finite associations of values to labels. We study a second-order type system over these operations, supporting both subtyping and polymorphism. Our approach unifies and extends previous notions of records, bounded quantification, record extension, and parametrization by row-variables. The general aim is to provide foundations for concepts found in object-oriented languages, within the framework of typed lambda-calculus.

1 Introduction

)bject oriented programming is based on record structures (called objects) intended as named :ollections of values (attributes) and functions (methods). Collections of objects form classes. A ubclass relation is defined on classes with the intention that methods work "appropriately" on all nember of subclasses of a given class. This property is important in software engineering because t permits after-the-fact extensions of system by subclasses, without having to modify the systems hemselves. The first object oriented language, Simula67, and most of the more recent ones are yped by using simple extensions of the type rules for Pascal-like languages, mainly involving a otion of subtyping. We are interested in more powerful type systems, particularly ones that ncorporate parametric polymorphism and integrate it smoothly with subtyping.

Several other type systems for record structures have been formulated recently, following the ame motivation. In [Car88], the basic notions of records types were defined in the context of a rst-order type system for fixed-size records. Then Wand [Wan87] introduced the concept of row-ariables while trying to solve the inference problem for records; this led to a system with extensible ecords and limited second-order typing. His system was later refined and shown to have principal ypes by [Rém89, JM88] and again [Wan89]. The resulting system provides a flexible integration f records types and Milner-style type inference [Mil78].

---

*This extended abstract appears in conjunction with John Mitchell's invited lecture at this meeting. The lecture is sequel to Luca Cardelli's invited lecture at the March, 1989 Workshop on Mathematical Foundations of Programming anguage Semantics. Both are based on joint work, with this lecture focusing on results obtained after the March onference. A complete paper, currently in preparation, will appear in the proceedings of Programming Language emantics meeting.

In a parallel development, [CW85] defined a full second-order extension of the system with fixed-size records, based on techniques from [Mit84]. In that system, a program can work polymorphically over all subtypes B of a given record type A, and it can preserve the "unknown" fields (the ones in B but not in A) of record parameters from input to output. However, some natural functions are not expressible. For example, by the nature of fixed-size records there is no way to add a field to a record and preserve all its unknown fields. Less obviously, a function that updates a record field (in the purely applicative sense of making a modified copy of it) is forced to remove all fields from the result which are not explicitly specified in the result type. Imperative update also requires a careful typing analysis.

In this paper we describe a second-order type system that allows more flexible manipulation of records. We believe this approach makes the presentation of record types more natural. The general idea is to extend a polymorphic type system with a notion of subtyping at all types. Record types are then introduced as specialized type constructions with some specialized subtyping rules. These new constructions interact well with the rest of the system. For example, row-variables fall out naturally from second-order type variables, and contravariance of function spaces and universal quantifiers mixes well with record subtyping. In moving to second-order typing we lose the principal type property of weaker type systems, in exchange for some additional expressiveness. However, we gain some perspective on the space of possible operations on records and record types. Since it is not yet clear where the bounds of expressiveness may lay, this perspective should prove useful for comparisons and further understanding. Although previous research provides significant hints [Rém89, JM88, OB88, Wan89], we leave it to future research to determine the precise fragment of our current system which admits automatic type inference.

In this short summary, we will only give an informal description of the type system and some of its properties. A formal presentation of the language and several semantic models are given in the full paper.

2 Record values

A record value is essentially a finite map from labels to values, where the values may belong to different types. Syntactically, a record value is a collection of fields, where each field is a labeled value. To capture the notion of a map, the labels in a given record must be distinct. Hence the labels can be used to identify the fields, and the fields should be regarded as unordered. This is the notation we use:

$\langle\rangle$	the empty record.
$\langle x = 3, y = \mathit{true}\rangle$	a record with two fields, labeled x and y, equivalent to $\langle y = \mathit{true}, x = 3\rangle$.

There are three basic operations on record values, *extension*, *restriction*, and *extraction*. They have the following basic properties.

Extension $\langle r|x = a\rangle$ adds a field of label x and value a to a record r, provided a field of label x is not already present. This restriction will be enforced statically by the type system. The additional brackets placed around the operator help to make the examples more readable; we also write $\langle r|x = a|y = b\rangle$ for $\langle\langle r|x = a\rangle|y = b\rangle$.

Restriction $r\backslash x$ removes the field of label x, if any, from the record r. We write $r\backslash xy$ for $(r\backslash x)\backslash y$.

Extraction $r.x$ extracts the value corresponding to the label x from the record r, provided a field having that label is present. This restriction will be enforced statically by the type system.

We have chosen these three operations because they seem to be the fundamental constituents of more complex operations. In particular, the extension operation is not required to check whether a new field is already present in a record: its absence is guaranteed statically. The restriction operation has the task of removing unwanted fields and fulfilling that guarantee. This separation of tasks has advantages for efficiency, and for static error detection since fields cannot be overwritten unintentionally by extension alone.

Here are some simple examples:

$\langle\langle x = 3\rangle \| y = true\rangle$	$=$	$\langle x = 3, y = true\rangle$	extension
$\langle x = 3, y = true\rangle\backslash y$	$=$	$\langle x = 3\rangle$	restriction (canceling y)
$\langle x = 3, y = true\rangle\backslash z$	$=$	$\langle x = 3, y = true\rangle$	restriction (no effect)
$\langle x = 3, y = true\rangle.x$	$=$	3	extraction
$\langle\langle x = 3\rangle \| x = 4\rangle$	invalid extension		
$\langle x = 3\rangle.y$	invalid extraction		

Some additional operators may be defined in terms of the ones above.

Renaming] $r[x \leftarrow y] \stackrel{\text{def}}{=} \langle r\backslash x | y = r.x\rangle$ changes the name of a record field.

Overriding $\langle r \leftarrow x = a\rangle \stackrel{\text{def}}{=} \langle r\backslash x | x = a\rangle$. If x is present in r, replace its value with one of a possibly unrelated type, otherwise extend r with $x = a$ (compare with [Wan89]). Given adequate type restrictions, this can be seen as an updating operator, or a method overriding operator. We write $\langle r \leftarrow x = a, y = b\rangle$ for $\langle\langle r \leftarrow x = a\rangle \leftarrow y = b\rangle$.

It is clear that any record may be constructed from the empty record using extension operations. In fact, it is convenient to regard the syntax for a record of many fields as an abbreviation for iterated extensions of the empty record, *e.g.*,

$$\langle x = 3, y = true\rangle \stackrel{\text{def}}{=} \langle\langle\langle\rangle | x = 3\rangle | y = true\rangle.$$

This approach to record values allows us to express the fundamental properties of records using combinations of simple operators of fixed arity, as opposed to n-ary operators. Hence we never have to use schemas with ellipses, such as $\langle x_1 = a_1, \ldots, x_n = a_n\rangle$, in our formal treatment.

Since $r\backslash x = r$ whenever r lacks a field of label x, we may write $\langle x = 3, y = true\rangle$ using any of the following expressions:

$$\langle\langle\rangle | x = 3 | y = true\rangle = \langle\langle\langle\rangle\backslash x | x = 3\rangle\backslash y | y = true\rangle = \langle\langle\rangle, x = 3, y = true\rangle$$

The latter forms match a similar definition for record types, given in the next section.

3 Record types

In describing operations on record values, we made positive assumptions of the form "a field of label x must occur in record r" and negative assumptions of the form "a field of label x must not occur in record r". These constraints will be verified statically by the type system. To accomplish this, record types must convey both positive and negative information. Positive information describes the fields that members of a record type must have, while negative information describes the fields the members of that type must not have. Within these constraints, the members of a record type may or may not have additional fields or lack additional fields. It is worth emphasizing that both positive and negative constraints restrict the elements of a type, hence increasing either kind of constraint will lead to smaller sets of values. The smallest amount of information is expressed by the "empty" record type $\langle\!\rangle$. The "empty" record type is empty only in that it places no constraints on its members – every record has type $\langle\!\rangle$, since all records have at least no fields and lack at least no fields. Here are some other examples.

$\langle\!\rangle$	the type of all records. Contains, *e.g.*, $\langle\rangle$ and $\langle x = 3\rangle$.
$\langle\!\rangle\backslash x$	the type of all records which lack a field labeled$x$. *E.g.*, $\langle\rangle$, $\langle y = true\rangle$, *but not* $\langle x = 3\rangle$.
$\langle\!\langle x{:}\ Int, y{:}\ Bool\rangle\!\rangle$	the type of all records which have at least fields labeled x and y, with values of types *Int* and *Bool*. *E.g.*,$\langle x = 3, y = true\rangle, \langle x = 3, y = true, z = str\rangle$ but not $\langle x = 3, y = 4\rangle, \langle x = 3\rangle$
$\langle\!\langle x{:}\ Int\rangle\!\rangle\backslash y$	the type of all records which have at least a field labeled x of type *Int*, and no field with label y. *E.g.*,$\langle x = 3, z = str\rangle$, but not $\langle x = 3, y = true\rangle$.

These examples illustrate that in general a record type is characterized by a finite collection o "positive" type fields, which are labeled types, and "negative" type fields, which are labels tha must not occur in any record of that type. For simplicity, we often say "fields" for "type field." The positive fields must have distinct labels and should be considered unordered. Negative field are also unordered and must be distinct in any "normalized" record type expression.

As with record values, we have three basic operations on record types.

> *Extension* $\langle\!\langle R|x{:}\ A\rangle\!\rangle$ This type denotes the collection obtained from R by adding x fields wit values in A in all possible ways (provided that none of the elements of R have x fields). Mo precisely, this is the collection of those records $\langle r|x = a\rangle$ such that r is in R and a is i A, provided that a positive type field x is not already present in R (this will be enforce statically). We sometime write $\langle\!\langle R|x{:}\ A|y{:}\ B\rangle\!\rangle$ for $\langle\!\langle\langle\!\langle R|x{:}\ A\rangle\!\rangle|y{:}\ B\rangle\!\rangle$.
>
> *Restriction* $R\backslash x$ This type denotes the collection obtained from R by removing the field (if any) from all its elements. More precisely, this is the collection of those records $r\backslash x$ suc that r is in R. We write $R\backslash xy$ for $(R\backslash x)\backslash y$.
>
> *Extraction* $R.x$ This is the type associated to label x in R, provided R has such a positi field. This provision will be enforced statically.

Again, several derived operators can be defined from these.

Renaming $R[x,y] \stackrel{\text{def}}{=} \langle R\backslash x | y = R.x \rangle$ changes the name of a record type field.

Overriding $\langle R \leftarrow x{:}\, A\rangle \stackrel{\text{def}}{=} \langle R\backslash x | x{:}\, A\rangle$ if a type field x is present in R, replaces it with a field x of type A, otherwise extends R. Given adequate type restrictions, this can be used to override a method type in a class signature (i.e. record type) with a more specialized one, to produce a subclass signature.

One crucial formal difference between these operators on types and the similar ones on values is that $\langle\rangle\backslash y \neq \langle\rangle$, since records belonging to the "empty" type may have y fields, whereas $\langle\rangle\backslash y = \langle\rangle$. In forming record types, one must always make a field restriction before a type extension, as illustrated in the examples below.

$\langle\langle x{:}\, Int\rangle\backslash y | y{:}\, Bool\rangle$	$=$	$\langle x{:}\, Int, y{:}\, Bool\rangle$	extension
$\langle x{:}\, Int, y{:}\, Bool\rangle\backslash y$	$=$	$\langle x{:}\, Int\rangle\backslash y$	restriction (canceling y)
$\langle x{:}\, Int, y{:}\, Bool\rangle\backslash z$	$=$	$\langle x{:}\, Int, y{:}\, Bool\rangle\backslash z$	restriction (no effect)
$\langle x{:}\, Int, y{:}\, Bool\rangle.x$	$=$	Int	extraction
$\langle\langle\rangle | x{:}\, Bool\rangle$	invalid extension		
$\langle x{:}\, Int\rangle.y$	invalid extraction		

It helps to read the examples in terms of the collections they represent. For example, the first example for restriction says that if we take the collection of records that have x and y (and possibly more) fields, and remove the y field from all the elements in the collection, then we obtain the collection of records that have x (and possibly more) but no y. In particular, we do not obtain the collection of records that have x and possibly more fields, because those would include y.

The way positive and negative information is formally manipulated is actually easier to understand if we regard record types as abbreviations, as we did for record values:

$$\langle x{:}\, Int, y{:}\, Bool\rangle \stackrel{\text{def}}{=} \langle\langle\langle\rangle\backslash x | x{:}\, Int\rangle\backslash y | y{:}\, Bool\rangle$$

Then, when considering $\langle y{:}\, Bool\rangle\backslash y$, we actually have $\langle\langle\rangle\backslash y | y{:}\, Bool\rangle\backslash y$. If we allow the outside positive and negative y labels to cancel, we are still left with $\langle\rangle\backslash y$. The inner y restriction reminds us that y fields have been eliminated from records of this type.

4 Summary

The record and record-type operations discussed in the preceding sections are incorporated in a typed calculus with subtyping. The subtyping rules for record types are essentially that every record type is a subtype of $\langle\rangle$, and the subtyping relation respects the type operations of extension, restriction and extraction. Writing $A <: B$ for *A is a subtype of B*, we have the following examples.

$\langle x{:}\, Int, y{:}\, Bool\rangle <: \langle\rangle$	
$\langle R | x{:}\, A\rangle <: \langle S | x{:}\, A\rangle$	if $R <: S <: \langle\rangle\backslash x$
$R\backslash x <: S\backslash x$	if $R <: S <: \langle\rangle$
$R.x <: S.x$	if $R <: S <: \langle x{:}\, A\rangle$

In general, a record type R will be a subtype of another record type S if every positive constraint (labeled field) associated with R is also a positive constraint imposed by S, and similarly for negative constraints (fields required to be absent). There are some subtleties. For example, $\langle R\backslash x | x\colon Int\rangle$ is not necessarily a subtype of R, and *never* a subtype of $R\backslash x$, even though this might seem consistent with the point of view expressed in [Car88], for example.

There are several variants of the typed calculus, each described in the full paper. In the basic calculus, any collection of records might reasonably be regarded as a type, as illustrated by a simple semantic model. One natural extension of the calculus is obtained by imposing the condition that whenever R is a type of records with x field (*i.e.*, whenever $R <: \langle x\colon A\rangle$ for some type A), we must have $R = \langle R\backslash x | x\colon R.x\rangle$. In category-theoretic terms, this equation says that R must be the cartesian product of $R\backslash x$ and $R.x$. This equation may fail if we allow arbitrary collections of records as types. For example,

$$R = \{\langle x = 1, y = true\rangle, \langle x = 0, y = false\rangle$$

is not the cartesian product of $R\backslash x = \{\langle y = true\rangle, \langle y = false\rangle\}$ and $R.x = \{0, 1\}$. By omitting certain sets of records, we may construct a semantic model in which every record type is the appropriate cartesian product.

A second extension of the basic calculus is obtained by adding the equational axiom

$$r = s : \langle\rangle$$

which equates all records of type $\langle\rangle$. This extensionality axiom makes good sense, since no operation of the calculus will allow us to distinguish between records r and s of type $\langle\rangle$. One importnat aspect of the equational theory is that equations must be regarded as triples, consisting of two expressions and a type, rather than pairs. Put another way, we may assert that records r and s are equal records of type R, but not equal when considered with respect to another record type S. In particular, the records $r ::= \langle x = 3\rangle$ and $s ::= \langle x = 4\rangle$ belong to types $\langle x\colon Int\rangle$ and $\langle\rangle$. Since we may extract the x field of any record belonging to type $\langle x\colon Int\rangle$, records r and s *are* distinguishable as elements of this type, and so we have

$$r \neq s\colon \langle x\colon Int\rangle.$$

However, since we cannot extract the x field of a record belonging to type $\langle\rangle$, we also have the equation $r = s\colon \langle\rangle$. Interpreting types as partial equivalence relations (sets together with equivalence relations), we may construct a semantic model in which every record type is a cartesian product and extensional equality is satisfied. However, we have not been able to construct a semantic model of extensional equality without restricting record types to cartesian products.

References

[Car88] L. Cardelli. A semantics of multiple inheritance. *Information and Computation*, 76:138–164, 1988. Special issue devoted to *Symp. on Semantics of Data Types*, Sophia-Antipolis (France), 1984.

[CW85] L. Cardelli and P. Wegner. On understanding types, data abstraction, and polymorphism. *Computing Surveys*, 17(4):471–522, 1985.

[JM88] L. Jategaonkar and J.C. Mitchell. ML with extended pattern matching and subtypes. In *Proc. ACM Symp. Lisp and Functional Programming Languages*, pages 198–212, July 1988.

[Mil78] R. Milner. A theory of type polymorphism in programming. *JCSS*, 17:348–375, 1978.

[Mit84] J.C. Mitchell. Coercion and type inference (summary). In *Proc. 11-th ACM Symp. on Principles of Programming Languages*, pages 175–185, January 1984.

[OB88] A. Ohori and P. Buneman. Type inference in a database language. In *Proc. ACM Symp. Lisp and Functional Programming Languages*, pages 174–183, July 1988.

[Rém89] D. Rémy. Typechecking records and variants in a natural extension of ML. In *16-th ACM Symposium on Principles of Programming Languages*, pages 60–76, 1989.

[Wan87] M. Wand. Complete type inference for simple objects. In *Proc. 2-nd IEEE Symp. on Logic in Computer Science*, pages 37–44, 1987. Corrigendum in *Proc. 3-rd IEEE Symp. on Logic in Computer Science*, page 132, 1988.

[Wan89] M. Wand. Type inference for record concatenation and simple objects. In *Proc. 4-nd IEEE Symp. on Logic in Computer Science*, pages 92–97, 1989.

Projections for Polymorphic Strictness Analysis

John Hughes
Department of Computing Science
University of Glasgow

ABSTRACT

We apply the categorical properties of polymorphic functions to compile-time analysis, specifically projection-based strictness analysis. First we interpret parameterised types as functors in a suitable category, and show that they preserve monics and epics. Then we define "strong" and "weak" polymorphism — the latter admitting certain projections that are not polymorphic in the usual sense. We prove that, under the right conditions, a weakly polymorphic function is characterised by a single instance. It follows that the strictness analysis of one simple instance of a polymorphic function yields results that apply to all. We show how this theory may be applied.

In comparison to earlier polymorphic strictness analysis methods, ours can apply polymorphic information to a particular instance very simply. The categorical approach simplifies our proofs, enabling them to be carried out at a higher level, and making them independent of the precise form of the programming language to be analysed.

1. Introduction

This is not a paper about category theory. Rather it is about using categorical ideas to develop better optimising compilers for functional languages. The categorical fact we use is that every first-order polymorphic function is a natural transformation [Rydeheard85]. This semantic property enables us to develop a semantic analysis method for polymorphic functions; it guides our intuitions and greatly simplifies our proofs, replacing structural inductions over terms by semantic arguments. These categorical proofs are not only at a higher level than the corresponding "syntactic" proofs would be, they are unspecific about the programming language in which polymorphic functions are expressed.

The particular semantic analysis technique we are interested in is backwards (or projection-based) strictness analysis. We prove a new result in this paper, that this kind of analysis is, in a sense, polymorphic. This confirms an earlier conjecture, and shows how the technique can be applied to first-order polymorphic functions.

The paper is organised as follows. In the next section, we review projection-based strictness analysis very briefly. In section 3 we introduce the types we will be working with: they are the objects of a category. We show that parameterised types are functors, with some very special properties. In section 4 we define strong and weak polymorphism: polymorphic functions in programming languages are strongly polymorphic, but we will need to use projections with a slightly weaker property. We prove that, under certain conditions, weakly polymorphic functions are characterised by any non-trivial instance. We can therefore analyse one monomorphic instance of a polymorphic function, and apply the results to every instance. In section 5 we choose a finite set of projections for each type, suitable for use in a practical compiler. We call these specially chosen projections *contexts*, and we show that contexts for compound types can be factored to facilitate application of the results of section 4. We give a number of examples of polymorphic strictness analysis. Finally in section 6 we discuss related work and draw some conclusions.

2. Projections for Strictness Analysis

Early strictness analysis methods could discover nothing informative about functions on lazy data-structures, and projection based strictness analysis was developed in an attempt to solve this problem. Recall that a *projection* is a function α such that

$$\alpha \cdot \alpha = \alpha$$
$$\alpha \leq \text{id}$$

The essential intuition is that a projection performs a certain amount of evaluation of a lazy data-structure. For example, the projection

$$\begin{aligned} &\alpha : \text{Nat} \times \text{Nat} \to \text{Nat} \times \text{Nat} \\ &\alpha\ (x, y) = (\bot, \bot) && \text{if } x = \bot \\ & = (x, y) && \text{if } x \neq \bot \end{aligned}$$

may be thought of as evaluating the first component of a pair, while

$$\begin{aligned} &\beta : \text{Nat} \times \text{Nat} \to \text{Nat} \times \text{Nat} \\ &\beta\ (x, y) = (\bot, \bot) && \text{if } x = \bot \text{ or } y = \bot \\ & = (x, y) && \text{otherwise} \end{aligned}$$

evaluates both. Now we can regard a function as β-strict — performing as much evaluation as β — if evaluating its argument with β before the call doesn't change its result. For example, the function

$$+ : \text{Nat} \times \text{Nat} \to \text{Nat}$$

evaluates both its arguments, and so

$$+ = + \cdot \beta$$

More generally, there may be parts of a function's argument that are evaluated only if certain parts of its result are evaluated — a function may evaluate more or less of its argument depending on context. For example,

$$\begin{aligned} &\text{swap} : \text{Nat} \times \text{Nat} \to \text{Nat} \times \text{Nat} \\ &\text{swap}\ (x, y) = (y, x) \end{aligned}$$

is not β-strict, but it *is* β-strict in a β-strict context since

$$\beta \cdot \text{swap} = \beta \cdot \text{swap} \cdot \beta$$

Thus, if both components of swap's result will be evaluated, then the components of its argument can be evaluated before the call without changing the meaning. We make the following definition:

Definition 2.1 Let f be a function and α, β be projections. We say f is α-strict in a β-strict context if

$$\beta \cdot f = \beta \cdot f \cdot \alpha$$

or equivalently,

$$\beta \cdot f \leq f \cdot \alpha$$

In this case we write

$$f : \beta \Rightarrow \alpha$$

Projections capture the notion of evaluating a component of a data-structure; to capture evaluation of a single value we must embed it in a "data-structure" with a single component, which we can think of as representing an unevaluated closure. Thus we think of a closure of type t as an element of $t_\bot$, and we "evaluate" it with the projection

$$
\begin{array}{lll}
S_t : t_\perp \to t_\perp & & \\
S_t \perp & = \perp & \\
S_t\ (\text{lift } x) & = \perp & \text{if } x=\perp \\
 & = \text{lift } x & \text{if } x\neq\perp
\end{array}
$$

(writing the lifted elements in the form lift x). Now, any function

$$f : s \to t$$

induces a function

$$f_\perp : s_\perp \to t_\perp$$

which behaves like f on elements of s, but maps the new $\perp$ to $\perp$. It is easy to show that f is strict if and only if

$$f_\perp : S_t \Rightarrow S_s$$

So strictness analysis can be performed by establishing facts of the form

$$f : \beta \Rightarrow \alpha$$

Usually f and β are given, and we want to find an α such that this property holds. We can always choose α to be id, but this is uninformative: id corresponds to performing no evaluation at all. We would like to find the smallest α with this property. In general this is impossible, but methods exists for finding quite small αs, for monomorphic functions. These methods are beyond the scope of this paper. We remark only that they depend crucially on choosing a *finite* set of projections for each type, so that recursive equations can be solved effectively. The interested reader will find a description of these methods, along with a much fuller presentation of projection based strictness analysis, in [Wadler87a].

3. Types and Functors

We work in a category $\mathbf{C}$ in which the objects are types, interpreted as domains, and the morphisms are continuous functions. The types we are interested in are:

$\mathbf{1}$	
$t_1 \oplus t_2$	where t_1, t_2 are types
$t_1 \otimes t_2$	where t_1, t_2 are types
$t_\perp$	where t is a type
$\mu t.T(t)$	where T(t) is a type, given that t is.

$\mathbf{1}$ is the one point type. $t_1 \oplus t_2$ is the coalesced sum, with elements inl x_1 and inr x_2 (where $x_i \in t_i$). $t_1 \otimes t_2$ is the smash product. We model separated sums and ordinary products using explicit lifting: $t_\perp$ has elements $\perp$ and lift x (where $x \in t$). (*Note*: separating lifting from the formation of sums and products simplifies the description of the choice of suitable contexts for each type, but it is not without its dangers. We intend that the programming language enforce the use of lifting whenever a product appears.)

$\mu t.T(t)$ is a recursive type, whose meaning is given by the recursive domain equation

$$\mu t.T(t) \quad = T(\mu t.T(t))$$

Note that $\mu t.T(t)$ has subdomains $\mathbf{1}$, $T(\mathbf{1})$, $T^2(\mathbf{1})$, ..., $T^n(\mathbf{1})$,.... It is useful to define functions to and from these subdomains.

Definition 3.1 For all types t, define

$$
\begin{array}{l}
\iota_t : t \to \mathbf{1} \\
\kappa_t : \mathbf{1} \to t \\
\perp_t : t \to t
\end{array}
$$

by

$$\begin{aligned} \iota_t\, x &= \bot \\ \kappa_t\, \bot &= \bot \\ \bot_t\, \bot &= \bot \end{aligned}$$

We will usually omit the subscripts on these functions. Note that $\bot = \kappa \cdot \iota$.

Now $T^n(\iota)$ maps $\mu t.T(t)$ onto its nth subdomain $T^n(\mathbf{1})$, and $T^n(\kappa)$ maps back again. Their composition, $T^n(\bot)$, is a projection on $\mu t.T(t)$. The following lemma expresses continuity and will be heavily used.

Lemma 3.2 For any $f : \mu t.T(t) \to s$,

$$f = \bigsqcup_n (f \cdot T^n(\bot))$$

Our type forming operators can be made into functors.

Definition 3.3 Let $f : A \to B$, $g : C \to D$ be strict functions. Define

$$\begin{aligned} & f_\bot : A_\bot \to B_\bot \\ & f \oplus g : A \oplus C \to B \oplus D \\ & f \otimes g : A \otimes C \to B \otimes D \end{aligned}$$

by

$$\begin{aligned} f_\bot\ \bot &= \bot \\ f_\bot\ (\text{lift } a) &= \text{lift } (f\ a) \\[1ex] f \oplus g\ (\text{inl } a) &= \text{inl } (f\ a) \\ f \oplus g\ (\text{inr } c) &= \text{inr } (g\ c) \\[1ex] f \otimes g\ (a,c) &= (f\ a, g\ c) \end{aligned}$$

Note that f and g must be strict for $f \oplus g$ to be well-defined. Let $\mathbf{C}_S$ be the category of domains and strict functions. Then it is easy to show

Lemma 3.4 $\cdot_\bot$, $\oplus$ and $\otimes$ are functors over $\mathbf{C}_S$:

$$\begin{aligned} & \cdot_\bot : \mathbf{C}_S \to \mathbf{C}_S \\ & \oplus, \otimes : \mathbf{C}_S \times \mathbf{C}_S \to \mathbf{C}_S \end{aligned}$$

We would like to make an arbitrary parameterised type $F(t_1 \ldots t_n)$ into a functor $F : {\mathbf{C}_S}^n \to \mathbf{C}_S$. To do so, we need only define the functor corresponding to a recursive type.

Definition 3.5 Let $\mathbf{D}$ be any category, and $G : \mathbf{D} \times \mathbf{C}_S \to \mathbf{C}_S$ be a functor. Then define

$$\mu G : \mathbf{D} \to \mathbf{C}_S$$

by

$$\begin{aligned} \mu G(d) &= \mu t.G(d,t) && \text{for objects} \\ \mu G(f) &= \textstyle\bigsqcup_i {G_f}^i(\bot) && \text{for morphisms} \end{aligned}$$

where

$$G_f(g) = G(f,g)$$

In practice we expect $\mathbf{D}$ to be ${\mathbf{C}_S}^n$ for some n, representing the parameters of the recursive type.

Lemma 3.6 (a) $G_{f\cdot g}{}^n(h\cdot j) = G_f{}^n(h) \cdot G_g{}^n(j)$

(b) $G_f{}^n(\perp) \cdot \mu G(g) = G_{f\cdot g}{}^n(\perp) = \mu G(f) \cdot G_g{}^n(\perp)$

(c) μG is a functor.

Proof

(a) By induction on n. The base case is trivial, and the step case is

$G_{f\cdot g}{}^{n+1}(h\cdot j)$
$= G(f\cdot g, G_{f\cdot g}{}^n(h\cdot j))$
= {induction hypothesis}
$G(f\cdot g, G_f{}^n(h) \cdot G_g{}^n(j))$
= {G is a functor}
$G(f, G_f{}^n(h)) \cdot G(g, G_g{}^n(j))$
$= G_f{}^{n+1}(h) \cdot G_g{}^{n+1}(j)$

(b) We prove the first equality; the second is similar.

$G_f{}^n(\perp) \cdot \mu G(g)$
$= G_f{}^n(\perp) \cdot \bigsqcup_i G_g{}^i(\perp)$
= {continuity}
$\bigsqcup_i G_f{}^n(\perp) \cdot G_g{}^i(\perp)$
= {since the chain is increasing}
$\bigsqcup_{i\geq n} G_f{}^n(\perp) \cdot G_g{}^i(\perp)$
$= \bigsqcup_{i\geq n} G_f{}^n(\perp) \cdot G_g{}^n(G_g{}^{i-n}(\perp))$
$= \bigsqcup_{i\geq n} G_{f\cdot g}{}^n(\perp \cdot G_g{}^{i-n}(\perp))$
$= \bigsqcup_{i\geq n} G_{f\cdot g}{}^n(\perp)$
$= G_{f\cdot g}{}^n(\perp)$

(c) Step 1: $\mu G(id) = id$

$\mu G(id) \quad = \bigsqcup_i G_{id}{}^i(\perp)$
$= \bigsqcup_i (id \cdot G_{id}{}^i(\perp))$
= {by continuity}
id

Step 2: $\mu G(f\cdot g) = \mu G(f) \cdot \mu G(g)$

From (a), we have

$G_{f\cdot g}{}^i(\perp) \quad = G_f{}^i(\perp) \cdot G_g{}^i(\perp)$

Taking limits,

$\mu G(f\cdot g) = \mu G(f) \cdot \mu G(g)$

End of proof

Now since parameterised types are combinations of these functors, the following theorem is straightforward.

Theorem 3.7 Every parameterised type $F(t_1 \ldots t_n)$ is a functor

$$F : \mathbf{C}_S{}^n \to \mathbf{C}_S$$

These functors have a number of pleasant properties. First note that we can extend the notion of a projection to $\mathbf{C}_S{}^n$.

Definition 3.8 A morphism α in $\mathbf{C}_S{}^n$ is a projection if

$$\alpha \cdot \alpha = \alpha$$
$$\alpha \leq id$$

(where $\mathbf{C}_S{}^n$ is ordered componentwise).

Lemma 3.9 Every parameterised type $F : \mathbf{C}_S{}^n \to \mathbf{C}_S$ maps projections to projections.

Proof

Immediate from the fact that F is a monotonic functor.

End of proof

Now recall that a morphism h is *epic* if f•h=g•h implies f=g, and *monic* if h•f=h•g implies f=g. It is easy to show

Lemma 3.10 If $f : A \to B$ and $g : C \to D$ are epic, then so are $f_\perp$, $f \oplus g$, and $f \otimes g$.

Lemma 3.11 If $f : A \to B$ and $g : C \to D$ are monic, then so are $f_\perp$, $f \oplus g$, and $f \otimes g$.

It is slightly harder to prove the corresponding results for recursive types.

Lemma 3.12 If $G : \mathbb{D} \times \mathbf{C}_S \to \mathbf{C}_S$ maps epics to epics, then so does μG.

Proof

Suppose h is epic, and

$$f \cdot \mu G(h) = g \cdot \mu G(h)$$

Then for any n,

$$f \cdot \mu G(h) \cdot G_{id}{}^n(\perp) = g \cdot \mu G(h) \cdot G_{id}{}^n(\perp)$$
$$\Rightarrow \{\text{Lemma 3.6(b)}\}$$
$$f \cdot G_h{}^n(\perp) = g \cdot G_h{}^n(\perp)$$
$$\Rightarrow \{\text{Lemma 3.6(a)}\}$$
$$f \cdot G_{id}{}^n(\kappa) \cdot G_h{}^n(\iota) = g \cdot G_{id}{}^n(\kappa) \cdot G_h{}^n(\iota)$$
$$\Rightarrow \{\text{h and } \iota \text{ are epic, so } G_h{}^n(\iota) \text{ is epic}\}$$
$$f \cdot G_{id}{}^n(\kappa) = g \cdot G_{id}{}^n(\kappa)$$
$$\Rightarrow f \cdot G_{id}{}^n(\kappa) \cdot G_{id}{}^n(\iota) = g \cdot G_{id}{}^n(\kappa) \cdot G_{id}{}^n(\iota)$$
$$\Rightarrow f \cdot G_{id}{}^n(\perp) = g \cdot G_{id}{}^n(\perp)$$

and, by continuity, f = g.

End of proof

Similarly we can prove

Lemma 3.13 If $G : \mathbb{D} \times \mathbf{C}_S \to \mathbf{C}_S$ maps monics to monics, then so does μG.

Now by structural induction, we can prove

Theorem 3.14 Every parameterised type maps epics to epics and monics to monics.

These pleasant properties will prove very useful in the following section.

4. Polymorphism

Let $F, G : \mathbf{C}_S{}^n \to \mathbf{C}_S$ be parameterised types, and let f be a polymorphic function whose type might be written in ML as

$$f : F(*,**,\ldots) \to G(*,**,\ldots)$$

Then it's well known that f is a natural transformation from F to G [Rydeheard85]:

$$f : F \to G$$

That is, f is a collection of instances $f_A : F(A) \to G(A)$, one for every object A of $\mathbb{C}_S{}^n$, such that for any morphism $\alpha : A \to B$ in $\mathbb{C}_S{}^n$, the following diagram commutes:

```
F(A) ------fA------> G(A)
 |                    |
 | F(α)               | G(α)
 |                    |
 V                    V
F(B) ------fB------> G(B)
```

Thus we can regard a polymorphic function as a collection of (strongly related) monomorphic instances.

Technical Note In general f_A and f_B need not be strict — they are $\mathbb{C}$-morphisms, not $\mathbb{C}_S$-morphisms. So this diagram includes morphisms in different categories. We need to embed $\mathbb{C}_S$ objects and morphisms into $\mathbb{C}$. We shall simply regard parameterised types as functors from $\mathbb{C}_S{}^n$ to $\mathbb{C}$ also, in the obvious way.

Definition 4.1 Let $F, G : \mathbb{C}_S{}^n \to \mathbb{C}_S$ be parameterised types. If f is a natural transformation from F to G, we say that f is *strongly polymorphic*.

Why introduce the word "strongly"? Unfortunately, some of the "polymorphic" primitives we wish to use do not satisfy this definition. Recall the projection

$$S_t : t_\perp \to t_\perp$$

introduced in section 2. We would like S to be a polymorphic projection, but consider $S_{\mathbf{2}}$, where $\mathbf{2}$ is the two-point domain $\mathbf{1}_\perp$. We would need in particular that

$$S_{\mathbf{2}} \cdot \perp_\perp = \perp_\perp \cdot S_{\mathbf{2}}$$

(where $\perp_\perp$ denotes the constant bottom function, lifted). But

$$(S_{\mathbf{2}} \cdot \perp_\perp) \text{ (lift (lift } \perp)) = S_{\mathbf{2}} \text{ (lift } \perp) = \perp$$
$$(\perp_\perp \cdot S_{\mathbf{2}}) \text{ (lift (lift } \perp)) = \perp_\perp \text{ (lift (lift } \perp)) = \text{lift } \perp$$

We shall have to weaken our definition of polymorphism slightly to admit S. We do so by requiring the diagram above to commute for a smaller class of functions α.

Definition 4.2 (Abramsky)
We say f is *⊥–reflecting* if $f\ x = \perp$ implies $x = \perp$.

Definition 4.3 $\mathbb{C}_T$ is the category of domains and strict–and–⊥–reflecting functions.

We will regard parameterised types as functors on $\mathbb{C}_T{}^n$ as well.

Definition 4.4 Let $F, G : \mathbb{C}_T{}^n \to \mathbb{C}_T$ be parameterised types. If f is a natural transformation from F to G, we say that f is *weakly polymorphic*.

It is easy to show that S is weakly polymorphic. It is also clear that strongly polymorphic functions are weakly polymorphic, and that functional programs (which cannot use S) are strongly polymorphic. The important projections id and ⊥ are also strongly polymorphic.

Technical Aside We have to be very careful that we only provide strongly polymorphic primitives to the functional programmer. Surprisingly,

$$fst_{A,B} : A \otimes B \to A$$

which selects the first component from a strict pair, is only weakly polymorphic. Fortunately

$$fst'_{A,B} : A_\perp \otimes B_\perp \to A$$

is strongly polymorphic. This is one reason for denying the programmer access to strict tuples.

We now embark on an extended argument to show that weakly polymorphic functions can, in certain circumstances, be characterised by one simple instance. Let F, G : $\mathbf{C}_r{}^n \to \mathbf{C}$ be parameterised types, suppose f, g : F→G are weakly polymorphic, and suppose that we establish that $f_A = g_A$ for some object A in $\mathbf{C}_r{}^n$. Then we will show that $f_t = g_t$ for all t, subject to two conditions. Firstly, A (which is a tuple of types) must not contain the type **1** — such instances contain too little information to characterise a function. Secondly, we must prove separately that f≤g. (This latter condition can be lifted for strongly polymorphic functions). We will use this theorem to show that the results of analysing the strictness of a single instance of a polymorphic function can be applied to all instances.

We make the argument in three major stages. First we show that if $f_A = g_A$ for some such A, then $f_{\mathbf{2}^*} = g_{\mathbf{2}^*}$, where X* denotes the object in $\mathbf{C}_r{}^n$ whose components are all X. Next we show that if $f_{\mathbf{2}^*} = g_{\mathbf{2}^*}$ then $f_{Nat^*} = g_{Nat^*}$, where Nat is the flat domain of natural numbers. (Nat can be expressed within our type language as $\mu n.\mathbf{2} \oplus n$). Finally we show that if $f_{Nat^*} = g_{Nat^*}$ then $f_t = g_t$ for every object t not containing **1**. Together these lemmata imply that if $f_t = g_t$ for any object t not containing **1**, then $f_t = g_t$ for all such objects. We then just have to show that, in this case, $f_t = g_t$ for objects containing **1** as well.

We begin the first stage of our argument with a definition.

Definition 4.5 For each type t except **1**, define

$$\begin{array}{lll} \Delta_t : t \to \mathbf{2} & & \\ \Delta_t\, x & = \perp & \text{if } x = \perp \\ & = \text{lift } \perp & \text{if } x \neq \perp \end{array}$$

Δ_t can be thought of as a definedness test.

Corollary 4.6 (a) Δ_t is the unique strict and ⊥–reflecting map from t to **2**.
(b) Δ_t is epic.

If $t = (t_1 \ldots t_n)$ is an object of $\mathbf{C}^n$ not containing **1**, we write Δ_t for the $\mathbf{C}_r{}^n$ morphism

$$(\Delta_{t1} \ldots \Delta_{tn}) : t \to \mathbf{2}^*$$

Δ_t is also epic.

Lemma 4.7 Let f, g : F→G be weakly polymorphic, and suppose that $f_A = g_A$ for some A not containing **1**. Then $f_{\mathbf{2}^*} = g_{\mathbf{2}^*}$.

Proof

$$\begin{array}{l} f_A = g_A \\ \Rightarrow G(\Delta_A) \cdot f_A = G(\Delta_A) \cdot g_A \\ \Rightarrow \{\text{by weak polymorphism}\} \\ \quad f_{\mathbf{2}^*} \cdot F(\Delta_A) = g_{\mathbf{2}^*} \cdot F(\Delta_A) \\ \Rightarrow \{\Delta_A \text{ is epic, so } F(\Delta_A) \text{ is epic}\} \\ \quad f_{\mathbf{2}^*} = g_{\mathbf{2}^*} \end{array}$$

End of proof

The second stage of our argument shows that if $f_{\mathbf{2}^*} = g_{\mathbf{2}^*}$ then $f_t = g_t$ for any tuple of *flat* domains t, and in particular for Nat*. We begin with a lemma:

Lemma 4.8 Let F be a parameterised type, t be a tuple of flat domains, and x, y $\in$ F(t). If $x \leq y$ and

$$F(\Delta_t)\ x = F(\Delta_t)\ y$$

then x=y.

Proof

By structural induction on F.

Case $F(t_1 \ldots t_n) = t_i$

Then x, y $\in$ t_i, $x \leq y$, and $\Delta_{ti}\ x = \Delta_{ti}\ y$. Since t_i is a flat domain, x and y must be equal.

Case $F(t_1 \ldots t_n) = A$ (a constant type)

Then $F(\Delta_t) = id_A$, so

$$x = F(\Delta_t)\ x = F(\Delta_t)\ y = y$$

Cases $F = G_\perp$, $F = G \oplus H$, $F = G \otimes H$

follow immediately from the inductive hypothesis.

Case $F = \mu G$

$$\mu G(\Delta_t)\ x = \mu G(\Delta_t)\ y$$

$\Rightarrow$ {apply $G_{id}{}^n(\iota)$ to each side, Lemma 3.6(b)}

$$G_{\Delta_t}{}^n(\iota)\ x = G_{\Delta_t}{}^n(\iota)\ y$$

$\Rightarrow$ {Lemma 3.6(a)}

$$G_{\Delta_t}{}^n(id)\ (G_{id}{}^n(\iota)\ x) = G_{\Delta_t}{}^n(id)\ (G_{id}{}^n(\iota)\ y)$$

$\Rightarrow$ {induction hypothesis and induction on n}

$$G_{id}{}^n(\iota)\ x = G_{id}{}^n(\iota)\ y$$

$\Rightarrow$ {apply $G_{id}{}^n(\kappa)$ to each side, Lemma 3.6(a)}

$$G_{id}{}^n(\perp)\ x = G_{id}{}^n(\perp)\ y$$

$\Rightarrow$ {taking limits}

$$x = y$$

End of proof

Lemma 4.9 Let f, g : F$\rightarrow$G be weakly polymorphic, and suppose that $f_{2*}=g_{2*}$, $f \leq g$, and t is a tuple of flat domains. Then $f_t = g_t$.

Proof

$$f_{2*} = g_{2*}$$

$\Rightarrow$ $f_{2*} \cdot F(\Delta_t) = g_{2*} \cdot F(\Delta_t)$

$\Rightarrow$ {weak polymorphism}

$$G(\Delta_t) \cdot f_t = G(\Delta_t) \cdot g_t$$

$\Rightarrow$ {$f \leq g$, previous lemma}

$$f_t = g_t$$

End of proof

This lemma applies in particular when t is Nat*.

For the third stage of our argument we need to introduce a function from Nat to every type. First note that the finite elements of every domain form a countable set. We can therefore define

Definition 4.10 For every type t other than **1**, let

$$\varepsilon_t : \mathrm{Nat} \rightarrow t$$

be a function that enumerates the finite elements of t, such that $\varepsilon_t\, x = \bot$ if and only if $x = \bot$.

Corollary 4.11 ε_t is continuous, strict and $\bot$–reflecting.

Lemma 4.12 ε_t is epic.
Proof

Suppose $f \cdot \varepsilon_t = g \cdot \varepsilon_t$. Then f and g agree at all finite elements of t, and so by continuity they agree at all elements of t. Therefore f = g, and ε_t is epic.

End of proof

If $t=(t_1 \ldots t_n)$ is an object of $\mathbf{C}_r{}^n$ not containing **1**, we will write ε_t for the (epic) $\mathbf{C}_r{}^n$ morphism

$$(\varepsilon_{t1} \ldots \varepsilon_{tn}) : \mathrm{Nat}^* \rightarrow t$$

Lemma 4.13 Let f, g : F→G be weakly polymorphic, and suppose that $f_{\mathrm{Nat}^*}=g_{\mathrm{Nat}^*}$. Then $f_t=g_t$ for every object t of $\mathbf{C}^n$ not containing **1**.

Proof

$$
\begin{aligned}
& f_{\mathrm{Nat}^*} = g_{\mathrm{Nat}^*} \\
\Rightarrow\ & G(\varepsilon_t) \cdot f_{\mathrm{Nat}^*} = G(\varepsilon_t) \cdot g_{\mathrm{Nat}^*} \\
\Rightarrow\ & \{\text{weak polymorphism}\} \\
& \quad f_t \cdot F(\varepsilon_t) = g_t \cdot F(\varepsilon_t) \\
\Rightarrow\ & \{\varepsilon_t \text{ is epic, so } F(\varepsilon_t) \text{ is epic}\} \\
& \quad f_t = g_t
\end{aligned}
$$

End of proof

Finally, we must show that if $f_t = g_t$ for every t not containing **1**, then it is also true for the rest.

Lemma 4.14 Let f, g : F→G be weakly polymorphic, and let $t=(t_1 \ldots t_n)$ be any object of $\mathbf{C}^n$. Let $s=(s_1 \ldots s_n)$ be defined by

$$
\begin{aligned}
s_i &= t_i && \text{if } t_i \neq \mathbf{1} \\
s_i &= \mathbf{2} && \text{if } t_i = \mathbf{1}
\end{aligned}
$$

Then s does not contain **1**, and if $f_s = g_s$ then $f_t = g_t$.

Proof

We define a $\mathbf{C}_r{}^n$ morphism $h=(h_1 \ldots h_n) : t \rightarrow s$ by

$$
\begin{aligned}
h_i &= id && \text{if } t_i \neq \mathbf{1} \\
h_i &= \kappa_2 && \text{if } t_i = \mathbf{1}
\end{aligned}
$$

It is clearly $\bot$–reflecting and monic. Now,

$$
\begin{aligned}
& f_s = g_s \\
\Rightarrow\ & f_s \cdot F(h) = g_s \cdot F(h) \\
\Rightarrow\ & \{\text{weak polymorphism}\} \\
& \quad G(h) \cdot f_t = G(h) \cdot g_t
\end{aligned}
$$

$\Rightarrow$ {h is monic, so G(h) is monic}
$f_t = g_t$

End of proof

Combining these results gives

Theorem 4.15 Let f, g : F→G be weakly polymorphic, with f≤g. If $f_A = g_A$ for any $\mathbb{C}_T^n$ object A not containing **1**, then $f_t = g_t$ for all $\mathbb{C}_T^n$ objects t.

Note that the restriction that A should not contain **1** is necessary: consider the two polymorphic functions

$$S_t : t_\perp \rightarrow t_\perp$$
$$B_t : t_\perp \rightarrow t_\perp$$
$$B_t\, x \qquad = \perp$$

(B is just a less polymorphic version of ⊥). Then B≤S and $B_{\mathbf{1}} = S_{\mathbf{1}}$, but it is not generally true that $B_t = S_t$.

Now let us apply this result to strictness analysis.

Definition 4.16 Let f : F→G, α : F→F, and β : G→G be weakly polymorphic. If f_t: $\beta_t \Rightarrow \alpha_t$ for all t, we write

$$f : \beta \Rightarrow \alpha$$

Corollary 4.17 Let f : F→G, α : F→F, and β : G→G be weakly polymorphic. If $f_A : \beta_A \Rightarrow \alpha_A$ for any $\mathbb{C}_T^n$ object A not containing **1**, then

$$f : \beta \Rightarrow \alpha$$

Proof

Consider the functions β • f and β • f • α. They are both weakly polymorphic, and β • f • α ≤ β • f. Now,

$f_A : \beta_A \Rightarrow \alpha_A$
$\Rightarrow \beta_A \cdot f_A \cdot \alpha_A = \beta_A \cdot f_A$
$\Rightarrow$ {previous theorem}
$\forall t.\ \beta_t \cdot f_t \cdot \alpha_t = \beta_t \cdot f_t$
$\Rightarrow f : \beta \Rightarrow \alpha$

End of proof

Corollary 4.18 Let f : F→G, α : F→F, and β : G→G be weakly polymorphic. Then f : β => α if and only if $f_{\mathbf{2}*} : \beta_{\mathbf{2}*} \Rightarrow \alpha_{\mathbf{2}*}$.

Since we can use the methods of [Wadler87a] to find a good $\alpha_{\mathbf{2}*}$ such that

$$f_{\mathbf{2}*} : \beta_{\mathbf{2}*} \Rightarrow \alpha_{\mathbf{2}*}$$

it follows that we can find a good polymorphic α, given polymorphic f and β, such that

$$f : \beta \Rightarrow \alpha$$

Of course, a polymorphic function may be used in a less polymorphic context. In these cases the following theorem is useful.

Theorem 4.19 Let $f : F \rightarrow G$ be strongly polymorphic, and $\alpha : F \rightarrow F$, $\beta : G \rightarrow G$ be weakly polymorphic projections such that

$$f : \beta \Rightarrow \alpha$$

Then for any projection $\gamma : t \Rightarrow t$,

$$f_t : \beta_t \cdot G(\gamma) \Rightarrow \alpha_t \cdot F(\gamma)$$

Proof

$$\begin{aligned}
& f : \beta \Rightarrow \alpha \\
\Rightarrow\ & \beta_t \cdot f_t \leq f_t \cdot \alpha_t \\
\Rightarrow\ & \beta_t \cdot f_t \cdot F(\gamma) \leq f_t \cdot \alpha_t \cdot F(\gamma) \\
\Rightarrow\ & \{f \text{ is strongly polymorphic}\} \\
& \beta_t \cdot G(\gamma) \cdot f_t \leq f_t \cdot \alpha_t \cdot F(\gamma) \\
\Rightarrow\ & f_t : \beta_t \cdot G(\gamma) \Rightarrow \alpha_t \cdot F(\gamma)
\end{aligned}$$

Note that f's strong polymorphism is necessary since γ might not be $\perp$–reflecting.
End of proof

5. Putting it into Practice

To apply these results in practice, we must choose a finite set of projections for each type. We call these specially chosen projections *contexts*, and describe our choice via a collection of inference rules.

Definition 5.1 The projection $\alpha : t \rightarrow t$ is a *context* for t if $\alpha\ \underline{cxt}\ t$ can be inferred using the rules below.

$$\frac{}{id_{\mathbf{1}}\ \underline{cxt}\ \mathbf{1}}$$

$$\frac{\alpha\ \underline{cxt}\ t}{\alpha_{\perp}\ \underline{cxt}\ t_{\perp}} \qquad \frac{\alpha\ \underline{cxt}\ t}{S \cdot \alpha_{\perp}\ \underline{cxt}\ t_{\perp}}$$

$$\frac{\alpha\ \underline{cxt}\ s \quad \beta\ \underline{cxt}\ t}{\alpha \oplus \beta\ \underline{cxt}\ s \oplus t}$$

$$\frac{\alpha\ \underline{cxt}\ s \quad \beta\ \underline{cxt}\ t}{\alpha \otimes \beta\ \underline{cxt}\ s \otimes t}$$

$$\frac{F(\alpha)\ \underline{cxt}\ T(t)\ [\alpha\ \underline{cxt}\ t]}{\mu\alpha.F(\alpha)\ \underline{cxt}\ \mu t.T(t)}$$

Clearly, for all types t we can infer

$$id_t\ \underline{cxt}\ t$$
$$\perp_t\ \underline{cxt}\ t$$

In fact, the set of contexts for each type forms a lattice. We can therefore map the domain of all projections over a type into the domain of contexts by mapping each projection to the least context greater than it. This mapping is continuous, and so we can use it to induce a continuous function on contexts from any continuous function on projections — we just apply the function to contexts viewed as projections, and then map the result back into the set of contexts as just described. A practical strictness analyser will of course compute with these induced operations on contexts. Since the induced operations always overestimate the true result it is safe to do so.

Note that the domain of contexts is not a subdomain of the domain of projections, because the least upper bound operation differs. In the domain of contexts for $t_\perp \otimes t_\perp$, we have

$$(S\otimes id) \sqcup (id\otimes S) = id\otimes id$$

but in the domain of projections this is not so: the left hand side maps the pair (lift $\perp$, lift $\perp$) to the pair ($\perp$, $\perp$), while the right hand side leaves it unchanged.

Examples of Contexts

Let us infer the contexts for a number of interesting types. First of all consider the type $\mathbf{2} = \mathbf{1}_\perp$. Since there is only one context for $\mathbf{1}$, we can infer that there are only two contexts for $\mathbf{2}$:

$$S_\mathbf{1} \cdot (id_\mathbf{1})_\perp$$
$$(id_\mathbf{1})_\perp$$

The former equals $\perp_\mathbf{2}$, and the latter equals $id_\mathbf{2}$.

Now consider the type Nat, which can be defined by

$$Nat = \mu n.\ \mathbf{2}\oplus n$$

Given that $\alpha\ \underline{cxt}\ n$, we can infer both $\perp_\mathbf{2}\oplus\alpha\ \underline{cxt}\ \mathbf{2}\oplus n$ and $id_\mathbf{2}\oplus\alpha\ \underline{cxt}\ \mathbf{2}\oplus n$, so there are two contexts for Nat:

$$\mu\alpha.\ \perp_\mathbf{2}\oplus\alpha$$
$$\mu\alpha.\ id_\mathbf{2}\oplus\alpha$$

These are equal to $\perp_{Nat}$ and id_{Nat} respectively. So as we would expect, the contexts chosen for Nat ignore the value of the natural, and indeed are just the two contexts that we have for every non-trivial type.

More interesting are the contexts for $Nat_\perp$ — i.e. closures of type Nat. Since there are two contexts for Nat, there are four for $Nat_\perp$:

$$(\perp_{Nat})_\perp$$
$$(id_{Nat})_\perp$$
$$S \cdot (\perp_{Nat})_\perp$$
$$S \cdot (id_{Nat})_\perp$$

$(\perp_{Nat})_\perp$ does not "evaluate" the closure, but discards its contents — this context was called ABS in [Wadler87a] and corresponds intuitively to ignoring the closure altogether. $(id_{Nat})_\perp$ is the identity on closures, and $S \cdot (\perp_{Nat})_\perp$ is the bottom projection. $S \cdot (id_{Nat})_\perp$ "evaluates" the closure and was called STR in [Wadler87a]. Thus these four contexts form the basic four-point domain of that paper:

```
     ID
    /  \
 STR    ABS
    \  /
     ⊥
```

Finally let us consider lazy lists. We must describe the list type using our chosen type formers: it is

$$List\ t = \mu l.\ \mathbf{2}\oplus(t_\perp\otimes l_\perp)$$

This is isomorphic to the more familiar

$$List\ t = \mu l.\ \mathbf{1}+(t\times l)$$

which uses separated sum and ordinary product, but note that our formulation uses lifting exactly where an implementation uses closures. As a result the contexts selected by our rules carry information directly useful to an implementation.

Since there are two contexts for **2**, and two contexts for $t_\perp$ and $l_\perp$ corresponding to each context for t and l, we would expect eight list contexts to correspond to each element context. Fortunately they can be expressed as compositions of just three of their number:

$$N = \mu\alpha.\ \perp_{\mathbf{2}} \oplus (id_\perp \otimes \alpha_\perp)$$
$$H = \mu\alpha.\ id_{\mathbf{2}} \oplus (S \otimes \alpha_\perp)$$
$$T = \mu\alpha.\ id_{\mathbf{2}} \oplus (id_\perp \otimes (S \cdot \alpha_\perp))$$

N is the projection that maps a nil at the end of a finite list to $\perp$ — an N-strict function (for example head) gives the same result for nil or $\perp$. It is not of great practical importance. H and T, however, capture the intuitive ideas of "head-strictness" and "tail-strictness", and were used for that purpose in our previous work. Now we can express the contexts for List t as follows, where β is any context for t.

$id \cdot List(\beta)$
$N \cdot List(\beta)$
$H \cdot List(\beta)$
$T \cdot List(\beta)$
$N \cdot H \cdot List(\beta)$
$T \cdot H \cdot List(\beta)$
$\perp$

(Note that since $T \cdot N = \perp$ there are only seven different contexts).

Factorising Contexts

Notice that contexts for List(t) can be "factored" into the form

$$\alpha \cdot List(\beta)$$

where α is independent of the element type and β is a context for t. Recall also that in Corollary 4.17 we showed that such a factorisation can help us to apply polymorphic knowledge about a function to analyse a particular instance. We will therefore generalise this factorisation to other types. We first need to define the notion of a polymorphic context, and a context for a tuple of types.

Definition 5.2 Let F be a parameterised type and $\alpha : F \to F$ be a projection. We say α is a *(polymorphic) context* for F, α $\underline{cxt}$ F, if α_t $\underline{cxt}$ F(t) for all t.

Definition 5.3 Let $t=(t_1 \ldots t_n)$ be a $\mathbb{C}^n$ object, and $\alpha=(\alpha_1 \ldots \alpha_n) : t \to t$ be a projection. We say α is a context for t, α $\underline{cxt}$ t, if α_i $\underline{cxt}$ t_i for each i.

Now given α, a context for F(t), we will factor it into $\alpha|F$, a (polymorphic) context for F, and α/F, a context for t. The factorisation will unfortunately not be exact, but we will contruct it so that

$$\alpha \le \alpha|F \cdot F(\alpha/F)$$

A strictness analyser may therefore factorise contexts freely, since the factorisation is certain to be an overestimate of the original context.

First we define the trivial cases of factorisation.

Definition 5.4 Let $F(t_1 \ldots t_n) = t_i$ and α be a context for F(t). Then

$$\alpha|F = id$$
$$\alpha/F = (\beta_1 \ldots \beta_n) \quad \text{where} \quad \beta_i = \alpha, \quad \beta_j = \perp \text{ if } i \neq j$$

Definition 5.5 Let $F(t_1 \ldots t_n) = A$ (a constant type), and α be a context for F(t). Then

$$\alpha|F = \alpha$$
$$\alpha/F = (\perp \ldots \perp)$$

Now let us cover the other non-recursive type-formers.

Definition 5.6 Let β be a context for G(t) and γ be a context for H(t). Then

$$\beta_\perp \mid G_\perp = (\beta \mid G)_\perp$$
$$(S \cdot \beta_\perp) \mid G_\perp = S \cdot (\beta \mid G)_\perp$$
$$(\beta \oplus \gamma) \mid (G \oplus H) = (\beta \mid G) \oplus (\gamma \mid H)$$
$$(\beta \otimes \gamma) \mid (G \otimes H) = (\beta \mid G) \otimes (\gamma \mid H)$$

$$\beta_\perp / G_\perp = \beta / G$$
$$(S \cdot \beta_\perp) / G_\perp = \beta / G$$
$$(\beta \oplus \gamma) / (G \oplus H) = (\beta \mid G) \sqcup (\gamma \mid H)$$
$$(\beta \otimes \gamma) / (G \otimes H) = (\beta \mid G) \sqcup (\gamma \mid H)$$

It remains to factorise recursive contexts. Recall that contexts for the type μs. G(t,s) take the form μb. H(β) where H(β) is a context for G(t,s) under the assumption that β is a context for s. Consider the factorisation of H(β) by G — we have

$$H(\beta) \mid G \ \underline{cxt}\ G$$
$$H(\beta) / G \ \underline{cxt}\ (t,s)$$

It is clear that H(β) | G cannot involve β, and that H(β) / G is a pair of contexts (τ,β) since β is the only available context for s. Therefore we define

Definition 5.7 Let G : $\mathbf{C}_S{}^n \times \mathbf{C}_S \to \mathbf{C}$ be a parameterised type, and let $\mu\beta$. H(β) be a context for μs. G(t,s). Then

$$\mu\beta.\ H(\beta) \mid \mu G = \mu\alpha.\ (H(\beta) \mid G) \cdot G(id,\alpha)$$
$$\mu\beta.\ H(\beta) / \mu G = \tau \qquad \text{where } (\tau,\sigma) = H(\beta) / G$$

Lemma 5.8 Let α be a context for F(t). Then

(a) $\alpha \mid F \ \underline{cxt}\ F$

(b) $\alpha / F \ \underline{cxt}\ t$

(c) $\alpha \leq (\alpha \mid F) \cdot F\,(\alpha / F)$

Proof

By structural induction on F.

End of proof

An Example of Analysis

We discuss the analysis of a simple polymorphic function — the well-known function append, with the type

$$append_t : (List\ t)_\perp \otimes (List\ t)_\perp \to List\ t$$

(Note that its arguments are passed lazily). We analyse such a function as follows: on encountering its definition, we enumerate all the polymorphic contexts for its result type, and analyse the function for each of them. This analysis uses the monomorphic techniques of [Wadler87a], at any convenient instance — say **2**. For example, we may discover that

$$append_{\mathbf{2}} : T_{\mathbf{2}} \Rightarrow (S_{List\ \mathbf{2}} \cdot (T_{\mathbf{2}})_\perp) \otimes (S_{List\ \mathbf{2}} \cdot (T_{\mathbf{2}})_\perp)$$

and Corollary 4.17 then allows us to conclude the polymorphic property

$$append : T \Rightarrow (S \cdot T_\perp) \otimes (S \cdot T_\perp)$$

These polymorphic properties are saved for later use.

Now, to analyse a call of append in a context β, we first factorise β by append's result type, List:

$$\beta \leq (\beta \mid \text{List}) \cdot \text{List}\ (\beta / \text{List})$$

$\beta \mid$ List is a polymorphic projection for List, and so there must be a saved fact

$$\text{append} : \beta \mid \text{List} \Rightarrow \alpha$$

for some polymorphic α. By Theorem 4.19 we can conclude

$$\text{append} : (\beta \mid \text{List}) \cdot \text{List}\ (\beta / \text{List}) \Rightarrow \alpha \cdot \text{List}\ (\beta / \text{List})$$

and so

$$\text{append} : \beta \Rightarrow \alpha \cdot \text{List}\ (\beta / \text{List})$$

since β approximates its factorisation. This completes the analysis of the call.

For example, suppose the result of append is passed to length. The length function is $T \cdot \text{List}(\bot)$ –strict — that is, it is tail strict and ignores the list elements. We can infer at once that

$$\text{append} : T \cdot \text{List}(\bot) \Rightarrow (S \cdot (T \cdot \text{List}(\bot))_\bot) \otimes (S \cdot (T \cdot \text{List}(\bot))_\bot)$$

That is, both arguments are strictly evaluated in a similar context.

A Difficulty

For some functions, this method gives poorer results than analysing each instance separately. For example, consider

$$\text{dup}_t : t_\bot \to t_\bot \otimes t_\bot$$
$$\text{dup}_t\ x = (x,x)$$

Analysis shows that

$$\text{dup} : S \otimes S \Rightarrow S$$

and so using Theorem 4.19 and factorisation we conclude that

$$\text{dup} : (S \cdot \alpha_\bot) \otimes (S \cdot \beta_\bot) \Rightarrow S \cdot (\alpha \sqcup \beta)_\bot$$

But note that if either $\alpha\ x = \bot$ or $\beta\ x = \bot$, then dup x is mapped to $(\bot,\bot)$ by the context on the left, because the product is strict. Our earlier monomorphic techniques take advantage of this to derive

$$\text{dup} : (S \cdot \alpha_\bot) \otimes (S \cdot \beta_\bot) \Rightarrow S \cdot (\alpha \,\&\, \beta)_\bot$$

where

$$\begin{array}{lll} (\alpha \,\&\, \beta)\ x & = \bot & \text{if } \alpha\ x = \bot \text{ or } \beta\ x = \bot \\ & = \alpha\ x \sqcup \beta\ x & \text{otherwise} \end{array}$$

It is not clear how to derive the same information from a polymorphic analysis. Future work will investigate factorisation taking into account the function to be analysed, since although it is not true that

$$(S \cdot \alpha_\bot) \otimes (S \cdot \beta_\bot) \leq (S \otimes S) \cdot ((\alpha \,\&\, \beta)_\bot \otimes (\alpha \,\&\, \beta)_\bot)$$

it *is* the case that

$$(S \cdot \alpha_\bot) \otimes (S \cdot \beta_\bot) \cdot \text{dup} \leq (S \otimes S) \cdot ((\alpha \,\&\, \beta)_\bot \otimes (\alpha \,\&\, \beta)_\bot) \cdot \text{dup}$$

6. Related Work and Conclusions

Strictness analysis of functional languages has attracted a great deal of interest over the past decade, beginning with Mycroft's seminal paper [Mycroft80]. Mycroft introduced the idea of using *abstract interpretation* for strictness analysis, and although his method was limited to first order functions operating on non-lazy data-structures it is the foundation for much later work. Mycroft worked with an untyped programming language, and this theme has been continued by Hudak and Young, who discovered a technique for analysing higher-order functions in this setting [Hudak86]. Others, however, have concentrated on exploiting a type system. Burn, Hankin and Abramsky developed an abstract interpretation of higher-order typed programs [BHA85], which Wadler extended to handle lazy lists [Wadler87b]. A common feature of this work is its emphasis on working in finite domains, in which fixed points can be calculated directly by iteration. Clack and Peyton-Jones invented a technique for finding such fixed points efficiently [Clack85] which has been extended to a wider class of domains by Martin and Hankin [Martin87]. Abstract interpretation has been applied to many other analysis problems, for example sharing analysis [Goldberg87] and abstract reference counting [Hudak87]. See [AH87] for a collection of recent work.

Abstract interpretation is a "forwards" analysis method; I became interested in backwards analysis because of its apparent ability to handle lazy data-structures well. I developed an ad-hoc analyser for first-order untyped languages that relied on algebraic simplification on infinite domains to analyse recursive functions over lazy data-structures [Hughes85]. More recently Hall and Wise have integrated a similar analyser with a program transformer [Hall87] — they also use algebraic simplification in infinite domains. Theories of backwards strictness analysis were developed based on continuations [Hughes87a], Scott-open sets [Dybjer87], and projections [Wadler87a]. The projection-based theory has inspired similar approaches to binding-time analysis [Launchbury87] and complexity analysis [Wadler88].

Backwards analysis, like forwards analysis, can take advantage of type information. An early backwards analyser that did so was developed by Wray: it handled a powerful polymorphic type system with higher-order functions, and moreover ran in an insignificant fraction of the total compile-time [Wray86]. However, it was not proved correct, and was unable to analyse functions on lazy data-structures usefully. Wadler and I used a monomorphic type system for our work on projection based analysis, and I have continued to do so in more recent work on analysing higher-order functions [Hughes87b] and performing sharing and life-time analysis backwards [Hughes89].

Monomorphically typed languages have proved amenable to compile-time analysis, whether forwards or backwards. Their advantage is that infinitely many different pieces of possible information may be divided into many finite sets, one per type. Thus an analyser gains the advantage of being able to represent infinitely many different possibilities, without the cost of needing to find one solution from an infinite set in any particular case. The disadvantage of basing an analyser on a monomorphic type system is that most practical functional languages are polymorphic.

Abramsky pointed the way to a solution of this problem by treating a polymorphic function as an infinite collection of monomorphic instances, and showing that some of the results of strictness analysis are common to all [Abramsky85]. Thus only one instance need by analysed — using monomorphic techniques — and the results can be applied to all. This idea is at the heart of my own work. However, Abramsky used a syntactic characterisation of polymorphism — the type inference rules — and therefore had to prove his result by structural induction over terms. I used the semantic characterisation of first order polymorphic functions as natural transformations to strengthen Abramsky's result in the first-order case, showing that an approximation to the abstract function of any instance can be calculated from that of the simplest [Hughes88]. Abramsky has shown that higher-order functions are natural transformations in a more complex category, and thereby proved a stronger version of his original result [Abramsky88]. In this paper I apply the same idea to projection analysis, and it is also finding other applications: Wadler is using naturality to derive theorems about higher-order functions for use in program transformation [Wadler89], and Sheeran is using Wadler's ideas to derive theorems about hardware descriptions [Sheeran89]. Launchbury has applied the results of this paper to polymorphic binding time analysis, based on his earlier work [Launchbury89].

In the future I hope to generalise the result of [Hughes88] to the abstract interpretation of higher-order functions. This is difficult because the function-space type former is not a covariant functor. Possible approaches are to use di-natural transformations, or to make the function type covariant by working in a more complex category as Abramsky and Wadler did. Unfortunately this can be done in several different ways, and it is not yet clear which way is best suited to this particular problem. In the longer term it would be interesting to make a connection with operational semantics, so as to attack a wider class of analysis problems.

The particular result in this paper is a new strictness analysis method for first-order polymorphic functions. A major advantage over previous methods is the ease with which polymorphic information can be applied to the call of a particular instance: neither my nor Abramsky's previous work can do so as smoothly as Theorem 4.19.

My more general thesis is that the categorical view is a potent tool for solving practical problems involving polymorphic functions. Evidence for its power is mounting.

Acknowledgements

I am grateful to John Launchbury for many useful discussions, and to Phil Wadler for his inspiration. This work was carried out under a grant from the UK Science and Engineering Research Council.

References

[Abramsky85] S. Abramsky, *Strictness Analysis and Polymorphic Invariance*, Workshop on Programs as Data Objects, Copenhagen, Springer LNCS 217, 1985.

[Abramsky88] S. Abramsky, *Notes on Strictness Analysis for Polymorphic Functions*, draft paper, 1988.

[AH87] S. Abramsky and C. L. Hankin (eds.) *Abstract Interpretation of Declarative Languages*, Ellis-Horwood, 1987.

[BHA85] G. L. Burn, C. L. Hankin, S. Abramsky, *The Theory of Strictness Analysis for Higher-order Functions*, Workshop on Programs as Data Objects, Copenhagen, Springer LNCS 217, 1985.

[Clack85] C. Clack and S. L. Peyton-Jones, *Strictness Analysis — a Practical Approach*, in IFIP Conference on Functional Programming Languages and Computer Architecture, Nancy, France, Springer LNCS 201, 1985.

[Dybjer87] P. Dybjer, *Computing Inverse Images*, in ICALP 1987.

[Goldberg87] B. Goldberg, *Detecting Sharing of Partial Applications in Functional Programs*, in IFIP Conference on Functional Programming Languages and Computer Architecture, Portland, Oregon, Springer LNCS 274, 1987.

[Hall87] C. Hall and D. S. Wise, *Compiling Strictness into Streams*, in ACM Symposium on Principles of Programming Languages, 1987.

[Hudak86] P. Hudak and J. Young, *Higher-order Strictness Analysis for Untyped Lambda Calculus*, in ACM Symposium on Principles of Programming Languages, 1986.

[Hudak87] P. Hudak, *Abstract reference counting*, in S. Abramsky and C. L. Hankin (eds.) *Abstract Interpretation of Declarative Languages*, Ellis-Horwood, 1987.

[Hughes85] J. Hughes, *Strictness Detection in Non-Flat Domains*, Workshop on Programs as Data Objects, Copenhagen, Springer Verlag 217, 1985.

[Hughes87a] J. Hughes, *Analysing Strictness by Abstract Interpretation of Continuations*, in S. Abramsky and C. L. Hankin (eds.) *Abstract Interpretation of Declarative Languages*, Ellis-Horwood, 1987.

[Hughes87b] J. Hughes, *Backwards Analysis of Functional Programs*, IFIP Workshop on Partial Evaluation and Mixed Computation, Bjørner, Ershov and Jones (eds.), North-Holland, 1987.

[Hughes88] J. Hughes, *Abstract Interpretation of First-order Polymorphic Functions*, Proc. Aspenæs Workshop on Graph Reduction, University of Gothenburg, 1988.

[Hughes89] J. Hughes, *Compile-time Analysis of Functional Languages*, Proc. Year of Programming Summer School on Declarative Programming, University of Texas, 1989 (to appear).

[Launchbury87] J. Launchbury, *Projections for Specialisation*, IFIP Workshop on Partial Evaluation and Mixed Computation, Bjørner, Ershov and Jones (eds.), North-Holland, 1987.

[Launchbury89] J. Launchbury, *Binding Time Aspects of Partial Evaluation*, Ph.D. thesis, Glasgow University, in preparation.

[Martin87] C. Martin and C. Hankin, *Finding Fixed Points in Finite Lattices*, IFIP Conference on Functional Programming Languages and Computer Architecture, Portland, Oregon, Springer LNCS 274, 1987.

[Mycroft80] A. Mycroft, *The Theory and Practice of Transforming Call-by-Need into Call-by-Value*, Proc. International Symposium on Programming, Springer LNCS 83, 1980.

[Rydeheard85] Tutorial on natural transformations, in Proc. Category Theory and Computer Programming, Springer LNCS 240, 1985.

[Sheeran89] M. Sheeran, *Categories for the Working Hardware Designer*, to appear in Proc. Int. Workshop on Hardware Specification, Verification, and Synthesis: Mathematical Aspects, Cornell, Springer LNCS, 1989.

[Wadler87a] P. Wadler and J. Hughes, *Projections for Strictness Analysis*, IFIP Conference on Functional Programming Languages and Computer Architecture, Portland, Oregon, Springer LNCS 274, 1987.

[Wadler87b] P. Wadler, *Strictness Analysis on Non-flat Domains (by Abstract Interpretation over Finite Domains)*, in S. Abramsky and C. L. Hankin (eds.) *Abstract Interpretation of Declarative Languages*, Ellis-Horwood, 1987.

[Wadler88] P. Wadler, *Strictness Analysis Aids Time Analysis*, in ACM Symposium on Principles of Programming Languages, 1988.

[Wadler89] P. Wadler, *Theorems for Free!*, to appear in IFIP Functional Programming Languages and Computer Architecture, London, Springer LNCS, 1989.

[Wray86] S. C. Wray, *Implementation and Programming Techniques for Functional Languages*, Ph.D. thesis, University of Cambridge, 1986.

A category-theoretic account of program modules

Eugenio Moggi
em@lfcs.edinburgh.ac.uk

LFCS, University of Edinburgh, EH9 3JZ Edinburgh, UK
(on leave from Università di Pisa)

Abstract

The type-theoretic explanation of modules proposed to date (for programming languages like ML) is unsatisfactory, in that it fails to reflect the distinction between compile-time, when type-expressions are evaluated, and run-time, when value-expressions are evaluated. This paper proposes a new explanation based on "programming languages as indexed categories" and illustrates, as an application, how ML should be extended to support higher order modules. The paper also outlines a methodology for a modular approach to programming languages, where programming languages (of a certain kind) are identified with objects in a 2-category and features are viewed as 2-categorical notions.

A critique to the [Mac86, HM88] account of ML-modules

The addition of module facilities to programming languages is motivated by the need to provide a better environment for the development and maintenance of large programs. Nowadays many programming languages include such facilities. Throughout the paper Standard ML (see [Mac85, HMM86, HMT87]) is taken as representative for these languages. The implementation of module facilities has been based mainly on an operational understanding. More recently a type-theoretic explanation of ML-modules has been proposed, based on a type theory with dependent types and a cumulative hierarchy of universes U_1 and U_2, i.e. $U_1 : U_2$ and $U_1 \subset U_2$ (see [Mac86, HM88]). This type-theoretic account of ML-modules is as follows:

- *ML-signatures* are elements of U_2 built from U_1 and types (i.e. elements of U_1) by the Σ-type constructor, while *ML-structures* are elements of ML-signature, namely tuples made of values (i.e. elements of types) and types

- *ML-functor signatures* are elements of U_2 of the form $\Pi s: sig_1.sig_2(s)$, where sig_1 and sig_2 are ML-signatures, while *ML-functors* are elements of ML-functor signatures, namely functions from ML-structures to ML-structures.

Unfortunately, the explanation of ML-functors as functions is *unsatisfactory*, in that it fails to capture the distinction between compile-time and run-time. The problems caused by this misunderstanding of ML-functors become apparent when trying to define *higher order modules* as done in XML (see [HM88]).

Example 0.1 Let τ be a type and $f\colon(\Pi x\colon\tau.U_1)$ be an ML-functor variable, which may occur in the body of an higher order module. Then two major problems arise:

- It is not clear what the meaning of $fM\colon U_1$ is, when the expression $M\colon\tau$ diverges. This may happen in ML, because functions can be defined by recursion.
- Type equality becomes undecidable as soon as recursive types are allowed (as in ML). For instance, if $\tau = \tau \to \tau$, then $fM_1 = fM_2\colon U_1 \iff M_1 = M_2\colon\tau$ and the second equality is undecidable, since it amounts to $\beta\eta$-conversion between untyped λ-terms.

In XML these problems are avoided by making everything total and decidable, but then the theory hardly captures the *essence* of ML. In ML they are avoided by banning higher order modules, so that the only ML-functors $\lambda x\colon\tau.M$ of ML-signature $\Pi x\colon\tau.U_1$ allowed are those where x is not free in M.

> This paper gives *a correct understanding of modules* for programming languages where there is a *syntactic distinction* between compile-time and run-time (as in ML, PASCAL, and ADA).

Syntactic distinction, more precisely, *phase distinction* is discussed by [Car88] in the context of Type Theory "...the execution of a program is carried out in two *phases*: a type-checking phase (*compile-time*) and an execution phase (*run-time*)." We reformulate the notion of phase distinction categorically and derive a definition of parametric program module consistent with such a distinction.

> The concrete outcome of our analysis is that ML-functors should be viewed as pairs of functions and not as functions (from pairs to pairs).

For instance, ML-functors from $\Sigma t_1\colon U_1.\tau_1(t_1)$ to $\Sigma t_2\colon U_1.\tau_2(t_2)$ are elements of

$$\Sigma f\colon U_1 \to U_2.(\Pi t_1\colon U_1.\tau_1(t_1) \to \tau_2(ft_1))$$

and not elements of $(\Sigma t_1\colon U_1.\tau_1(t_1)) \to (\Sigma t_2\colon U_1.\tau_2(t_2))$, as in [Mac86, HM88]. We will also show, by exhibiting a toy language called HML, that higher order modules are consistent with other features of programming languages, provided the correct notion of ML-functor is used.

Pragmatic issues and general methodology

If the objective of this paper were to be the description of a toy language like HML (see Section 5), then one may fairly ask why it should be necessary to go through indexed categories, the Grothendieck construction and even 2-categories. As a matter of fact, HML is fully described in an unpublished manuscript (see [Mog86]), which relies only on a bit of type theory, operational semantics and common sense. Although that manuscript describes also a "category of modules", at that time we did not realise that it is the

Grothendieck construction applied to a suitable indexed category, nor did we feel necessary to see things in such a way.

However, the objective of this paper is to propose a general methodology for studying programming languages, and to explain how the compile-time/run-time distinction and program modules may fit into it (in Section 3 we also briefly discuss other issues). From this perspective Category Theory is particularly appropriate. The need for a general methodology can be better appreciated by recalling a failed attempt by Bob Harper, John Mitchell and the author to develop a type theory combining dependent types, computational lambda-calculus, HML and capable of accounting for ML plus higher order modules with sharing constraints. Such an ambitious project failed (at least temporarily), since it produced several big formal systems whose properties were very difficult to establish, even though each system was the combination of simple and well-understood ideas.

> The failure of such an attempt highlights a deep-rooted problem of programming language theory (and other areas): the lack of *modularity*.

More precisely researchers have been focusing on mathematically clean and easy to understand toy languages, where particular features are isolated and investigated in depth, but they have neglected the problem of combining features. In a modular approach the key concept to investigate should be the "addition of features to a language" rather than investigating specific "toy languages" (i.e. languages with only a few features).

Syntax-independent view. The first ingredient of our methodology is to abstract as much as possible from the concrete presentation of a language. This is standard practice in categorical logic, where theories are identified with categories (having certain additional structure). We follow a similar paradigm for programming languages. In particular, we propose to identify a programming language where some of the expressions are evaluated at compile-time and the others at run-time with an indexed category $\mathcal{C}: \mathcal{B}^{op} \to \mathbf{Cat}$ (see Section 2 and Section 6 for a discussion on possible improvements). Intuitively:

- the *compile-time part* is represented by the base category $\mathcal{B}$, e.g. ML types expressions become morphisms in $\mathcal{B}$, while
- the *run-time part* is represented by the fiber categories, e.g. ML program expressions become morphisms in one of the fibers $\mathcal{C}[X]$.

Program modules. If programming languages are indexed categories (as outlined above), then it is natural to expect that program modules should live in a category where the compile-time and run-time part are combined. There is a standard construction, due to Grothendieck, which transforms an indexed category $\mathcal{C}: \mathcal{B}^{op} \to \mathbf{Cat}$ into a fibration $\mathcal{FC}: \mathcal{GC} \to \mathcal{B}$. We take $\mathcal{GC}$ as the *category of modules* for the programming language $\mathcal{C}$ (see Section 4). Once we have established the correct link between programming language concepts and category-theoretic notions, it is mainly a matter of letting standard category-theoretic machinery do the rest, e.g. tell us what it means to have higher order modules.

2-categorical approach. The second ingredient of our methodology is ways in which to combine features. This is a very difficult problem to solve in generality, because it depends on the way features *interact*. For instance:

- they could be *unrelated*, like products and coproducts, or
- one feature could be defined in terms of the other, e.g. the definition of function space relies on having products.

In this paper we focus on "adding one feature *on top* of another". The motivating example is to make precise the idea of a notion of computation which respects the compile-time/run-time distinction, or more formally of a monad (i.e. a notion of computation) over an indexed category (which captures the compile-time/run-time distinction). A simpler example of such a combination is the notion of topological group, which amounts to a group in the category of topological spaces. The strategy suggested by the second example is to look for a category of topological spaces (the first feature) and generalise the definition of group (the second feature) over a set to that of group over an object in a category (with finite products). By applying (mutatis mutandis) the same strategy to the first example, we have to generalise the definition of monad over a category to that of monad over an object in a 2-category (see Section 1) and show that indexed categories form a 2-category (see Section 2). Summarising, the key ideas of the 2-categorical approach to programming languages are:

- programming languages (with certain features) are objects of a 2-category $\mathcal{C}$ (which will depend on the features one is interested in)
- an additional feature is an instance in $\mathcal{C}$ of a 2-categorical concept.

Section 3 discusses other features of programming languages and how they might be added on top of the compile-time/run-time distinction.

1 Preliminaries on 2-categories

Both categories and $\mathcal{B}$-indexed categories can be viewed as objects of suitables 2-categories, **Cat** and **ICat**($\mathcal{B}$) respectively. This view is particularly useful for giving definitions involving $\mathcal{B}$-indexed categories (and proving their properties) by analogy with categories. In fact, it is just a matter of rephrasing familiar concepts, like monad or adjunction, in the formal language of 2-categories. We recall the definitions of 2-category, 2-functor and 2-natural transformation, i.e. the **Cat**-enriched analogue of category, functor and natural transformation (see [Kel82]).

Definition 1.1 *A* **2-category** $\mathcal{C}$ *is a* **Cat**-*enriched category, i.e.*

- *a class of objects* $\mathrm{Obj}(\mathcal{C})$
- *for every pair of objects* c_1 *and* c_2 *a category* $\mathcal{C}(c_1, c_2)$
- *for every object* c *an object* $id_c^{\mathcal{C}}$ *of* $\mathcal{C}(c, c)$ *and for every triple of objects* c_1, c_2 *and* c_3 *a functor* $comp^{\mathcal{C}}_{c_1,c_2,c_3}$ *from* $\mathcal{C}(c_1, c_2) \times \mathcal{C}(c_2, c_3)$ *to* $\mathcal{C}(c_1, c_3)$ *satisfying the associativity and identity axioms*

- $comp(\_, comp(\_, \_)) = comp(comp(\_, \_), \_)$
- $comp(id, \_) = \_ = comp(\_, id)$

Notation 1.2 An object f of $C(c_1, c_2)$ is called a **1-morphism**, while an arrow σ is called a **2-morphism**. We write $\_;\_$ for $comp(\_, \_)$

$$c_1 \ \begin{array}{c} \xrightarrow{f_1} \\ \Downarrow \sigma_1 \\ \xrightarrow[f_1']{} \end{array} \ c_2 \ \begin{array}{c} \xrightarrow{f_2} \\ \Downarrow \sigma_2 \\ \xrightarrow[f_2']{} \end{array} \ c_3 \ \stackrel{-;-}{\longmapsto} \ c_1 \ \begin{array}{c} \xrightarrow{f_1;f_2} \\ \Downarrow \sigma_1;\sigma_2 \\ \xrightarrow[f_1';f_2']{} \end{array} \ c_3$$

and $\_\cdot\_$ for composition of 2-morphisms

$$c_1 \ \begin{array}{c} \xrightarrow{f_1} \\ \Downarrow \sigma_1 \\ \longrightarrow \\ \Downarrow \sigma_2 \\ \xrightarrow[f_2]{} \end{array} \ c_2 \ \stackrel{-\cdot-}{\longmapsto} \ c_1 \ \begin{array}{c} \xrightarrow{f_1} \\ \Downarrow \sigma_1\cdot\sigma_2 \\ \xrightarrow[f_2]{} \end{array} \ c_2$$

Example 1.3 The canonical example of a 2-category is **Cat** itself (see [Mac71]):

- the objects are categories
- the 1-morphisms are functors and $\_;\_$ is functor composition,
- the 2-morphisms are natural transformations and $\_;\_$ and $\_\cdot\_$ are respectively horizontal and vertical composition of natural transformations.

Definition 1.4 *A* **2-functor** F *from* C_1 *to* C_2 *is a mapping*

$$c \ \begin{array}{c} \xrightarrow{f} \\ \Downarrow \sigma \\ \xrightarrow[f']{} \end{array} \ c' \text{ in } C_1 \ \stackrel{F}{\longmapsto} \ Fc \ \begin{array}{c} \xrightarrow{Ff} \\ \Downarrow F\sigma \\ \xrightarrow[Ff']{} \end{array} \ Fc' \text{ in } C_2$$

which commutes with identities, $\_;\_$ *and* $\_\cdot\_$.

Definition 1.5 *If* F_1 *and* F_2 *are 2-functors from* C_1 *to* C_2*, then a* **2-natural transformation** τ *from* F_1 *to* F_2 *is a family* $\langle F_1c \xrightarrow{\tau_c} F_2c \mid c \in \mathrm{Obj}(C_1)\rangle$ *of 1-morphisms in* C_2 *s.t.*

$$\begin{array}{c} c \\ \Big\downarrow \overset{\sigma}{\Rightarrow} \Big\downarrow \\ c' \end{array} \quad \Longrightarrow \quad \begin{array}{ccc} F_1c & \xrightarrow{\tau_c} & F_2c \\ \Big\downarrow \overset{F_1\sigma}{\Rightarrow} \Big\downarrow & & \Big\downarrow \overset{F_2\sigma}{\Rightarrow} \Big\downarrow \\ F_1c' & \xrightarrow{\tau_{c'}} & F_2c' \end{array}$$

i.e. the functors $\tau_c; F_2\_$ *and* $F_1\_;\tau_{c'}$ *from* $C_1(c, c')$ *to* $C_2(F_1c, F_2c')$ *are equal for every* c *and* c' *in* $\mathrm{Obj}(C_1)$.

Adjunctions are the basic tool to define data-types, while monads are used to model computations (see [Mog89]). Their definition can be rephrased in the language of 2-categories and most of their properties can be proved in such a formal setting (see [Str72]), so these standard tools can be applied in a different 2-category, e.g. that of indexed categories (see Section 2).

Definition 1.6 *Let c and c′ be objects of a 2-category C.*

- *A* **monad** *over c is a triple (T, η, μ) s.t.*

$$c \xrightarrow{T} c \qquad \begin{array}{c} c \\ id_c \big\downarrow \overset{\eta}{\Rightarrow} \big\downarrow T \\ c \end{array} \qquad \begin{array}{c} c \\ T;T \big\downarrow \overset{\mu}{\Rightarrow} \big\downarrow T \\ c \end{array}$$

$(T; \mu) \cdot \mu = (\mu; T) \cdot \mu$ *and* $(T; \eta) \cdot \mu = \mathrm{id}_T = (\eta; T) \cdot \mu$.

- *An* **adjunction** *from c to c′ is a quadruple (F, G, η, ϵ) s.t.*

$$c \underset{G}{\overset{F}{\rightleftarrows}} c' \qquad \begin{array}{c} c \\ id_c \big\downarrow \overset{\eta}{\Rightarrow} \big\downarrow F;G \\ c \end{array} \qquad \begin{array}{c} c' \\ G;F \big\downarrow \overset{\epsilon}{\Rightarrow} \big\downarrow id_{c'} \\ c' \end{array}$$

$(\eta; F) \cdot (F; \epsilon) = \mathrm{id}_F$ *and* $(G; \eta) \cdot (\epsilon; G) = \mathrm{id}_G$.

It is obvious from the definition above, that the 2-categorical notions of monad and adjunction are preserved by 2-functors and that in the 2-category **Cat** they amount to familiar definitions.

2 Indexed categories and programming languages

In this section we define the 2-category $\mathbf{ICat}(\mathcal{B})$ of $\mathcal{B}$-indexed categories. Indexed categories model only one feature of a strongly typed programming language, namely that expressions are partitioned into two groups, those evaluated at compile-time and those evaluated at run-time, and that compile-time expressions do not depend on run-time ones. Section 3 will discuss how to model other features by additional structure over an indexed category.

Indexed categories do not enforce the distinction between compile-time and run-time, they simply allow it. The situation resembles that of hyperdoctrines (see [See83, See84]), where the indexed category structure allows the distinction between types and formulas in first order logic, as well as the identification of types and formulas advocated by Martin-Löf type theory (by providing additional structure, which enforces such identification).

The general definition of indexed category is fairly complicated, since it involves the notion of canonical isomorphism. However, for representing languages it is more appropriate to use a stricter definition of $\mathcal{B}$-indexed category (e.g. see [See84, See87]), namely a functor from $\mathcal{B}^{op}$ to **Cat**, where $\mathcal{B}$ is a small category and **Cat** is the category of small categories and functors.

Definition 2.1 *Given a small category $\mathcal{B}$, the 2-category* $\mathbf{ICat}(\mathcal{B})$ *of $\mathcal{B}$-indexed categories is defined as follows:*

- *an object (indexed category) is a functor $C: \mathcal{B}^{op} \to \mathbf{Cat}$*

$$\begin{array}{ccc} X & & C[X] \\ f\uparrow & \stackrel{C}{\longmapsto} & \downarrow C[f] \\ Y & & C[Y] \end{array}$$

- *a 1-morphism (indexed functor) from C_1 to C_2 is a natural transformation $F: C_1 \dot{\to} C_2$, i.e. a family $\langle F[X]: C_1[X] \to C_2[X] | X \in \mathcal{B}\rangle$ of functors s.t.*

$$\begin{array}{ccccc} X & & C_1[X] & \xrightarrow{F[X]} & C_2[X] \\ f\uparrow & \Longrightarrow & C_1[f]\downarrow & & \downarrow C_2[f] \\ Y & & C_1[Y] & \xrightarrow[F[Y]]{} & C_2[Y] \end{array}$$

- *Given 1-morphisms F_1 and F_2 from C_1 to C_2, a 2-morphism (indexed natural transformation) from F_1 to F_2 is a family $\langle \sigma[X]: F_1[X] \dot{\to} F_2[X] | X \in \mathcal{B}\rangle$ of natural transformations s.t.*

$$\begin{array}{ccccc} X & & C_1[X] & \underset{\longrightarrow}{\overset{\longrightarrow}{\Downarrow \sigma[X]}} & C_2[X] \\ f\uparrow & \Longrightarrow & C_1[f]\downarrow & & \downarrow C_2[f] \\ Y & & C_1[Y] & \underset{\longrightarrow}{\overset{\longrightarrow}{\Downarrow \sigma[Y]}} & C_2[Y] \end{array}$$

  i.e. for every $f: Y \to Y$ in $\mathcal{B}$ the natural transformations $C_1[f]; \sigma[Y]$ and $\sigma[X]; C_2[f]$ are equal.

The definition of monad and adjunction for $\mathcal{B}$-indexed categories are particular instances of the 2-categorical definitions. The following proposition characterises an adjunction in $\mathbf{ICat}(\mathcal{B})$ as a family of adjunctions in $\mathbf{Cat}$ satisfying the *Beck-Chevalley condition*.

Proposition 2.2 ([PS78]) *Given a $\mathcal{B}$-indexed functor G from C_2 to C_1, an adjunction (F, G, η, ϵ) from C_1 to C_2 amounts to have a family*

$$\langle (F[b], G[b], \eta[b], \epsilon[b]) | b \in \mathcal{B}\rangle$$

satisfying the following properties:

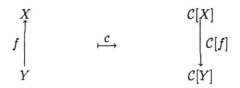

loc $(F[b], G[b], \eta[b], \epsilon[b])$ *is an adjunction from* $C_1[b]$ *to* $C_2[b]$, *for every* $b \in \mathcal{B}$

BC *for every* $f: b_2 \to b_1$ *in* $\mathcal{B}$ *the functors* $C_1[f]; F[b_2]$ *and* $F[b_1]; C_2[f]$ *are equal and the natural transformation* $(\eta[b_1]; C_1[f]; F[b_2]) \cdot (F[b_1]; C_2[f]; \epsilon[b_2])$ *from* $C_1[f]; F[b_2]$ *to* $F[b_1]; C_2[f]$ *is the identity over* $C_1[f]; F[b_2]$.

Remark 2.3 The Beck-Chevalley condition in [PS78] requires only that the natural transformation $(\eta[b_1]; C_1[f]; F[b_2]) \cdot (F[b_1]; C_2[f]; \epsilon[b_2])$ is a *canonical isomorphism*, but we have adopted a *strict* notion of indexed category, where canonical isomorphisms are identities.

3 Intermezzo

At this point we review the category-theoretic structures used for interpreting some typed λ-calculi and discuss the additional structures needed to model various features of programming languages.

- Hyperdoctrines model the proof theory of intuitionistic first order logic (see [See83]). They are indexed-categories $C: \mathcal{B}^{op} \to \mathbf{Cat}$, where morphisms in the base correspond to terms and morphisms in the fibers correspond to derivations. Moreover, the base $\mathcal{B}$ has finite products, the fibers $C[X]$ are bicartesian closed, the functors $C[f]$ preserve such structure and for every first projection $\pi_1: b_1 \times b \to b_1$ in $\mathcal{B}$ the functor $C[\pi_1]$ has right and left adjoints, corresponding to universal and existential quantification over b, satisfying the Beck-Chevalley condition.

- Locally cartesian closed categories model intuitionistic type theory with *equality types* (see [See84]). They amount to *identifying* the two levels of an hyperdoctrine, intuitively propositions and types are identified.

- Contextual categories, class of display maps, D-categories provide *equivalent* accounts of dependent types (see [Car78, Tay87, Erh88, HP87]). Unlike the approach based on locally cartesian closed categories, they give a general category-theoretic understanding of dependent types.

- PL Categories model the higher order lambda calculus, or equivalently the proof theory of higher order intuitionistic propositional calculus (see [See87]). They are hyperdoctrines with an object $\Omega \in \mathcal{B}$ (the type of propositions) s.t. the set of objects of $C[X]$ is $\mathcal{B}(X, \Omega)$, and for any $X \in \mathcal{B}$ a distinguished exponential Ω^X (the type of predicates over X).

- Monads can be used to model *notions of computation* (see [Mog89]).

Other features of programming languages (beside the distinction between compile-time and run-time) can be modelled as follows.

- A distinguished object Ω in the base category corresponds to the *kind* of all types, while exponentials Ω^X allow the interpretation of higher order type-constructors.

- Computations at *run-time* are modelled by a monad in the 2-category $\mathbf{ICat}(\mathcal{B})$. Since monads are a 2-category concept, it is clear how to define monads over indexed categories.

- Data-types (like products, sums, functional types, ...) are modelled by the usual adjunctions but in the 2-category **ICat**(B) instead of **Cat**.

  This is not quite right, because function spaces (and dependent products) are given via an adjunction with parameter. At the best of the author's knowledge adjunctions with parameter cannot be reformulated 2-categorically, although the definitions of all data-types mentioned above can be given (by hand) for indexed category.

- Polymorphic types are modelled like universal quantifiers (in Hyperdoctrines), while abstract data-types are modelled like existential quantifiers (see [MP88]).

Remark 3.1

- The requirement $\mathrm{Obj}(\mathcal{C}[X]) = \mathcal{B}(X, \Omega)$ for the kind Ω of all types is not always justified in relation to programming languages. For instance, in ML there are type and type schema. Types correspond to elements of Ω, while type schemas correspond to objects in the fiber categories. In Section 5 we introduce a language which does not identify types and type schemas. The inclusion of types into type schemas is modelled by an object $t \in \mathcal{C}[\Omega]$ (the generic type), so that a type expression $f: X \to \Omega$ (with a free variable of kind X) corresponds to the type schema $\mathcal{C}[f](t)$ in $\mathcal{C}[X]$. When $\mathrm{Obj}(\mathcal{C}[X]) = \mathcal{B}(X, \Omega)$, the generic type is simply $\mathrm{id}_\Omega \in \mathcal{C}[\Omega]$.

- A general understanding of dependent types is essential for correctly explaining dependent types in the category of modules (defined below) in terms of dependent types in the base and fiber categories. For instance in ML dependent types appear in a rather restricted form, namely polymorphic and product types, but there are non-trivial dependencies at the level of ML-signatures.

  The semantics of dependent types is based on a special kind of indexed categories (fibrations), where it is possible to go back and forth from one level to the other (see the definition of D-category in [Erh88]). Such a possibility of moving back and forth contradicts the separation between compile-time and run-time, because run-time expressions should not be allowed at compile-time. The correct way of having dependent types in a programming language $\mathcal{C}$, without violating the separation between compile-time and run-time, is to have a structure of D-category over the base category and/or the fibers, rather than requiring $\mathcal{C}$ itself to be a D-category.

4 The category of modules

The 2-category **ICat**($\mathcal{B}$) is isomorphic to the 2-category of split $\mathcal{B}$-fibrations (see [Ben85]). Since $\mathcal{B}$-fibrations are functors with codomain $\mathcal{B}$ satisfying certain additional properties, the 2-category of $\mathcal{B}$-fibrations is a 2-subcategory of **Cat**$\downarrow\mathcal{B}$ and the 2-embedding $\mathcal{F}$, mapping a $\mathcal{B}$-indexed category $\mathcal{C}$ to the corresponding $\mathcal{B}$-fibration $\mathcal{FC}: \mathcal{GC} \to \mathcal{B}$, can be viewed as a 2-functor from **ICat**($\mathcal{B}$) to **Cat**$\downarrow\mathcal{B}$.

For our purposes we need only to define the 2-functor $\mathcal{G}$ from **ICat**($\mathcal{B}$) to **Cat**, mapping a programming language $\mathcal{C}$ to its category of modules $\mathcal{GC}$. In Section 5 we will define the category of modules for HML.

Definition 4.1 *The 2-functor $\mathcal{G}$ from* $\mathbf{ICat}(\mathcal{B})$ *to* $\mathbf{Cat}$ *is defined as follows:*

- *if C is an indexed category, then $\mathcal{G}C$ is the category s.t.*

$$\begin{array}{lcl} \langle X_1, c_1\rangle & & \textit{where } X_1 \in \mathcal{B} \textit{ and } c_1 \in C[X_1] \\ \langle f, g\rangle \Big\downarrow & & \textit{where } f \in \mathcal{B}(X_1, X_2) \textit{ and } g \in C[X_1](c_1, C[f]c_2) \\ \langle X_2, c_2\rangle & & \textit{where } X_2 \in \mathcal{B} \textit{ and } c_2 \in C[X_2] \end{array}$$

  identity over $\langle X, c\rangle$ is the pair $\langle \mathrm{id}_X, \mathrm{id}_c\rangle$ and composition of $\langle f_1, g_1\rangle$ and $\langle f_2, g_2\rangle$ is the pair $\langle f_1; f_2, g_1; C[f_1]g_2\rangle$

- *if F is an indexed functor from C_1 to C_2, then $\mathcal{G}F$ is the functor from $\mathcal{G}C_1$ to $\mathcal{G}C_2$ s.t.*

$$\begin{array}{ccccc} \langle X, c\rangle & & & \langle X, F[X]c\rangle & \\ \langle f, g\rangle \Big\downarrow & \textit{in } \mathcal{G}C_1 & \overset{\mathcal{G}F}{\longmapsto} & \langle f, F[X]g\rangle \Big\downarrow & \textit{in } \mathcal{G}C_2 \\ \langle X', c'\rangle & & & \langle X', F[X']c'\rangle & \end{array}$$

- *if F_1 and F_2 are indexed functors from C_1 to C_2 and τ is an indexed natural transformation from F_1 to F_2, then $\mathcal{G}\tau$ is the natural transformation from $\mathcal{G}F_1$ to $\mathcal{G}F_2$ s.t.*

$$\langle X, c\rangle \quad \textit{in } \mathcal{G}C_1 \quad \overset{\mathcal{G}\tau}{\longmapsto} \quad \langle X, F_1[X]c\rangle \xrightarrow{\langle \mathrm{id}_X, \tau[X]_c\rangle} \langle X, F_2[X]c\rangle \quad \textit{in } \mathcal{G}C_2$$

Remark 4.2 After defining the general construction which maps a programming language, viewed as an indexed category C, to its category $\mathcal{G}C$ of modules, we can investigate how additional structure on $\mathcal{G}C$ depends on (is induced by) additional structure on C and/or the base category $\mathcal{B}$. For instance, an indexed monad (T, η, μ) over C, which corresponds to a notion of run-time computation, induces a monad over $\mathcal{G}C$ by simply taking its image w.r.t. the 2-functor $\mathcal{G}$, more precisely $(\mathcal{G}T)(\langle b, c\rangle) = \langle b, T[b]c\rangle$. Another interesting structure, which deserves further investigation, is type dependency in the category of modules. However, its treatment is too complex to be included here.

5 A toy example: HML

In this section we define a calculus, HML (for higher-order ML), similar to higher-order lambda calculus $F\omega$ (see [BMM88]). We could have chosen $F\omega$, but HML is the *minimal extension* of simply typed lambda calculus with *type variables* whose category of modules is cartesian closed.

In HML the kind of all types, Ω, is closed only w.r.t. $\rightarrow$, like in simply typed lambda calculus, but there is a superset of Ω, the set of type schemas (which, however, is not a kind), closed w.r.t. $\rightarrow$ and quantification over any kind. The distinction between types and type schemas is typical of ML (where type schemas are closed w.r.t. quantification over the kind Ω of all types, but are not closed w.r.t. $\rightarrow$ like in HML) and it is easily captured by the indexed category structure. In fact, types correspond to elements of an object Ω in the base category, while type schemas correspond to objects in the fiber categories. We require kinds and type schemas to be closed w.r.t. products only for simplifying the category-theoretic view of the language. HML is given by a set of rules for deriving well-formation and equality judgements.

- Kinds, k, are defined by the following BNF

$$\mathrm{k} \in \mathrm{kind} \quad ::= \quad \Omega \mid 1 \mid \mathrm{k}_1 \times \mathrm{k}_2 \mid \mathrm{k}_1 \rightarrow \mathrm{k}_2$$

  We write Δ for a constructor context, i.e. a sequence $v_1{:}\,\mathrm{k}_1, \ldots, v_m{:}\,\mathrm{k}_m$ where the variables v_i are all different.

  Equality of kinds, $\mathrm{k}_1 = \mathrm{k}_2$, is just syntactic equality.

- Constructors, $\Delta \vdash u{:}\,\mathrm{k}$, are defined by the rules in Table 1. In other words u is a λ-term with pairing, projections and constants 1, $\times$ and $\rightarrow$; as common practice, we write $u_1 \rightarrow u_2$ instead of $\rightarrow u_1 u_2$ (and similarly for $\times$).

  Constructor equality, $\Delta \vdash u_1 = u_2{:}\,\mathrm{k}$, is the congruence relation generated by the equations in Table 1, i.e. $\beta\eta$-conversion.

- Type schemas, $\Delta \vdash \sigma$, are defined by the rules in Table 1, i.e. they are the superset of types generated by binary products, functions spaces and quantifications over any kind.

  We write $\Delta; \Gamma$ for a term context, where Δ is a constructor context and Γ is a sequence $x_1{:}\,\sigma_1, \ldots, x_n{:}\,\sigma_n$ s.t. $\Delta \vdash \sigma_j$ are type schemas and the variables x_i are all different (and different from those declared in Δ).

  Type schema equality, $\Delta \vdash \sigma_1 = \sigma_2$, is the congruence relation generated by constructor equality.

- Terms, $\Delta; \Gamma \vdash e{:}\,\sigma$, are defined by the rules in Table 2. In other words e is a λ-term with pairing, projections and polymorphic abstraction and application.

  Term equality, $\Delta; \Gamma \vdash e_1 = e_2{:}\,\sigma$, is the congruence relation generated by constructor equality and the equations in Table 2, i.e. $\beta\eta$-conversion of polymorphic λ-terms.

The calculus HML can be viewed as an indexed category, $\mathcal{C}{:}\,\mathcal{B}^{op} \rightarrow \mathbf{Cat}$, according to a standard term-model construction (see [See87]), where constructors (up to constructor equality) are morphisms in the base category and terms are morphisms in the fiber categories. More precisely:

- The base category $\mathcal{B}$ is defined as follows:

  - objects are kinds k
  - morphisms from k_1 to k_2 are equivalence classes $[v{:}\,\mathrm{k}_1 \vdash u{:}\,\mathrm{k}_2]$ of constructors, i.e.

$$\{v{:}\,\mathrm{k}_1 \vdash u'{:}\,\mathrm{k}_2 | v{:}\,\mathrm{k}_1 \vdash u = u'{:}\,\mathrm{k}_2\}$$

  - $[v{:}\,\mathrm{k}_1 \vdash u_1{:}\,\mathrm{k}_2]$ followed by $[v{:}\,\mathrm{k}_2 \vdash u_2{:}\,\mathrm{k}_3]$ is substitution of v by u_1 in u_2, i.e.

$$[v{:}\,\mathrm{k}_1 \vdash u_2[v{:}= u_1]{:}\,\mathrm{k}_3]$$

- If k is an object of $\mathcal{B}$, then the category $\mathcal{C}[\mathrm{k}]$ is defined as follows:
  - objects are equivalence classes $[v{:}\,\mathrm{k} \vdash \sigma]$ of type schemas $v{:}\,\mathrm{k} \vdash \sigma$
  - morphisms from $[v{:}\,\mathrm{k} \vdash \sigma_1]$ to $[v{:}\,\mathrm{k} \vdash \sigma_2]$ are equivalence classes $[v{:}\,\mathrm{k}; x{:}\,\sigma_1 \vdash e{:}\,\sigma_2]$ of terms, i.e.

$$\{v{:}\,\mathrm{k}; x{:}\,\sigma_1' \vdash e'{:}\,\sigma_2' | v{:}\,\mathrm{k} \vdash \sigma_i = \sigma_i' \text{ and } v{:}\,\mathrm{k}; x{:}\,\sigma_1 \vdash e = e'{:}\,\sigma_2\}$$

  - $[v{:}\,\mathrm{k}; x{:}\,\sigma_1 \vdash e_1{:}\,\sigma_2]$ followed by $[v{:}\,\mathrm{k}; x{:}\,\sigma_2 \vdash e_2{:}\,\sigma_3]$ is substitution of x by e_1 in e_2, i.e.

$$[v{:}\,\mathrm{k}; x{:}\,\sigma_1 \vdash e_2[x{:}= e_1]{:}\,\sigma_3]$$

- If f is a morphism from k_1 to k_2 in $\mathcal{B}$, say $[v{:}\,\mathrm{k}_1 \vdash u{:}\,\mathrm{k}_2]$, then $\mathcal{C}[f]$ is the functor from $\mathcal{C}[\mathrm{k}_2]$ to $\mathcal{C}[\mathrm{k}_1]$ which substitutes v by u, i.e.
  - $[v{:}\,\mathrm{k}_2 \vdash \sigma]$ is mapped to $[v{:}\,\mathrm{k}_1 \vdash \sigma[v{:}= u]]$ and
  - $[v{:}\,\mathrm{k}_2; x{:}\,\sigma_1 \vdash e{:}\,\sigma_2]$ is mapped to $[v{:}\,\mathrm{k}_1; x{:}\,\sigma_1[v{:}= u] \vdash e[v{:}= u]{:}\,\sigma_2[v{:}= u]]$

Proposition 5.1 *The indexed category $\mathcal{C}{:}\,\mathcal{B}^{op} \to$ **Cat** corresponding to HML satisfies most of the properties of a PL category (see [See87]).*

- *The category $\mathcal{B}$ is cartesian closed (because kinds are closed w.r.t. products and function spaces).*
- *For every object k in $\mathcal{B}$ the category $\mathcal{C}[\mathrm{k}]$ is cartesian closed (because type schemas are closed w.r.t. products and function spaces) and for each morphism f in $\mathcal{B}$ the functor $\mathcal{C}[f]$ preserves the cartesian closed structure, i.e. the indexed category $\mathcal{C}$ is cartesian closed.*
- *For every first projection $\pi_1{:}\,\mathrm{k}_1 \times \mathrm{k} \to \mathrm{k}_1$ in $\mathcal{B}$ the functor $\mathcal{C}[\pi_1]$ has a right adjoints satisfying the Beck-Chevalley condition (because type schemas are closed w.r.t. quantification over any kind k).*

The category $\mathcal{GC}$ of modules for HML is defined as follows:

- objects are pairs $\langle \mathrm{k}, [v{:}\,\mathrm{k} \vdash \sigma]\rangle$, denoted by $[\Sigma v{:}\,\mathrm{k}.\sigma]$, where k is a kind and $v{:}\,\mathrm{k} \vdash \sigma$ is a type schema
- morphisms from $[\Sigma v{:}\,\mathrm{k}_1.\sigma_1]$ to $[\Sigma v{:}\,\mathrm{k}_2.\sigma_2]$ are pairs

$$\langle [v{:}\,\mathrm{k}_1 \vdash u{:}\,\mathrm{k}_2], [v{:}\,\mathrm{k}_1; x{:}\,\sigma_1 \vdash e{:}\,\sigma_2[v{:}= u]]\rangle$$

  denoted by $[v{:}\,\mathrm{k}_1; x{:}\,\sigma_1 \vdash \langle u, e\rangle{:}\,\Sigma v{:}\,\mathrm{k}_2.\sigma_2]$, where $v{:}\,\mathrm{k}_1 \vdash u{:}\,\mathrm{k}_2$ is a constructor and $v{:}\,\mathrm{k}_1; x{:}\,\sigma_1 \vdash e{:}\,\sigma_2[v{:}= u]$ is a term
- $[v{:}\,\mathrm{k}_1; x{:}\,\sigma_1 \vdash \langle u_1, e_1\rangle{:}\,\Sigma v{:}\,\mathrm{k}_2.\sigma_2]$ followed by $[v{:}\,\mathrm{k}_2; x{:}\,\sigma_2 \vdash \langle u_2, e_2\rangle{:}\,\Sigma v{:}\,\mathrm{k}_3.\sigma_3]$ is substitution of v by u_1 and of x by e_1, i.e. it is the pair

$$[v{:}\,\mathrm{k}_1; x{:}\,\sigma_1 \vdash \langle u_2[v{:}= u_1], e_2[v{:}= u_1, x{:}= e_1]\rangle{:}\,\Sigma v{:}\,\mathrm{k}_3.\sigma_3]$$

Remark 5.2 The explanation of ML-modules in terms of the category of modules for HML goes as follows:

- an ML-signature corresponds to an object in the category of modules, for instance

$$\textbf{signature}\ sig = (\textbf{sig}\ \text{type}\ v,\ \text{val}\ x{:}\,\sigma\ \textbf{end})$$

  corresponds to the object $[\Sigma v{:}\,\Omega.\sigma]$

- an ML-structure of signature sig corresponds to an element of $[\Sigma v{:}\,\Omega.\sigma]$, for instance

$$\textbf{structure}\ S = (\textbf{struct}\ \text{type}\ v = u,\ \text{val}\ x{:}\,\sigma = e\ \textbf{end})$$

  corresponds to the morphism $[v{:}\,1; x{:}\,1 \vdash \langle u, e\rangle{:}\,\Sigma v{:}\,\Omega.\sigma]$ from the terminal object $[\Sigma v{:}\,1.1]$ to $[\Sigma v{:}\,\Omega.\sigma]$

- an ML-functor corresponds to a morphism in the category of modules, for instance

$$\begin{array}{l}\textbf{functor}\ F(S{:}\,sig_1){:}\,sig_2 = \\ \quad\textbf{struct} \\ \qquad\text{type}\ v = u[v{:=}\,S.v], \\ \qquad\text{val}\ x{:}\,\sigma_2 = e[v{:=}\,S.v, x{:=}\,S.x] \\ \quad\textbf{end}\end{array}$$

  corresponds to the morphism $[v{:}\,\Omega; x{:}\,\sigma_1 \vdash \langle u, e\rangle{:}\,\Sigma v{:}\,\Omega.\sigma_2]$ from $[\Sigma v{:}\,\Omega.\sigma_1]$ to $[\Sigma v{:}\,\Omega.\sigma_2]$.

While the category-theoretic explanation of ML-signatures and ML-structures matches the type-theoretic intuition, there is a major difference in the understanding of ML-functors.

> For us it is essential that ML-types (including those in the body of an ML-functor) do not depend on values, otherwise it would not be possible to associate a morphism in $\mathcal{GC}$ to an ML-functor. On the other hand, such property is ignored in the type-theoretic account given in [Mac86, HM88].

We claim that HML has *higher order modules*.

Proposition 5.3 *The category $\mathcal{GC}$ is cartesian closed, more precisely:*

- *the terminal object* $[\Sigma v{:}\,1.1]$
- *the product* $c_1 \times c_2$ *is* $[\Sigma v{:}\,\mathrm{k}_1 \times k_2.(\sigma_1[v_1{:=}\,\pi_1(v)] \times \sigma_2[v_2{:=}\,\pi_2(v)])]$
- *the exponential* ${c_2}^{c_1}$ *is* $[\Sigma v{:}\,\mathrm{k}_1 \to k_2.(\forall v_1{:}\,\mathrm{k}_1.\sigma_1 \to \sigma_2[v_2{:=}\,vv_1])]$

where c_i *is* $[\Sigma v_i{:}\,\mathrm{k}_i.\sigma_i]$.

6 Conclusions

We have proposed an abstract view of programming languages as objects of a 2-category and modelled the compile-time/run-time distinction by an indexed category. We believe that such an approach is potentially useful, and in Section 3 we have indicated possible applications. However, the 2-categorical framework seems not to be expressive enough to reformulate important concepts like adjunctions with parameter (used to define function spaces). Finding a better framework is a challenge for the pure category-theorist.

Indexed categories should also capture the idea of evaluating constant value expressions at compile-time. However, the presence of dependent types (necessary to deal with ML-sharing constraints) may require a partial evaluation of run-time expressions for the purpose of type-checking. We expect that a $\mathcal{B}$-indexed functor should capture the idea of partial evaluation of run-time expressions, namely the domain of the functor would correspond to the view of the language at compile-time (where equality of programs is rather weak, but decidable), the codomain would correspond to the view of the language at run-time (here equality of programs may not be decidable), while the functor itself relates these two views.

The 2-categorical framework applies to theories as well, by replacing Lawvere's view of theories as categories with theories as objects of a 2-category. So it seems that a modular approach to programming languages can go hand in hand with a modular approach to logics (after all most of the ideas in Section 3 for modelling programming language features are borrowed from categorical logic).

Acknowledgements

All my thanks to Dave MacQueen, who introduced me to ML-modules, AT&T, for financing my visit at Bell Labs in 1986, and Bob Harper for many discussions on ML. Thanks also to Luca Cardelli, Joseph Goguen, Tony Hoare and John Mitchell for discussions and to Michael Mendler, Douglas Gurr and James McKinna for comments on a previous draft.

References

[Ben85] J. Benabou. Fibred categories and the foundation of naive category theory. *Journal of Symbolic Logic*, 50, 1985.

[BMM88] K. Bruce, A. Meyer, and J. Mitchell. The semantics of second order polymorphic lambda calculus. *Information and Computation*, 73(2/3), 1988.

[Car78] J. Cartmell. *Generalized Algebraic Theories and Contextual Categories.* PhD thesis, University of Oxford, 1978.

[Car88] L. Cardelli. Phase distinction in type theory. Draft 4/1/88, DEC SRC, 1988.

[Erh88] T. Erhard. A categorical semantics of constructions. In *3rd LICS Conf.* IEEE, 1988.

[HM88] R. Harper and J. Mitchell. The essence of ML. In *15th POPL.* ACM, 1988.

[HMM86] R. Harper, D. MacQueen, and R. Milner. Standard ML. Technical Report ECS-LFCS-86-2, Edinburgh Univ., Dept. of Comp. Sci., 1986.

[HMT87] R. Harper, R. Milner, and M. Tofte. The semantics standard ML. Technical Report ECS-LFCS-87-36, Edinburgh Univ., Dept. of Comp. Sci., 1987.

[HP87] J.M.E. Hyland and A.M. Pitts. The theory of constructions: Categorical semantics and topos-theoretic models. In *Proc. AMS Conf. on Categories in Comp. Sci. and Logic (Boulder 1987)*, 1987.

[Kel82] G.M. Kelly. *Basic Concepts of Enriched Category Theory.* Cambridge University Press, 1982.

[Mac71] S. MacLane. *Categories for the Working Mathematician.* Springer Verlag, 1971.

[Mac85] D. MacQueen. Modules for standard ML. *Polymorphism*, 2, 1985.

[Mac86] D. MacQueen. Using dependent types to express modular structures. In *13th POPL.* ACM, 1986.

[Mog86] E. Moggi. The cartesian closed category of signatures and functors and its description in Martin-Löf type theory. Manuscript, Bell Laboratories, 1986.

[Mog89] E. Moggi. Computational lambda-calculus and monads. In *4th LICS Conf.* IEEE, 1989.

[MP88] J.C. Mitchell and G.D. Plotkin. Abstract types have existential type. *ACM Trans. on Progr. Lang. and Sys.*, 10(3), 1988.

[PS78] R. Pare and D. Schumacher. Abstract families and the adjoint functor theorems. In P.T. Johnstone and R. Pare, editors, *Indexed Categories and their Applications*, volume 661 of *Lecture Notes in Mathematics.* Springer Verlag, 1978.

[See83] R.A.G. Seely. Hyperdoctrines, natural deduction and the beck condition. *Zeitschr. f. math. Logik und Grundlagen d. Math.*, 29, 1983.

[See84] R.A.G. Seely. Locally cartesian closed categories and type theory. *Math. Proc. Camb. Phil. Soc.*, 95, 1984.

[See87] R.A.G. Seely. Categorical semantics for higher order polymorphic lambda calculus. *Journal of Symbolic Logic*, 52(2), 1987.

[Str72] R. Street. The formal theory of monads. *Journal of Pure and Applied Algebra*, 2, 1972.

[Tay87] P. Taylor. *Recursive Domains, Indexed Category Theory and Polymorphism.* PhD thesis, University of Cambridge, 1987.

$$v \quad \Delta \vdash v{:}\,\mathrm{k} \qquad v{:}\,\mathrm{k} \text{ in } \Delta$$

$$\text{unit} \quad \Delta \vdash 1{:}\,\Omega$$

$$\text{prod} \quad \Delta \vdash \times{:}\,\Omega \to \Omega \to \Omega$$

$$\text{fun} \quad \Delta \vdash \to{:}\,\Omega \to \Omega \to \Omega$$

$$1\mathrm{I} \quad \Delta \vdash *{:}\,1$$

$$\times\mathrm{I} \quad \frac{\Delta \vdash u_1{:}\,\mathrm{k}_1 \qquad \Delta \vdash u_2{:}\,\mathrm{k}_2}{\Delta \vdash \langle u_1, u_2\rangle{:}\,\mathrm{k}_1 \times \mathrm{k}_2}$$

$$\times\mathrm{E} \quad \frac{\Delta \vdash u{:}\,\mathrm{k}_1 \times \mathrm{k}_2}{\Delta \vdash \pi_i(u){:}\,\mathrm{k}_i}$$

$$\to\mathrm{I} \quad \frac{\Delta, v{:}\,\mathrm{k}_1 \vdash u{:}\,\mathrm{k}_2}{\Delta \vdash (\lambda v{:}\,\mathrm{k}_1.u){:}\,\mathrm{k}_1 \to \mathrm{k}_2}$$

$$\to\mathrm{E} \quad \frac{\Delta \vdash u{:}\,\mathrm{k}_1 \to \mathrm{k}_2 \qquad \Delta \vdash u_1{:}\,\mathrm{k}_1}{\Delta \vdash u(u_1){:}\,\mathrm{k}_2}$$

$$1.\eta \quad \frac{\Delta \vdash u{:}\,1}{\Delta \vdash * = u{:}\,1}$$

$$\times.\beta \quad \frac{\Delta \vdash u_1{:}\,\mathrm{k}_1 \qquad \Delta \vdash u_2{:}\,\mathrm{k}_2}{\Delta \vdash \pi_i(\langle u_1, u_2\rangle) = u_i{:}\,\mathrm{k}_i}$$

$$\times.\eta \quad \frac{\Delta \vdash u{:}\,\mathrm{k}_1 \times \mathrm{k}_2}{\Delta \vdash \langle \pi_1(u), \pi_2(u)\rangle = u{:}\,\mathrm{k}_1 \times \mathrm{k}_2}$$

$$\to.\beta \quad \frac{\Delta, v{:}\,\mathrm{k}_1 \vdash u_2{:}\,\mathrm{k}_2 \qquad \Delta \vdash u_1{:}\,\mathrm{k}_1}{\Delta \vdash (\lambda v{:}\,\mathrm{k}_1.u_2)(u_1) = u_2[v{:}= u_1]{:}\,\mathrm{k}_2}$$

$$\to.\eta \quad \frac{\Delta \vdash u{:}\,\mathrm{k}_1 \to \mathrm{k}_2}{\Delta \vdash (\lambda v{:}\,\mathrm{k}_1.u(v)) = u{:}\,\mathrm{k}_1 \to \mathrm{k}_2}$$

$$\text{type} \quad \frac{\Delta \vdash \tau{:}\,\Omega}{\Delta \vdash \tau}$$

$$\times \quad \frac{\Delta \vdash \sigma_1 \qquad \Delta \vdash \sigma_2}{\Delta \vdash \sigma_1 \times \sigma_2}$$

$$\to \quad \frac{\Delta \vdash \sigma_1 \qquad \Delta \vdash \sigma_2}{\Delta \vdash \sigma_1 \to \sigma_2}$$

$$\forall \quad \frac{\Delta, v{:}\,\mathrm{k} \vdash \sigma}{\Delta \vdash \forall v{:}\,\mathrm{k}.\sigma}$$

Table 1: Constructors, constructor equality and type schemas

$$x \quad \Delta;\Gamma \vdash x{:}\,\sigma \qquad x{:}\,\sigma \text{ in } \Gamma$$

$$1\mathrm{I} \quad \Delta;\Gamma \vdash *{:}\,1$$

$$\times\mathrm{I} \quad \frac{\Delta;\Gamma \vdash e_1{:}\,\sigma_1 \qquad \Delta;\Gamma \vdash e_2{:}\,\sigma_2}{\Delta;\Gamma \vdash \langle e_1, e_2\rangle{:}\,\sigma_1 \times \sigma_2}$$

$$\times\mathrm{E} \quad \frac{\Delta;\Gamma \vdash e{:}\,\sigma_1 \times \sigma_2}{\Delta;\Gamma \vdash \pi_i(e){:}\,\sigma_i}$$

$$\to\mathrm{I} \quad \frac{\Delta;\Gamma, x{:}\,\sigma_1 \vdash e{:}\,\sigma}{\Delta;\Gamma \vdash (\lambda x{:}\,\sigma.e){:}\,\sigma_1 \to \sigma_2}$$

$$\to\mathrm{E} \quad \frac{\Delta;\Gamma \vdash e{:}\,\sigma_1 \to \sigma_2 \qquad \Delta;\Gamma \vdash e_1{:}\,\sigma_1}{\Delta;\Gamma \vdash e(e_1){:}\,\sigma_2}$$

$$\forall\mathrm{I} \quad \frac{\Delta, v{:}\,\mathrm{k};\Gamma \vdash e{:}\,\sigma}{\Delta;\Gamma \vdash (\Lambda v{:}\,\mathrm{k}.e){:}\,\forall v{:}\,\mathrm{k}.\sigma} \; v \notin \mathrm{FV}(\Gamma)$$

$$\forall\mathrm{E} \quad \frac{\Delta;\Gamma \vdash e{:}\,\forall v{:}\,\mathrm{k}.\sigma \qquad \Delta \vdash u{:}\,\mathrm{k}}{\Delta;\Gamma \vdash e(u){:}\,\sigma[v{:}= u]}$$

$${:}\text{-eq} \quad \frac{\Delta;\Gamma \vdash e{:}\,\sigma_1 \qquad \Delta \vdash \sigma_1 = \sigma_2}{\Delta;\Gamma \vdash e{:}\,\sigma_2}$$

$$1.\eta \quad \frac{\Delta;\Gamma \vdash e{:}\,1}{\Delta;\Gamma \vdash * = e{:}\,1}$$

$$\times.\beta \quad \frac{\Delta;\Gamma \vdash e_1{:}\,\sigma_1 \qquad \Delta;\Gamma \vdash e_2{:}\,\sigma_2}{\Delta;\Gamma \vdash \pi_i(\langle e_1, e_2\rangle) = e_i{:}\,\sigma_i}$$

$$\times.\eta \quad \frac{\Delta;\Gamma \vdash e{:}\,\sigma_1 \times \sigma_2}{\Delta;\Gamma \vdash \langle \pi_1(e), \pi_2(e)\rangle = e{:}\,\sigma_1 \times \sigma_2}$$

$$\to.\beta \quad \frac{\Delta;\Gamma, x{:}\,\sigma_1 \vdash e_2{:}\,\sigma_2 \qquad \Delta;\Gamma \vdash e_1{:}\,\sigma_1}{\Delta;\Gamma \vdash (\lambda x{:}\,\sigma_1.e_2)(e_1) = e_2[x{:}= e_1]{:}\,\sigma_2}$$

$$\to.\eta \quad \frac{\Delta;\Gamma \vdash e{:}\,\sigma_1 \to \sigma_2}{\Delta;\Gamma \vdash (\lambda x{:}\,\sigma_1.e(x)) = e{:}\,\sigma_1 \to \sigma_2}$$

$$\forall.\beta \quad \frac{\Delta, v{:}\,\mathrm{k};\Gamma \vdash e{:}\,\sigma \qquad \Delta \vdash u{:}\,\mathrm{k}}{\Delta;\Gamma \vdash (\lambda v{:}\,\mathrm{k}.e)(u) = e[v{:}= u]{:}\,\sigma[v{:}= u]}$$

$$\forall.\eta \quad \frac{\Delta;\Gamma \vdash e{:}\,\forall v{:}\,\mathrm{k}.\sigma}{\Delta;\Gamma \vdash (\Lambda v{:}\,\mathrm{k}.e(v)) = e{:}\,\forall v{:}\,\mathrm{k}.\sigma}$$

$$=\text{-eq} \quad \frac{\Delta;\Gamma \vdash e_1 = e_2{:}\,\sigma_1 \qquad \Delta \vdash \sigma_1 = \sigma_2}{\Delta;\Gamma \vdash e_1 = e_2{:}\,\sigma_2}$$

Table 2: Terms and term equality

A NOTE ON CATEGORICAL DATATYPES

G.C.Wraith
Department of Mathematics, University of Sussex

Abstract

It is shown how Hagino's categorical datatypes can be expressed in the polymorphic typed λ-calculus. This gives a way of passing from a description of a datatype in terms of its universal properties, to a representation in terms of λ-expressions.

A Summary of Hagino's categorical datatypes

Hagino's ideas for a more categorical programming language involve two sorts of type declarations:

```
product X(t) =              sum X(t) =
    d_1 : X → e_1,              c_1 : e_1 → X,
    ..............              ..............
    d_n : X → e_n               c_n : e_n → X
end                         end
```

The notation is mine, not Hagino's. The symbol **X** stands for a type constructor identifier, and **t** for a possibly empty list of covariant type parameters. The body of the declaration is a possibly empty list of destructor/constructor declarations. These play a special role in the definition of **X**, and correspond to the canonical projections/injections of a product/coproduct. The e_k are type expressions in which the definiend **X** and the members of **t** may occur covariantly. The type declaration would have the following effects:

1) **X** is added to the type constructor environment.
2) The destructors/constructors are added to the value environment.
3) A value, also denoted by **X** of type

$$(t \to t') \to X(t) \to X(t')$$

expressing the functoriality (or strength) of **X** is added to the value environment.

If the expressions e_k do not contain **X**, we simply have products (i.e. records) and sums (i.e. unions). We must interpret recursive products and sums using the notions of terminal coalgebra and initial algebra of an endofunctor **A**. In the product case **A(X)** is given by the product of the e_k's and in the sum case by the sum of the e_k's.

Recall that an **A**-coalgebra is a pair (Y, η) where

$$\eta : Y \to A(Y)$$

and that a map $f: Y \to Y'$ defines a map of A-coalgebras from (Y,η) to (Y',η') if $f;\eta' = \eta; A(f)$. Here $f;g$ denotes the map f followed by the map g. Dually an A-algebra is a pair (X,s) where

$$s : A(X) \to X$$

and a map $f : X \to X'$ defines a map of A-algebras from (X,s) to (X',s') if $s; f = A(f); s'$.

We will denote a terminal A-coalgebra by

$$(\nu X.A(X), d)$$

and an initial A-algebra by

$$(\mu X.A(X), c)$$

when they exist (in which case they are unique up to isomorphism). The maps **d** and **c** must be isomorphisms. To see why, consider the A-algebra

$$(A(\mu X.A(X)), A(c))$$

Let $u : \mu X.A(X) \to A(\mu X.A(X))$ be the unique A-algebra map, so that

$$c; u = A(u); A(c).$$

Since we have an A-algebra map from $(A(\mu X.A(X)), A(c))$ to $(\mu X.A(X), c)$ by **c**, it follows from the uniqueness that

$$u; c = Id(\mu X.A(X))$$

whence

$$c; u = A(u); A(c) = A(Id(\mu X.A(X))) = Id(A(\mu X.A(X)))$$

so that **c** is an isomorphism with inverse **u**. An entirely dual argument shows that **d** is an isomorphism.

It may make these notions more familiar to think of μ and ν as least and greatest fixedpoint operators. However this would hide the fact that **c** and **d** are isomorphisms not equalities. It is unfortunate that in the past types have been treated as elements of an algebra (leading to the idea of recursive equations for types) rather than as objects of a category, for which the notion of equality is meaningless. We can only compare types via values (i.e. maps) which mediate between them. It is the fact that $\mu X.A(X)$ is the underlying object of an initial A-algebra which is the important aspect. That it should be a fixedpoint (up to isomorphism) is in some sense an accident. Besides, thinking in terms of equations obscures the categorical duality between μ and ν. We can cope with mutually recursive product/sum type declarations by using parameters in the usual way.

To summarize, Hagino's categorical programming language is based on categories with:

1) finite products - terminal object **1**
   binary products $\times$
2) finite coproducts - initial object $\emptyset$
   binary sums **+**
3) exponentials $\rightarrow$
4) terminal coalgebras ν
5) initial algebras μ

So for example **1** is given by

```
product unit = end
```

and **list(α)** $= \mu X.(1+\alpha\times X)$, the type of finite lists, by

```
sum list(α) =
    nil : unit → list,
    cons : α×list → list
end
```

and **stream(α)** $= \nu X.\alpha\times X$, the type of infinite lists, by

```
product stream(α) =
    head : stream → α,
    tail : stream → stream
end
```

Similarly we have **game** $= \mu X.$**list(X)** given by

```
sum game = moves : list(game) → game end
```

and **cascade(α)** $= \nu X.\alpha\times$**list(X)**, the type of finitely branching, possibly infinite α-labelled trees, given by

```
product cascade(α) =
    label : cascade → α,
    shower : cascade → list(cascade)
end
```

We could describe this system as a polymorphically typed combinatory algebra, but we forebear to go into the details, which can be had from [Hagino 1]. In this reference, Hagino shows that his calculus, which I will refer to as Cλ, is strongly normalizing.

Note that we have made a sharp distinction between lists (they must be finite) and streams (they must be infinite). They are quite different sorts of type constructor. Because we have strong normalization there is no possibility of fudging the distinction with normal order evaluation. Indeed we have decoupled types from evaluation order. Of course, the **print** function cannot be represented in Cλ. Nevertheless, we still have a wealth of functions for manipulating infinite data structures.

We shall show that Cλ can be interpreted as a fragment of 2Tλ, the polymorphic typed λ-calculus. This gives another way of proving strong normalization, and explains why the various "strange" constructions of standard types in 2Tλ, which have been discovered over the years by various authors, really work. It also explains why recursive type definitions and fixedpoint operators for values are redundant for practical purposes in 2Tλ. This in turn answers a question posed in [Reynolds 2, page 113] about principal type schemes and type inference for "infinite recursively defined" types. Essentially, the 2Tλ expressions which encode functions for handling infinite data structures in Cλ use a continuation style of representation; the continuation types arise from the use of quantified type variables in 2Tλ.

2Tλ The polymorphic typed λ-calculus

We will use capital letters for types, lowercase for values. We suppose that we start with a given set of type variables, and a given set of value variables.

Types are formed as follows:

1) A type variable **X** is a type.
2) If **S** and **T** are types so is **S→T**.
3) If **X** is a type variable and **T** a type, **ΠX.T** is a type.

We treat α-convertible expressions, that is those that differ by a consistent change of bound variables, as denoting the same thing. When a type expression **A** is thought of as parametrized by a type variable **X**, we shall write **A(X)** to emphasize its dependence on **X**, and we shall write **A(T)** for the expression obtained by substituting the type expression **T** for **X** in **A**, with appropriate change of bound variables to avoid capture. A type with no free variables is called a closed type.

Values are defined as follows:

1) A value variable is a value.
2) If **f** and **g** are values, the 'application' **f g** is a value.
3) If **f** is a value and **T** is a type then the 'instance' **f T** is a value.
4) If **f** is a value and **x** is a value variable and **T** is a type then the 'value abstraction' **λx∈T.f** is a value.
5) If **f** is a value and **X** is a type variable, not occurring free in the type of a free variable of f, then the 'type abstraction' **ΛX.f** is a value.

A typing is a relation **f:T** between values **f** and types **T** satisfying the rules :

1) If $f : S \to T$ and $g : S$ then $(f\ g) : T$.
2) If $f : \Pi X.S(X)$ then $(f\ T) : S(T)$.
3) If $f : S$ then $\lambda x \in T.f : T \to S$.
4) If $f : T$ then $\Lambda X.f : \Pi X.T$.

We may define inductively the relations

$$\mathbf{co}(A,X) \qquad \mathbf{contra}(A,X)$$

between types **A** and type variables **X**, expressing the idea that **A(X)** depends covariantly or contravariantly on **X**.

1) $\mathbf{co}(X,X)$.
2) If **Y** is a variable distinct from **X** then $\mathbf{co}(Y,X)$ and $\mathbf{contra}(Y,X)$.
3) If $\mathbf{co}(S,X)$ and $\mathbf{contra}(T,X)$ then $\mathbf{co}(T \to S,X)$ and $\mathbf{contra}(S \to T,X)$.
4) If $\mathbf{co}(S,X)$ then $\mathbf{co}(\Pi Y.S,X)$.
5) If $\mathbf{contra}(S,X)$ then $\mathbf{contra}(\Pi Y.S,X)$.

If $\mathbf{co}(A(X),X)$ then we can define the strength of **A** as a value $\langle A \rangle$ with the closed type

$$\langle A \rangle : \Pi X.\Pi Y.(X \to Y) \to A(X) \to A(Y)$$

by induction. We omit the details. We abbreviate $\langle A \rangle\, X\, Y\, f$ to $A(f)$ to fit in with categorical practice, with apologies.

Similarly, if $\mathbf{contra}(A(X),X)$ we have

$$\langle A \rangle : \Pi X.\Pi Y.(X \to Y) \to A(Y) \to A(X)$$

giving the action of **A** as a contravariant functor.

Now suppose that **A(X)** is covariant in **X**. We make the definition:

$$\mu X.A(X) = \Pi X.(A(X) \to X) \to X$$

and we define:

$$\begin{aligned} \varphi &= \Lambda X.\lambda f \in A(X) \to X.\lambda w \in (\mu Y.A(Y)).w\ X\ f \\ &: \Pi X.(A(X) \to X) \to (\mu Y.A(Y)) \to X \end{aligned}$$

and

$$\begin{aligned} \mathbf{s} &= \lambda h \in A(\mu Y.A(Y)).\Lambda X.\lambda f \in (A(X) \to X).f\ (A(\varphi\ X\ f)\ \ h) \\ &: A(\mu X.A(X)) \to (\mu X.A(X)) \end{aligned}$$

The following proposition has been known for some time:

Proposition

$(\mu X.A(X), \iota)$ is a weakly initial A-algebra. That is, for any A-algebra (S, c) we have a value $f : \mu X.A(X) \to S$ such that $\iota ; f$ β-reduces to $A(f) ; c$.

Proof : Simply take f to be $\varphi\, S\, c : \mu X.A(X) \to S$.

What I believe is new, is that we have a dual result. We make the definitions :

$$\nu X.A(X) = \Pi Y.(\Pi X.(X \to A(X)) \to X \to Y) \to Y$$

where Y is not a free variable of $A(X)$.

$$\phi = \Lambda X.\lambda f \in X \to A(X).\lambda x \in X.\Lambda Y.\lambda h \in (\Pi Z.(Z \to A(Z)) \to Z \to Y).h\, X\, f\, x$$
$$: \Pi X.(X \to A(X)) \to X \to (\nu Y.A(Y))$$

$$\eta = \lambda w \in (\nu X.A(X)).w\, A(\nu X.A(X))\ \Lambda X.\lambda f \in X \to A(X).\lambda x \in X.A(\phi\, X\, f)\ (f\, x)$$
$$: (\nu X.A(X)) \to A(\nu X.A(X))$$

Proposition

$(\nu X.A(X), \eta)$ is a weakly terminal A-coalgebra. That is, for any A-coalgebra (S, d) there is a value $f : S \to \nu X.A(X)$ such that $f ; \eta$ β-reduces to $d ; A(f)$.

Proof : Just take $f = \phi\, S\, d : S \to \nu X.A(X)$.

Neglecting for the moment the question of whether we have actual rather than weak initiality and terminality, we can argue for the "poly" versions of the standard types as follows:

If $\emptyset$ and 1 denote empty and unit types respectively

$$\emptyset - \mu X.\emptyset = \Pi X.(\emptyset \to X) \to X - \Pi X.1 \to X - \Pi X.X$$

$$1 - \mu X.1 = \Pi X.(1 \to X) \to X - \Pi X.X \to X$$

If $\times$ and $+$ denote product and sum, then using exponential adjointness

$$X \to Y - \mu Z.X \to Y = \Pi Z.(X \to Y \to Z) \to Z - \Pi Z.(X \to Y \to Z) \to Z$$

$$X+Y - \mu Z.(X+Y) = \Pi Z.((X+Y) \to Z) \to Z$$
$$- \Pi Z.((X \to Z) \to (Y \to Z)) \to Z$$
$$- \Pi Z.(X \to Z) \to (Y \to Z) \to Z$$

Note that if we make the definition

$$\Sigma X.A(X) = \Pi Y.(\Pi X.A(X) \to Y) \to Y$$

where **Y** is not a free type variable of **A(X)**, then we could have

$$\nu X.A(X) = \Sigma X.X \to (X \to A(X))$$

which perhaps brings out the duality better. $(X \times -, X \to -)$ is an adjoint pair dual to the adjoint pair $(- \to X, - \to X)$.

Various "magic formulae" have long been known for representing certain datatypes and functions for manipulating them in terms of the λ-calculus. For example:

$$\mathbf{pair} = \lambda x.\lambda y.\lambda f.f\ x\ y$$

is a classic. What we have now is a piece of machinery for producing such representations from categorical specifications of the datatype in terms of its universal properties. First describe the datatype in Hagino's categorical language using **sum** and **product** declarations. Then translate into Cλ, then into 2Tλ, and finally throw away the type information. Of course, this gives in principle a way of compiling Hagino's language, once one has chosen a particular mechanism for evaluating λ-terms. Whether there exist more direct methods that bypass the translation into λ-terms would seem to be an interesting question (especially in the light of the Categorical Abstract Machine), suggesting that we should search for simple operational semantics of Cλ.

By way of example, here is the result of applying this translation scheme to get a representation of **hd** and **tl** for streams:

Define:

```
fst = λp.p (λx.λy.x)          snd = λp.p (λx.λy.y)
hd = λs.s (λf.λa.fst (f a))
tl = λs.s (λf.λa.λg.g f (snd (f a)))
```

These definitions are got by stripping types from:

```
fst = ΛX.ΛY.λp∈X×Y.p X (λx∈X.λy∈Y.x)
snd = ΛX.ΛY.λp∈X×Y.p Y (λx∈X.λy∈Y.y)
hd = ΛX.λs∈stream(X).s X (ΛY.λf∈Y→X×Y.λa∈Y.fst X Y (f a))
tl = ΛX.λs∈stream(X).s stream(X) (
        ΛY.λf∈Y→X×Y.λa∈Y.ΛZ.λg∈(ΠW.(W→X×Y)→W→Z).
          g Y f (snd X Y (f a))          )
```

which in turn are got from decoding **hd** and **tl** from the η expression for the case $A(Y) = X \times Y$ in $\mathbf{stream}(X) = \nu Y.X \times Y$. I hope this gives the general idea of the process.

It is the thesis of this note that by a categorical model of 2Tλ we should mean something in which the type constructors we have described

$$\emptyset \quad 1 \quad + \quad \times \quad \to \quad \mu \quad \nu$$

are interpreted as genuine limits and colimits, that is as initial object, terminal object, sum, product, exponential, initial algebra, terminal coalgebra, respectively. This means that as well as identifying values with their normal forms for β-reduction, we must also impose η-rules in the model. The η-rules are

$$\lambda x \in T.(f\ x) = f \qquad x \text{ not free in } f$$

$$\varphi\ \mu X.A(X)\ \varepsilon = \lambda x \in \mu X.A(X).x$$

$$\psi\ \nu X.A(X)\ \eta = \lambda x \in \nu X.A(X).x$$

using the notation above. For example, the η-rule for $\emptyset$ (take $A(X)=X$) asserts that the values

$$\lambda x \in (\Pi X.X).x \qquad\qquad \lambda x \in (\Pi X.X).x\ (\Pi X.X)$$

of type $(\Pi X.X) \to (\Pi X.X)$ are to be identified. It is not clear to me whether Coquand and Breazu-Tannen's minimal extensional model or Hyland's PER model satisfy these conditions. Extensionality (**1** is a generator) would seem an unnecessarily strong condition for a categorical model in general. P. Freyd, in e-mail notes on "Structors" has pointed out that we have more than simply weakly initial/terminal objects. We have objects that come equipped with canonical maps which are preserved by composition with other maps. This is a notion stronger than what has hitherto been called weak initiality/terminality.

We see that Cλ corresponds to a fragment of 2Tλ in which Π appears only either at the outermost level or concealed within $\times$, $+$, $\emptyset$, **1**, μ or ν. This fragment already contains the functorial strengths $\langle A \rangle$ without explicit recursion; instances of this phenomenon (e.g. **listrec**) have been noted before [Reynolds 2]. It is a consequence of the fact that universal constructions are automatically functorial. The implicit recursion in the functorial types gives rise to implicitly recursive values.

We have type inference for this fragment, because if we simply omit the Π quantifiers at the outermost level we are left with a free algebraic theory of types, generated by the binary operation $\to$ and the user named type constructors. The unification theorem asserts that in a free algebraic theory (in the sense of Lawvere, i.e. a category with finite products in which all the objects are finite powers of a basic object) any finite diagram having a cone has a limiting cone. A value expression in the fragment we are considering gives rise to such a finite diagram, with vertices representing type constraints between subexpressions, and arrows representing substitutions of expressions for variables. A typing corresponds to a cone on the diagram, and a principle type scheme to a limiting cone. This is really a consequence of "taming" quantification by adjoining constant type constructors to replace certain patterns of occurrence of Π anywhere but at the outermost level.

I would like to thank Andy Pitts for much useful advice at Sussex, and Anders Kock for the same at Aarhus. Martin Hyland convinced me at Trondheim of the usefulness of PER models.

References

I am grateful to the referee for pointing out references to the work of Mendler and Constable.

K.B.Bruce, A.R.Meyer (1984)
The semantics of second order polymorphic lambda-calculus.
SLNCS 173 pp 131-144

J.Bell (1988)
Toposes and Local set Theories.
Oxford University Press

T. Coquand, V.Breazu-Tannen (1988)
Extensional Models of Polymorphism.
T.C.S. 59 pp 85-114

Jon Fairbairn (1985)
Design and Implementation of a simple typed language based on the lambda-calculus.
University of Cambridge
Technical Report No. 75

T. Hagino (1987)
A Typed Lambda Calculus with Categorical Type Constructors.
Preprint.

T. Hagino (1987)
A Categorical Programming Language.
Thesis. Edinburgh University.

D.B.MacQueen, R.Sethi, G.Plotkin (1984)
An Ideal Model for Recursive Polymorphic Types.
11-th Annual ACM Symposium on the Principles of Programming Languages.

P. Mendler (1987)
Inductive Definitions in type Theory.
Thesis - Cornell University.

Mendler, Constable (1985)
Recursive Definitions in Type Theory.
LNCS 193 pp 61-78

J.C.Reynolds (1984)
Polymorphism is not Set-Theoretic.
SLNCS 173 pp 145-156

J.C.Reynolds (1985)
Three approaches to type structure.
TAPSOFT. SLNCS 185 pp 97-138

A.Pitts (1987)
Polymorphism is Set-Theoretic Constructively.
Preprint. University of Sussex.

A Set Constructor for Inductive Sets in Martin-Löf's Type Theory

Kent Petersson* Dan Synek*

1 Introduction

An important construction in programming languages is the definition facility for inductive data types [Hoa75]. One way to understand this construction is as a very general type constructor for trees. Trees are used for many purposes in computer science, one important example is syntax trees for representing phrases in languages. It is therefore vital that such a type constructor, together with its proof rules, should be available in a programming logic such as Martin-Löf's intuitionistic type theory [Mar82, Mar84]. In this paper, we will define a set constructor that one could use for defining many inductive data types.

The rest of the paper is organized as follows: We first explain the wellorder set constructor W introduced by Martin-Löf in [Mar82]. This set constructor is the least solution of a particular parameterized fixed point set equation. And since the parameters of this equation are closely related to the different parts of a single ML datatype definition [Mil84], we continue to discuss the correspondence between the wellorder constructor in Type Theory and the datatype constructor in ML.

The wellorder constructor is not sufficient when we want to define a family of mutually dependent inductive sets. We therefore introduce a set constructor for such sets. This set constructor, Tree, is explained in the following section and its formal rules are given.

In section 3, we show that all wellorders could be defined by the tree set constructor and then we discuss a slight variant of the tree constructor. Finally some examples are presented.

1.1 An introduction to the wellorder set constructor

With the wellorder set constructor we can construct many different sets of trees and to characterize a particular set we must provide information about two things,

- the different ways the trees may be formed, and
- for each way to form a tree which parts it consists of.

To provide this information, the wellorder set constructor W has two arguments:

1. The *constructor set B*.

*Programming Methodology Group, Dept. of Computer Sciences, Univ. of Göteborg and Chalmers, S-412 96 Göteborg, Sweden.

2. The *index family* C.

Given a constructor set B and an index family C over B, we can form a wellorder $\mathsf{W}(B, C)$ (two other notations are $(\mathsf{W}x : B)\, C(x)$ and $\mathsf{W}_{x:B}C(x)$).

One way of understanding the wellorder set constructor is to see it as a solution of an inductive definition of a particular form. So instead of solving a fixed point set equation every time we want to introduce a new set, as we do in domain theory [Sco82] we look at one particular equation of a general form and parameterize it with the information that differ between different inductive definitions. Since we just have one equation, we do not need to introduce the notions of fixed point set constructor, set operator and strictly positive set operator into the theory. Dybjer has furthermore shown [Dyb87] that all sets defined by a fixed point set operator could, in an extensional version of type theory, be represented by a wellorder.

The set $\mathsf{W}(B, C)$ is a solution to the equation

$$X \cong \sum_{y:B} C(y) \to X \tag{1}$$

In other words, isomorphic to the least fixed point of the strictly positive set operator

$$(X) \sum_{y:B} C(y) \to X$$

So $\mathsf{W}(B, C)$ is a fixed point to a *particular* (strictly positive) set operator.

If we have an element b_1 in the set B, that is, if we have a particular form we want the tree to have, and if we have a function c_1 from $C(b_1)$ to $\mathsf{W}(B, C)$, that is if we have a collection of subtrees, we may form the tree $\mathsf{sup}(b_1, c_1)$. We visualize this element in figure 1.

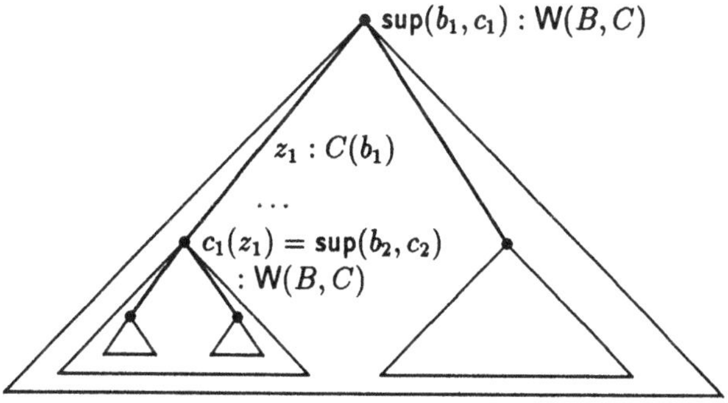

Figure 1: An element of a wellorder

We now give the formal rules for the wellorder set constructor and then an informal comparison between sets introduced using the wellorder set constructor and ML datatypes.

W - formation

$$\frac{B : \mathsf{set} \qquad C(y) : \mathsf{set}\ [y : B]}{\mathsf{W}(B, C) : \mathsf{set}}$$

W - introduction

$$\frac{b : B \qquad c(z) : \mathsf{W}(B,C)\ [z : C(b)]}{\mathsf{sup}(b,c) : \mathsf{W}(B,C)}$$

The non-canonical constant **wrec** is computed according to the computation rule:

$$\mathsf{wrec}(\mathsf{sup}(b,c),f) \ \rightarrow\ f(b,c,(x)\mathsf{wrec}(c(x),f))$$

The computation rule for **wrec** justifies the elimination and equality rules:

W - elimination

$$\frac{\begin{array}{l} D(x) : \mathbf{set}\ \ [x : \mathsf{W}(B,C)] \\ w : \mathsf{W}(B,C) \\ f(y,z,u) : D(\mathsf{sup}(y,z)) \\ \quad [y : B, z(v) : \mathsf{W}(B,C)\ [v : C(y)], u(v) : D(z(v))\ [v : C(y)]] \end{array}}{\mathsf{wrec}(w,f) : D(w)}$$

W - equality

$$\frac{\begin{array}{l} D(x) : \mathbf{set}\ \ [x : \mathsf{W}(B,C)] \\ b : B \\ c(y) : \mathsf{W}(B,C)\ \ [y : C(b)] \\ f(y,z,u) : D(\mathsf{sup}(y,z)) \\ \quad [y : B, z(v) : \mathsf{W}(B,C)\ [v : C(y)], u(v) : D(z(v))\ [v : C(y)]] \end{array}}{\mathsf{wrec}(\mathsf{sup}(b,c),f) = f(b,c,(x)\mathsf{wrec}(c(x),f) : D(\mathsf{sup}(b,c))}$$

Let us see how we can represent the set introduced by an ML datatype definition of the form[1]

$$\textbf{datatype}\ A = b_1\ \textbf{of}\ C_1\ \mid \cdots \mid b_n\ \textbf{of}\ C_n$$

First, we define a set B containing the constructor names $b_1, \ldots, b_n$. Then we would like to construct a family of sets C over B in such a way that $C(b_i)$ expresses the information in the ML type C_i. Let us for a moment restrict the form of C_i to a cartesian product $\underbrace{A * \cdots * A}_{n}$. But instead of using a cartesian product where the elements are decomposed by selectors (projection functions), we use a function from a set of selector names to A. That is, instead of A^n we use $\{c_1, \ldots, c_n\} \rightarrow A$. We can now, for each constructor, express the information in the ML type C_i as the set $C(b_i) = \{c_{b_i1}, \ldots, c_{b_in_{b_i}}\}$. Then we can construct the wellorder $\mathsf{W}(B,C)$, which is a set in Type Theory that represents the ML type.

Let us give some examples. An ML datatype definition such as

[1]Note the similarity between the ML definition

$$A = b_1 \text{ of } \{c_{b_1 1}, \ldots, c_{b_1 n_{b_1}}\} \rightarrow A \mid \cdots \mid b_m \text{ of } \{c_{b_m 1}, \ldots, c_{b_m n_{b_m}}\} \rightarrow A$$

and the equation

$$A \ \cong \sum_{y:\{b_1,\ldots,b_m\}} \{c_{y1}, \ldots, c_{yn_y}\} \rightarrow A$$

where $\sum$ is the *disjoint sum* of a family of sets and | is ML:s "sum" operator.

$$\textbf{datatype } BinTree = leaf \mid node \textbf{ of } BinTree{*}BinTree$$

will be represented as follows. The set of constructors is $\{leaf, node\}$. There is no selector for the constructor *leaf* because it has no components, and the set of selectors for *node* is a two element set, for example $\{left, right\}$ where the two elements give names to the two components. Putting this together, we can see that the wellorder[2]

$$BinTree \equiv \mathsf{W}(\{leaf, node\}, (x)\mathsf{case}(x, leaf : \{\}, node : \{left, right\}))$$

represents the ML-type above. The element $\mathsf{sup}(leaf, (x)\mathsf{case}_{\{\}}(x))$ represents the ML object *leaf* and $\mathsf{sup}(node, (x)\mathsf{case}(x, left : \hat{t_1}, right : \hat{t_2}))$ represents $node(t_1, t_2)$ if $\hat{t_1}$ represents t_1 and $\hat{t_2}$ represents t_2. Notice that we need an extensional version of type theory to insure that all elements of the form $\mathsf{sup}(leaf, z)$, where $z(x) : BinTree\ [x : \{\}]$, are equal to $\mathsf{sup}(leaf, (x)\mathsf{case}_{\{\}}(x))$.

It comes perhaps as a surprising fact that the wellorder set constructor is not constrained to datatypes of the limited form above. We can also allow constant sets to be used in the cartesian product C_i. This is made possible by viewing the constant part of C_i as part of the constructor. Let us see how this works by adding a natural number to the nodes in the example above.

$$\textbf{datatype } BinTree = leaf \mid node \textbf{ of } N{*}BinTree{*}BinTree$$

The set of constructors now becomes $\{leaf\} + \mathsf{N}$, the selector set $B(\mathsf{inl}(leaf))$ is the empty set and $B(\mathsf{inr}(n))$ is the set $\{left, right\}$. Let us give an ML-like definition to show what we have done:

$$\textbf{datatype } BinTree = leaf \mid node\ N \textbf{ of } BinTree{*}BinTree$$

The intuition is that we have a different node constructor for each natural number. This is possible for the wellordering since the constructor set may be infinite. The result is the wellorder

$$\mathsf{W}(\{leaf\} + \mathsf{N}, (x)\mathsf{when}(x, (y)\{\}, (y)\{left, right\}))$$

which represents the ML datatype. The ML-value *leaf* is represented by $\mathsf{sup}(\mathsf{inl}(leaf), \mathsf{case}_{\{\}})$ and $node(n, t_1, t_2)$ is represented by $\mathsf{sup}(\mathsf{inr}(n), (x)\mathsf{case}(x, \hat{t_1}, \hat{t_2}))$, if $\hat{t_1}$ represents t_1 and $\hat{t_2}$ represents t_2. The method outlined above can also be extended to definitions where C_i is of the form $K \rightarrow T$ and where K is a constant.

As an example of a function from an inductive set, we define the function that adds all numbers in the binary trees we defined above. In ML such a function could be defined as

$$\begin{array}{lll} \textbf{fun} & f(leaf) & = 0 \\ \mid & f(node(n, t_1, t_2)) & = n + f(t_1) + f(t_2); \end{array}$$

and in Type Theory it could be defined as

$$f(w) = \mathsf{wrec}(w, (x, y, z)\mathsf{when}(x, (u)0, (u)u + z(left) + z(right)))$$

[2]In order to define many families of sets we need a set of sets, a universe. When we use it here, we will identify sets and elements of the universe and also use a slightly changed syntax for case expressions.

2 The Tree Set Constructor

When trying to mimic mutually recursive inductive definitions with the wellorder set constructor one encounters a problem since it is not always the case that all components of an element belong to the same set as the element itself. For example if we want to represent the types defined in ML by:

$$\begin{array}{lll} \textbf{datatype}\ \mathit{Odd} & = & \mathit{sO}\ \textbf{of}\ \mathit{Even} \\ \textbf{and}\qquad \mathit{Even} & = & \mathit{zeroE} \mid \mathit{sE}\ \textbf{of}\ \mathit{Odd}; \end{array}$$

we can not do this directly using wellorders, we have to introduce a more complicated set constructor.

The constructor should produce a family of sets instead of one set as the wellorder set constructor did. In order to do this, we introduce a *name set*, which is a set of names of the mutually defined sets in the inductive definition. A suitable choice of name set for the example above would be $\{Odd, Even\}$. Instead of having one set of constructors B and one index family C over B, as we had in the wellorder case, we now have one constructor set and one index set for each element in A. The constructors form a family of sets B, where $B(x)$ is a set for each x in A and the index set form a family of sets C where $C(x, y)$ is a set for each x in A and y in $B(x)$. Furthermore, since the parts of a tree now may come from different sets, we introduce a function d which provides information about this; $d(x, y, z)$ is an element of A if $x : A$, $y : B(x)$ and $z : C(x, y)$. We call this element the ***component set name***.

The family of sets $\mathsf{Tree}(A, B, C, d)$ is a representation of the family of sets introduced by a collection of inductive definitions, for example an ML datatype definition. It could also be seen as a solution to the equation

$$\mathcal{T} \cong (x) \sum_{y:B(x)} \prod_{z:C(x,y)} \mathcal{T}(d(x, y, z))$$

where $\mathcal{T}$ is a family of sets over A and $x : A$. This equation could be interpreted as a possibly infinite collection of ordinary set equations, one for each $a : A$.

$$\begin{array}{rcl} \mathcal{T}(a_1) & \cong & \displaystyle\sum_{y:B(a_1)} \prod_{z:C(a_1,y)} \mathcal{T}(d(a_1, y, z)) \\ \mathcal{T}(a_2) & \cong & \displaystyle\sum_{y:B(a_2)} \prod_{z:C(a_2,y)} \mathcal{T}(d(a_2, y, z)) \\ & \vdots & \end{array}$$

Or, if we want to express the tree set constructor as the least fixed point of a set function operator.

$$\mathsf{Tree}(A, B, C, d) \cong \mathsf{FIX}((\mathcal{T})(x) \sum_{y:B(x)} \prod_{z:C(x,y)} \mathcal{T}(d(x, y, z)))$$

Comparing this equation with the equation for the wellorder set constructor, we can se that it is a generalization in that the non-dependent function set, "$\rightarrow$", has become a set of dependent functions, "$\prod$". This is a natural generalization since we are now defining a family of sets instead of just one set and every instance of the family could be defined in terms of every one of the other instances. It is the function d that expresses this relation.

2.1 Rules

The formation rule for the set of trees is:

$$\frac{\begin{array}{l} A : \mathsf{set} \\ B(x) : \mathsf{set} \;\; [x : A] \\ C(x,y) : \mathsf{set} \;\; [x : A, y : B(x)] \\ d(x,y,z) : A \;\; [x : A, y : B(x), z : C(x,y)] \\ a : A \end{array}}{\mathsf{Tree}(A,B,C,d,a) : \mathsf{set}}$$

The different parts have the following intuitive meaning:

- A, the name set, is a set of names for the mutually dependent sets.
- $B(x)$, the constructor set, is a set of names for the clauses defining the set x.
- $C(x,y)$, the index set, is a set of names for selectors of the parts in the clause y in the definition of x.
- $d(x,y,z)$, the component set name, is the name of the set corresponding to the selector z in clause y in the definition of x.
- a determines a particular instance of the family of sets.

Understood as a set of syntax-trees generated by a grammar, the different parts have the following intuitive meaning:

- A is a set of non-terminals.
- $B(x)$ is a set of names for the alternatives defining the non-terminal x.
- $C(x,y)$, is a set of names for positions in the sequence of non-terminals in the clause y in the definition of x.
- $d(x,y,z)$, is the name of the non-terminal corresponding to the position z in clause y in the definition of x.
- a is the startsymbol.

In order to reduce the notational complexity, we write $\mathcal{T}(a)$ instead of $\mathsf{Tree}(A,B,C,d,a)$ in the rest of the paper.

Introduction rule:

$$\frac{\begin{array}{l} a : A \\ b : B(a) \\ c(z) : \mathcal{T}(d(a,b,z)) \;\; [z : C(a,b)] \end{array}}{\mathsf{tree}(a,b,c) : \mathcal{T}(a)}$$

Intuitively:

- a is the name of one of the mutually dependent sets.
- b is one of the constructors of the set a.

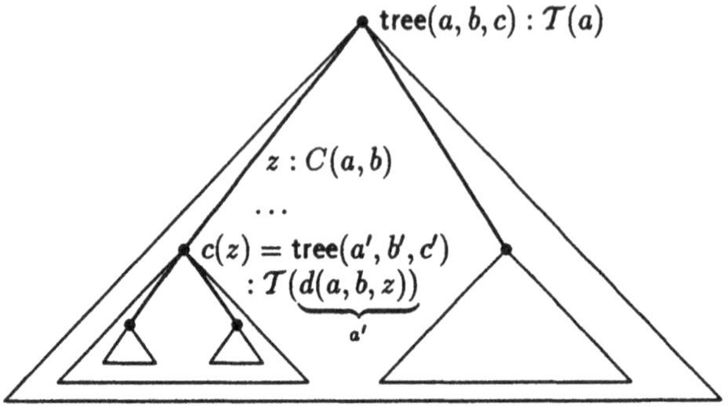

Figure 2: An element of a tree

- c is a function from $C(a,b)$ to a tree. This function defines the different parts of the element.

The element $\mathsf{tree}(a,b,c)$ in the set $\mathcal{T}(a)$ corresponds to the tree in figure 2, where $C(a,b) = \{z_1,\ldots,z_n,\ldots\}$ and $c(z_i) : \mathcal{T}(d(a,b,z_i))$.

Elimination rule:

$$\frac{\begin{array}{l} D(x,t) : \mathsf{set} \;\; [x : A, t : \mathcal{T}(x)] \\ a : A \\ t : \mathcal{T}(a) \\ f(x,y,z,u) : D(x, \mathsf{tree}(x,y,z)) \\ \qquad [x : A, y : B(x), z(v) : \mathcal{T}(d(x,y,v))\,[v : C(x,y)], \\ \qquad\; u(v) : D(d(x,y,v), z(v))\,[v : C(x,y)]] \end{array}}{\mathsf{treerec}(t,f) : D(a,t)}$$

The non-canonical constant treerec has the following computation rule:

$$\mathsf{treerec}(\mathsf{tree}(a,b,c),f) \;\; \rightarrow \;\; f(a,b,c,(x)\mathsf{treerec}(c(x),f))$$

This is reflected in the equality rule:

$$\frac{\begin{array}{l} D(x,t) : \mathsf{set} \;\; [x : A, t : \mathcal{T}(x)] \\ a : A \\ b : B(a) \\ c(z) : \mathcal{T}(d(a,b,z)) \;\; [z : C(a,b)] \\ f(x,y,z,u) : D(x, \mathsf{tree}(x,y,z)) \\ \qquad [x : A, y : B(x), z(v) : \mathcal{T}(d(x,y,v))\,[v : C(x,y)], \\ \qquad\; u(v) : D(d(x,y,v), z(v))\,[v : C(x,y)]] \end{array}}{\mathsf{treerec}(\mathsf{tree}(a,b,c),f) = f(a,b,c,(x)\mathsf{treerec}(c(x),f)) : D(a, \mathsf{tree}(a,b,c))}$$

3 Relation to the Wellorder Set Constructor

A wellorder set $\mathsf{W}(B,C)$ can be seen as an instance of a Tree set. We get the wellorders by defining a family of trees over a set with only one element. If we make the definitions:

$$\mathsf{W}(B,C) \;\; = \;\; \mathsf{Tree}(\{\mathsf{e}\},(x)B,(x,y)C(y),(x,y,z)\mathsf{e},\mathsf{e})$$

$$\begin{aligned} \mathsf{sup}(b,c) &= \mathsf{tree}(\mathsf{e},b,c) \\ \mathsf{wrec}(t,f) &= \mathsf{treerec}(t,(x,y,z,u)f(y,z,u)) \end{aligned}$$

where $\{\mathsf{e}\}$ is the set consisting of the element e. Then we can derive the rules for wellorders from the rules for trees as follows:

Formation rule If we assume that the premisses of the wellorder formation rule hold, that is, if we assume

$B : \mathsf{set}$
$C(y) : \mathsf{set}\ \ [y : B]$

we can infer

$\{\mathsf{e}\} : \mathsf{set}$
$((x)B)(x) : \mathsf{set}\ \ [x : \{\mathsf{e}\}]$
$((x,y)C(y))(x,y) : \mathsf{set}\ \ [x : \{\mathsf{e}\}, y : B]$
$((x,y,z)\mathsf{e})(x,y,z) : \{\mathsf{e}\}\ \ [x : \{\mathsf{e}\}, y : B, z : C(y)]$
$\mathsf{e} : \{\mathsf{e}\}$

and then, by the Tree-formation rule, get

$\mathsf{Tree}(\{\mathsf{e}\},(x)B,(x,y)C(y),(x,y,z)\mathsf{e},\mathsf{e}) : \mathsf{set}$

which is the same as

$\mathsf{W}(B,C) : \mathsf{set}$

and also the conclusion of the formation rule. So we have proved that the formation rule holds for the definition of wellorders in terms of trees.

Introduction rule Assume

$b : B$
$c(z) : \mathsf{W}(B,C)\ \ [z : C(b)]$

From the last assumption we get

$c(z) : \mathsf{Tree}(\{\mathsf{e}\},(x)B,(x,y)C(y),(x,y,z)\mathsf{e},\mathsf{e})\ \ [z : C(b)]$

It then follows that

$\mathsf{e} : \{\mathsf{e}\}$
$b : ((x)B)(\mathsf{e})$
$c(z) : \mathsf{Tree}(\{\mathsf{e}\},(x)B,(x,y)C(y),(x,y,z)\mathsf{e},((x,y,z)\mathsf{e}))(\mathsf{e},b,z)\ \ [z : ((x,y)C(y))(b)]$

and, from the Tree-introduction rule,

$\mathsf{tree}(\mathsf{e},b,c) : \mathsf{Tree}(\{\mathsf{e}\},(x)B,(x,y)C(y),(x,y,z)\mathsf{e},\mathsf{e})$

which is the same as

$\mathsf{sup}(b,c) : \mathsf{W}(B,C)$

The elimination and equality rules could be proved in the same way.

4 A Derived Version of the Tree Set Constructor

In this section we give a short description of an internal definition of Tree, called Tree_I, It is defined using the W set constructor. In [Syn89] we give the formal definition and prove that it is correct.

The construction of Tree_I is done in two steps. First we define a set constructor Ap, such that for all elements in $\mathsf{Tree}(A,B,C,d,a)$ there is a corresponding element in the approximation $\mathsf{Ap}(A,B,C)$. The set $\mathsf{Ap}(A,B,C)$ is an instance of the well order set constructor and does not simulate $\mathsf{Tree}(A,B,C,d,a)$ fully, i.e. there are elements in $\mathsf{Ap}(A,B,C)$ which have no correspondence in $\mathsf{Tree}(A,B,C,d,a)$. We then define a property $\mathsf{Ok}(A,C,d,a,t)$, where t is an element in $\mathsf{Ap}(A,B,C)$, which is true iff there is a corresponding element in $\mathsf{Tree}(A,B,C,d,a)$.

The set which simulates $\mathsf{Tree}(A,B,C,d,a)$ is the set of all x in $\mathsf{Ap}(A,B,C)$ such that $\mathsf{Ok}(A,C,d,a,t)$ is true. Since we do not have subsets in pure type theory we define Tree_I as a disjoint union of $\mathsf{Ok}(A,B,C,d,a)$ over $\mathsf{Ap}(A,B,C)$.

$$\mathsf{Tree}_I(A,B,C,d,a) \equiv \sum_{t:\mathsf{Ap}(A,B,C)} \mathsf{Ok}(A,C,d,a,t)$$

Instead of viewing the rules for Tree as primitive we can now prove them for Tree_I. The proof object of the introduction rule is the internal version of tree and the proof object of the elimination rule is the internal version of treerec.

5 A Variant of the Tree Set Constructor

We will in this section introduce a slight variant of the tree set constructor. Instead of having information in the element about what instance of the family a particular element belongs to, we move this information to the recursion operator. We call the new set constructor Tree′, the new element constructor tree′ and the new recursion operator treerec′. The formation rule for Tree′ is exactly the same as for Tree, but the other rules are slightly modified.

Introduction rule:

$$\frac{\begin{array}{l} a : A \\ b : B(a) \\ c(z) : \mathsf{Tree}'(A,B,C,d,d(a,b,z)) \;\; [z : C(a,b)] \end{array}}{\mathsf{tree}'(b,c) : \mathsf{Tree}'(A,B,C,d,a)}$$

Elimination rule:

$$\frac{\begin{array}{l} D(x,t) : \mathsf{set} \;\; [x : A, t : \mathsf{Tree}'(A,B,C,d,x)] \\ a : A \\ t : \mathsf{Tree}'(A,B,C,d,a) \\ f(x,y,z,u) : D(x,\mathsf{tree}'(y,z)) \\ \quad [x : A, y : B(x), z(v) : \mathsf{Tree}'(A,B,C,d,d(x,y,v))\,[v : C(x,y)], \\ \quad u(v) : D(d(x,y,v),z(v))\,[v : C(x,y)]] \end{array}}{\mathsf{treerec}'(d,a,t,f) : D(a,t)}$$

We leave the formulation of the equality rule to the reader. Notice that we in the first version of the tree sets can view the constructor tree as a family of constructors, one for each $a : A$. In the variant we have one constructor for the whole family, but instead we get a family of recursion operators, one for each a in A.

6 Examples of Different Tree Sets

6.1 Even and odd numbers

Consider the following datatype definition in ML:

datatype *Odd* = *sO* **of** *Even*
and *Even* = *zeroE* | *sE* **of** *Odd*;

and the corresponding grammar:

<odd> ::= s_O(<even>)
<even> ::= 0_E | s_E(<odd>)

If we want to define a set with elements corresponding to the phrases defined by this grammar (and if we consider <odd> as startsymbol), we can define $OddNrs = \mathsf{Tree}(A, B, C, d, a)$ where:

$$A = \{Odd, Even\}$$

$$a = Odd$$

$$\begin{aligned} B(Odd) &= \{s_O\} \\ B(Even) &= \{zero_E, s_E\} \\ \text{i.e. } B &= (x)\mathsf{case}(x, Odd : \{s_O\}, Even : \{zero_E, s_E\}) \end{aligned}$$

$$\begin{aligned} C(Odd, s_O) &= \{pred_O\} \\ C(Even, zero_E) &= \{\} \\ C(Even, s_E) &= \{pred_E\} \\ \text{i.e. } C &= (x, y)\mathsf{case}(x, \\ &\qquad\qquad Odd : \{pred_O\}, \\ &\qquad\qquad Even : \mathsf{case}(y, zero_E : \{\}, s_E : \{pred_E\})) \end{aligned}$$

$$\begin{aligned} d(Odd, s_O, pred_O) &= Even \\ d(Even, s_E, pred_E) &= Odd \\ \text{i.e. } d &= (x, y, z)\mathsf{case}(x, \\ &\qquad\qquad Odd : Even, \\ &\qquad\qquad Even : \mathsf{case}(y, zero_E : \mathsf{case}_{\{\}}(z), s_E : Odd)) \end{aligned}$$

The element $s_E(s_O(zero_E))$ is represented by

$$2_E = \mathsf{tree}(Even, s_E, (x)\mathsf{tree}(Odd, s_O, (x)\mathsf{tree}(Even, zero_E, (x)\mathsf{case}_{\{\}}(x)))$$

and $s_O(s_E(s_O(0_E)))$ is represented by

$$3_O = \mathsf{tree}(Odd, s_O, (x)2_E)$$

We get the set of even numbers by just changing the "startsymbol"

$$EvenNrs = \mathsf{Tree}(A, B, C, d, Even)$$

and we can define a mapping from even or odd numbers to ordinary natural numbers by:

$$\begin{aligned} tonat(w) &= \mathsf{treerec}(w, \\ &\qquad (x,y,z,u)\,\mathsf{case}(x, \\ &\qquad\qquad Odd : \mathsf{succ}(u(pred_O)), \\ &\qquad\qquad Even : \mathsf{case}(y, \\ &\qquad\qquad\qquad zero_E : 0, \\ &\qquad\qquad\qquad s_E : \mathsf{succ}(u(pred_E)))))) \\ tonat(w) &: \mathsf{N}\ [v : \{Odd, Even\}, w : \mathsf{Tree}(A,B,C,d,v)] \end{aligned}$$

6.2 An infinite family of sets

In ML, and all other programming languages with some facility to define mutually inductive types, one can only introduce finitely many new datatypes. A family of sets in type theory, on the other hand, could range over infinite sets and the tree set constructor therefore could introduce families with infinitely many instances. In this section we will give an example where the name set is infinite.

The problem is to define a set $\mathsf{Array}(A,n)$, whose elements are lists with exactly n elements from the set A. If we make a generalization of ML's datatype construction to dependent types this type could be defined as:

$$\begin{aligned} \mathsf{Array}(E,0) &= \mathsf{empty} \\ \mathsf{Array}(E,s(n)) &= \mathsf{add\ of}\ E * \mathsf{Array}(E,n)) \end{aligned}$$

The corresponding definition with the tree set constructor is:

$$\mathsf{Array}(E,n) = \mathsf{Tree}'(\mathsf{N},B,C,d,n)$$

where

$$\begin{aligned} B(n) &= \mathsf{natrec}(n, \{\mathsf{nil}\}, (x,y)E) \\ C(n,x) &= \mathsf{natrec}(n, \{\}, (x,y)\{\mathsf{tail}\}) \\ d(n,x,y) &= \mathsf{natrec}(n, \mathsf{case}_{\{\}}(y), (z,u)z) \end{aligned}$$

We can then define:

$$\begin{aligned} \mathsf{empty} &= \mathsf{tree}'(\mathsf{nil}, \mathsf{case}_{\{\}}) \\ \mathsf{add}(e,l) &= \mathsf{tree}'(e,l) \end{aligned}$$

as the elements. Notice that we in this example have used the variant of the tree constructor we introduced in section 5.

7 Related Work

Several different approaches to the introduction of inductive sets in type theory have already been suggested:

- Using the wellorder set constructor in type theory.
- Introducing a fixed point set constructor [CM85, Dyb87].
- Introducing a method to add new rules [Bac87].

We have seen that the wellorder set constructor is not general enough such that all inductive sets that we need are instances of it. The fixed point set constructor, on the other hand, is very complicated. Several different collections of proof rules have been suggested, and we think it is hard, or at least much harder than for most other sets, to realize the correctness of the rules. One reason why the rules are so complicated is because they do not follow the pattern from other set constructors. There is also one premise that requires a substantial modification of the whole theory (the subtype relation between sets) and another of a very "syntactical nature" (the notion of strictly positive set operator). Furthermore, if one wants to define mutually recursive sets, the fixed point construction seems as inadequate as the wellorder constructor and has to be extended to give fixed points to operators over families of sets.

The third approach is not sufficient for our purposes because it is not formalized within the Type Theory formalism and we want to see the set constructor for inductive sets as a formal entity which we could reason about using the rules of the theory. We want, for example, to be able to make a specification of a set and then formally derive an implementation using the proof rules of Type Theory.

8 Conclusion

We have presented a set constructor in the framework of Per Martin-Löf's type theory. It is a natural generalization of the wellorder set constructor to families of sets. The families of mutually dependent sets that can be introduced using the tree set constructor occur frequently in computer science, for example when representing abstract syntax trees.

An interesting application of the tree type constructor is for specifying parsers. In [Chi87], Chisholm proves the correctness of a parser for a small language. However, since he introduces a new set for the particular parse trees of his language, he has no concept of a grammar or any specification of a parser in general.

Another application is to represent the context sensitive parts of a language using an infinite family of sets. A method like this have already been used in the specification of Algol68 [vWea74] where a two level grammar was used.

Acknowledgement

We would like to thank Peter Dybjer and Per Martin-Löf for their comments on an earlier version of this paper.

References

[Bac87] Roland Backhouse. On the meaning and construction of the rules in Martin-Löf's theory of types. In *Proceedings of the Workshop on General Logic, Edinburgh*, Laboratory for the Foundations of Computer Science, University of Edinburgh, February 1987.

[Chi87] P. Chisholm. Derivation of a Parsing Algorithm in Martin-Löf's theory of types. *Science of Computer Programming*, 8:1–42, 1987.

[CM85] Robert L. Constable and Nax P. Mendler. Recursive Definitions in Type Theory. In *Proceedings of the LICS-Conference, Brooklyn, N. Y., Lecture Notes in Computer Science*, Springer-Verlag, June 1985.

[Dyb87] Peter Dybjer. Inductively Defined Sets in Martin-Löf's Type Theory. 1987. Presented at the Workshop on General Logic, Edinburgh.

[Hoa75] C. A. R. Hoare. Recursive Data Structures. *International Journal of Computer and Information Sciences*, 4(2):105 - 132, 1975.

[Mar82] Per Martin-Löf. Constructive Mathematics and Computer Programming. In *Logic, Methodology and Philosophy of Science, VI, 1979*, pages 153–175, North-Holland, 1982.

[Mar84] Per Martin-Löf. *Intuitionistic Type Theory*. Bibliopolis, Napoli, 1984.

[Mil84] R. Milner. Standard ML Proposal. *Polymorphism: The ML/LCF/Hope Newsletter*, 1(3), January 1984.

[Sco82] Dana Scott. Domains for Denotational Semantics. In *Automata, Languages and Programming, 9th Colloquium*, pages 577–613, Springer-Verlag, LNCS 140, July 1982.

[Syn89] Dan Synek. *Deriving Rules for Inductive Sets in Martin-Löf's Type Theory*. Technical Report, Programming Methodology Group, Dept. of Computer Science, Chalmers University of Technology, S–412 96 Göteborg, 1989. In preparation.

[vWea74] A. van Wijngarden et al. Revised report on the Algorithmic Language, ALGOL 68. *Acta Informatica*, 5:1–236, 1974.

Independence Results for Calculi of Dependent Types

Thomas Streicher
Fakultät für Mathematik und Informatik
Univ. Passau, D-8390, West Germany

Abstract. Based on a categorical semantics for impredicative calculi of dependent types we prove several independence results. Especially we prove that there exists a model where all syntactical concepts can be interpreted with one exception: in the model the strong sum of a family of propositions indexed over a proposition need not be a proposition again. The method of proof consists of restricting the set of propositions in the well-known **ω-Set** model due to E. Moggi.

1. Basic Concepts

Starting from previous work by R. Seely [Se1] on the categorical semantics of the logical part of Martin-Löf's intuitionistic type theory and work on the categorical semantics of polymorphically typed λ-calculus by Moggi and Hyland, see [Pi] and again R.Seely [Se2], a categorical semantics of impredicative calculi of dependent types (e.g. the Huet-Coquand calculus of constructions and extensions thereof) combining the features of Martin-Löf's intuitionistic type theory and polymorphically typed λ-calculus has emerged. The categorical framework for the semantics is explained in the author's Thesis [Str] and [Hy,Pi].

We just repeat here the definition of the kind of structures used for modelling impredicative calculi of dependent types.

Let $\mathbb{C}$ be a locally cartesian closed category, i.e. a category with finite limits and right adjoints to all pullback functors, which we know to model the logical part of Martin-Löf's intuitionistic type theory, see [Se1].

A *full* subcategory of $\mathbb{C}$ is given by a class $\mathbb{D}$ of *display maps*, i.e. a class of morphisms in $\mathbb{C}$ such that

(i) pullbacks of display maps along *arbitrary* morphisms in $\mathbf{C}$ exist and

(ii) any pullback of a display map is a display map again .

Notation To indicate that $f : B \to A$ is a display map we write $f : B$ ---|> A.
By $\mathbf{D}/A$ we denote the full subcategory of $\mathbf{C}/A$ whose objects are just the display maps with codomain A.

The full subcategory of $\mathbf{C}$ given by $\mathbf{D}$ is often called the *category of propositions* and display maps are often called *families of propositions*.

Any locally cartesian closed category models the concept of products of families of types as for any morphism f the pullback functor f^* is guaranteed to have a right adjoint Π_f. But from that it does not yet follow that the product of a family of propositions (indexed over an arbitrary type) is a proposition again. Therefore we say that $\langle \mathbf{C}, \mathbf{D} \rangle$ admits *impredicative universal quantification* iff it satisfies the following condition.

($\forall$) If $f : B \to A$ is a morphism in $\mathbf{C}$ and $g : C$ ---|> B is a display map (i.e. $g \in \mathbf{D}$) then $\Pi_f(g)$ is a display map (i.e. $\Pi_f(g) \in \mathbf{D}$), too.

It is most natural to claim that the full subcategory described by $\mathbf{D}$ has all finite products, i.e. has a terminal object and all binary products. This claim can be expressed in an elementary way by the following two conditions.

(Term) For any object A of $\mathbf{C}$ the morphism id_A is a display map.

(Prod) If $a : A$ --|> I and $b : B$ --|> I are display maps and $p : P \to A$ and $q : P \to B$ is a pullback cone for a and b in $\mathbf{C}$ then the morphism $a \circ p \ (= b \circ q)$ is a display map, too.

Obviously, if $\langle \mathbf{C}, \mathbf{D} \rangle$ satisfies ($\forall$), (Term) and (Prod) then for any object I of $\mathbf{C}$ the category $\mathbf{D}/I$ is a full sub-cartesian-closed-category of $\mathbf{C}/I$ as $\mathbf{D}/I$ inherits products from $\mathbf{C}/I$ and it follows already from ($\forall$) that for any $f : A$ ---|> I and $g : B$ ---|> I the exponential object $f \Rightarrow g = \Pi_f f^* g \in \mathbf{D}$.

One may notice that the notion of a category with display maps $\langle \mathbf{C}, \mathbf{D} \rangle$ satisfying ($\forall$), (Term) and (Prod) coincides with a notion introduced quite recently by Meseguer, see [Me].

Two further first order concepts, sums of families of types and identity types, which are typical for Martin-Löf's type theory, can be interpreted in any locally cartesian closed category by composition of morphisms and by fibrewise diagonal maps, see [Se1].

As the statement that objects (of the same type) are identical is usually considered as a proposition it is natural to assume that identity types are propositions. The following condition expresses this assumption in categorical language.

(Id) For any morphism $f : B \to A$ in $\mathbb{C}$ the diagonal morphism $\Delta_f \in \mathbf{D}$, where Δ_f is the mediating arrow in the diagram below

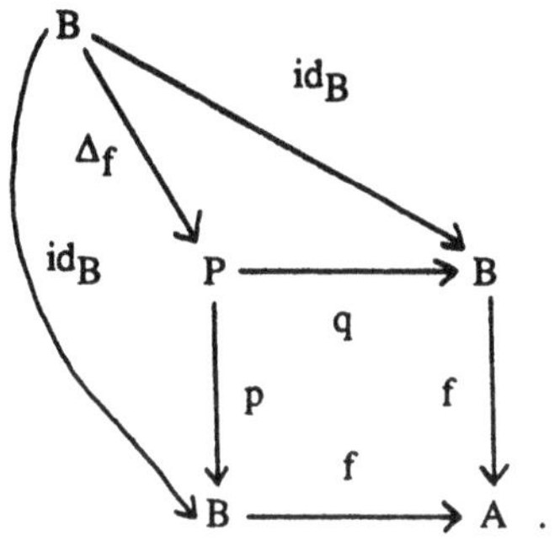

But for certain "philosophical" reasons one can take the position that statements which claim that two propositions are identical do not correspond to propositions but to something like "meta-propositions". Therefore some people may consider the statement of identity of two objects to be a proposition only if the type of the underlying objects is a proposition. The following condition expresses this weaker assumption in categorical terms.

(P-Id) If $f \in \mathbf{D}$ then $\Delta_f \in \mathbf{D}$.

If one assumes that the sum of a family of propositions *indexed over an arbitrary type* is a proposition then by Girard's paradox any proposition is inhabited. According to the principle of propositions as types any proposition must be considered as provable which makes the logical part of such a calculus inconsistent. In presence of identity types the whole calculus would be trivialized then. But it is quite natural to claim (and actually it holds in all natural models we know) that the sum of a family of propositions *indexed over a proposition* is a proposition again without forcing all propositions to be inhabited.

This can be expressed in categorical language by the following condition.

(Sum) If $f : B$ ---|> A and $g : C$ ---|> B then $\Sigma_f g = f \circ g \in \mathbf{D}$.

Now we are going to prove some lemmas relating some of the conditions above to *internal completeness* of the full subcategory given by $\mathbf{D}$.

Lemma 1 If $\mathbb{C}$ is a category with finite limits (not necessarily locally cartesian closed) and $\mathbf{D}$ is a class of display maps satisfying the conditions (Term), (Id-P) and (Sum) then

(i) if $a : A$ ---|> I and $b : B$ ---|> I and $a \circ f = b$ then $f \in \mathbf{D}$ and

(ii) for any object I of $\mathbb{C}$ the category $\mathbf{D}/I$ has all finite limits and the embedding of $\mathbb{D}/I$ into $\mathbb{C}/I$ preserves and reflects all finite limits.

Proof:

(i) : Assume that $a : A$ ---|> I and $b : B$ ---|> I are display maps and $f : B \rightarrow A$ is an arbitrary morphism in $\mathbb{C}$ with $a \circ f = b$.

Consider the following diagram where all rectangles are assumed to be pullbacks.

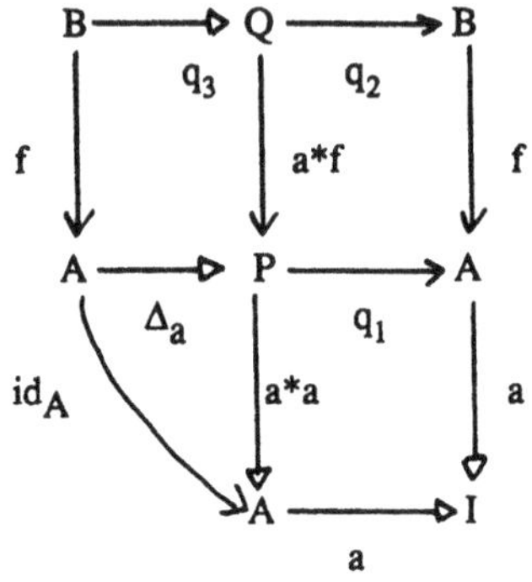

The pullback of f along $q_1 \circ \Delta_a$ can be chosen to be f as $q_1 \circ \Delta_a = id_A$.

$a^*a \circ a^*f$ is a display map as it is a pullback of the display map $a \circ f = b$ along a.

As q_3 is a pullback of the map Δ_a along a^*f and by (Id-P) Δ_a is a display map we have that q_3 is a display map.

Now $f = id_A \circ f = a^*a \circ \Delta_a \circ f = a^*a \circ a^*f \circ q_3$. Thus f is the composition of two display maps and thus by (Sum) we conclude that f is a display map, too.

(ii) : As (Term) guarantees that $\mathbb{D}/I$ inherits the terminal object from $\mathbb{C}/I$ it remains to prove that $\mathbb{D}/I$ has pullbacks and inherits them from $\mathbb{C}/I$.

Let $a : A$ ---|> I, $b : B$ ---|> I and $c : C$ ---|> I be display maps and $f : B \rightarrow A$, $g : C \rightarrow A$ be morphisms in $\mathbb{C}$ with $a \circ f = b$ and $a \circ g = c$ then by (i) of Lemma 1 f and g are display maps.

Let $p : P \rightarrow B$ and $q : P \rightarrow C$ be a pullback cone for f and g in $\mathbb{C}$. As f and g are display maps p and q must be display maps, too, as display maps are stable under pullbacks.

By (Sum) it follows that $a \circ f \circ p = a \circ g \circ q$ is a display map.

It follows immediately that the pair of morphisms $p : a \circ f \circ p \rightarrow b$ and $q : a \circ g \circ q \rightarrow c$ is a pullback cone for $f : b \rightarrow a$ and $g : c \rightarrow a$ in $\mathbb{D}/I$.

end of Lemma 1

Corollary 1 If $\mathbb{C}$ is a category with finite limits and $\mathbb{D}$ is a class of display maps in $\mathbb{C}$ satisfying the conditions (Term), (Id) and (Sum) then for arbitrary morphisms $a : A \rightarrow I$, $f : B \rightarrow A$ in $\mathbb{C}$ one has that $a \circ f \in \mathbb{D}$ implies $f \in \mathbb{D}$.

Proof : We proceed just as in the proof of (i) of Lemma 1 with the following exception : although the morphism a need not be a display map (Id) guarantees that Δ_a is a display map.

end of Corollary 1

As any locally cartesian closed category $\mathbb{C}$ has all finite limits Lemma 1 and Corollary 1 are valid especially for all locally cartesian closed categories. This implies that any class $\mathfrak{D}$ of display maps satisfying the conditions (Term), (Id-P) and (Sum) also satisfy the following property.

(Pb) If $f : B \to A$ and $g : C \to A$ are morphisms in $\mathbb{C}$ and $p : P \to B$, $q : P \to C$ is a pullback cone for f and g in $\mathbb{C}$
then for all display maps a : A ---|> I with $a \circ f$, $a \circ g \in \mathfrak{D}$
$a \circ f \circ p$ $(= a \circ g \circ q) \in \mathfrak{D}$, too.

Remark. For any locally cartesian closed category $\mathbb{C}$ the notion of a class $\mathfrak{D}$ of diplay maps satisfying conditions ($\forall$), (Term) and (Pb) coincides with the notion of a *full, internally complete subcategory of* $\mathbb{C}$, as introduced by J. Benabou in his (unpublished) work on the foundations of naive category theory using Grothendieck's concept of a fibred category.

Next we shall prove a lemma stating that for any category $\mathbb{C}$ with pullbacks and any full sub-left-exact category represented by a class $\mathfrak{D}$ of display maps any morphism in the subcategory is a display map.

Lemma 2 Let $\mathbb{C}$ be a category having all pullbacks and $\mathfrak{D}$ a class of display maps satisfying the conditions (Term) and (Pb) then
for all display maps a : A ---|> I , b : B ---|> I and arbitrary morphisms f in $\mathbb{C}$
$a \circ f = b$ implies $f \in \mathfrak{D}$.

Proof : Consider the diagram of the proof of Lemma 1.

The pair of morphisms f and q_3 is a pullback cone for Δ_a and a*f in $\mathbb{C}$. The morphism a*a is a display map as a is a display map and display maps are stable under pullbacks. The morphism $a^*a \circ \Delta_a = id_A$ is a display map by (Term) and $a^*a \circ a^*f$ is a display map as it is the pullback of the display map $a \circ f$ along a and display maps are stable under pullbacks. By an application of (Pb) we conclude that $a^*a \circ \Delta_a \circ f$ is a display map. But as $a^*a \circ \Delta_a = id_A$ we conclude that f is a display map, too.

end of Lemma 2

This shows that any family of morphisms of the subcategory given by $\mathfrak{D}$ is also a family of display maps. Now the question is whether any family of display maps is also a family of morphisms in the subcategory (which is expressed in categorical terms by the condition (Sum)). We shall show that the answer to this question is *negative* even under rather strong assumptions on the class $\mathfrak{D}$ of display maps.

So one of the main theorems we shall prove in the next section is that there exist full, internally complete subcategories of locally cartesian closed categories such that the representing class $\mathfrak{D}$ of display maps is *not* closed under composition.

Furthermore all the models we shall consider will have the following property which allows to represent any display map by a so called classifying map.

(Gen) There exists a display map generic : Proof ---|> Prop such that for any display map a : A ---|> I there exists a (not necessarily unique) classifying map $\chi_a : I \rightarrow Prop$ with a isomorphic to χ_a*generic in $\mathfrak{D}/I$ (i.e. a is obtainable as a pullback of the display map generic along some morphism, e.g. χ_a).

This property arises from the study of the calculus of constructions where propositions have an ambiguous nature. Propositions can be considered as ***objects** of type Prop* and as *propositional **types***. Any Prop-valued morphism $\chi : I \rightarrow Prop$ corresponds to the family of propositions χ*generic and any display map can be obtained in this way as guaranteed by the axiom (Gen).

Of course, classifying morphism represent families of propositions only up to isomorphism (as pullbacks are defined only up to isomorphism) and there may be several different classifying morphisms for one and the same family of propositions. This non-uniqueness causes some problems for the interpretation of $\forall$ in the calculus of constructions. These problems are avoided by using the framework of *doctrines of constructions* built on Cartmell's notion of contextual categories, see [Str]. Furthermore all *natural* models of the kind we have defined above, such as realizability models or domain-theoretic models, can be transformed into a doctrine of constructions by considering *pairs* $\langle a, \chi \rangle$ where a is a display map corresponding to a family of propositions and χ is a morphism classifying a instead of considering display maps alone.

2. Some Independence Results Obtained by Modifications of Realizability Models

Starting from unpublished work by E. Moggi on realizabilty models for polymorphically typed λ-calculus I have shown in my Thesis (independent from a few other researchers at the same time) how these results for the polymorphically typed λ-calculus can be extended to construct realizability models for the calculus of constructions introduced by Coquand and Huet.

We shall now define the well known realizability model for a calculus of dependent types extending both the calculus of constructions and Martin-Löf's intuitionistic type theory.

Definition 1 The category **ω-Set** of *realizability sets* and *realizable morphisms* (due to Moggi) is given by the following data :

Objects : a *realizability set* $\underline{X}$ is given as a pair $\langle X, \Vdash_X \rangle$
where X is a set and $\Vdash_X \subseteq \omega \times X$ such that
for all $x \in X$ there exists $n \in \omega$ such that $n \Vdash_X x$

Morphisms : a *realizable morphism* from $\underline{X}$ to $\underline{Y}$ is a function $f : X \to Y$
such that for some $n \in \omega$ one has that
for all $x \in X$ and $m \Vdash_X x$ the computation $\{n\}m$ terminates and $\{n\}m \Vdash_Y f(x)$
(we say that "n realizes f" and write $n \Vdash f$).

Composition and identity morphisms are inherited from **Set**.

As the class $\mathcal{P}$ of *families of propositions* we take the class of *locally discrete families* consisting of all realizable morphisms $f : \underline{Y} \to \underline{X}$ such that for all $y, y' \in Y$ with $f(y) = f(y')$ and all $n \in \omega$ with $n \Vdash_Y y$ and $n \Vdash_Y y'$ one has that $y = y'$.

end of Definition 1

In my Thesis [Str] (chapter 2) I have proven the following theorem.

Theorem 1 ω-Set is a locally cartesian closed category and $\mathcal{P}$ is a class of display maps closed under arbitrary pullbacks satisfying the conditions $(\forall)$, (Gen), (Term), (Id) and (Sum).

end of Theorem 1

A generic morphism for $\mathcal{P}$ can be obtained in the following way.

Definition 2

Let $\underline{Prop} = \langle Prop, \Vdash_{Prop} \rangle$ where Prop = PERω, the set of partial equivalence relations on ω (i.e. transitive and symmetric, not necessarily reflexive relations on ω), and $\Vdash_{Prop} = \omega \times PER\omega$.

Let $\underline{Proof} = \langle Proof, \Vdash_{Proof} \rangle$ where $Proof = \{\langle R, M \rangle \mid R \in PER\omega \text{ and } M \in \omega_{/R}\}$ and $\Vdash_{Proof} = \{\langle n, \langle R, M \rangle \rangle \mid \langle R, M \rangle \in Proof \text{ and } n \in M\}$.

generic : $\underline{Proof} \to \underline{Prop}$ is defined as the projection on the first component, i.e. for all $\langle R, M \rangle \in$ Proof $generic(\langle R, M \rangle) = R$, which is realized by any total recursive function, e.g. $\Lambda n.0$.

end of Definition 2

We now shall prove several independence results by *restricting* Prop to certain subsets. We define sets Prop-3 $\subseteq$ Prop-2 $\subseteq$ Prop-1 $\subseteq$ Prop = PERω and associate with them generic morphisms $generic_i$: $\underline{Proof\text{-}i} \to \underline{Prop\text{-}i}$:

- $\underline{Prop\text{-}i} = \langle Prop\text{-}i, \Vdash_{Prop\text{-}i} \rangle$ where $\Vdash_{Prop\text{-}i} = \omega \times Prop\text{-}i$

- $\underline{Proof\text{-}i} = \langle Proof\text{-}i, \Vdash_{Proof\text{-}i} \rangle$

where $\text{Proof-i} = \{\langle R, M\rangle \mid R \in \text{Prop-i} \text{ and } M \in \omega_{/R}\}$ and
$\Vdash_{\text{Proof-i}} = \{\langle n, \langle R, M\rangle\rangle \mid \langle R, M\rangle \in \text{Proof-i} \text{ and } n \in M\}$

$\text{generic}_i : \underline{\text{Proof-i}} \rightarrow \underline{\text{Prop-i}}$
is the projection on the first component, i.e. $\text{generic}_i(\langle R, M\rangle) = R$ for all $\langle R, M\rangle \in$ Proof-i, and is realized by any total recursive function, e.g. $\Lambda n.0$.

Definition 3

A subset M of ω is called a *functional class* iff there exists a *partial* function f from ω to ω such that $M = \{n \in \omega \mid \text{graph}(f) \subseteq \text{graph}(\{n\})\}$, called *representative for* M. For every functional class M the (uniquely determined) representative for M is denoted by rep(M).

$R \in \text{PER}\omega$ is an element of *Prop-1* iff any $M \in \omega_{/R}$ is a functional class and for all $M' \in \omega_{/R}$ dom(rep(M)) = dom(rep(M')) where dom denotes the domain of definition of a partial function.

$R \in \text{PER}\omega$ is an element of *Prop-2* iff R is empty or there exists an ω-Set $\underline{X}$ such that for all $n, m \in \omega$ $n R m$ iff for all $x \in X$ and $k \Vdash_X x$ $\{n\}(k)$ and $\{m\}(k)$ are defined and equal.

$R \in \text{PER}\omega$ is an element of *Prop-3* iff R is non-empty and an element of *Prop-2* .

For i, $1 \le i \le 3$, $\mathcal{P}_i = \{ \chi^*\text{generic}_i \mid \chi$ is a morphism of ω-Set with codomain $\underline{\text{Prop-i}}$ } is a class of display maps which obviously is stable under pullbacks along arbitrary morphism of ω-Set .

end of Definition 3

Now we have defined enough notions to be able to formulate the following theorem.

Theorem 2

(1) $\langle \omega\text{-Set}, \mathcal{P}_1\rangle$ satisfies ($\forall$), (Gen), (Term), (Pb) and (Id), but *not* (Sum).

(2) $\langle \omega\text{-Set}, \mathcal{P}_2\rangle$ satisfies ($\forall$), (Gen), (Term), (Prod) and (Id), but *neither* (Pb) *nor* (Sum).

(3) $\langle \omega\text{-Set}, \mathcal{P}_3\rangle$ satisfies ($\forall$), (Gen), (Term), (Prod), but no one of (Pb), (Id-P) and (Sum).

Proof : Obviously, by the definition of $\mathcal{P}_1, \mathcal{P}_2, \mathcal{P}_3$ the condition (Gen) is satisfied.
Next we shall show that ($\forall$) is satisfied for $\mathcal{P}_1, \mathcal{P}_2, \mathcal{P}_3$.
Let $f : \underline{Y} \rightarrow \underline{X}$ be a morphism in ω-Set and let $g \in \mathcal{P}_1$ be classified by $\chi : \underline{Y} \rightarrow \underline{\text{Prop-i}}$. Then $\Pi_f g$ is classified by the morphism $\varphi : \underline{X} \rightarrow \underline{\text{Prop-i}}$ which maps any $x \in X$ to the partial equivalence relation $\varphi(x)$ in $\mathcal{P}_i$ defined as follows.

$n\ \varphi(x)\ n'$ iff for all $y \in Y$, $m, m' \in \omega$ if $f(y) = x$, $m \Vdash_Y y$ and $m' \Vdash_Y y$
then $\Lambda k.\{n\}(\langle m, k\rangle)\ \chi(y)\ \Lambda k.\{n'\}(\langle m', k\rangle)$.

It can be seen immediately that for any i, $1 \le i \le 3$, the ω-sets corresponding to partial equivalence relations in Prop-i contain a terminal object (as $\omega \times \omega$ is in Prop-i) and are closed under finitary products (if f and g are representatives then the partial function h satisfying $h(2 \cdot n) = f(n)$ and $h(2 \cdot n + 1) = g(n)$ represents the pair of elements represented by f and g, respectively).

Now we shall prove that none of $\mathcal{P}_1, \mathcal{P}_2, \mathcal{P}_3$ is closed under composition. Therefore it is sufficient to give composable morphisms f and g in $\mathcal{P}_3$ such that $f \circ g$ is *not* in $\mathcal{P}_1$.

Let $\underline{X} = \langle \{f : \omega \to \omega : f \text{ is total recursive}\}, \{\langle n, f\rangle : \{n\} = f\}\rangle$ and $\underline{Y} = \langle Y_1 \cup Y_2, \{\langle n, f\rangle : \text{graph}(f) \subseteq \text{graph}(\{n\})\}\rangle$ where $Y_1 = \{ f : \{2\} \cdot \omega \cup \{1\} \to \omega : f(2n) = 0 \text{ for all } n \in \omega \}$ and $Y_2 = \{ f : \{2\} \cdot \omega \cup \{3\} \to \omega : f(2n) > 0 \text{ for some } n \in \omega \}$. Let f be the unique ω-**Set** morphism from $\underline{X}$ to the terminal object $\underline{1}$ and let g be the unique morphism from $\underline{Y}$ to $\underline{X}$ realized by $\Lambda n.\Lambda m. \{n\}(2m)$. Obviously both f and g are in $\mathcal{P}_3$.

Now we shall prove by a continuity argument that $\underline{Y}$ cannot be classified by a morphism from $\underline{1}$ to Prop-1, i.e. that $f \circ g$ is not in $\mathcal{P}_1$. More precisely we shall show that $\underline{Y}$ does *not* satisfy the following *separability condition* which is true for all ω-sets classified by a morphism from $\underline{1}$ to Prop-1.

Separability Condition

For all $y_1, y_2 \in Y$ with $y_1 \neq y_2$ there exists a morphism $\varphi : \underline{Y} \to \underline{\omega}$ in ω-**Set** such that $\varphi(y_1) = 0$ and $\varphi(y_2) = 1$, where $\underline{\omega} = \langle \omega, \{\langle n, n\rangle : n \in \omega\}\rangle$.

Let $y_1, y_2 \in Y_1$ with $y_1(1) = 0$ and $y_2(1) = 1$. Suppose that there exists a separating morphism $\varphi : \underline{Y} \to \underline{\omega}$ with $\varphi(y_1) = 0$ and $\varphi(y_2) = 1$. Let n_0 be a realizer for φ. As any code for a total recursive function is a realizer for some object of Y, n_0 codes an effective operation on total recursive functions. Let n_1 be a code for the total recursive function which maps any number to 0 and n_2 be a code for the total recursive function which maps 1 to 1 and all other numbers to 0.

Obviously n_1 is a realizer for y_1 and n_2 is a realizer for y_2. Thus $\{n_0\}(n_1) = 0$ and $\{n_0\}(n_2) = 1$. By the well known Kreisel-Lacombe-Shoenfield Theorem stating the continuity of all effective operations on total recursive functions there exist $m_1, m_2 \in \omega$ such that for any $m \in \omega$ coding a total recursive function it holds that if for all $k < m_1$ one has $\{m\}(k) = \{n_1\}(k)$ then $\{n_0\}(m) = \{n_0\}(n_1)$ and if for all $k < m_2$ one has $\{m\}(k) = \{n_2\}(k)$ then $\{n_0\}(m) = \{n_0\}(n_2)$.

Let n_3, n_4 be codes of a total recursive functions such that

- $\{n_3\}(k) = \{n_1\}(k)$ and $\{n_4\}(k) = \{n_2\}(k)$ for $k < \max(m_1, m_2)$ and
- $\{n_3\}(k) = \{n_4\}(k) = 1$ for $k \geq \max(m_1, m_2)$.

Obviously n_3 and n_4 realize the same object of Y_2 and therefore $\{n_0\}(n_3) = \{n_0\}(n_4)$. But on the other hand for all $k < m_1$ one has $\{n_3\}(k) = \{n_1\}(k)$ and for all $k < m_2$ one has $\{n_4\}(k) = \{n_2\}(k)$ which implies $\{n_0\}(n_3) = \{n_0\}(n_1) = 0$ and $\{n_0\}(n_4) = \{n_0\}(n_2) = 1$ by the conditions on m_1 and m_1 stated above. Thus $\{n_0\}(n_3) \neq \{n_0\}(n_4)$.

But as $\{n_0\}(n_3)$ and $\{n_0\}(n_4)$ cannot be both equal and different the separating morphism φ cannot exist.

That $\mathcal{P}_1$ satisfies the condition (Pb) follows from the fact that $\mathcal{P}_1$ satisfies the condition (Prod) and that for any $R \in$ Prop-1 and partial equivalence relation R' with $\omega_{/R'} \subseteq \omega_{/R}$ one has that $R' \in$ Prop-1, too.

$\mathcal{P}_2$ and $\mathcal{P}_3$ cannot satisfy (Pb) as by assuming pullbacks in $\mathcal{P}_3/\mathbb{1}$ we can construct a type with cardinality 2 whereas all types in $\mathcal{P}_2/\mathbb{1}$ and $\mathcal{P}_3/\mathbb{1}$ are either empty, singletons or infinite.

For $\mathcal{P}_1$ and $\mathcal{P}_2$ the condition (Id) is satisfied as Prop-2 contains the empty relation and $\omega \times \omega$. Thus for an arbitrary morphism $f: \underline{Y} \to \underline{X}$ in **ω-Set** the morphism Δ_f is classified by the morphism δ_f which maps any pair $\langle y, y \rangle$ with $y \in Y$ to $\omega \times \omega$ and all other pairs $\langle y_1, y_2 \rangle$ with $y_1, y_2 \in Y$, $f(y_1) = f(y_2)$ and $y_1 \neq y_2$ to the empty relation.

For $\mathcal{P}_3$ the condition (P-Id) does not hold as for any display map in $\mathcal{P}_3$ every fibre is inhabited, which is obviously false for Δ_f if f is a display map such that one of its fibres contains more than one element.

end of Theorem 2

Next we shall prove that one cannot prove the independence of the conditions (Term) and (Prod) from the conditions (∀) and (Gen) by realizability models.

Theorem 3 If $\mathcal{P}$ is a class of display maps in **ω-Set** stable under pullbacks along arbitrary morphisms and satisfying (∀) and (Gen) then $\mathcal{P}$ also satisfies the conditions (Term) and (Prop).

Proof: **ω-Set** contains as objects the empty realizability set $\underline{0} = \langle \{\}, \{\} \rangle$ and the discrete realizability set $\underline{2} = \langle \{0, 1\}, \{\langle 0, 0 \rangle, \langle 1, 1 \rangle\} \rangle$ containing two elements.

The terminal object is obtained by taking the product of the (unique) family of propositions indexed over $\underline{0}$, which by (∀) is guaranteed to be a proposition. A family of propositions indexed over $\underline{0}$ exists, because there exists a (unique) classifying morphism χ with domain $\underline{0}$ by (Gen).

The cartesian product of propositions A and B is obtained as the product of the family of propositions indexed over $\underline{2}$ such that the fibre over 0 is A and the fibre over 1 is B. This product is guaranteed to be a proposition by (∀) and the family of propositions we used exists by (Gen).

end of Theorem 3

Next we prove that there exists a full subcategory of **ω-Set** represented by a class of display maps $\mathcal{D}$ satisfying the conditions (Sum), (Id), (Term), (Gen) and instead of the condition (∀) expressing impredicative universal quantification the following *weaker* condition which guarantees the product of a family of propositions to be a proposition *only* if the indexing set of the family under consideration is a proposition.

(Sub-Π) For $f, g \in \mathcal{D}$ with $f \circ g$ defined one has that $\Pi_f g \in \mathcal{D}$.

A class $\mathfrak{D}$ of display maps in a locally cartesian closed category $\mathfrak{C}$ satisfying the conditions (Sum), (Id), (Term), (Gen) and (Sub-Π) is said to represent an *internal full sub-locally-cartesian-closed-category of* $\mathfrak{C}$.

Definition 4 A morphism $f : \underline{Y} \to \underline{X}$ is *source finite* iff for any $x \in X$ the inverse image of x under f is finite, i.e. card({ $y \in Y : f(y) = x$ }) is finite. **end of Definition 4**

Theorem 4 Let $\mathfrak{F}$ be the class of source finite morphisms in ω-Set.
Obviously $\mathfrak{F}$ is stable under pullbacks along arbitrary morphisms. Furthermore $\mathfrak{F}$ satisfies the conditions (Gen), (Sub-Π), (Term), (Id) and (Sum), but not (∀).

Proof : As one can easily find a *set* Fin of finite realizability sets such that any finite realizability set $\underline{X}$ is isomorphic to some $\underline{Y}$ in Fin by an isomorphism pair whose both components are realized by Λn.n, condition (Gen) is satisfied for $\mathfrak{F}$.

As the empty set and any singleton set is a finite realizability set (Term) and (Id) are satisfied.

(Sum) and (Sub-Π) are satisfied as the disjoint sum and the product of a family of finite realizability sets indexed over a finite realizability set are finite realizability sets themselves.

But condition (∀) is not satisfied as the product of a family of *finite* realizability sets indexed over an *infinite* realizability set need not be finite in general, e.g. $\underline{\omega} \to \underline{2}$ (the product of the constant family indexed over $\underline{\omega}$ with value $\underline{2}$) contains all recursive sets and there are infinitely many of them.

end of Theorem 4

Remark. Instead of $\mathfrak{F}$ one can choose the class $\mathfrak{PF}$ of all source finite morphisms contained in $\mathfrak{P}$. Then Theorem 4 holds with $\mathfrak{F}$ replaced by $\mathfrak{PF}$.

After having established a number of positive and negative answers to questions concerning the independence of concepts w.r.t realizability models we want to indicate an open question.

Arising from the things said above we suggest as a most concise and most "category-theoretic" formulation of the notion of model for a rather strong extension of the calculus of constructions the following one: a locally cartesian closed category $\mathfrak{C}$ together with a class of display maps $\mathfrak{D}$ stable under pullbacks along arbitrary morphisms in $\mathfrak{C}$ such that

(i) the full subcategory given by $\mathfrak{D}$ is *internally complete*, i.e. $\mathfrak{D}$ satisfies the conditions (Term), (Pb), (∀), and the category given by $\mathfrak{D}$ is *internal*, which is guaranteed by (Gen).

(ii) $\mathfrak{D}$ is closed under composition , i.e. (Sum).

Condition (Sum) can be intuitively interpreted as the constraint that the subcategory as given by $\mathfrak{D}$ is *not only* full w.r.t. morphisms, *but also* "full w.r.t. families of propositions". This can be seen from the following observation : if an object A is in the subcategory, i.e. $!_A : A \to 1$ is in $\mathcal{D}$, and $b : B$ ---|> $A \in \mathcal{D}$, i.e. represents a family of objects of the subcategory, then by (Sum) we have $!_B = !_A \circ b \in \mathcal{D}$ and therefore B is in the subcategory, too, and the display map b is a morphism of the subcategory. (Sum) just states this constraint relative to arbitrary contexts.

From conditions (i) and (ii) it follows that for all objects I of $\mathcal{C}$ the comma category $\mathcal{D}/I$ is a locally cartesian closed category and the embedding of $\mathcal{D}/I$ into $\mathcal{C}/I$ preserves and reflects the structure of locally-cartesian-closedness. That the embedding preserves finite limits is an immediate consequence of internal completeness, i.e. condition (i). The only crucial condition which has to be checked is the following.

(Sub-lcc) For all morphisms f, g in $\mathcal{C}$ and display maps a in $\mathcal{D}$
if $a \circ f$ and $a \circ f \circ g$ are in $\mathcal{D}$ then $a \circ \Pi_f g$ is in $\mathcal{D}$, too.

This follows very easily from (Sum): if a is in $\mathcal{D}$ and $a \circ f$ and $a \circ f \circ g$ are in $\mathcal{D}$ then by Corollary 1 it follows that f and g are display maps, by ($\forall$) also $\Pi_f g \in \mathcal{D}$ (although condition (Sub-Π) would be sufficient) and finally by (Sum) we have that $a \circ \Pi_f g$ is in $\mathcal{D}$.

Now it is natural to ask whether the implication holds in the reverse direction, too, i.e. whether (Sub-lcc) does imply (Sum) under the assumption of the conditions (Term), (Pb), ($\forall$) and (Gen). From (1) of Theorem 2 we already know that these conditions alone do *not* imply (Sum).

This motivates the following conjecture :

There exists a locally cartesian closed category $\mathcal{C}$ *and a class* $\mathcal{D}$ *of display maps, such that the full subcategory given by* $\mathcal{D}$ *is internal and internally complete and for any object* I *in* $\mathcal{C}$ *the comma category* $\mathcal{D}/I$ *is locally cartesian closed and* $\mathcal{D}$ *is* <u>*not*</u> *closed under composition.*

The problem does not seem to be very important from the point of view of the categorical semantics of calculi of dependent types since the condition (Sub-lcc) does not admit an easy and intuitive explanation in terms of a rule of a theory of dependent types. But the problem seems to be interesting at least from the point of view of *fibrational category theory* . It asks the question whether a weak and a strong notion of full, internal, internally complete sub-locally-cartesian- closed-category coincide in a fibrational framework..

A first guess would be that $\mathcal{P}_1$ solves the problem. But one can show that condition (Sub-lcc) is not satisfied for $\mathcal{P}_1$, which proves that (Sub-lcc) is independent from internal completeness.

To see that we can adapt the counterexample for (Sum) given in the proof of Theorem 2. The morphism $g : \underline{Y} \to \underline{X}$ is not a morphism in $\mathcal{P}_1/1$ (as $!_{\underline{Y}} : \underline{Y} \to 1$ is not in $\mathcal{P}_1$) but it is isomorphic to $\Pi_h k$ for some morphisms h and k in $\mathcal{P}_1/1$ (let h be classified by χ which maps $\lambda n.0$ to $\{\langle n,m\rangle : \{n\}(0) = \{m\}(0) = 0\}$ and all other objects of X to $\{\langle n,m\rangle : \{n\}(0) = \{m\}(0) = 1\}$ and let k be classified by the morphism which constantly maps every argument to $\{\langle n,m\rangle :$ $\{n\}(0)$ and $\{m\}(0)$ are both defined and equal $\}$).

So in order to prove the conjecture above one needs a new idea which is still missing.

3. Summary

We have proven some independence results for impredicative calculi of dependent types by *modifying* the realizability model due to Hyland and Moggi in several ways. Our main emphasis has been to prove that the condition (Sum) is independent from all other concepts.

We think this emphasis is justified as a model where the condition (Sum) holds but the condition (P-Id) and therefore also the condition (Pb) are not satisfied is provided by the domain-like model introduced by Hyland and Pitts in [Hy,Pi] since in that model any type is inhabited.

So we finally give a listing of all known dependencies and independencies between the conditions (P-Id), (Sum) and (Pb) for locally cartesian closed categories $\mathbb{C}$ with a class $\mathfrak{D}$ of display maps stable under arbitrary pullbacks satisfying the conditions (Term), (Prod), ($\forall$) and (Gen), i.e. full internal subcategories of locally cartesian closed categories having all internal products.

(1) There exists a model where (Sum), (Pb) and (P-Id) are all false (as shown by (3) of Theorem 2).

(2) (Sum) + (Pb) is consistent as shown by the well known realizability model.

(3) (P-Id) implies neither (Pb) nor (Sum) (as shown by (2) of Theorem 2).

(4) (Pb) implies (P-Id) (follows from Lemma 2), but (Pb) does not imply (Sum) (follows from (1) of Theorem 2).

(5) (Sum) does not imply (P-Id) as shown by the domain-like models and therefore also does not imply (Pb).

(6) (Sum) + (P-Id) implies (Pb) (as shown by Lemma 1) and therefore is equivalent to (Sum) + (Pb).

Acknowledgements

I want to thank Martin Hyland and Andy Pitts for making unpublished work available to me and discussing with me the problems of categorical semantics of calculi of dependent types. Especially I want to thank Pino Rosolini from whom I learned that the subset Prop-1 of PERω provides a model for the calculus of constructions.

Finally I want to thank Bart Jacobs for his careful reading of a draft and the suggestion of a lot of stylistic improvements.

References

[Hy,Pi] M. Hyland, A. Pitts *The Theory of Constructions : Categorical Semantics and Topos-Theoretic Models* preprint, to appear in the Proceedings of the Conference on *Categories in Computer Science and Logic* , Contemporary Mathematics, Amer. Math.Soc., 1988.

[Me] J. Meseguer *Universe Models and Initial Model Semantics for the Second Order Polymorphic Lambda Calculus* abstract in "Abstracts of papers presented to A.M.S.", Issue 58, Vol. 9, Num. 4, 1988.

[Pi] A. Pitts *Polymorphism is Set-Theoretic, Constructively* Proc.of the Conference on Category Theory and Computer Science, Edinburgh 1987, SLNCS 283, 1987.

[Se1] R. Seely *Locally Cartesian Closed Categories and Type Theory* . In Math. Proc. Cambridge Phil. Soc. 95, 1984.

[Se2] R. Seely *Categorical Semantics for Higher Order Polymorphic Lambda Calculus* Journal of Symbolic Logic, 1987.

[Str] T. Streicher *Correctness and Completeness of a Categorical Semantics of the Calculus of Constructions* Ph.D.Thesis, Univ. of Passau, 1988.

Quantitative Domains, Groupoids and Linear Logic

by

Paul Taylor

Department of Computing,
Imperial College of Science and Technology,
180 Queen's Gate,
London SW7 2BZ
01 589 5111 ext 5057
pt@doc.ic.ac.uk

Abstract

We introduce the notion of a ***candidate*** for "multiple valued universal constructions" and define ***stable functors*** (which generalise functors with left adjoints) in terms of factorisation through candidates. There are many mathematical examples, including the ***Zariski spectrum*** of a ring (as shown by Diers [81]) and the ***Galois group*** of a polynomial, but we are mainly interested in Berry's [78] ***minimum data property***. In fact we begin with a completely non-mathematical example.

The aim is to find domain models in which terms of the typed or polymorphic λ-calculus are interpreted as stable functors. We study Girard's ***quantitative domains*** [85], in which information is represented by a collection of *tokens* from a universe of tokens for a particular type, and there is no restriction on the ability of different tokens to co-exist or on the number of occurrences of a particular token. This idea may be used to code ***parallelism*** (with no suppression of duplicated output) or ***accounted resources***.

Unfortunately Girard did not fully describe the function-spaces, which should be equipped with the "***Berry order***"; this turns out to mean that function-tokens must have "internal symmetries". It is our purpose to describe the smallest cartesian closed category with these function-spaces which contains **Set** (the simplest non-trivial quantitative domain, with one token which may appear arbitrarily often) as an object.

The natural way of presenting this is as a new interpretation of ***Linear Logic*** given by **group** (and more generally groupoid) actions. These stand in the same relation to quantitative domains as ***coherence spaces*** do to qualitative domains, and there is a kind of coherence between group(oid) elements. By a similar analysis of stable functors we obtain an ***of course*** operation. Finally, our (generalised) quantitative domains themselves form a domain of this kind with ***rigid comparisons*** as morphisms, and hence we have a ***type of types***.

1. Stable Functors

1.1. The Minimum Data Property

Imagine marking examination scripts in history, the question set being

What were the causes of the Second World War?

Of course it is not our business to pass judgement on this question ourselves, but merely to imagine the process whereby the examiner awards marks for what the student has written. In programming terms, she executes a function from *scripts* to *numbers* which is defined by her *mark-sheet.*

This is what a typical student wrote:

The Germans invaded Poland.
The Japanese invaded Manchuria.
The Italians bombed Pearl Harbor.
Marie-Antoinnette said "let them eat cake."

He obviously gets marks for the first two assertions but not the last two: the third because it is false (or a confusion) and the last because it is irrelevant (noise).

Perhaps the examiner had the following mark-sheet:

Japanese invaded Manchuria.	*10 marks*
Japanese bombed Pearl Harbor.	*5 marks*
Germans invaded Poland.	*5 marks*
Germans invaded Czechoslovakia.	*10 marks*
Italians invaded Abyssinia.	*10 marks*

and so she gave this student 15 marks.

Suppose our student had just written

Germans invaded Poland.

and so got 5 marks. Obviously writing less than this (*i.e.* omitting any of these three words) would have got him none at all, and adding comments about Marie-Antoinette, King Canute or his history teacher would only have wasted time. So these three words are *the least part of the script which would gain these marks.* The same is true of each of the other four lines of the mark-sheet. Observe that (i) although the Polish marks and those for Pearl Harbor are interchangeable for the purpose of awarding the final grade, they accumulate rather than become identified, and (ii) it's no good trying to match half of one pattern (Pearl Harbor) with half of another (the Italians).

Now let us formalise this a bit[1]. The marking process is a function S which takes a script X and returns a number of marks SX. Suppose that the script X includes the assertion that the Germans invaded Poland; then five marks $Y = 5$ are *distinguishably* included in the total SX, by a function $w : Y \to SX$. (If it also included the assertion about Pearl Harbor there would be a *different* function $w' : Y \to SX$.) Let X_0 be the one-line script containing just this assertion, then again $u : Y \to SX_0$, and since this is a sub-script, $f : X_0 \to X$. We don't write $X_0 \subset X$ because parts of the pattern X_0 might be counted twice, for instance if

$X_0 =$

Germans invaded Poland.
Germans invaded Czechoslovakia.

and

$X =$

The Germans invaded Poland and Czechoslovakia.

then the function is not mono on "Germans invaded".

The minimal sub-script X_0 has a certain "universal property". Suppose $g : X' \to X$ is another sub-script which also wins *these same* marks, so that $v : Y \to SX'$ *and* $Sg \circ v = Sf \circ u$, then X' contained the one-line script X_0, *i.e.* $h : X_0 \to X'$ such that[2] $v = Sh \circ u$ and $g \circ h = f$. (h is not mono, for the same reason as f).

Let us write this definition in its general form.

Definition 1.1.1 Let $S : \mathcal{X} \to \mathcal{Y}$ be a functor. Then the map $u : Y \to SX_0$ in $\mathcal{Y}$ is said to be a ***candidate*** for [a universal map from the object Y to the functor] S if for any triple of maps $v : Y \to SX'$ in $\mathcal{Y}$ and $f : X_0 \to X$ and $g : X' \to X$ in $\mathcal{X}$, such that the square

[1] The reader with little mathematical background should concentrate on section 2 and ignore the remainder of the paper.

[2] This is *left handed* composition, contrary to my personal habits, so that later $A \vdash B$ will be a left action of A and a right action of B, not *vice versa*.

$$\begin{array}{ccc} Y & \xrightarrow{\ u\ } & SX_\circ \\ {\scriptstyle v}\big\downarrow & & \big\downarrow{\scriptstyle Sf} \\ SX' & \xrightarrow{\ Sg\ } & SX \end{array}$$

commutes, there is a *unique* $h : X \to X''$ such that *both* triangles

$$\begin{array}{ccc} Y & \xrightarrow{\ u\ } & SX_\circ \\ {\scriptstyle v}\big\downarrow & \swarrow{\scriptstyle Sh} & \\ SX' & & \end{array} \qquad \text{and} \qquad \begin{array}{ccc} & & X_\circ \\ & \swarrow{\scriptstyle h} & \big\downarrow{\scriptstyle f} \\ X' & \xrightarrow[\ g\ ]{} & X \end{array}$$

commute. Note that $X_\circ$, as well as u, Y and $SX_\circ$, is part of the data defining the candidate. The word candidate is used to signify that this is one of many (for given Y), and derives from the special case of coproduct candidates (this term appears in [Lamarche 88]).

There are numerous mathematical[3] examples of this idea, and we only give a few representatives; the first is typical of the special case of functors with left adjoint.

Example 1.1.2 Let S be the forgetful functor from $\mathcal{X} = \mathbf{Gp}$ to $\mathcal{Y} = \mathbf{Set}$ and $Y \in \mathcal{Y}$. Then $u : Y \to SX_\circ$ is a candidate iff $X_\circ$ is the *free group* on Y and u is the inclusion of generators. In this case, there is only one candidate (up to unique isomorphism) for S for given Y. □

Example 1.1.3 Let $\mathcal{X} = \mathbf{IntDom}$, the category of integral domains and *mono*morphisms, and $\mathcal{Y} = \mathbf{CRng}$, the category of commutative rings and homomorphisms. Then $u : Y \to SX_\circ$ is a (quotient) candidate iff $X_\circ$ is the quotient of Y by a *prime ideal*; any particular ring Y may have many quotient candidates. In this case (as in the previous one) $f = g \circ h$ is a corollary of $Sh \circ u = v$. □

Example 1.1.4 Let p be a polynomial with integer coefficients. Let $\mathcal{X} = \mathbf{Fld}[p]$ be the category of fields in which p splits into linear factors and S be its inclusion in $\mathcal{Y} = \mathbf{Fld}$, the category of all fields (and homomorphisms). Then $u : Y \to SX_\circ$ is a candidate iff $X_\circ$ is the ***splitting field*** for p over Y. In this case there is only one candidate for given Y, but it has many automorphisms; indeed they form the ***Galois group*** of p over Y. □

Example 1.1.5 Consider a program S consisting of several parallel processes which merge their output indiscriminantly ***without suppression of duplication***; for instance "parallel or" would output t *twice* on input $\langle \mathtt{t}, \mathtt{t} \rangle$. Suppose that on input X (a bag of tokens from A) its output includes an instance $Y = \{j\}$ of a token[4] $j \in B$. This has come from a particular process, which itself has pursued a ***sequential*** execution path, involving certain "hurdles" which amount to reading and matching a pattern $X_\circ$ in X; moreover if X had contained only this pattern, Y would still have been output. The candidate is the function $u : Y \to SX_\circ$ which identifies this instance of the token j in the output, but there may have been many other ways in which this or other parallel processes could have generated j, but identified by different u's.

[3] All prerequisites from mainstream pure mathematics will be found in, for example, [Cohn 77] or [Lang 65].

[4] The reason for the convention $j \in B$ will become clear in section 2.3.

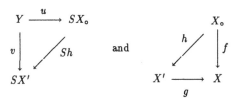

1.2. Factorisation

Stable functors generalise functors with left adjoints, and can be characterised as functors which acquire adjoints whenever they are restricted to slices (down-sets or principal lower sets). The following definition is, however, the most useful.

Definition 1.2.1 A functor $S : \mathcal{X} \to \mathcal{Y}$ is ***stable*** if every map $w : Y \to SX$ factors as $Sf \circ u$ with $u : Y \to SX_\circ$ a candidate and $f : X_\circ \to X$. By definition of candidacy, it is immediate that this factorisation is unique up to unique isomorphism.

Example 1.2.2 [Vickers] Let $\mathcal{X}$ be the category of complete Boolean algebras and *frame monomorphisms* and $\mathcal{Y}$ be the category of frames and homomorphisms, with $S : \mathcal{X} \to \mathcal{Y}$ the forgetful functor. Then S is stable.

Definition 1.2.3 A ***wide pullback*** is a diagram $d : \mathcal{I} \to \mathcal{X}$ (or its limit, according to context) such that $\mathcal{I}$ has a terminal vertex (and has a set rather than a proper class of vertices and edges). Thus ordinary (binary) pullbacks are wide pullbacks, and any cofiltered limit diagram is equivalent to a wide pullback, but equalisers, binary products and terminal objects are not.

Exercise 1.2.4 Show that stable functors preserve wide pullbacks and monos. □

Suppose we are given a functor $S : \mathcal{X} \to \mathcal{Y}$; how might we prove that it is stable? We may consider *all* possible factorisations $w = Sf' \circ u'$; these form a category whose morphisms are as illustrated:

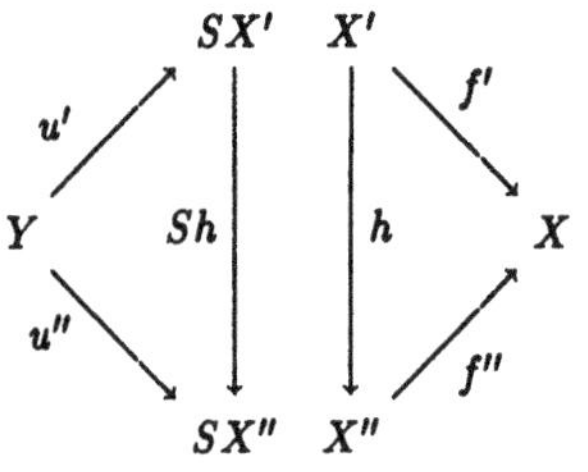

Clearly the factorisation is through a candidate iff the corresponding object of this category is *initial*. We know that stable functors must preserve wide pullbacks, and the idea of the converse is to take the wide pullback of all possible factorisations. The problem is that we are only allowed to do this with a *set* of them, and so we need the following:

Definition 1.2.5 $S : \mathcal{X} \to \mathcal{Y}$ satisfies the ***solution set condition*** if for every $w : Y \to SX$ the (possibly large) category of factorisations has a *small full* cofinal subcategory. This means that for any given factorisation $Sf'' \circ u''$ there is a factorisation $Sf' \circ u'$ in the subcategory and a morphism h with $Sh \circ u' = u''$ and $f'' \circ h = f'$, and that between any two factorisations in the subcategory there is only a set of morphisms h.

There are important examples where this condition fails, but in the cases which are of interest to us in this paper it will hold automatically (essentially because we shall be using toposes).

Theorem 1.2.6 Let $S : \mathcal{X} \to \mathcal{Y}$ be any functor, where the category $\mathcal{X}$ has wide pullbacks. Then the following are equivalent:
(α) S is stable, *i.e.* the factorisation property holds;
(β) S has a left adjoint on each slice;
(γ) S preserves wide pullbacks and satisfies the solution set condition. □

We're going to be interested in stable functors in this paper, but we want to assume something else of them (and will modify the terminology accordingly).

Definition 1.2.7 A functor is ***continuous*** if it preserves filtered colimits.

Definition 1.2.8 An object $X \in \mathcal{X}$ is ***finitely presentable*** if the functor $\mathsf{Hom}_{\mathcal{X}}(X, -) : \mathcal{X} \to \mathbf{Set}$ is continuous. This means that if we have $f : X \to \mathsf{colim}^{\uparrow} Y^{(r)}$ then for some (non-unique) r_o and $g : X \to Y^{(r_o)}$ we have $f = \nu^{(r_o)} \circ g$, where $\nu^{(r_o)} : Y^{(r_o)} \to \mathsf{colim}^{\uparrow} Y^{(r)}$ belongs to the colimiting cocone. The corresponding property for lattices is called ***compactness***.

The following exercise serves as a good test of the reader's understanding of the important concepts of candidates, continuity and finite presentability, although it doesn't use stability.

Exercise 1.2.9 Suppose $u : Y \to SX_o$ is a candidate, where S is continuous, and Y is finitely presentable. Then X_o is also finitely presentable. □

1.3. Bags and Power Series

Now we shall turn our attention to the domains of specific interest to us.

Definition 1.3.1 Let A be a set. A ***bag*** of elements of A is an assignment of an abstract set X_i (its ***multiplicity***) to each element $i \in A$. Abstractly, a bag is represented by a multiplicity function $X_- : A \to \mathbf{Set}$ or by a display $x : X = (\bigcup_{i \in A} X_i) \to A$. Hence a finite bag may be written as an unordered list (with repetition) whose terms are from A.

Definition 1.3.2 $\mathbf{Set}^A$ denotes the category of bags of elements of A. There are three ways of seeing its morphisms: (α) as an A-indexed family of functions $f_i : X_i \to Y_i$; (β) as a natural transformation $f : X_- \to Y_-$ between functors $A \rightrightarrows \mathbf{Set}$, or ($\gamma$) as functions $f : X \to Y$ such that $y \circ f = x$.

Exercise 1.3.3 Show that a bag is finitely presentable (Definition 1.2.9) iff it is finite, *i.e.* all elements of A have finite multiplicity and all but finitely many of them have zero multiplicity. The notation $\vec{n}$ will be used for a finite bag, where n_i is the multiplicity of $i \in A$. □

Girard's quantitative domains were the categories of the form $\mathbf{Set}^A$; we shall begin by investigating stable functors between these, but will find that more complicated categories are needed for functor-spaces.

Lemma 1.3.4 Let $\vec{n}$ be a finite bag. The ***representable functor***

$$(-)^{\vec{n}} : \mathbf{Set}^A \to \mathbf{Set} \qquad \text{by} \quad X \mapsto \mathbf{Set}^A(\vec{n}, X) = \prod_i (X_i)^{n_i}$$

(which acts on morphisms by composition) is stable and continuous.

Proof Continuity is immediate from the Exercise, and actually there is a left adjoint. I claim that $u : Y \to (X_o)^{\vec{n}}$ is a candidate iff its exponential transpose[5] $e : Y \times \vec{n} \to X_o$ is an isomorphism. For in

$$\begin{array}{ccc}
Y & \xrightarrow{\;u = \ulcorner e \urcorner\;} & (\vec{n})^{\vec{n}} \\
{\scriptstyle v = \ulcorner h \circ e \urcorner}\downarrow & \swarrow{\scriptstyle h \circ -} & \downarrow{\scriptstyle f \circ -} \\
(X')^{\vec{n}} & \xrightarrow[\;g \circ -\;]{} & (X)^{\vec{n}}
\end{array}$$

it is clear that h exists and is uniquely determined by v iff e is invertible. The factorisation of $w : Y \to X^{\vec{n}}$ is $(f \circ -) \circ u$ where $u = \ulcorner \mathrm{id} \urcorner$ and f is the exponential transpose of w. □

[5] Actually $Y \times \vec{n}$ is a copower, i.e. Y-fold coproduct, not a product.

Girard's power-series expansion amounts to a sum of representable functors ("monomials"), so we have to show that we can compute sums pointwise. For coherence spaces this is not possible (directly resulting in the non-representability of *parallel or*) and so we have to make crucial use of the idea that quantitative domains admit arbitrary sums *without suppression of duplication*. Categorically, this amounts to the following two properties:

Proposition 1.3.5 Sums in $\mathbf{Set}^A$ are disjoint and universal.

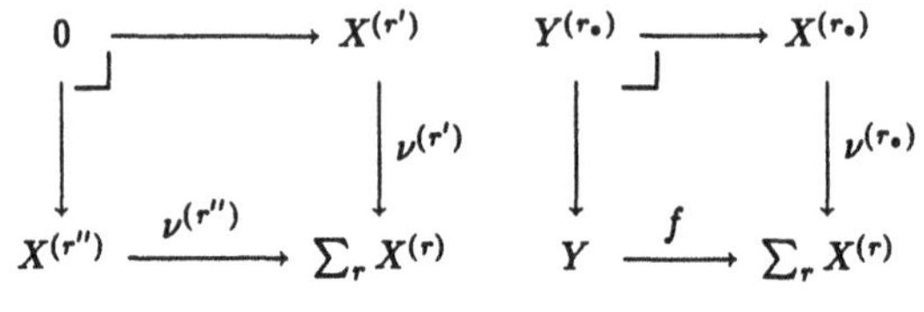

Disjoint means that when we form the intersection of different inclusions ($r' \neq r''$) we get the initial object 0. ***Universal*** means that when we form the pullback of the coproduct diagram against an arbitrary map f, we get another coproduct diagram: $Y \cong \sum_r Y^{(r)}$. In the special case of the empty coproduct (the initial object) this is equivalent to ***strictness***, *i.e.* any map $Y \to 0$ is an isomorphism. □

We defer to section 1.5 the proof that we may compute sums pointwise, and merely state that any bag $(c_{\vec{n}})$ of finite bags gives rise to a stable functor:

$$SX = \sum_{\vec{n}} c_{\vec{n}} \times X^{\vec{n}}$$

Now we shall prove the converse: every stable functor $\mathbf{Set}^A \to \mathbf{Set}$ is (isomorphic to one) of this form. The following lemma will be the source of the complication (creeds) which we shall meet later.

Lemma 1.3.6 Let $S : \mathbf{Set}^A \to \mathbf{Set}$ be a continuous stable functor and $u : 1 \to S\vec{n}$ be a candidate. Suppose that $h : \vec{n} \to \vec{n}$ in $\mathbf{Set}^A$ is such that $Sh \circ u = u$. Then $h = \mathrm{id}$.

Proof The notation $\vec{n}$ is justified because we know that it is finite. Let $X = A$ be the bag in which each element of A occurs just once (so $X_i = 1$); there is a unique map $f : \vec{n} \to X$ in $\mathbf{Set}^A$, *i.e.* X is the ***terminal object***. Then $Sf \circ u = Sf \circ u$, so there is a unique $h : \vec{n} \to \vec{n}$ in $\mathbf{Set}^A$ with $Sh \circ u = u$ and $f \circ h = f$; clearly both the given h and also the identity will do, so by uniqueness $h = \mathrm{id}$. □

Theorem 1.3.7 Every stable functor $\mathbf{Set}^A \to \mathbf{Set}$ is a power-series.

Proof For each [isomorphism class of] finite bag $\vec{n}$, let $c_{\vec{n}}$ be the set of equivalence classes of candidates $u : 1 \to S\vec{n}$, where we identify u with $Sh \circ u$ for automorphisms $h : \vec{n} \cong \vec{n}$. If $w \in SX$, then this is a map $w : 1 \to SX$, which factorises as $Sf \circ u$, and isomorphic factorisations belong to the same equivalence class; hence w corresponds to a unique element of $c_{\vec{n}}$ together with a function $f \in X^{\vec{n}}$. □

Girard proved this result on the additional assumption that stable functors are to preserve equalisers; but by Lemma 1.3.6 this assumption is actually a consequence of having a terminal object. However Lamarche has shown (privately) that *evaluation does not preserve equalisers*, and so this stronger notion of stability does not lead to a cartesian closed category. We shall in fact see that for higher types there is no terminal object, and so this lemma breaks down.

Exercise 1.3.8 Let $S : \mathcal{X} \to \mathcal{Y}$ be a stable functor between categories with equalisers. Show that S preserves equalisers iff Lemma 1.3.6 holds, *i.e.* for every candidate $u : Y \to SX$ and automorphism $h : X \cong X$, we have $Sh \circ u = u \Rightarrow h = \mathrm{id}$.

In this case, commutativity of the second triangle in Definition 1.1.1 is automatic, and we are reduced to Diers' original definition of diagonal universality (candidacy); *cf.* Example 1.1.3.

The power-series expansion was really as far as Girard developed this topic. We shall proceed by insisting on the notion of stable functor, but one may alternatively take the point of view that the power-series is itself the essential feature. This is the subject of work in progress by Lamarche [89] on polynomial algebras, which is also the idea behind [Joyal 87].

1.4. Cartesian Transformations

Morphisms between bags (in the category $\mathbf{Set}^A$) are "colour-preserving" (1.3.2γ) functions $X \to Y$; *i.e.* the set X_i of elements of kind i is taken to the set Y_i. Although the copies of i are regarded as being alike, we nevertheless consider as different functions which affect them differently; in particular the "switch" function on the bag $\{\bullet, \bullet\}$ is not the same as the identity.

With this in mind we look for the morphisms of the stable function-space, which we call

$$[\mathcal{X} \to \mathcal{Y}]$$

If $S, T : \mathbf{Set}^A \rightrightarrows \mathbf{Set}^B$ are two stable functors, for each A-bag X there are B-bags SX and TX, and a morphism $\phi : S \to T$ (in particular) gives functions $\phi X : SX \to TX$. In the special case where $A = B = 1$ and $S = T = (-)^2$, the identity and switch functions $(-)^2 \to (-)^2$ are regarded as being different.

Recall that stable functors preserve (wide) pullbacks. Since we're aiming to describe a cartesian closed category, we need in particular that the evaluation functor $\mathsf{ev} : [\mathcal{X} \to \mathcal{Y}] \times \mathcal{X} \to \mathcal{Y}$ be stable. There is a particular pullback square which it must preserve, for any given morphisms $\phi : S \to T$ in $[\mathcal{X} \to \mathcal{Y}]$ and $f : X' \to X$ in $\mathcal{X}$, namely

$$\begin{array}{ccc} \langle S,X'\rangle & \xrightarrow{\langle\phi,\mathrm{id}\rangle} & \langle T,X'\rangle \\ {\scriptstyle\langle\mathrm{id},f\rangle}\downarrow & \lrcorner & \downarrow{\scriptstyle\langle\mathrm{id},f\rangle} \\ \langle S,X\rangle & \xrightarrow{\langle\phi,\mathrm{id}\rangle} & \langle T,X\rangle \end{array} \qquad\qquad \begin{array}{ccc} SX' & \xrightarrow{\phi X'} & TX' \\ {\scriptstyle Sf}\downarrow & \lrcorner & \downarrow{\scriptstyle Tf} \\ SX & \xrightarrow{\phi X} & TX \end{array}$$

giving the square on the right. This leads to the

Definition 1.4.1 A ***cartesian transformation*** $\phi : S \to T$ is an assignment of a function $\phi X : SX \to TX$ in $\mathcal{Y}$ to each $X \in \mathcal{X}$ making the right-hand square above a *pullback* for each $f : X' \to X$. If the square merely commutes for each f, we say ϕ is ***natural***.

Our main concern is to investigate the relationship between cartesian transformations and functions between bags of finite bags. This relationship is a close one, but it is not as direct as Girard would have us believe, as the following example shows:

Example 1.4.2 The squaring functor $(-)^2 : \mathbf{Set} \to \mathbf{Set}$ is stable, being represented by the singleton bag $\{\ulcorner 2 \urcorner\}$. The identity and switch functions $(-)^2 \to (-)^2$ are ***distinct*** cartesian transformations from this functor to itself, whereas a singleton bag has only one endofunction (the identity). □

This shows that the token $\vec{n}$ which stands for the representable functor has "internal structure" which we must take into account. This becomes relevant when we consider higher-order functions. For instance, in a large public examination, many examiners must be employed to mark the scripts, and so there must be an "examiner of examiners" (moderator) to ensure fairness. The moderator would observe the behaviour of the examiners on typical scripts (say on the minimal ones we discussed). Instead of searching for particular words in scripts, the moderator would test for the inclination of the examiner to award marks for particular phrases. This has two important consequences.

First, the stable order is capable of detecting *lazy* behaviour, *i.e.* when the process outputs without reading its input. For instance there may be "free marks" awarded to all candidates irrespective of performance; one marker may award these marks automatically, even to students who failed to return scripts, whilst another may require to see a (blank) sheet of paper before giving the marks. The moderator, of course, will detect this behaviour. Phrasing this more computationally, the stable order can detect that a process has read its input, whilst the pointwise order can only detect that it has not.

Second, although this is not something we can see in the examples given, there are "internal symmetries" of tokens at higher order. A first order function may be searching for two occurrences of a (zero-order) token x in its input; its output will then be the *square* of the number of occurrences (this is a *representable functor*). Multiples of this functor are also stable, but the functor "$\frac{1}{2}X^2$" (which returns the set[6] of *unordered pairs*, or two-element bags) is not (by Lemma 1.3.6). However if we test this functor, we can look for occurrences of the pattern "match $x^{(1)}, x^{(2)}$", which is isomorphic to the pattern "match $x^{(2)}, x^{(1)}$", and we *are* permitted to count these two patterns (which necessarily occur together) as one if we wish. This illustrates that higher-order pattern-matching is inherently more complicated, and may help the reader to grasp the behaviour of creeds in section 3.

Returning to candidates and cartesian transformations the correspondence is summed up abstractly by the following result:

Lemma 1.4.3 Let $S, T : \mathcal{X} \to \mathcal{Y}$ be functors and $\phi : S \to T$ a cartesian transformation. Then $u' : Y \to SX$ in $\mathcal{Y}$ is a candidate for S iff $u = \phi X \circ u' : Y \to TX$ is a candidate for T. Conversely, if S is stable and postcomposition with the natural transformation ϕ preserves candidates then ϕ is cartesian.

Proof The following diagram was discovered independently by Lamarche (with Y atomic); we shall use it three times.

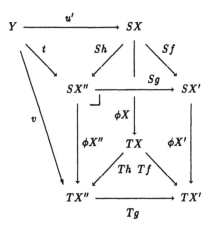

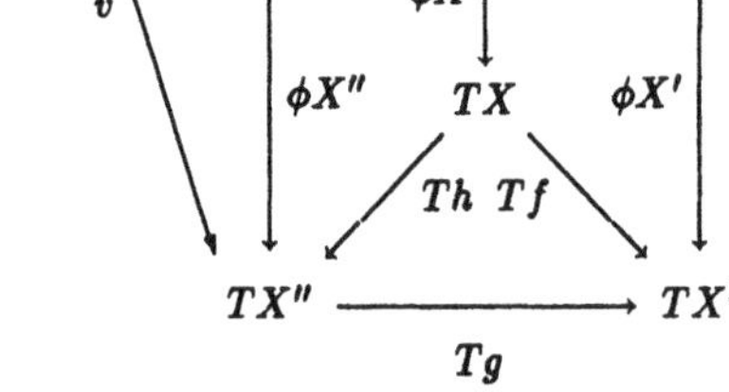

We are given f, g and either t or v.

[⇒] Given v, let t mediate the pullback and define h by candidacy of u'; then it mediates for u. Conversely h determines $t = Sh \circ u'$.

[6] "$\frac{1}{2}$" does not mean numerical division here, but quotient by the natural action of the group of order 2.

[$\Leftarrow$] Given t, put $v = \phi X'' \circ t$ and define h by candidacy of u, so that $f = g \circ h$. Then t and $Sh \circ u'$ both mediate for the pullback. But any h which makes the diagram commute mediates for u.

verse] Use stability of S to factorise one of the sides of an arbitrary commutative square as Sf o[eon Define h by candidacy of u and put $t = Sh \circ u'$; conversely t determines h by candidacy of u'.

In all three parts we have to use *naturality* of ϕ, *i.e.* the commutation of the square faces of the prism. □

Cartesian transformations therefore correspond to functions between sets of candidates. In the examination example, the candidates correspond to items in the mark-sheet, *i.e.* pairings of a minimal answer with the number of marks it gains. A cartesian transformation $\phi : S \to T$ from one examiner S to another T is determined by a function which assigns to each line of the first mark-sheet a line of the second. In particular T gives marks for at least as many things as S does, whatever the nature of ϕ; however some of the marks which S gave might become collapsed. Thus S might end up giving fewer marks than T, unless ϕ is mono.

However although T may be more willing to award marks than S, she never awards the same marks for less information. (This is the effect of cartesianness.)

Exercise 1.4.4 Show that every cartesian transformation into a representable functor $S \to (-)^{\vec{n}}$ is an isomorphism, so that representables are **atomic**. Moreover the automorphisms of $(-)^{\vec{n}}$ are given by composition with an automorphism of $\vec{n}$ [Hint: Yoneda]. □

Remark 1.4.5 Lamarche has studied the similar situation with $\mathcal{M}$, the category of sets and *monomorphisms*, instead of **Set**. In this case, the representable functors, $\mathcal{M}(\vec{n}, -)$, are still atomic, but they are not the same as the powering functors $(-)^{\vec{n}} = \mathbf{Set}^A(\vec{n}, -)$. Indeed as functors $\mathcal{M} \to \mathcal{M}$, $X \times X \cong \mathcal{M}(1, X) + \mathcal{M}(2, X)$, where the first inclusion is the diagonal $\Delta : X = \mathcal{M}(1, X) \to X \times X$, which is *not* cartesian as a transformation between functors **Set** $\to$ **Set**. The second component consists of the (ordered) *unequal* pairs.

1.5. Power Series Revised

We shall now reformulate Girard's result to give a precise characterisation of the stable function-space $[\mathbf{Set}^A \to \mathbf{Set}]$. First we show that we can form sums of stable functors.

Proposition 1.5.1 Suppose ($\mathcal{X}$ and) $\mathcal{Y}$ have disjoint universal coproducts, and let $S^{(r)} : \mathcal{X} \to \mathcal{Y}$ be stable functors. Then the pointwise coproduct $S : X \mapsto \sum_r S^{(r)} X$ is a stable functor, the inclusion $\nu^{(r_\circ)} : S^{(r_\circ)} \to \sum_r S^{(r)}$ is cartesian and this is the coproduct in $[\mathcal{X} \to \mathcal{Y}]$.

Proof We use the diagram to show (a) that $\nu^{(r)}$ is cartesian, (b) how to factorise maps using (c) "sums" of candidates, and (d) that $\sum_r S^{(r)}$ is the coproduct in $[\mathcal{X} \to \mathcal{Y}]$.

[a] Suppose $\nu^{(r_\circ)} X \circ t = \sum_r S^{(r)} g \circ v$. Define $Y^{(r)}$ by making the left face a pullback. Then we have $Y^{(r)} \to S^{(r)} X$ and $Y^{(r)} \to Y \xrightarrow{t} S^{(r_\circ)} X$, so this factors through the intersection, which (for $r \neq r_\circ$) is 0 by disjointness. Then $Y^{(r)} \cong 0$ by strictness and $Y \cong Y^{(r_\circ)}$ by universality, so $v^{(r_\circ)}$ is (essentially) the required mediator. Uniqueness is easy.

[b] Given $w : Y \to \sum_r S^{(r)} X$, factorise its pullback $(\nu^{(r)} X)^* w = S^{(r)} f^{(r)} \circ u^{(r)} : Y^{(r)} \to S^{(r)} X$ with $u^{(r)} : Y^{(r)} \to S^{(r)} X^{(r)}$ a candidate for $S^{(r)}$ and $f^{(r)} : X^{(r)} \to X$. Define $v^{(r)}$ using cartesianness and u out of the coproduct (by universality).

[c] Using candidacy of $u^{(r)}$ we have diagonal fill-ins $h^{(r)} : S^{(r)} X^{(r)} \to S^{(r)} X'$ at the back, and the required fill-in is their sum $h = \sum_r h^{(r)}$ at the front.

[d] Given a cocone $S^{(r)} \to T$ over the diagram, there is a unique *natural* mediator τ: we have to show it's cartesian. If $w : Y \to \sum_r S^{(r)} X$ and $Y \to TX'$ make the diagram commute, we find $v^{(r)}$ by cartesianness of the cocone and $v = \sum_r v^{(r)}$ as their sum, using universality of sums again. □

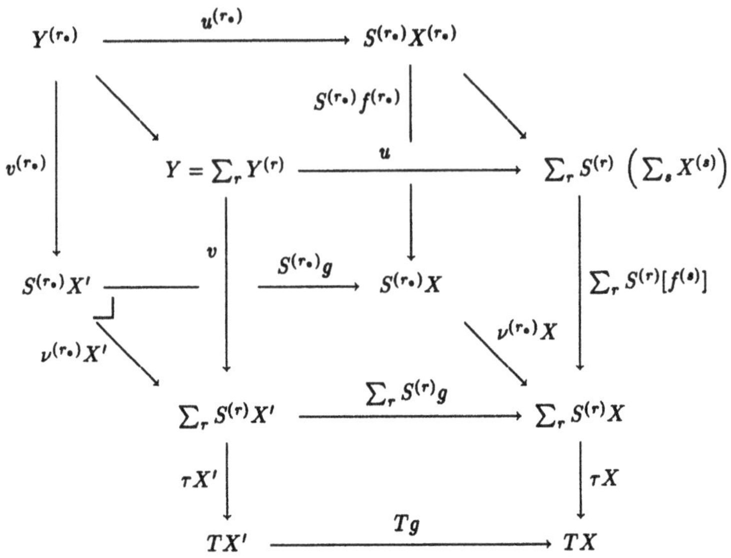

To explain this result in the terms of the original example, suppose S' is the behaviour of the history examiner and S'' that of the geography examiner. The function $w : Y \to S'X + S''X$ means that certain specific marks were awarded in the two examinations, $Y' \subset Y$ of them in history and $Y'' \subset Y$ in geography. Only the part $X' \to X$ of the script was actually considered worthwhile history (and this got the Y' marks), and similarly $X'' \to X$ in geography, so altogether $X' + X'' \to X$ was what earned the student the Y marks.

Exercise 1.5.2 Suppose $\mathcal{X}$ has binary coproducts and that $S : \mathcal{X} \to \mathcal{Y}$ and $T : \mathcal{X} \to \mathcal{Z}$ are stable. Show that $\langle S, T\rangle : \mathcal{X} \to \mathcal{Y} \times \mathcal{Z}$ is stable. In what sense are binary coproducts necessary? □

We can now characterise cartesian transformations between stable functors $\mathbf{Set}^A \to \mathbf{Set}$ precisely, but to describe this we have to modify the power-series representation. First we reformulate Lemma 1.3.6.

Lemma 1.5.3 Let $S : \mathbf{Set}^A \to \mathbf{Set}$ be a stable functor. For each finite bag $\vec{n}$, let $\sigma_{\vec{n}}$ be the set of candidates $u : 1 \to S\vec{n}$. This carries a ***locally faithful*** action of the group $\mathsf{Aut}(\vec{n})$ on the left, by

$$h \cdot u = Sh \circ u$$

whilst the representable $X^{\vec{n}}$ carries an action on the right:

$$f \cdot h = f \circ h$$

Proof We shall discuss group actions at length in section 2. An action is locally faithful if $\forall h, u\,.\,(h \cdot u = u \Rightarrow h = 1)$: this property was shown in Lemma 1.3.6. □

Definition 1.5.4 Suppose the group $\mathsf{Aut}(\vec{n})$ acts on the left of $\sigma_{\vec{n}}$ and the right of $X^{\vec{n}}$. Then we can form their ***tensor product***, $X^{\vec{n}} \otimes_{\mathsf{Aut}(\vec{n})} \sigma_{\vec{n}}$, which consists of set of pairs $\langle f, u\rangle$ with $f \in X^{\vec{n}}$ and $u \in \sigma_{\vec{n}}$ subject to the equivalence relation that $\langle f \circ h^{-1},\ Sh \circ u\rangle = \langle f, u\rangle$.

Theorem 1.5.5 Let $S : \mathbf{Set}^A \to \mathbf{Set}$ be a stable functor and $\sigma_{\vec{n}}$ be the set of candidates $u : 1 \to S\vec{n}$. Then

$$SX \cong \sum_{\vec{n}} X^{\vec{n}} \otimes_{\mathsf{Aut}(\vec{n})} \sigma_{\vec{n}}$$

Let $T : \mathbf{Set}^A \to \mathbf{Set}$ be another stable functor corresponding to $(\tau_{\vec{n}})$. Then cartesian transformations $\phi : S \to T$ correspond bijectively to families of functions

$$\phi_{\vec{n}} : \sigma_{\vec{n}} \to \tau_{\vec{n}}$$

which *respect the action* of $\mathsf{Aut}(\vec{n})$. Hence the stable function-space $[\mathbf{Set}^A \to \mathbf{Set}]$ is equivalent to the category of *locally faithful actions of the groupoid* whose components are $\mathsf{Aut}(\vec{n})$.

Proof

[$\cong$] The proof of the isomorphism runs the same as Theorem 1.3.7, except that we have taken the quotient by the equivalence relation at a later stage. Notice we have also swapped the order of the product.

[$\Rightarrow$] By Lemma 1.4.3, every cartesian transformation ϕ restricts to a function on candidates which we may write as $\phi_{\vec{n}}$ (in fact this is the restriction of the component $\phi\vec{n}$ of ϕ to the subset $\sigma_{\vec{n}} \subset S\vec{n}$). Moreover naturality of ϕ with respect to automorphisms $h : \vec{n} \cong \vec{n}$ implies that the action is respected.

[$\Leftarrow$] For $\langle f, u\rangle \in SX$, so that $u : 1 \to S\vec{n}$ is a candidate and $f : \vec{n} \to X$, put $\phi X\langle f, u\rangle = \langle f, \phi_{\vec{n}}(u)\rangle$. This is independent of the choice from the equivalence class because $\phi^{\vec{n}}$ respects the action of $\mathsf{Aut}(\vec{n})$. We have to show that ϕ is natural: let $g : X \to X_1$; then $Sg\langle f, u\rangle = \langle g \circ f,\ u\rangle$, so

$$\phi X_1\big(Sg\langle f, u\rangle\big) = \big\langle g \circ f,\ \phi_{\vec{n}}(u)\big\rangle = Tg\big\langle f,\ \phi^{\vec{n}}(u)\big\rangle = Tg\big(\phi X\langle f, u\rangle\big)$$

as required. Cartesianness then follows from the converse of Lemma 1.4.3. □

Corollary 1.5.6 $[\mathbf{Set}^A \to \mathbf{Set}]$ does not have a terminal object (unless $A = 0$). □

Clearly we must study actions of groups and groupoids in detail, replacing Girard's set

$$\mathsf{Int}(A) = \mathsf{Ob}(\mathbf{Set}^A)_f$$

with the groupoid whose components are $\mathsf{Aut}(\vec{n})$ for $\vec{n} \in \mathsf{Int}(A)$. Moreover these actions must, it would appear, be locally faithful. However, as this was the result of Lemma 1.3.6, which depended on the existence of a terminal object, this is invalidated (for higher types) by the Corollary, and so we have to introduce a more flexible concept (creeds).

2. Groups and Linear Logic

2.1. Group Actions and the Multiplicative Fragment

This section describes the groupoid interpretation of linear logic in an informal way. We leave the reader to work out the interpretation of the linear λ-calculus [Lafont]; this is more or less obvious, but inadequate as it involves equations between terms which are *objects* of categories. We shall concentrate on group theory and use the characterisation of stable functors as our correctness criterion.

Let us say a word here about cardinality. In everything we do the groups are finite, but the groupoids will be countable (but recursively enumerable): infinitely many distinct finite groups may occur, each infinitely often, and different occurrences of the same group may carry different creeds. It doesn't really harm matters to assume that the actions are also countable, but on the other hand it doesn't help either.

One of the virtues of category theory is that it describes not only *collections* of mathematical objects (and their morphisms), but also in many cases the objects *themselves*. In particular, a ***group***, which is of course a set with an associative binary operation (∘), an identity and inverses[7], is (the morphism-set of) a category with one object in which every morphism is invertible. For purposes of calculation, the classical definition is more appropriate, but for theoretical use the abstract version simplifies matters, as we shall see. In fact we shall find it convenient to use both definitions interchangeably.

Definition 2.1.1 Let A be a group and X a set. A ***right action*** A on X is a function $\cdot : X \times A \to X$ such that

$$x \cdot \mathsf{id} = x \qquad \text{and} \qquad x \cdot (a \circ b) = (x \cdot a) \cdot b$$

for all $x \in X$ and $a, b \in A$. Suppose Y also carries an action of A; then a function $f : X \to Y$ ***respects the action*** of A, or is ***equivariant***, if

$$f(x \cdot a) = (fx) \cdot a$$

for all $x \in X$ and $a \in A$ (where $\cdot$ on the left is the action on X and on the right is that on Y). Similarly ***left action***.

Lemma 2.1.2 Right actions of A are isomorphic to left actions of A^{op} (the group with the same elements as A — written $\hat{a}$ for clarity — but the opposite composition: $\hat{a}\hat{b} = \widehat{ba}$).

Proof Exercise. [Hint: $\hat{a} \cdot x = x \cdot a$] □

The value of the categorical definition lies in the following:

Proposition 2.1.3 The category of right A-actions and equivariant functions is isomorphic to the functor-category $\mathbf{Set}^{A^{\mathrm{op}}}$, whose objects are functors $A^{\mathrm{op}} \to \mathbf{Set}$ (***presheaves on*** A) and whose morphisms are natural transformations. Similarly the category of left A-actions is isomorphic to $\mathbf{Set}^{A}$.

Proof The "single object" is taken by the functor to the set X; what's interesting is what happens to the morphisms: they are taken to the corresponding automorphisms of the set. That naturality and equivariance are the same is an elementary (but important) **exercise**. □

We shall introduce a few more classical ideas about permutation actions later, but our main point is to show how we can use groups to interpret the multiplicative fragment of linear logic, *i.e.* that involving only ⊗, ⅋ and $^\perp$, together with the identity (axiom) and the cut rule. In fact in this interpretation, ⊗ = ⅋. The interpretation will be in some ways similar to Girard's *coherence space* interpretation; in particular neither of them has any obvious direct connection with logic ("*linear logic is not necessarily logic*").

The notation

$$A_{(1)}, A_{(2)}, \ldots, A_{(k)} \vdash B_{(1)}, B_{(2)}, \ldots, B_{(l)}$$

informally means that we can deduce *at least one* of the conclusions $B_{(j)}$ using *some of* the hypotheses $A_{(i)}$. Better, it means that we have a particular proof, not specified in the notation. In these non-logical interpretations, there is usually at least one "proof" for any hypotheses and conclusions, so it is really a specific "proof-object" (whatever that is) that we're interested in.

In classical logic, the $A_{(i)}$'s may be permuted (the technical word is exchanged) and if two hypotheses are actually (different occurrences of) the same formula we may identify them (contraction). This means that the string $A_{(1)}, A_{(2)}, \ldots, A_{(k)}$ is a (finite) *set*. The same holds for the $B_{(j)}$'s, and in fact there is a further possibility (structural rule), namely to add further hypotheses to the set (weakening).

[7] The inverses are a bit misleading as part of the exposition, and produce red herrings like $A \cong A^{\mathrm{op}}$ by $a \mapsto \hat{a}^{-1}$. In fact the *strictly linear* parts (*i.e.* this section) can be done with general categories, but when we return to stable functors in the next section, we shall need invertibility (but see Proposition 3.5.4).

The point of *linear logic* is to forbid contraction and weakening, so that we have a proof which uses the hypotheses *exactly once each*. The *set* $\{A_{(1)}, A_{(2)}, ..., A_{(k)}\}$ now becomes a *bag* since we may now have duplicates but may still exchange. The effect of this is that there are now two different conjunctions ($\otimes$ and $\&$) and two different disjunctions ($\parr$ and $\oplus$). The power of intuitionistic logic is regained by the "of course" operator $!A$, which manufactures as many copies of the hypothesis A as we need. The reader is referred to any of the recent work of Girard or Lafont for further descriptions of the system.

When we say that *linear logic is not necessarily logic* we mean that what we call "proofs" may bear no resemblance to ordinary deduction. This is the case with the group-action interpretation.

Definition 2.1.4 A ***proof*** of

$$A_{(1)}, A_{(2)}, ..., A_{(k)} \vdash B_{(1)}, B_{(2)}, ..., B_{(l)}$$

is a set P together with an action of each $A_{(i)}$ on the left, and each $B_{(j)}$ on the right, *i.e.*

$$a_{(i)} \cdot p \qquad \text{and} \qquad p \cdot b_{(j)}$$

for each $p \in P$, $a_{(i)} \in A_{(i)}$ and $b_{(j)} \in B_{(j)}$, which *commute*, *i.e.*

$$\begin{aligned} a_{(i)} \cdot (a_{(i')} \cdot p) &= a_{(i)} \cdot (a_{(i')} \cdot p) \\ (a_{(i)} \cdot p) \cdot b_{(j)} &= a_{(i)} \cdot (p \cdot b_{(j)}) \\ (p \cdot b_{(j)}) \cdot b_{(j')} &= (p \cdot b_{(j)}) \cdot b_{(j')} \end{aligned}$$

for $i \neq i'$ and $j \neq j'$.

When $k = l = 1$, a proof of $A \vdash B$ is a functor $P : A \times B^{\mathrm{op}} \to \mathbf{Set}$, which also called a ***profunctor*** from A to B. This numerical restriction is unimportant because of the

Exercise 2.1.5 A simultaneous action of $A_{(1)}, ..., A_{(k)}$ on the left and $B_{(1)}, ..., B_{(l)}$ on the right of X is the same as an action of $A^{\mathrm{op}}_{(1)} \times ... \times A^{\mathrm{op}}_{(k)} \times B_{(1)} \times ... \times B_{(l)}$ on the right. □

What justifies the use of the word "proof" for these very un-proof-like objects is the fact that the sequent rules for the (linear) logical connectives are sound, although with (single) groups we shall only be able to interpret $\otimes$, $\parr$ and ${}^{\perp}$.

Proposition 2.1.6 For groups C and D, let $C \otimes D = C \parr D = C \times D$, the cartesian product. Also, let ${}^{\perp}C = C^{\mathrm{op}}$, the opposite group. Then the rules

$$\frac{\vec{A}, C, D \vdash \vec{B}}{\vec{A}, C \otimes D \vdash \vec{B}}\mathcal{L}\otimes \qquad\qquad \mathcal{R}\parr\frac{\vec{A} \vdash C, D, \vec{B}}{\vec{A} \vdash C \parr D, \vec{B}}$$

and

$$\frac{\vec{A} \vdash C, \vec{B}}{\vec{A}, {}^{\perp}C \vdash \vec{B}}\mathcal{L}^{\perp} \qquad\qquad \mathcal{R}^{\perp}\frac{\vec{A}, C \vdash \vec{B}}{\vec{A} \vdash {}^{\perp}C, \vec{B}}$$

are interpreted as isomorphisms. The rules

$$\frac{\vec{A}, C \vdash \vec{B} \qquad \vec{A}', D \vdash \vec{B}'}{\vec{A}, \vec{A}', C \parr D \vdash \vec{B}, \vec{B}'}\mathcal{L}\parr \qquad\qquad \mathcal{R}\otimes\frac{\vec{A} \vdash C, \vec{B} \qquad \vec{A}' \vdash D, \vec{B}'}{\vec{A}, \vec{A}' \vdash C \otimes D, \vec{B}, \vec{B}'}$$

are interpreted by a product.

Proof We have dealt with the first four rules in the Exercise. For the last two, if P and Q are the actions corresponding to the proofs above the line, $P \times Q$ corresponds to that below; the action is

$$a_{(i)} \cdot \langle p, q\rangle = \langle a_{(i)} \cdot p, q\rangle, \quad a_{(i')} \cdot \langle p, q\rangle = \langle p, a_{(i')} \cdot q\rangle, \quad \langle p, q\rangle \cdot b_{(j)} = \langle p \cdot b_{(j)}, q\rangle, \quad \langle p, q\rangle \cdot b_{(j')} = \langle p, q \cdot b_{(j')}\rangle$$

and

$$\langle c, d\rangle \cdot \langle p, q\rangle = \langle c \cdot p, d \cdot q\rangle \qquad \text{or} \qquad \langle p, q\rangle \cdot \langle c, d\rangle = \langle p \cdot c, q \cdot d\rangle$$

for $\mathcal{L}\parr$ and $\mathcal{R}\otimes$ respectively. □

2.2. Identity and Cut

So far we have given definitions and constructions but no concrete examples of actions. The most important example was found by Cayley, and we shall use it to interpret the identity (axiom) $A \vdash A$: we take $P = A$, *i.e.* just the set of elements of the group, where $a \cdot p = a \circ p$ and $p \cdot a = p \circ a$. (This same idea occurs in category theory as the Yoneda lemma.) We take this opportunity to introduce some more general notation; we shall also often drop the $\circ$ when composing group elements.

Definition 2.2.1 Let A be a group and H any subgroup, written $H \leq A$. For $x \in A$, $xH = \{xh : h \in H\}$ is a ***left coset*** and $Hx = \{hx : h \in H\}$ is a ***right coset***. Write $A_{/H}$ for the set of left cosets and ${}_{H\backslash}A$ for the set of right cosets.

Definition 2.2.2 Let X carry a right action of A and $x \in X$. Then the ***orbit*** of x is the set

$$x \cdot A = \{x \cdot a : a \in A\}$$

and if this is the whole of X we call the action ***transitive***. The ***stabiliser*** of x is the subgroup

$$\mathsf{Stab}_A(x) = \{a \in A : x \cdot a = x\}$$

The action of A on X is called ***faithful*** if

$$\forall a \in A \,.\, (\forall x \in X \,.\, x \cdot a = x) \Rightarrow a = \mathsf{id}$$

but the condition we had in Lemma 1.5.4 was

$$\forall a \in A \,.\, (\exists x \in X \,.\, x \cdot a = x) \Rightarrow a = \mathsf{id} \qquad i.e. \quad \forall a \in A \,.\, \forall x \in X \,.\, (x \cdot a = x \Rightarrow a = \mathsf{id})$$

which we call ***locally faithful***[8]. Alternatively, the action is locally faithful at x iff $\mathsf{Stab}_A(x) = \{\mathsf{id}\}$; of course if $\mathsf{Stab}_A(x) = A$ then x is a ***fixed point***.

Proposition 2.2.3 Let A be a group. Then
(a) For any subgroup H, ${}_{H\backslash}A$ carries a transitive right action given by $Hx \cdot a = H(xa)$ and $A_{/H}$ carries a transitive left action given by $a \cdot xH = (ax)H$.
(b) Every action of A on a set X is uniquely expressible as the disjoint union of actions on cosets.

Proof Checking that [a] is well-defined is an easy standard exercise. For [b] we decompose X as a disjoint union of orbits, on each of which the action is transitive. With $H = \mathsf{Stab}_A(x)$, the right action of A on ${}_{H\backslash}A$ is isomorphic to that on $x \cdot A$ by $Ha \leftrightarrow x \cdot a$. □

This gives us a concrete representation of actions.

Exercise 2.2.4 Show that $\mathsf{Stab}_A(x \cdot a) = a^{-1}\,\mathsf{Stab}_A(x)\,a$; this is called a ***conjugate*** subgroup.

Notation 2.2.5 Write $\mathsf{Conj}(A)$ for the set of conjugacy classes of subgroups of a group A. For an action of A on X and $[H] \in \mathsf{Conj}(A)$, write $X^{[H]}$ for the number (set) of orbits whose stabiliser belongs to the conjugacy class $[H]$, so that

$$X \cong \sum_{[H] \in \mathsf{Conj}(A)} X^{[H]} \times {}_{H\backslash}A$$

We also need ***two-sided cosets***:[9] if $H \leq A^{\mathsf{op}} \times B$ we write $A_{/H\backslash}B$ for the set of objects $aHb \stackrel{\text{def}}{=} H\langle \hat{a}, b\rangle$ with $a \in A$ and $b \in B$. This has left action of A and right action of B by

$$a' \cdot aHb = a'aHb \qquad \text{and} \qquad aHb \cdot b' = aHbb'$$

and so is an (atomic) proof of $A \vdash B$. As a special case of this, write

$$\Delta = \{\langle \hat{a}^{-1}, a\rangle : a \in A\} \leq A^{\mathsf{op}} \times A$$

for the ***diagonal subgroup***; then $A_{/\Delta\backslash}A$ is (isomorphic to) the set A with its obvious (***regular***) two-sided action.

[8] The term *semi-regular* is used in group theory: Cayley's action is the *regular* one: *cf.* simple and semi-simple in ring theory. I am grateful to Steve Linton for his remarks on standard terminology.
[9] In computational group theory, a *double* coset is of the form HaK.

Proposition 2.2.6
(a) For a group A,

$$A_{/\Delta\backslash}A \quad \text{is a proof of} \quad A \vdash A$$

(b) The rule

$$\frac{\vec{A} \vdash C, \vec{B} \qquad \vec{A}', C \vdash \vec{B}'}{\vec{A}, \vec{A}' \vdash \vec{B}, \vec{B}'}\text{Cut}$$

is interpreted by

$$P,\ Q \quad \mapsto \quad P \otimes_C Q$$

where $P \otimes_C Q$ is the set of pairs $\langle p, q\rangle$ with $p \in P$ and $q \in Q$ subject to the equivalence relation $\langle p \cdot c, q\rangle = \langle p, c \cdot q\rangle$ (Definition 1.5.4), and the action of $\vec{A}$ *etc.* is as in Proposition 2.1.5.

Proof It is an exercise to show that the actions respect the equivalence relation. □

Observe that this rule is the same as $\mathcal{R}\otimes$ (and $\mathcal{L}\invamp$) but with $C \otimes {}^\perp C$ (respectively ${}^\perp C \invamp C$) deleted. In simpler terms, this amounts to the rule

$$\frac{C \vdash C}{\vdash}$$

which, semantically, turns an action of C on both sides of P into a set without any action, using the ***condensation***

$$p \mapsto [p] = \{c^{-1} \cdot p \cdot c : c \in C\}$$

which forms the quotient by an equivalence relation. With *linear* (now in the traditional sense of Linear *Algebra*) instead of *permutation* representations of groups, we may express $[p]$ as an "average"

$$[p] = \frac{1}{|C|} \sum_{c \in C} c^{-1} \cdot p \cdot c$$

(*cf.* Maschke's theorem; [Lang] Theorem 18.1.1). This is essentially what Girard does in his "C*-algebra" interpretation of linear logic [89].

Exercise 2.2.7 Show that every equivariant function $f : {}_{H\backslash}A \to {}_{J\backslash}A$ is onto, and is of the form $Hx \mapsto Jfx$ where $H \subset f^{-1}Jf$. Suppose that $K : \mathbf{Q}$ is a normal field extension with Galois group A; show that the category of intermediate fields and homomorphisms (not necessarily commuting with the inclusion in K) is dual to this category of transitive actions.

2.3. Groupoids and the Additive Fragment

The multiplicative fragment of Linear Logic is very inexpressive: we need to extend it with the ***additive*** connectives, & and ⊕. In our interpretation, they become identified, and in fact are interpreted as a *disjoint sum* of groups.

But what *is* a disjoint sum of groups? Just what it says, and certainly not their coproduct in **Gp**. It is at this point that we see the value of the abstract (categorical) definition of a group.

Definition 2.3.1 A ***groupoid*** is a category in which every morphism is invertible.

Lemma 2.3.2 Every groupoid is equivalent to a unique bag of groups.

Proof A bag of groups is a ***skeletal*** groupoid, *i.e.* one in which if two objects are isomorphic (*quā* objects of the category, not just that their automorphism groups are isomorphic in **Gp**) then they are equal. Given any groupoid, we choose (arbitrarily) one object in each component and form the full subcategory; this is a skeletal groupoid. Any object is isomorphic to a unique chosen object, but not uniquely so, hence we have also to choose a particular isomorphism. By pre-composing morphisms with this chosen isomorphism and post-composing with its inverse, we can define a functor from the whole groupoid to its skeleton, and this provides an equivalence. The bag of groups is easily seen to be unique up to isomorphism, although (the pair of functors defining) the equivalence is certainly not unique. □

As we said, both the abstract and the concrete definitions are useful, even for connected groupoids. A good example is the ***Fundamental Group*** of a (path-connected) topological space, which consists of the loops at a given basepoint under concatenation. The abstract version of this is the groupoid of paths between any two points, which enables us to extend the definition to the non-connected case. The functoriality of this construction is more easily seen abstractly, whereas for calculations in algebraic topology the additional copies of the same group serve no useful purpose. The same will apply in our case.

We shall in future use A to denote a group*oid* and write A_i for a typical component, which we shall consider to be a group (skeletal). Note that the $i \in I$ correspond to components and not objects of the groupoid. We shall also drop the usual shorthand $X \in C$ for an *object* of a category, instead using $a \in A$ to mean that a is a *morphism* of A; this generalises $a \in A_i$ for an element of a group. This is consistent with seeing a skeletal groupoid as its set of morphisms with *partial* composition.

Just as a groupoid is a bag of groups, A_i, so a groupoid action is a bag of sets, X^i (one for each group). The elements $a \in A_i$ act on the $x \in X^i$ in the corresponding set, and local faithfulness is defined in the obvious way. (Contrast this with Definition 2.1.4, in which all the $A_{(i)}$ acted on the *same* set X.) Moreover an equivariant function $f : X \to Y$ is a bag of functions $f^i : X^i \to Y^i$ each respecting the action of the A_i. The abstract definition proves its worth in the following result, which unifies 1.3.2 and 2.1.3.

Lemma 2.3.3 Let A be a groupoid. The category of right A-actions is isomorphic to the category $\mathbf{Set}^{A^{op}}$ of functors from A^{op} to $\mathbf{Set}$ (or presheaves on A) and natural transformations between them. Similarly left actions form the category $\mathbf{Set}^A$. □

If we add groups to a bag, it can still act on the same family of sets, on the understanding that the missing sets are empty. Thus if X carries an action of C then $X + 0$ carries an action of $C + D$. This is "vectorial addition" in the sense that the set X^i is "oriented" by the group A_i. In particular we shall write ${}_{H\backslash}A_i$ for the "unit vector", *i.e.* action of the groupoid A on the family of sets which are empty in all components except i, where it is the set of right cosets of $H \leq A_i$.

Lemma 2.3.4 ${}_{H\backslash}A_i$ is ***atomic***: it has no proper subobject as an object of $\mathbf{Set}^{A^{op}}$. Every object of this category may be expressed uniquely as a coproduct of atoms.

Proof Clearly

$$X \cong \sum_{i \in I} \sum_{\substack{[H] \in \\ \mathrm{Conj}(A_i)}} X^{i,[H]} \times {}_{H\backslash}A_i$$

whilst it is an exercise to show that if $f : X \to Y$ is an equivariant function between two such decompositions, each component of X is mapped *onto* some unique component of Y. □

Of course we shall now interpret $C \,\&\, D = C \oplus D = C + D$ as the disjoint union of groupoids, but certain other things have to be generalised about the earlier interpretation. We have to replace the product of groups and of sets by the product of groupoids and families of sets; this means that we form the product of the indexing sets, $I \times J$, and for each $\langle i, j \rangle$ form the product of the groups, $A_i \times B_j$, or sets, $X^i \times Y^j$. Observe that the sum and product of skeletal groupoids are again skeletal.

Proposition 2.3.5 The rules

$$\frac{\vec{A}, C \vdash \vec{B}}{\vec{A}, C \,\&\, D \vdash \vec{B}}\mathcal{L}1\& \qquad \mathcal{L}2\&\frac{\vec{A}, D \vdash \vec{B}}{\vec{A}, C \,\&\, D \vdash \vec{B}}$$

and

$$\frac{\vec{A} \vdash \vec{B}, C}{\vec{A} \vdash \vec{B}, C \oplus D}\mathcal{R}1\oplus \qquad \mathcal{R}2\oplus\frac{\vec{A} \vdash \vec{B}, D}{\vec{A} \vdash \vec{B}, C \oplus D}$$

are interpreted by $X \mapsto X + 0$ or $X \mapsto 0 + X$, whilst the rules

$$\frac{\vec{A}, C \vdash \vec{B} \qquad \vec{A}, D \vdash \vec{B}}{\vec{A}, C \oplus D \vdash \vec{B}}\mathcal{L}\oplus \qquad \text{and} \qquad \mathcal{R}\&\frac{\vec{A} \vdash C, \vec{B} \qquad \vec{A} \vdash D, \vec{B}}{\vec{A} \vdash C \,\&\, D, \vec{B}}$$

are interpreted by $X, Y \mapsto X + Y$. □

The interpretation of Cut is also more complicated: if X is a proof of $\vdash C$ and Y of $C \vdash$ then we have to reinterpret

$$X \otimes_C Y = \sum_i X^i \otimes_{C_i} Y_i$$

which is already reminiscent of Theorem 1.5.5. The categorically-minded will observe that this is a colimit or coend over a diagram whose type is a groupoid (namely C).

Before we return to our discussion of stable functors, let us first complete the interpretation of the linear connectives by giving the units.

Exercise 2.3.6 Show that with $\mathbf{1} = \bot$ interpreted as the one-element group and $\mathbf{0} = \top$ as the empty groupoid, the rules

$$\mathcal{L}\bot\frac{}{\bot \vdash} = \frac{}{\vdash \mathbf{1}}\mathcal{R}\mathbf{1} \qquad \mathcal{L}\mathbf{1}\frac{\vec{A} \vdash \vec{B}}{\vec{A}, \mathbf{1} \vdash \vec{B}} = \frac{\vec{A} \vdash \vec{B}}{\vec{A} \vdash \bot, \vec{B}}\mathcal{R}\bot$$

$$\mathcal{L}\mathbf{0}\frac{}{\vec{A}, \mathbf{0} \vdash \vec{B}} = \frac{}{\vec{A} \vdash \top, \vec{B}}\mathcal{R}\top \qquad \text{(no rule } \mathcal{L}\top = \mathcal{R}\mathbf{0})$$

are sound. [Hint: use a singleton singleton for the first two, and an empty bag for the third.] □

3. Quantitative Domains

3.1. Linear Functors

We have already *defined* a proof of $\vdash A$ to be an object of $\mathbf{Set}^{A^{op}}$, so more generally what is the relationship between proofs (actions)

$$A_{(1)}, \ldots, A_{(k)} \vdash B_{(1)}, \ldots, B_{(l)}$$

and stable functors

$$\mathbf{Set}^{A^{op}_{(1)} \times \ldots \times A^{op}_{(k)}} \to \mathbf{Set}^{B^{op}_{(1)} \times \ldots \times B^{op}_{(l)}}$$

(without loss of generality, $k = l = 1$)? Suppose that P is a proof of $A \vdash B$, *i.e.* an action of A on the left and B on the right, and let $X \in \mathbf{Set}^{A^{op}}$, *i.e.* an action of A on the right. We may form the tensor product

$$SX = X \otimes_A P \qquad i.e. \qquad (SX)^j = \sum_i X^i \otimes_{A_i} P_i^j$$

and hence define a functor $S : \mathbf{Set}^{A^{op}} \to \mathbf{Set}^{B^{op}}$.

Now S is not stable, but $S^{op} : (\mathbf{Set}^{A^{op}})^{op} \to (\mathbf{Set}^{B^{op}})^{op}$ is!

Lemma 3.1.1 S has a right adjoint and preserves cofiltered limits.

Proof The right adjoint is given by

$$Y \mapsto Y/_A P \stackrel{\text{def}}{=} \{x : P \to_B Y\} \qquad \text{with } x \cdot a : p \mapsto x \cdot (a \cdot p)$$

Preservation of cofiltered limits uses the fact that the intersection of a descending sequence of non-empty finite sets is non-empty, which depends on our use of *finite* groups. □

Exercise 3.1.2 Show that if $S : \mathbf{Set}^{A^{op}} \to \mathbf{Set}^{B^{op}}$ has a right adjoint then $S \cong - \otimes_A P$ for some $P : A \vdash B$. □

The problem is that S does not preserve pullbacks (we already have the counterexample $\frac{1}{2}X^2$), but it does preserve a wider class of squares:

Definition 3.1.3 A square in $\mathbf{Set}^{A^{op}}$ is called ***sur-cartesian*** if it commutes and the mediator to the pullback is an epimorphism.

Exercise 3.1.4 By considering the decomposition in Lemma 2.3.4, show that a commutative square is sur-cartesian iff the underlying square of atoms is a pullback. Hence show that S preserves sur-cartesian squares. □

Since not every two-sided action (profunctor) gives rise to a *stable* functor, we have to modify the theory. We have (at least) three options:

(a) Replace *stability* with this weaker notion of preserving sur-cartesian squares. Joyal [87] has studied this, proving similar results to ours (in particular uniqueness in the definition of candidacy is dropped) but unfortunately not quite in sufficient generality to give a cartesian closed category. However he does also develop a theory with vector spaces instead of sets, to which we have only alluded.

(b) Restrict the *morphisms* to monos, forcing sur-cartesian=cartesian to restore stability: Lamarche has shown (privately) that domains of the form $\mathcal{M}(A)$ and stable functors form a cartesian closed category, where $\mathcal{M}(A)$ is the category of locally faithful A-actions and monomorphisms.

(c) Restrict the *objects* to those for which the functor *does* preserve pullbacks: this is what we choose to do.

The interested reader should be able to adapt our results to cases (a) and (b), which are somewhat simpler. Joyal's ideas probably lead to other models of Linear Logic worthy of study.

Definition 3.1.5 A functor is ***linear*** if its restriction to slices has adjoints on both sides; we write $S : \mathcal{X} \multimap \mathcal{Y}$.

The term is justified *semantically* by the representation $S \cong - \otimes_A P$, which says that a linear functor performs "matrix multiplication" on atoms. The *syntactic* justification lies in the proof-theoretic principle of using each hypothesis exactly once. We cannot say that S has a (global) right adjoint, because $Y/_A P$ may not lie in the chosen full subcategory.

What help is this in classifying *stable* functors? As Girard put it, it is that *every* stable functor can be made linear if only we change the domain. Indeed, comparing

$$SX = X \otimes_A P = \sum_i X^i \otimes_{A_i} P_i^j \qquad \text{with} \qquad SX \cong \sum_{\vec{n}} X^{\vec{n}} \otimes_{\mathrm{Aut}(\vec{n})} \sigma_{\vec{n}}$$

we see that we can write

$$SX \cong\, !X \otimes_{!A} \sigma$$

if we make the following

Definition 3.1.6

(a) $!A$ is the groupoid of finite A-actions $(\vec{n})$ and isomorphisms; as a bag of groups its components are $\mathrm{Aut}(\vec{n})$.

(b) $!X$ is the right $!A$-action with $(!X)^{\vec{n}} = X^{\vec{n}}$ (*i.e.* $\mathrm{Hom}(\vec{n}, X)$, *cf.* Lemma 1.5.3).

In the language of category theory [Mac Lane, Chapter VI], $!-$ is a comonoid in the monoidal category with linear functors as morphisms, and the category with stable functors is the coKleisli category [Lafont]. The fact that

$$[\mathcal{X} \to \mathcal{Y}] \simeq [!\mathcal{X} \multimap \mathcal{Y}]$$

is a non-trivial achievement, because we have reduced a *complex binary* operator $\to$ to a *simple* binary operator $[- \multimap -]$ (the linear function-space) and a complex *unary* operator $!$.

Observe that *evaluation is linear in the functor*. This means that although a process may have to read its input many times to match the parts of a pattern, the pattern (a token of the function) need itself only be read once. Girard noticed that many λ-definable operations are actually linear.

3.2. Creeds

We have already seen that for a set (or groupoid) A, the groupoid of all finite objects of $\mathbf{Set}^{A^{op}}$ is involved in $!A$, but this is further complicated by local faithfulness. In order to state and prove the theorem correctly, we must therefore make a definition of quantitative domains which is sufficiently complex to account for these phenomena.

Definition 3.2.1 A ***creed*** on a group A is a sub*set* $\Gamma \subset A$ which is closed under inverses, powers and conjugation (*not* multiplication: that would make it a normal subgroup, which is too strong). A creed on a groupoid is a creed on each component group. An action of A on X is ***locally faithful to*** Γ if $\forall x \in X \,.\, \mathsf{Stab}_A(x) \subset \Gamma$, or equivalently $\forall x \in X, a \in A \,.\, (a \cdot x = x \Rightarrow a \in \Gamma)$.

Exercises 3.2.2

(a) Let A be a groupoid equivalent to the bag of groups (A_i) (*cf.* Lemma 2.3.2), and let $\Gamma_i \subset A_i$ be a family of subsets. Using the chosen isomorphisms this can be extended to a family of subsets of the automorphisms of the other objects. Show that this extension is *independent of the choice of isomorphisms* iff each Γ_i is closed under conjugation by elements of A_i. [Hint: *cf.* Exercise 2.2.4.]

(b) Show that every subset of a group *contains* a largest creed, to which it is equivalent as a way of defining local faithfulness. [Hint: $\mathsf{Stab}_A(x)$ might be a cyclic subgroup.] □

Definition 3.2.3 A ***quantitative type*** is a groupoid A equipped with a creed Γ, *i.e.* a creed Γ_i on each component group A_i. The corresponding ***quantitative domain***, $\mathsf{QD}(A,\Gamma)$, is the category of locally Γ-faithful right A-actions and equivariant functions. Note that $\mathsf{QD}(A,\Gamma)$ does not have a terminal object unless $\Gamma = A$, so Lemma 1.3.6 fails, but everything else carries over.

Lemma 3.2.4 $\mathcal{X} = \mathsf{QD}(A,\Gamma) \subset \mathbf{Set}^{A^{op}}$ is closed under

(i) backwards-arrows (isotomic): if $Y \in \mathcal{X}$ and $f : X \to Y$ in $\mathbf{Set}^{A^{op}}$ then $X \in \mathcal{X}$,
(ii) morphisms (full): if $X, Y \in \mathcal{X}$ and $f : X \to Y$ in $\mathbf{Set}^{A^{op}}$ then $f \in \mathcal{X}$,
(iii) isomorphisms (replete): special case of (i) with $f : Y \cong X$,
(iv) coproducts: if $X, Y \in \mathcal{X}$ and $Z = X + Y$ in $\mathbf{Set}^{A^{op}}$ then $Z \in \mathcal{X}$,
(v) filtered colimits: if $X = \mathsf{colim}^{\dagger} X^{(r)}$ in $\mathbf{Set}^{A^{op}}$ with $X^{(r)} \in \mathcal{X}$ then $X \in \mathcal{X}$.
(vi) representable objects: ${}_1\backslash A_i \in \mathcal{X}$ (in fact ${}_H\backslash A_i \in \mathcal{X}$ iff $H \subset \Gamma$).

Conversely, if $\mathcal{X} \subset \mathbf{Set}^{A^{op}}$ is a subcategory with these closure conditions then there is a unique creed Γ with $\mathsf{QD}(A,\Gamma) = \mathcal{X}$.

Proof [i] If $f : X \to Y$ then the stabilisers of points of X are no bigger than those in Y. [ii] Definition. [iv] Lemma 2.3.4. [v] Exercise. [vi] The representable objects are locally faithful actions because their stabilisers are $\{1\} \subset \Gamma$. [$\Leftarrow$] Let $\Gamma_i = \bigcup\{H \leq A_i : {}_H\backslash A_i \in \mathcal{X}\}$. □

Corollary 3.2.5 $\mathsf{QD}(A,\Gamma) \subset \mathbf{Set}^{A^{op}}$ is also closed under binary products with, and exponentiation by, arbitrary objects of $\mathbf{Set}^{A^{op}}$. Also equalisers, but not coequalisers. □

We have chosen a definition in terms of a *unary predicate* on group(oid) elements, *viz.* membership of a creed. One could alternatively define a (reflexive, symmetric) *binary* predicate

$$a \frown b \iff \exists i \,.\, a, b \in A_i \wedge ab^{-1} \in \Gamma_i$$

making an analogy with ***coherence spaces*** [GLT, chapter 8]. We recover the creed by $\Gamma = \{a \in A : \mathsf{id} \frown a\}$. A pairwise coherent subset of a groupoid which is also closed under the (partial) ***Mal'cev operation***

$$\mu(a,b,c) = ab^{-1}c$$

is the same thing as a right coset of a subgroup $H \subset \Gamma_i$, and this is a *token* of a locally Γ-faithful A-action. An object of a *quantitative* domain is a set of coherent Mal'cev-closed sets. In contrast, an object of a *qualitative* domain is just a coherent set. The reason is that in qualitative domains we are controlling sums, whilst in quantitative domains we need to control quotients.

We have to generalise the power-series expansion and the interpretation of linear logic. We shall abuse notation by writing

$$(A,\Gamma)\,\&\,(B,\Delta) \qquad \text{as} \qquad (A\,\&\,B,\Gamma\,\&\,\Delta)$$

and similarly for the other connectives. This *is* an abuse, because $!A$ (in particular) depends on Γ as well as A, because its objects are *locally* Γ*-faithful* finite right actions. Obviously $!\Gamma$ depends on A.

Exercise 3.2.6 Every stable functor $S : \mathsf{QD}(A,\Gamma) \to \mathsf{QD}(B,\Delta)$ has a power-series expansion similar to Theorem 1.4.9, except for the local faithfulness condition, and so extends to a functor (not necessarily stable) $\mathbf{Set}^{A^{op}} \to \mathbf{Set}^{B^{op}}$. □

The units are given by

$$\mathsf{QD}(\bot) = \mathbf{Set} = \mathsf{QD}(1) \qquad \mathsf{QD}(\top) = 1 = \mathsf{QD}(0)$$

whilst the additives are both given as before by disjoint unions:

$$(A,\Gamma)\,\&\,(B,\Delta) = (A,\Gamma)\oplus(B,\Delta) = (A+B,\Gamma+\Delta)$$

where $A+B$ is the disjoint union and $\Gamma+\Delta$ is similarly the family in which Γ_i corresponds to A_i and Δ_j to B_j.

Exercise 3.2.7 Show that

$$\left[\mathsf{QD}(A,\Gamma) \multimap \mathsf{QD}(B_1,\Delta_1)\times\mathsf{QD}(B_2,\Delta_2)\right] \simeq \left[\mathsf{QD}(A,\Gamma)\multimap \mathsf{QD}(B_1\,\&\,B_2,\Delta_1\,\&\,\Delta_2)\right]$$

and

$$\left[\mathsf{QD}(A_1,\Gamma_1)\times\mathsf{QD}(A_2,\Gamma_2)\multimap\mathsf{QD}(B,\Delta)\right] \simeq \left[\mathsf{QD}(A_1\oplus A_2,\Gamma_1\oplus\Gamma_2)\multimap\mathsf{QD}(B,\Delta)\right]$$

where $[\mathcal{X} \multimap \mathcal{Y}]$ is the linear function-space: we have yet to define the linear connective $(A,\Gamma)\multimap(B,\Delta)$. Show also that Proposition 2.3.5 remains valid. □

Recall that in coherence spaces, $\oplus$ and $\otimes$ have simple descriptions as sums and products of graphs, whereas $\&$ and ⅋ are defined *via* linear negation. In our case, linear negation commutes with $\oplus$, so $\& = \oplus$, but (unlike for groups) it does not commute with $\otimes$. We define

$$(A,\Gamma)\otimes(B,\Delta) = (A\times B,\Gamma\times\Delta)$$

in the obvious way, and it is easy to check this gives a creed.

Exercise 3.2.8 Describe $\mathsf{QD}\left((A,\Gamma)\otimes(B,\Delta)\right)$ and show that it is a topological product.

3.3. Creeds and Negation

Now it was Lemma 1.3.6 which was the source of our difficulties, so let us look carefully at how the problem arose. Because of the existence of the terminal object, it was possible to satisfy

$$\exists X, f : \vec{n} \to X \;\; . \;\; f \circ h = f$$

for arbitrary $h \in \mathsf{Aut}(\vec{n})$. In general, if h satisfies this property, we call it an ***annihilable automorphism***. (In the study of linear functors, we shall just have $h = a \in A_i \cong {}_1\backslash\mathsf{Aut}(A_i)$.)

Lemma 3.3.1 $a \in A_i$ is annihilable iff $a \in \Gamma_i$.

Proof [$\Leftarrow$] Put $X = {}_{\langle a\rangle}\backslash A_i$ where $\langle a\rangle \le A_i$ is the cyclic subgroup generated by a. [$\Rightarrow$] Any annihilating f factors through this. □

It is then clear that we must let $!\Gamma^{\vec{n}} \subset \mathsf{Aut}(\vec{n})$ be the set of annihilable automorphisms. We shall see shortly that $!(A,\Gamma) \multimap (B,\Delta)$ gives the function-space. However the basic difficulty actually lies in the linear negation, which we want to satisfy

$$[\mathsf{QD}(A,\Gamma) \multimap \mathbf{Set}] \simeq \mathsf{QD}({}^{\perp}A, {}^{\perp}\Gamma)$$

since $\mathbf{Set} = \mathsf{QD}(\perp)$ and $A \multimap \perp = {}^{\perp}A$.

Lemma 3.3.2 ${}^{\perp}A = A^{\mathrm{op}}$ and ${}^{\perp}\Gamma = \{\hat{a} : \forall k \,.\, a^k \in \Gamma \Rightarrow a^k = 1\}$.

Proof The question is when $SX = X \otimes_A P$ is stable, for a left action of A on P; without loss of generality this is atomic, so $P = A_{i/H}$ where $H \le A_i$. An element of $X \otimes_A P$ is then a pair $\langle x, aH \rangle$ where $x \in X_i$ and $aH \in A_{i/H}$; but this is identified with $\langle x \circ a, H \rangle$, so we may write it as xaH. The (potential) candidate is $H = \mathsf{id}H \in {}_{1\backslash}A_i \otimes_{A_i} A_{i/H}$ and the factorisation is $xH = Sx \circ H$.

The problem is uniqueness of the diagonal a in the definition of the candidacy of H.

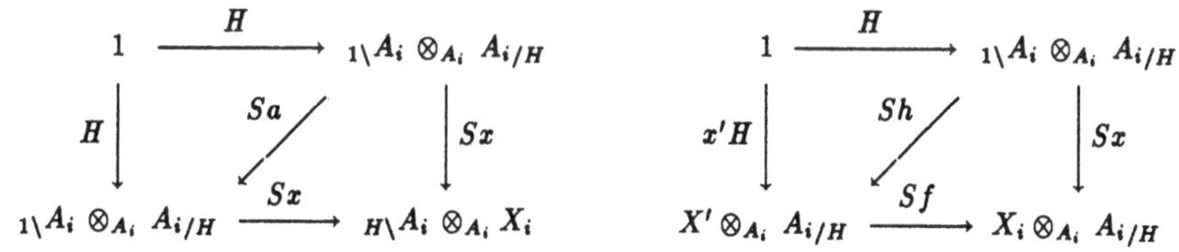

If $a = \mathsf{id}$ is to be the *unique* diagonal in the diagram on the left for all x, we must have

$$\frac{a \in H \qquad \exists x \in X \,.\, x \circ a = x}{a = \mathsf{id}}$$

Conversely, if the right-hand square commutes then $xH = fx'H$ and so there is some $a_1 \in H$ with $x = fx'a_1$; in fact $h_1 = x'a_1$ is a typical diagonal. If $a_2 = a_1 a$ also satisfies this then the rule makes $a_1 = a_2$. So we have shown that the rule is necessary and sufficient.

By lemma 3.3.1, the rule says that

$$H \subset \{a \in A_i : a \in \Gamma_i \Rightarrow a = \mathsf{id}\}$$

However H, being a subgroup, is closed under powers, so we can strengthen this to the given definition of ${}^{\perp}\Gamma_i$. On the other hand, cyclic subgroups show that this is the most we can do. It is also easy to show that ${}^{\perp}\Gamma$ is closed under inverses and conjugation. □

We do not in general have ${}^{\perp\perp}(A,\Gamma) = (A,\Gamma)$ (Exercise 3.6.4). The other two binary connectives, ⅋ and $\multimap$, are related by

$${}^{\perp}(A,\Gamma) \otimes {}^{\perp}(B,\Delta) = {}^{\perp}((A,\Gamma) ⅋ (B,\Delta)) \qquad \text{and} \qquad (A,\Gamma) \multimap (B,\Delta) = {}^{\perp}(A,\Gamma) ⅋ (B,\Delta)$$

where

$$(\Gamma ⅋ \Delta)_{ij} = \{\langle a, b \rangle : \forall k \,.\, (a^k \in \Gamma_i \wedge b^k \in \Delta_j) \vee (\mathsf{id} \ne a^k \in \Gamma_i) \vee (\mathsf{id} \ne b^k \in \Delta_j)\}$$

and

$$(\Gamma \multimap \Delta)_{ij} = \{\langle \hat{a}, b \rangle : \forall k \,.\, (a^k \in \Gamma_i \Rightarrow b^k \in \Delta_j) \wedge (a^k \in \Gamma_i \wedge b^k = \mathsf{id} \Rightarrow a^k = \mathsf{id})\}$$

so

$$((C,\Theta) \otimes (A,\Gamma)) \multimap (B,\Delta) = (C,\Theta) \multimap ((A,\Gamma) \multimap (B,\Delta))$$

Lemma 3.3.3 With this definition we have

$$\left[\mathsf{QD}(A,\Gamma) \multimap \mathsf{QD}(B,\Delta)\right] \simeq \mathsf{QD}\left((A,\Gamma) \multimap (B,\Delta)\right)$$

Proof Put $SX = X \otimes P$ with $P = A_{i/K\backslash}B_j$ for some subgroup $K \le A_i^{\mathrm{op}} \times B_j$. We have to show that the action of B on SX is locally faithful to Δ and the diagonals for the potential candidate

$K \in {}_1\backslash A_i \otimes A_{i/K}\backslash B_j$ are unique. A typical element of SX is xKb where $x \in X_i$ and $b \in B_j$, but since b is an isomorphism we may assume $b = \mathsf{id}$. Local faithfulness to Δ amounts to the rule

$$\frac{\langle \hat{a}, b\rangle \in K \qquad \exists x \in X \,.\, x \circ a = x}{b \in \Delta_j}$$

and candidacy, as before, to the rule

$$\frac{\langle \hat{a}, \mathsf{id}\rangle \in K \qquad \exists x \in X \,.\, x \circ a = x}{a = \mathsf{id}}$$

Again $\exists x \in X \,.\, x \circ a = x \iff a \in \Gamma_i$ and the rules may be strengthened to $\langle \hat{a}, b\rangle^k$. These two rules correspond exactly to the two parts of the expression for $\big((A,\Gamma) \multimap (B,\Delta)\big)$. □

Finally, we leave the identity

$$\left[\mathsf{QD}(A,\Gamma) \to \mathsf{QD}(B,\Delta)\right] \simeq \mathsf{QD}\left(!(A,\Gamma) \multimap (B,\Delta)\right)$$

as an exercise: the argument is exactly analogous to Lemmas 3.3.2&3. We have already shown that evaluation is not just stable but linear, so we have completed the proof of the

Theorem 3.3.4 Quantitative types model linear logic, and quantitative domains and stable functors form a cartesian closed category. □

Exercise 3.3.5 State and prove the adjunctions between $\otimes$ and $\multimap$ and between & and $\to$.

3.4. Rigid Adjunctions and the Type of Types

Following standard practice with domain models of polymorphism, we shall use the following to interpret dependent types:

Definition 3.4.1 $\Phi : \mathcal{X} \to \mathcal{Y}$ is a ***rigid comparison*** if it has a right adjoint $\Theta : \mathcal{Y} \to \mathcal{X}$ and the unit $\eta : \mathsf{id}_{\mathcal{X}} \to \Theta\Phi$ and counit $\varepsilon : \Phi\Theta \to \mathsf{id}_{\mathcal{Y}}$ are cartesian.

Theorem 3.4.2 Rigid comparisons are comonadic.

Proof Suppose $\alpha : Y \to \Phi\Theta Y$ is a coalgebra, so $\varepsilon Y \circ \alpha = \mathsf{id}$ — we don't even need the other equation $\Phi\Theta\alpha \circ \alpha = \nu Y \circ \alpha$ because it will follow automatically! Form the pullback

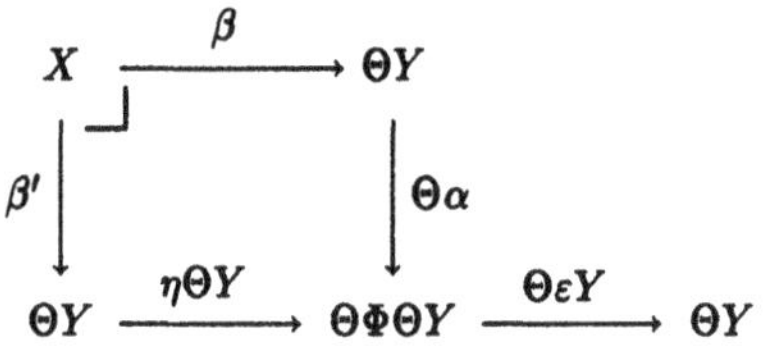

and since $\Theta\varepsilon Y \circ \Theta\alpha = \mathsf{id} = \Theta\varepsilon Y \circ \eta\Theta Y$ we have $\beta = \beta'$. We shall show that its adjoint transpose, $\tilde{\beta} = \varepsilon Y \circ \Phi\beta : \Phi X \to Y$, is a coalgebra isomorphism. Since Φ preserves pullbacks and ε is cartesian, the left-hand diagram below is a pullback:

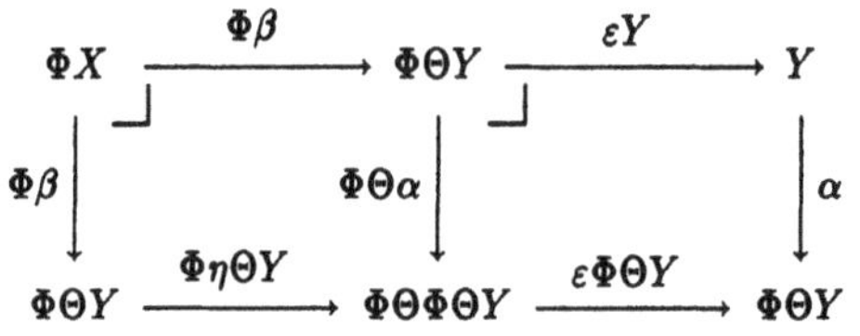

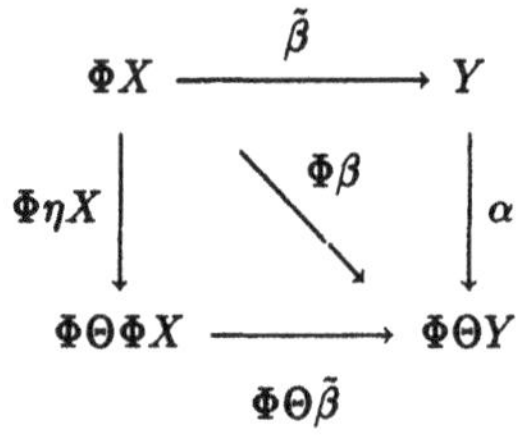

The lower composite is an identity, so $\tilde{\beta}$ is an isomorphism of objects. However the right-hand diagram now commutes, so in fact it is a coalgebra isomorphism. In categorical jargon we have now shown that the Eilenberg-Moore comparison functor (which is always full and faithful) is essentially surjective. □

Definition 3.4.3 A ***local isomorphism*** $\vartheta : (A,\Gamma) \to (B,\Delta)$ between quantitative types is a full and faithful functor $\vartheta : A \to B$ which preserves and reflects creeds. In terms of bags of groups, $\vartheta = (f, \vartheta_i)$ where $f : I \to J$ is an arbitrary function and $\vartheta_i : A_i \cong B_{f(i)}$ are group isomorphisms such that $\vartheta_i(\Gamma_i) = \Delta_{f(i)}$. An automorphism $\vartheta : B \cong B$ is called ***inner*** if there is some family $b_j \in B_j$ such that $\vartheta_j : B \mapsto b_j{}^{-1} b\, b_j$.

Proposition 3.4.4 Each local isomorphism $\vartheta : A \to B$ gives rise to a rigid comparison $\Phi : \mathsf{QD}(A,\Gamma) \to \mathsf{QD}(B,\Delta)$ by

$$\Phi(X)^j = \sum_{f(i)=j} X^i$$
$$\Theta(Y)^i = Y^{f(i)}$$
$$\Psi(X)^j = \prod_{f(i)=j} X^i$$

where the actions are translated *via* ϑ_i and $\Phi \dashv \Theta \dashv \Psi$. Conversely, every rigid comparison arises uniquely up to isomorphism in this way, where $\vartheta = \varepsilon B$. Moreover cartesian transformations between rigid comparisons correspond to postcomposition with inner automorphisms and so are invertible.

Proof By $B \in \mathsf{QD}(B,\Delta)$ we mean the right action on itself, *i.e.* $B^j = {}_1\backslash B_j$. Since any equivariant function into B^j is invertible (Exercise 1.4.4), $\varepsilon B : \hat{A} \stackrel{\text{def}}{=} \Phi\Theta B \to B$ must be of the form of a local isomorphism, where $\hat{A} \cong \sum_i B^{f(j)}$ is naturally a groupoid and $\hat{\Gamma}$ is induced in the obvious way. We have to show that $\mathsf{QD}(A,\Gamma)$, which we already know to be equivalent to the category of coalgebras, is equivalent to $\mathsf{QD}(\hat{A},\hat{\Gamma})$. But if $\alpha : Y \to \Phi\Theta Y$ is a coalgebra, the elements $y \in Y^j$ already carry an action of B_j, whilst the function α (where $\varepsilon_Y \circ \alpha = \mathsf{id}$) corresponds to a choice of $i \in f^{-1}(j)$. The remaining details are left to the reader. □

Notice that the 2-structure is induced in some way by 1-automorphisms. This is in contrast to the continuous analogue, *cf.* [Hyland-Pitts], where it corresponds to homomorphisms of models of theories and is therefore essential. If we ignore it, it not difficult to see that the category of quantitative domains and rigid comparisons (or quantitative types and local isomorphisms) is equivalent to a quantitative domain of the form $\mathsf{QD}(V,1)$. The wastage committed by this, though less, still seems comparable to our original complaint against Girard (Example 1.4.2), but we shall indicate later what the effect actually is (Remark 3.5.3).

Definition 3.4.5 V denotes the groupoid of finite groups with creeds (G,Δ) and group isomorphisms $\vartheta : G \cong G'$ which preserve and reflect creeds, *i.e.* $\vartheta(\Delta) = \Delta'$. As a bag of groups, V has one component for each isomorphism class of finite groups with creeds, and the component group is the group of creed-preserving automorphisms.

Corollary 3.4.6 $(V,1)$ is a ***quantitative type of types.*** □

For $T \in \mathsf{QD}(V,1)$ we shall write $[T]$ for the corresponding quantitative *type*, *i.e.* groupoid with creed. Since the action of V on T is locally faithful, the components of T are of the form ${}_1\backslash \mathsf{Aut}(G,\Delta)$ and correspond to components (G,Δ) of $[T]$.

It is now easy to code the quantifier-free types of the Girard & Reynolds' System F and of Coquand & Huet's Theory of Constructions (for details of the method, see [Hyland-Pitts]). Indeed we have

Exercise 3.4.7 Show that the following are linear functors:

$$\oplus, \& : (V,1) \oplus (V,1) \multimap (V,1) \qquad \otimes, \parr, \multimap : (V,1) \otimes (V,1) \multimap (V,1)$$

$$^{\perp} : (V,1) \multimap (V,1) \qquad ! : !(V,1) \multimap (V,1)$$

□

3.5. Dependent Sums and Products

Before attempting to compute dependent (sums and) products of quantitative domains we need a more explicit description of dependent types. Fix a domain of variation $\mathcal{X} = \mathsf{QD}(A,\Gamma)$ and write $\mathcal{Q}$ for a functor which to each object $X \in \mathcal{X}$ assigns a quantitative domain $\mathcal{Q}(X)$ and to each morphism $f : X' \to X$ of $\mathcal{X}$ a rigid comparison $\mathcal{Q}(f)_! : \mathcal{Q}(X') \to \mathcal{Q}(X)$ with $\mathcal{Q}(f)_! \dashv \mathcal{Q}(f)^* \dashv \mathcal{Q}(f)_*$. Using the "type of types", $\mathcal{Q}$ corresponds to a stable functor $\mathsf{QD}(A,\Gamma) \to \mathsf{QD}(V,1)$ and hence to a proof $Q : !(A,\Gamma) \vdash (V,1)$ such that

$$\mathcal{Q}(X) \simeq \mathsf{QD}\left([!X \otimes_{!A} Q]\right)$$

Such a proof is a sum of atoms, and so we shall concentrate on the case where it is the atom $Q = \mathsf{Aut}(\vec{m})_{/K\backslash}\mathsf{Aut}(G,\Delta)$. More generally, we shall abuse notation by using K to index components of Q.

Exercise 3.5.1 Let $K \leq A^{\mathrm{op}} \times B$. Show that the action of B on $A_{/K\backslash}B$ is locally faithful iff $K \to A^{\mathrm{op}} \times B \to A^{\mathrm{op}}$ is mono; we write $K_o \leq A$ for the image, so $K_o \cong K^{\mathrm{op}}$. This means that K defines a ***partial homomorphism*** $\kappa : A^{\mathrm{op}} \rightharpoonup B$ with support K_o. Then if B has a right action on an object G, there is an induced left action of A on $G \times A_{/K}$. □

Lemma 3.5.2 $[!X \otimes_{!A} Q]$ is the (non-skeletal) groupoid with
(i) objects $\vec{x}K\vartheta$ for $\vec{x} : \vec{m} \to X$ and $\vartheta \in \mathsf{Aut}(G,\Delta)$,
(ii) morphisms $\langle\phi, g\rangle : \vec{x}K\vartheta \to \vec{x}'K\vartheta'$ where $\vec{x}K\vartheta\phi = \vec{x}'K\vartheta'$ and
(iii) composition $\langle\phi,g\rangle\langle\phi',g'\rangle = \langle\phi\phi', \phi'(g)g'\rangle$.
If $f : X' \to X$ in $\mathsf{QD}(A,\Gamma)$,
(iv) the local isomorphism $[!f \otimes Q]$ is given by precomposition with f, *i.e.* $\vec{x}K\vartheta \mapsto f\vec{x}K\vartheta$ and $\langle\phi,g\rangle \mapsto \langle\phi,g\rangle$.
Equivalently, the component groups are G with creed Δ and are indexed by a choice of representatives for the classes $\vec{x}K_o$ but $[!f \otimes Q]$ involves a renormalisation of this choice, which gives an element of K and hence an automorphism of G.

□

Remark 3.5.3 This action of $K_o \leq \mathsf{Aut}(\vec{m})$ on G gives rise to a (split) ***group extension*** $G : K$ which is the set of pairs $\langle k, g\rangle$ with $\langle k,g\rangle\langle k',g'\rangle = \langle kk', \phi'(g)g'\rangle$ where $\langle\hat{k},\phi\rangle, \langle\hat{k}',\phi'\rangle \in K$. Observe that K_o is a subgroup of $G : K$ (this is the meaning of "split"), but in the more general case where we take account of the 2-structure of the category of domains by admitting *pseudo*functors and general (non-split) fibrations, we obtain general group extensions where G is a kernel and K_o is only a quotient. Alternatively the data may be coded as an *arbitrary* (creed-preserving) groupoid homomorphism.

If we attempt to perform the Grothendieck construction to interpret dependent sums, we find that we *never* get a quantitative domain (except in the constant case: the binary product).

Proposition 3.5.4 Let C be the category obtained by adding to the groupoid A the group $G : K$ for each atom of Q and the hom-set $\mathsf{Hom}(A_i, G : K) = m^i$, where m^i is the i-component of $\vec{m}$ and carries the obvious left action of $G : K$ (*via* $\mathsf{Aut}(\vec{m})$) and right action of A_i. Then the total category $\Sigma X : \mathcal{X} \,.\, \mathcal{Q}(X)$ is embedded as a subcategory of $\mathbf{Set}^{C^{\mathrm{op}}}$ with the same closure properties as in Lemma 3.2.4.

Proof Corresponding to $\langle X,Y\rangle$ is a presheaf on C which extends that (*viz.* X) on A. The value at $G : K$ is $\Sigma\vec{x} : X^{\vec{m}} \,.\, Y^{\vec{x}K}$ with $\langle\vec{x},y\rangle \cdot \langle k,g\rangle = \langle\vec{x}\cdot k, y\cdot g\rangle$, and for $r \in m^i$ in the other hom-set, $r : \langle\vec{x},y\rangle \mapsto x^r$. Verification is left to the reader. □

We chose not to develop this paper with general presheaf categories because the action of non-invertible $\mathcal{C}$-maps on candidates is not defined. Nevertheless we only need to consider candidates $u : Y \to SX$ where Y is a *generator*. Corresponding to the two kinds of generator for the total category are two essential kinds of candidate for sections of the display map, but the first is useless because we always have the unique candidate $\langle \mathsf{id}, ? \rangle : \langle {}_1\backslash A_i, 0 \rangle \to \langle {}_1\backslash A_i, S({}_1\backslash A_i) \rangle$. The other kind is of the form

$$u : \langle \vec{m}, {}_1\backslash G_{[K]} \rangle \to \langle \vec{n}, S(\vec{n}) \rangle$$

where $\vec{n}$ is finite by Exercise 1.2.9.

Notation 3.5.5 $\vec{\mu} : \vec{m} \to \vec{n}$ denotes the underlying map of u in $\mathcal{X}$. For $\vec{x} : \vec{m} \to X$, we write

$$\vec{x}^{\vec{\mu}} = \{ f : \vec{n} \to X : f \circ \vec{\mu} \in \vec{x} K_\circ \}$$

In particular for $\vec{x} = \vec{\mu}$, $\mathsf{Aut}(\vec{\mu}) = \vec{\mu}^{\vec{\mu}} = \{ f : f \circ \vec{\mu} \in \vec{\mu} K_\circ \} \subset \mathsf{Aut}(\vec{n})$; this carries the creed $\Gamma_{\vec{\mu}} = \Gamma_{\vec{n}} \cap \mathsf{Aut}(\vec{\mu})$.

Lemma 3.5.6 The set $\sigma_{\vec{\mu}}^{[K]}$ of candidates of the above form carries an action of $\mathsf{Aut}(\vec{\mu})$ on the left which is locally faithful to $\Gamma_{\vec{\mu}}$, and an action of G on the right which is locally faithful to Δ. Likewise $\vec{x}^{\vec{\mu}}$ carries a right action of $\mathsf{Aut}(\vec{\mu})$. □

Proposition 3.5.7 Every object S of the dependent product $\Pi X : \mathsf{QD}(A, \Gamma) \,.\, \mathcal{Q}(X)$ is of the form

$$(SX)^{\vec{x}K} \cong \sum_{\vec{\mu}} \vec{x}^{\vec{\mu}} \otimes_{\mathsf{Aut}(\vec{\mu})} \sigma_{\vec{\mu}}^{[K]}$$

where K ranges over the copies of subgroups corresponding to atoms of Q and $\vec{x} : \vec{m} \to X$. Conversely every such power series is an object of the product.

Proof As in Lemma 1.3.7 and Theorem 1.5.4, an element of $(SX)^{\vec{x}K}$ is a (vertical) map $\langle X, {}_1\backslash G_{\vec{x}K} \rangle \to \langle X, SX \rangle$. By the ophorizontal-vertical factorisation, this corresponds to a map $\langle \vec{m}, {}_1\backslash G_K \rangle \to \langle X, SX \rangle$ over $\vec{x}' : \vec{m} \to X$ (for any $\vec{x}' \in \vec{x}K$). Using stability of $X \mapsto \langle X, SX \rangle$, we factorise this into a candidate $u : \langle \vec{m}, {}_1\backslash G_K \rangle \to \langle \vec{n}, S(\vec{n}) \rangle$ and $\langle f, Sf \rangle$ for $f : \vec{n} \to X$. Considering alternative factorisations yields the given tensor product. The converse is an exercise. □

Theorem 3.5.8 Quantitative domains admit dependent products and hence model System F and the Theory of Constructions.

Proof The components of the quantitative type are of the form

$$\left(\mathsf{Aut}(\vec{\mu}), ! \Gamma_{\vec{\mu}} \right) \multimap (G, \Delta)$$

and there is such a component for each component K of Q and each $\vec{\mu} : \vec{m} \to \vec{n}$. □

3.6. Some Calculations

We have not described $!\Gamma$ explicitly, but the following sketch should serve as a guide to the serious group theory addict.

Exercises 3.6.1

(a) Write $\vec{n} = \sum_{i,[H]} {}_H\backslash A_i \times n_{i,[H]}$, where H ranges over the *conjugacy classes* of subgroups of A_i contained in Γ_i.

(b) Show that

$$\mathsf{Aut}(\vec{n}) = \prod_{i,[H]} \left(A_i / \bigcap [H] \right) \wr \mathsf{Symm}\left(n_{i,[H]} \right)$$

where $P \wr Q$ denotes the **wreath product**, which is the split extension of B^N by the implicit action of Q on the set N.

(c) Hence $h \in \mathsf{Aut}(\vec{n})$ can be written as $h = h_0 \circ \pi$, where π is a permutation of isomorphic atoms and h_0 acts on individual atoms.
(d) Hence h can be written as a product of commuting terms each of the form $h_0 \circ \pi$ in which $\pi = (12 \ldots k)$ is a cycle (possibly $k = 1$) of atoms with the same i and $[H]$.
(e) Suppose $\vec{n} = {}_H\backslash A_i \times k$ and $h = \langle h_1, h_2, \ldots, h_k \rangle \circ (12 \ldots k)$. Then there is an object $X = {}_K\backslash A_i \in \mathsf{QD}(A, \Gamma)$ and a map $f : \vec{n} \to X$ with $f \circ h = f$ iff the subgroup of A_i generated by H and the product $h_1 h_2 \ldots h_k$ (is contained in K which) is contained in Γ_i.
(f) Hence $(!\Gamma)^{\vec{n}} \subset \mathsf{Aut}(\vec{n})$ consists of those $h = h_0 \circ \pi$ such that for every cycle of π, the subgroup of A_i generated by H and the product of the h_x in the cycle is contained in Γ_i. □

From this we may recover our preliminary results.

Example 3.6.2 $^{\perp}!1$ is the groupoid consisting of the finite permutation groups $\mathsf{Symm}(n)$ once each, with the creed $\{1\} \subset \mathsf{Symm}(n)$. (*cf.* Lemmas 1.3.6 and 1.5.4) □

Example 3.6.3 $^{\perp}!^{\perp}!1$ is the groupoid with components

$$\prod_n \mathsf{Symm}(n) \wr \mathsf{Symm}(m_n) \qquad \text{for } (m_n) \text{ finite}$$

with the creed $\prod_n \mathsf{Symm}(n)^{m_n}$. (*cf.* Corollary 1.5.6) □

Example 3.6.4 Let $A = 2 \wr 2$ (the dihedral group of order eight) and $\Gamma = 2^2$; this is for instance one of the components of $^{\perp}!^{\perp}!1$. Let $\vec{n} = \{ A_{/1} \}$, so $\mathsf{Aut}(\vec{n}) = A$. Then $\{1\} \cup A \setminus (!\Gamma)^{\vec{n}}$ has five elements, but is not closed under powers. The creed $^{\perp}!\Gamma$ has three elements and is not closed under multiplication. Finally, $^{\perp\perp}\Gamma$ has six elements. □

Exercise 3.6.5 Show that $\mathsf{Aut}(A, \Gamma) \leq \mathsf{Aut}(^{\perp}A, {}^{\perp}\Gamma)$, but the inclusion may be strict.

Since the formula 3.6.1b involves quotients and not subgroups, the groups generated from **Set** using ! and $^{\perp}$ involve only the symmetric groups with binary and wreath product (because the alternating groups are simple). This means that our original claim to have found the *smallest* cartesian closed category including **Set** as an object is false. However from 3.5.5, we construct $\mathsf{Aut}(\vec{\mu})$ using subgroups, so it seems reasonable to suppose that the following is true (but only group junkies should attempt to prove it).

Conjecture 3.6.6 Every quantitative type is a subtype (in the sense implicit in Definition 3.4.3) of the interpretation of some type of System F or the Theory of Constructions.

Conclusions

Finally we might ask about polymorphic types such as $\Pi\alpha \,.\, \alpha$. This is very disappointing, because it involves the group $G : \mathsf{Aut}(G, \Delta)$ for each isomorphism class of finite groups and creeds. Moggi's "uniformity property" (which holds for the coherence space model) also fails. It seems that Stable Domain Theory has not lived up to its early promise of giving "minimal" models of polymorphism, but we should not therefore consider it to have been a dead end: we have profited by the discovery of Linear Logic, which has shown that (and how) Intuitionistic Logic and Cartesian Closed Categories are not as simple as we once thought.

I would like to express my appreciation for the deep interest shown in this work by Steven Vickers and François Lamarche. John Horton Conway was the source of my amateur fascination for Finite Group Theory.

Bibliography

M. Barr and R. Diaconescu
[80] Atomic toposes, *Journal of Pure and Applied Algebra* **17** (1980) 1–24

M. Barr and C. Wells
[85] *Toposes, triples and theories*, Springer Gr. d. math. W. **278**

G. Berry
[78] Stable models of typed lambda calculi, *Automata, Languages and Programming* (Udine, July 1978), Springer Lecture Notes in Computer Science **62**, 62–90

P.M. Cohn
[77] *Algebra*, Wiley, 2 vols (frequently reprinted)

Y. Diers
[81] Some spectra relative to functors, *JPAA* **22** (1981) 57–74

J.Y. Girard
[85] Normal functors, power series and lambda calculus, *Ann. P.&A. Logic*, 1986
[86] The system F of variable types, fifteen years later, *Theoretical Computer Science* **45** (1986) 159–192
[87] Linear Logic, *TCS* **50** (1987) 1–102
[88] *Towards a geometry of interaction*, in [Gray & Scedrov]
[89] Geometry of interaction I: interpretation of system F, *ASL meeting* (Padova, August 1988), to appear

J.Y. Girard and Y. Lafont
[87] Linear logic and lazy computation, *TAPSOFT '87* (Pisa), Springer LNCS **250** (1987) II 52–66

J.Y. Girard, ***translated and with appendices by*** **Y. Lafont and P. Taylor**
[89] *Proofs and Types*, CUP Cambridge Tracts in Theoretical Computer Science **7**

J.W. Gray and A. Scedrov, editors
[89] *Categories in computer science and logic* (Boulder, June 1987), American Mathematical Society Contemporary Mathematics, to appear

J.M.E. Hyland and A.M. Pitts
[87] *The theory of constructions: categorical semantics and topos-theoretic models*, in [Gray & Scedrov]

A. Joyal
[87] Foncteurs analytiques et espèces de structures, *Combinatoire énumérative* (Montréal, 1986), Springer L.N. Mathematics **1234** (1987) 126–159

Y. Lafont
[88] The categorical abstract machine, *TCS* **59** (1988) 157–180

F. Lamarche
[87] *A simple model of the theory of constructions*, in [Gray & Scedrov]
[88] *Modelling polymorphism with categories*, Ph.D. thesis, McGill University.
[89] *Domains and infinitary algebras*, seminar

S. Lang
[65] *Algebra*, Addison-Wesley (frequently reprinted)

R.A.G. Seely
[87] *Linear logic, *-autonomous categories and co-free alebras*, in [Gray & Scedrov]

P. Taylor
[85] Internal completeness of categories of domains, *Category Theory and Computer Programming* (Guildford, September 1985), Springer LNCS **240** (1986) 449–465
[88] *An algebraic approach to stable domains*, submitted to JPAA
[89] *The trace factorisation and cartesian closure for stable categories*, manuscript 70pp.

Graded Multicategories of Polynomial-time Realizers

R.A.G. Seely*
Department of Mathematics and Computer Science
John Abbott College, Ste. Anne de Bellevue
and
McGill University, Montréal, Québec

Abstract

We present a logical calculus which imposes a grading on a sequent-style calculus to account for the runtime of the programmes represented by the sequents. This system is sound for a notion of polynomial-time realizability. An extension of the grading is also considered, giving a notion of "dependant grades", which is also sound. Furthermore, we define a notion of closed graded multicategory, and show how the structure of polynomial-time realizers has that structure.

0 Introduction

In [4], a restricted notion of realizability is defined, a special case of which is polynomial-time realizability: this is like Kleene's original realizability, save for three features. First, closed atomic formulae are realized only by realizers that express a reason for the "truth" (or provability) of the formula, unlike Kleene's system which only reflects the fact that the formula is provable. Second, open formulae are treated as the corresponding closed formulae with all free variables universally quantified simultaneously. (There is a difference between the quantifiers $\forall\langle\xi,\eta\rangle$ and $\forall\xi\forall\eta$.) And third, the realizers code polynomial-time ("p-time") functions, rather than arbitrary recursive functions.

In [4], only the p-time realizability of single formulae is discussed—in [5] these notions are extended to logical rules, to give a sequent calculus that is sound for p-time realizability. This sequent calculus is much like Gentzen's formulation for intuitionist logic, with three main points of difference, which we summarize again here. First, a sequent of the form $A, B \longrightarrow C$ is interpreted as if it were a formula $A \supset (B \supset C)$, rather than $(A \wedge B) \supset C$ (which would be the Gentzen interpretation). As was shown in [4], these are not equivalent. Indeed, if $\Vdash (A \wedge B) \supset C$ then $\Vdash A \supset (B \supset C)$, but not conversely.

*This work was done following a visit to Monash University and is based on the joint work done there by J. N. Crossley, G.L. Mathai, and the author, who wishes to express his thanks to those two for their many kindnesses during his visit.

Research partially supported by a grant from Le Fonds F.C.A.R., Québec.

($\Vdash A$ means "A is realizable"; similarly $e \Vdash A$ means "e realizes A".) Second, among the structure rules we keep thinning, but drop exchange and contraction, (roughly the opposite of Girard's linear logic [7].) Again, it was shown in [4] that one could have $\Vdash A \supset (B \supset C)$ without having $\Vdash B \supset (A \supset C)$.

Our structure is then a closed multicategory [9,10] with finite products and coproducts: $\supset$ gives the internal hom but this is not a hom for the product structure given by $\wedge$. We do not have a tensor $\otimes$—the comma in the sequent notation takes that role—but if we did, it would not be symmetric. Furthermore, it would not satisfy the expected axiom $A, B \longrightarrow A \otimes B$, for we could then deduce $\Vdash A \otimes B \equiv A \wedge B$, which is false. However, $\Vdash A, B \longrightarrow A \wedge B$ is valid (take $C = A \wedge B$ in the result quoted above).

Third, we extend the usual notion of "sequent" by the addition of variable declarations. A variable declaration $(x :)$ delimits the scope of a free variable x in a sequent, and gives finer control over the functional aspect of realizing open formulae. For example, suppose B is a closed formula and $A(x)$ contains exactly x as free variable. It is *not* the case that $\Vdash B \supset A(x)$ implies $\Vdash B \supset \forall \xi A(\xi)$, according to the definitions of [4]. The point is that such an implication involves an exchange in the order of the realizers m, a standard numeral realizing x, and $b \Vdash B$—in the first m comes first, while in the second b does. If we can shift the functional dependence so that m only enters when wanted then this problem will disappear. The (valid) rule corresponding to our discussion then becomes

$$\frac{\Gamma, (x :) \longrightarrow A(x)}{\Gamma \longrightarrow \forall \xi A(\xi)}$$

where $(x :)$ indicates the beginning of the scope of the variable x, (its end being the end of the sequent), Γ being a finite sequence of formulae (and possibly other variable declarations). Such variable declarations should be thought of as the eponymous statements in computer programmes; as in programming, a variable may only appear locally, and so we wish to capture that locality in the functionality of our realizers. In the above rule, both lines in fact are realized by the same realizer.

In this note, we shall modify the calculus presented in [5] by the introduction of grading: instead of the usual implication $A \supset B$, we shall have a countable set of different "graded" implications, $A \overset{k}{\supset} B$, which indicate a bound on the degree of the runtime of the relevant algorithm. These grades will, initially, be natural numbers, as in the graded λ-calculus of [11], but we shall see that this requirement leads to a weakening of the system of [5], and so at the end of this note we shall present a variant in which the grades may be natural-number-valued functions, dependant on "previously realized data".

Aside The use of dependent grades is really the point of the exercise, since it is the ability to capture instances of "dependent realizability" that distinguishes [4,5] from [11]. For example, in [4] there is an example (my "Favorite Example") that occurs frequently. In the notation of this paper, it can be given in these three flavours:

$$(x :), (y :) \longrightarrow x^{y+1} = x^y \cdot x \tag{1}$$
$$(y :), (x :) \longrightarrow x^{y+1} = x^y \cdot x \tag{2}$$
$$(x, y :) \longrightarrow x^{y+1} = x^y \cdot x \tag{3}$$

(These examples may also be presented as quantified formulae.) Of course, there is no way that Examples 1, 3 can be realizable in polynomial time, but with dependent realizers, Example 2 *is* realizable in polynomial time. However, in the system of this paper, with integer grades, we cannot realize Example 2 because the second realizer (of $(x :) \longrightarrow x^{m+1} = x^m \cdot x$) depends on the first realizer (the m realizing y) for the degree of its runtime. However, at the time of this writing, [5] is still in preparation, and so I have thought it best to concentrate on the integer-graded system for the time being. This system is in some ways an extension of [11], from which it derives its inspiration, but there are some differences, as the reader will see.

Analagous to the graded implication, we shall have graded universal quantifiers $\overset{k}{\forall}\xi A(\xi)$. We mentioned in [5] that the usual quantifier $\forall\xi A(\xi)$ could be thought of as $(x :) \supset A(x)$; analagously $\overset{k}{\forall}\xi A(\xi)$ may be thought of as $(x :) \overset{k}{\supset} A(x)$. Further, the sequents in [5] are directly related to the notion of implication, and so in the graded system we shall have graded sequents: for example $A, B, C \overset{\mathbf{k}}{\longrightarrow} D$ will be interpreted as $A \overset{k_1}{\supset} (B \overset{k_2}{\supset} (C \overset{k_3}{\supset} D))$, where $\mathbf{k} = \langle k_1, k_2, k_3 \rangle$.

Finally, we shall define a notion of graded multicategory, which extends Lambek's notion of a multicategory by associating with each morphism an index or grade in a coherent manner. In fact, the graded multicategorical structure suitable for the graded sequent calculus described above will have much more structure, *e.g.* it will be "graded closed" since it has a "graded internal hom" given by the graded implication. Of course, all this is directly analagous to the ungraded case, and follows the paradigm case of a Gentzen multicategory [10]. (Also, in the intended example the grading is filtered—if a morphism has grade $\mathbf{k}$ then it also has grade $\mathbf{k}'$, for any $\mathbf{k}' \geq \mathbf{k}$, in a suitable ordering. However, it is not clear this should be part of the general definition.)

One final point: for simplicity, in this note we shall not consider the structure of the natural numbers, and so shall make no reference to the non-logical axioms needed to describe that structure. Of course, this is a serious oversight, but that structure (essentially Buss's axioms ([1,2], see also [3]) together with the expected equality axioms) is discussed in [5], as well as in [11] *mutatis mutandis.*

1 Polynomial-time realizers

We recall the basic definitions from [4] and [5]—for full details, the reader should refer to those papers.

1.1 Realizers

We assume a formal language $\mathcal{L}$ with constants including 0, function letters including $+$, $\cdot$, and *exp* (exponentiation), and predicate letters including $=$ (equality). As indicated in the Introduction, $\mathcal{L}$ is modified by replacing $\supset$ with $\overset{k}{\supset}$ and $\forall$ with $\overset{k}{\forall}$ (for all k) in the formation rules for formulae. Realizers are taken as (codes of) pairs $e = \langle e_m, e_t \rangle$, where e_m is (the code of) a Turing machine and e_t is (the code of) a polynomial: e acts as e_m except that it turns off after $e_t(|\mathsf{x}|)$ steps, where $|\mathsf{x}|$ is the length of the input. In the following, "function" means such a realizer, unless otherwise specified. (This includes constants, as functions of 0 variables.)

We shall write $e(f)$ to denote $\{e_m\}(f)$, $\{\ \}$ denoting the Kleene bracket.

Definition 1 *A realizer* $e = \langle e_m, e_t \rangle$ *is said to have grade* k *if the logarithm of the degree of* e_t *is* $\leq k$*; that is, if*

$$deg(e_t) \leq 2^k.$$

For some of the function symbols, in particular, $+, \cdot$, but not *exp*, there is associated a realizer e_f with the function symbol f, and likewise for some of the predicate symbols, including $=$, a realizer e_P is associated with P. Terms are realized by this definition:

Definition 2 • *If* t *is closed:*

1. *If* t *is a constant* c, *with realizer* e_c, *then* $e_c \Vdash t$.
2. *If* $t = f(t_1, \ldots, t_n)$, $e_i \Vdash t_i$ *for* $i = 1, \ldots, n$, e_f *is the realizer of* f, *then* $e_f(e_1, \ldots, e_n) \Vdash t$.

- *If* t *contains free variables* $x_1, \ldots, x_n$, *then* $e \Vdash t$ *iff* e *is (the code of) a function that, for standard numerals* $m_1, \ldots, m_n$, *gives* $e(m_1, \ldots, m_n) \Vdash t(m_1, \ldots, m_n)$.

Definition 3 *Closed atomic formulae are realized by the realizers of predicate symbols: if* $e_i \Vdash t_i$ *(for* $i = 1, \ldots, n$*) and* e_P *is the realizer of* P, *then* $e_P(e_1, \ldots, e_n) \Vdash P(t_1, \ldots, t_n)$.

Finally, we modify Kleene's original inductive definition of $e \Vdash A$ as in [5], with the addition of grading to account for the graded connective $\overset{k}{\supset}$ and the graded quantifier $\overset{k}{\forall}$.

Definition 4 $e \Vdash A$ *in the following situations:*

- *If* A *is closed:*
  1. $\bot$ *is never realized.*
  2. A *is* $B \wedge C$, $e = \langle e_0, e_1 \rangle$, $e_0 \Vdash B$ *and* $e_1 \Vdash C$.
  3. A *is* $B \vee C$, $e = \langle e_0, e_1 \rangle$, *and either* e_0 *is 0 and* $e_1 \Vdash B$ *or* e_0 *is not 0 and* $e_1 \Vdash C$.
  4. A *is* $B \overset{k}{\supset} C$, *the grade of* e *is* k, *and, for all* f, *if* $f \Vdash B$ *then* $e(f)$ *is defined and* $e(f) \Vdash C$.
  5. A *is* $\exists \xi B(\xi)$, $e = \langle e_0, e_1 \rangle$, e_0 *is a standard numeral and* $e_1 \Vdash B(x := e_0)$, *(where* $B(x := e_0)$ *denotes the substitution of* e_0 *for* x *in* B.*)*
  6. A *is* $\overset{k}{\forall} \xi B(\xi)$, *the grade of* e *is* k, *and for all standard numerals* k, $e(k) \Vdash B(x := k)$.
- *If* A *is open with free variables exactly* $x_1, \ldots, x_n$ *and for all standard numerals* $m_1, \ldots, m_n$,

$$e(m_1, \ldots, m_n) \Vdash A(x_1 := m_1, \ldots, x_n := m_n).$$

1.2 The sequent calculus

In the logical calculus we shall develop, it will be convenient to assume that all formulae are "homogeneous" in their occurrences of free variables: in forming a compound formula $A\Diamond B$, for any connective $\Diamond$, we shall suppose A and B have exactly the same free variables, which then also occur in $A\Diamond B$. This may be done without loss in expressive power by a liberal use of "dummy free variables"; perhaps the simplest technical way to do this is to add new function symbols to $\mathcal{L}$ corresponding to projections (—see [12] for example, where first order logic with equality is handled this way.) Note that in quantifying a formula, exactly one free variable disappears—*viz.* the one quantified.

Definition 5 *A graded sequent* $\Gamma \xrightarrow{\mathbf{k}} A$ *consists of a usual sequent* $\Gamma \longrightarrow A$ *together with a finite sequence* $\mathbf{k}$ *(of natural numbers) of the same length as the sequence* Γ. *Recall from [5] that* Γ *is a finite sequence of formulae and variable declarations, where a variable declaration* $(x :)$ *must precede all formulae in which* x *appears. Furthermore, a variable may only be declared once within the sequent.*

Remarks

1. There may be formulae within the scope of x in which x does not appear—such formulae will be treated as if they have 'dummy' occurrences of x.
2. We can declare several variables simultaneously via pairing: $(\mathbf{x} :)$ will mean $(\langle x_1, \ldots, x_n\rangle :)$, where $\mathbf{x} = x_1, \ldots, x_n$. These variables can also be declared sequentially: $(x_1 :), (x_2 :), \ldots, (x_n :)$. These forms of declaration are not equivalent.

Definition 6 *Our theory consists of all sequents generated* [1] *from the following axioms by the following rules:*

Logical Axioms

1. $(\mathbf{x} :), A \xrightarrow{(0,0)} A$ *(where* $\mathbf{x}$ *lists all free variables of* A.*)*
2. $(\mathbf{x} :), A, B \xrightarrow{(0,0,1)} A \wedge B$ *(where* $\mathbf{x}$ *lists all free variables of* $A \wedge B$.*)*
3. $(\mathbf{x} :), \overset{k}{\forall}\xi A(\xi), (x :) \xrightarrow{(0,0,k)} A(x)$ *(where* $\mathbf{x}$ *lists all free variables of* A *other than* x.*)*

Structural Rules

(grade filtering) $$\frac{\Gamma \xrightarrow{\mathbf{k}} A}{\Gamma \xrightarrow{\mathbf{k}'} A}$$

where $\mathbf{k}' \geq \mathbf{k}$ *in the sense that each coordinate* $k_i \geq k'_i$.

[1] As outlined by Definition 7.

(thinning)
$$\frac{\Gamma, \Delta \xrightarrow{\mathbf{k}} A}{\Gamma, B, \Delta \xrightarrow{\mathbf{k}'} A}$$

where $\mathbf{k}'$ *is the sequence* $\mathbf{k}$ *with a 0 inserted in the position corresponding to the position of* B *in the sequent.*

(curry)
$$\frac{\Gamma, (\langle x, y\rangle :), \Delta \xrightarrow{\mathbf{k}} A}{\Gamma, (x :), (y :), \Delta \xrightarrow{\mathbf{k}'} A} \qquad \frac{\Gamma, (\langle y, x\rangle :), \Delta \xrightarrow{\mathbf{k}} A}{\Gamma, (x :), (y :), \Delta \xrightarrow{\mathbf{k}'} A}$$

where $\mathbf{k}'$ *is the sequence* $\mathbf{k}$ *with a 1 inserted in the position corresponding to the position of* $(x :)$ *in the conclusion-sequent.*

(cut)
$$\frac{\Delta, B, \Theta \xrightarrow{(\mathbf{k}_1, k, \mathbf{k}_2)} A \quad \Gamma \xrightarrow{\mathbf{l}} B}{\Delta, \Gamma', \Theta \xrightarrow{(\mathbf{k}_1', \mathbf{l}' + k, \mathbf{k}_2)} A}$$

where

- Γ' *is* Γ *with the variable declarations for* B *omitted, (see Remarks following),*
- $\mathbf{l}'$ *is* $\mathbf{l}$ *with the entries corresponding to variable declarations for* B *deleted,*
- $\mathbf{k}_1'$ *is* $\mathbf{k}_1$ *with the entries corresponding to the variable declarations for* B *augmented by adding 1 + the corresponding entries from* $\mathbf{l}$ *(*i.e. *those deleted to get* $\mathbf{l}'$*), and*
- $\mathbf{k}_1', \mathbf{l}' + k$ *means add* k *to the* last *entry in the sequence* $\mathbf{k}_1', \mathbf{l}'$.

Logical Rules

$(\overset{k}{\supset}$ L)
$$\frac{\Delta, B, \Theta \xrightarrow{(\mathbf{k}_1, k, \mathbf{k}_2)} C \qquad \Gamma \xrightarrow{\mathbf{l}} A}{\Delta, (A \overset{h}{\supset} B), \Gamma', \Theta \xrightarrow{(\mathbf{k}_1', 0, \mathbf{l}' + k + h, \mathbf{k}_2)} C}$$

where Γ', $\mathbf{k}'$, $\mathbf{l}'$ *are as in* (cut).

$(\overset{k}{\supset}$ R)
$$\frac{\Gamma, A \xrightarrow{(\mathbf{k}, k)} B}{\Gamma \xrightarrow{\mathbf{k}} (A \overset{k}{\supset} B)}$$

$(\wedge$ L)
$$\frac{\Gamma, A, \Delta \xrightarrow{\mathbf{k}} C}{\Gamma, (A \wedge B), \Delta \xrightarrow{\mathbf{k}'} C} \qquad \frac{\Gamma, B, \Delta \xrightarrow{\mathbf{k}} C}{\Gamma, (A \wedge B), \Delta \xrightarrow{\mathbf{k}'} C}$$

where $\mathbf{k}'$ *is the sequence* $\mathbf{k}$ *with no less than 1 in the position corresponding to the position of* $A \wedge B$ *in the conclusion-sequent.*

$(\wedge$ R)
$$\frac{\Gamma \xrightarrow{\mathbf{k}} A \quad \Gamma \xrightarrow{\mathbf{l}} B}{\Gamma \xrightarrow{max'(\mathbf{k}, \mathbf{l})} A \wedge B}$$

where $max'(\mathbf{k}, \mathbf{l})$ *means take the maximum at each coordinate, and where the last entry is no less than 1.*

$$(\vee\ \mathbf{L})\quad \frac{\Gamma, A, \Delta \xrightarrow{\mathbf{k}} C \quad \Gamma, B, \Delta \xrightarrow{\mathbf{l}} C}{\Gamma, (A \vee B), \Delta \xrightarrow{max'(\mathbf{k},\mathbf{l})} C}$$

where $max'(\mathbf{k},\mathbf{l})$ *means take the maximum at each coordinate, and where the entry in the* $A \vee B$ *position is no less than 1.*

$$(\vee\ \mathbf{R})\quad \frac{\Gamma \xrightarrow{\mathbf{k}} A}{\Gamma \xrightarrow{\mathbf{k}'} A \vee B} \qquad \frac{\Gamma \xrightarrow{\mathbf{k}} B}{\Gamma \xrightarrow{\mathbf{k}'} A \vee B}$$

where $\mathbf{k}'$ *is the sequence* $\mathbf{k}$ *with the last position no less than 1.*

$$(\overset{k}{\forall}\ \mathbf{R})\quad \frac{\Gamma, (x :) \xrightarrow{(\mathbf{k},k)} A(x)}{\Gamma \xrightarrow{\mathbf{k}} \overset{k}{\forall} \xi A(\xi)}$$

$$(\exists\ \mathbf{L})\quad \frac{\Gamma, (x :), A, \Delta \xrightarrow{(\mathbf{k},k,l,\mathbf{l})} B}{\Gamma, \exists \xi A(\xi), \Delta \xrightarrow{(\mathbf{k},max'(k,l),\mathbf{l})} B}$$

where $max'(k,l)$ *means* $max(k,l,1)$.

$$(\exists\ \mathbf{R})\quad \frac{\Gamma \xrightarrow{\mathbf{k}} A}{\Gamma \xrightarrow{\mathbf{k}'} \exists \xi A(\xi)}$$

where $\mathbf{k}'$ *means* $\mathbf{k}$ *with the last entry no less than 1.*

Remarks: There are several restrictions on these rules, as discussed in [5]. Some of the restrictions are no longer relevant, because they are subsumed by the grading. However, the following restrictions remain:

(thinning) B must be a formula whose variables are declared in Γ.

(cut) The variable declarations for B must be identical in Γ and in Δ, and must occur at the beginning of Γ. In the conclusion of the rule, these declarations are dropped from Γ to give Γ'.

($\overset{k}{\supset}$ L) Similar restrictions on variable declarations to those above in (cut).

($\exists$ L) Implicit in the syntax is that x occurs free only in A, not in Δ nor in B.

($\exists$ R) Γ must begin with the declaration $(x :)$, where x is the variable quantified.

Definition 7 *A derivation of a graded sequent consists of a finite tree such that each branch ends in an axiom and each step is one of the rules of inference given in Definition 6 above, or is a substitution instance of such an axiom or rule.*

We remark here that by a "substitution instance" of a graded sequent $\Gamma \xrightarrow{\mathbf{k}} A$ we mean

- the replacement throughout the sequent of a free variable x, say, occurring in the sequent, by a *realizable* term $t(\mathbf{x})$ which has only new free variables $\mathbf{x}$, not occurring in the sequent,
- the replacement of the declaration $(x :)$ by the simultaneous declaration $(\mathbf{x} :)$, and

- the replacement of $\mathbf{k}$ by the sequence $\mathbf{k}'$, which differs from $\mathbf{k}$ by adding to the coordinate in the $(x :)$ position the grade of the realizer of t of least grade, (or 1 if that is larger.)

A substitution instance of a rule is defined similarly—take the corresponding substitution instances of the sequents involved in the rule.

Note that we only allow substitution instances of realizable terms.

1.3 Graded realizability

As discussed in the Introduction, we shall treat realizability for sequents as if the sequents consisted of a successive introduction of premisses, *i.e.* a nested sequence of implications. Within these successive hypotheses, a variable declaration amounts, in effect, to another such hypothesis. Finally, a rule is realized by a function (not necessarily p-time, however) that assigns a realizer of the conclusion to a simultaneously-presented tuple of realizers of the premisses of the rule. These points are summarized in the following definitions:

Definition 8 *For a graded sequent* $\Gamma \xrightarrow{\mathbf{k}} A$, *we define* $e \Vdash \Gamma \xrightarrow{\mathbf{k}} A$ *inductively:*

1. $e \Vdash B \xrightarrow{k} A$ *iff* $e \Vdash B \overset{k}{\supset} A$.
2. $e \Vdash (x :) \xrightarrow{k} A$ *iff* $e \Vdash \overset{k}{\forall}\xi A(\xi)$.
3. $e \Vdash B, \Gamma \xrightarrow{(k,\mathbf{k})} A$ *iff e is (the code of) a function of grade k which to any* $b \Vdash B$, *produces an output* $e(b) \Vdash \Gamma \xrightarrow{\mathbf{k}} A$.
4. $e \Vdash (x :), \Gamma \xrightarrow{(k,\mathbf{k})} A$ *iff e is (the code of) a function of grade k which to any standard numeral m, produces an output* $e(m) \Vdash \Gamma(x := m) \xrightarrow{\mathbf{k}} A(x := m)$.

Definition 9 *Given a rule of the form*

$$\frac{P_1, \ldots, P_n}{C}$$

where $P_i (i = 1, \ldots, n)$, C, *are sequents, we say f realizes the rule iff f is a function (not necessarily p-time) so that for realizers* $e_i \Vdash P_i$, $f(e_1, \ldots, e_n) \Vdash C$.

Proposition 1 *Each of the rules of Definition 6 is realizable.*

Proof In each case we shall define the realizer by giving an equation for the "fully evaluated form". Lower case letters will represent realizers of the corresponding formulae given by upper case letters. We shall write $e(\gamma)$ for $e(a_1)(a_2)\ldots(a_n)$ when $\Gamma = A_1, A_2, \ldots, A_n$. Variable declarations are realized by standard numerals m. Formulae involving functions ($A \overset{k}{\supset} B$ or $\overset{k}{\forall}\xi A(\xi)$, as appropriate) will be realized by p; e, f will denote realizers of the premisses of the rule. Finally we shall denote the realizer of the rule by the same name as the rule.

- $(\text{thin})(e)(\gamma)(b)(\delta) = e(\gamma)(\delta)$
- $(\text{curry})(e)(\gamma)(m)(m')(\delta) = e(\gamma)\langle m, m'\rangle(\delta)$
- $(\text{cut})\langle e, f\rangle(\delta)(\gamma)(\theta) = e(\delta)(f(\gamma))(\theta)$
- $(\stackrel{k}{\supset} \text{L})\langle e, f\rangle(\delta)(p)(\gamma)(\theta) = e(\delta)(p(f(\gamma)))(\theta)$
- $(\stackrel{k}{\supset} \text{R})(e)(\gamma)(a) = e(\gamma)(a)$
- $(\wedge \text{ L})(e)(\gamma)\langle a, b\rangle(\delta) = e(\gamma)(a)(\delta)$
- $(\wedge \text{ L})'(e)(\gamma)\langle a, b\rangle(\delta) = e(\gamma)(b)(\delta)$
- $(\wedge \text{ R})\langle e, f\rangle(\gamma) = \langle e(\gamma), f(\gamma)\rangle$
- $(\vee \text{ L})\langle e, f\rangle(\gamma)\langle i, a\rangle(\delta) = \begin{cases} e(\gamma)(a)(\delta) & \text{if } i = 0 \\ f(\gamma)(a)(\delta) & \text{if } i \neq 0 \end{cases}$
- $(\vee \text{ R})(e)(\gamma) = \langle 0, e(\gamma)\rangle$
- $(\vee \text{ R})'(e)(\gamma) = \langle 1, e(\gamma)\rangle$
- $(\stackrel{k}{\forall} \text{R})(e)(\gamma)(m) = e(\gamma)(m)$
- $(\exists \text{ L})(e)(\gamma)\langle m, a\rangle(\delta) = e(\gamma)(m)(a)(\delta)$
- $(\exists \text{ R})(e)(\gamma) = \langle m, e(m)(\gamma)\rangle$ where m is the realization of the variable declaration $(x :)$ in Γ.

Of course, now we must justify the grading given in the rules. For the most part this is based on that in [11], but reflecting the difference in our definition of grading, which makes it unnecessary to go up a grade when adding a routine of low degree runtime to a given routine. Notice in particular the frequent references to "not less than 1"—these are to account for the pairing – unpairing operations—which in [11] require a "+ 1" instead.

The main point here is our grading of cut, which we now illustrate with some examples.

First, consider the simplest case, "composition":

$$\frac{B \xrightarrow{k} A \quad C \xrightarrow{l} B}{C \xrightarrow{k+l} A}$$

This grading is the result of the way polynomial-time functions are composed: the runtime of the composite is given by composing the runtimes (see [4,5,11].)

Now, let's extend this one step:

$$\frac{e \Vdash D, B \xrightarrow{k_1, k} A \quad f \Vdash C \xrightarrow{l} B}{g \Vdash D, C \xrightarrow{k_1, l+k} A}$$

The programme g is as follows. First we compute $g(d)$, for $d \Vdash D$. This is: compute $e(d)$, (this has a runtime of grade k_1, by assumption); output $e(d)$ composed with f, (this is the previous algorithm, and amounts to just providing the given f with a tail end. Notice that we are not yet running this, so we have merely added some constant to the runtime, and have not changed the grade.)

Next we compute $g(d)(c)$ for $c \Vdash C$. This is where we actually run the composition above, and so as we saw earlier, this step has grade $l + k$.

Next, consider a simple case with variable declarations:

$$\frac{e \Vdash (x :), B \xrightarrow{\langle k_1, k \rangle} C \quad f \Vdash (x :) \xrightarrow{l} B}{g \Vdash (x :) \xrightarrow{k_1 + l + 1 + k} C}$$

(Notice how the grading instructions work here: we have, in the notation of Definition 6, that $\mathbf{k}_1 = k_1$, $\mathbf{l} = l$, $\mathbf{l}' = \langle\rangle$, $\mathbf{k}'_1 = \langle k_1 + l + 1 \rangle$, and $\mathbf{k}'_1, \mathbf{l}' + k = \langle k_1 + l + 1 + k \rangle$, as shown.)

The point about this instance of cut is that now we have an instance of an implicit *contraction* rule at play here: we have lost an hypothesis $(x :)$ in the conclusion (because we cannot declare variables more than once in a sequent). This will require us to step up one degree to access an appropriate Universal Turing Machine (UTM) which can perform the required joint application/composition operation for us. So, the algorithm $g(m)$, for $m \Vdash (x :)$, is to calculate both $e(m) \Vdash B \xrightarrow{k} C$ and $f(m) \Vdash B$, plug these into the UTM for grade k to compose the results and get a realizer for C. (Notice we actually run this, we do not just output the instructions "compose $e(m)$, $f(m)$". Calculating $e(m)$ takes runtime of grade k_1, calculating $f(m)$ takes runtime of grade l, and since $e(m)$ itself has grade k, using the UTM pushes us up a grade to give an additional grade of $k + 1$ (see [11]), as shown above.)

Now an induction on the structure of the sequents in the cut rule, similar to that above (going from the simple composition to the one-step extension), gives the general grading formula. As an example, consider:

$$\frac{e \Vdash A_1, (y :), (x :), A_2, B \xrightarrow{\langle k_1, k_2, k_3, k_4, k \rangle} C \quad f \Vdash (x :), (z :), D \xrightarrow{\langle l_1, l_2, l_3 \rangle} B}{g \Vdash A_1, (y :), (x :), A_2, (z :), D \xrightarrow{\langle k_1, k_2, k_3 + l_1 + 1, l_2, k_5 + l_3 \rangle} C}$$

The crucial step is the third, where $m_x \Vdash (x :)$ is read, $e(a_1)(m_y)(m_x)$ and $f(a_1)(m_y)(m_x)$ are calculated, plugged into an appropriate UTM, and so composed to produce a realizer of $A_2, (z :), D \longrightarrow C$. (Note that if we wish, we can always bring hypotheses to the right hand side, via ($\overset{k}{\supset}$ R) and ($\overset{k}{\forall}$ R).)

Theorem 1 *The sequent calculus given in Definition 6 is sound with respect to realizability.*

Proof All that remains to be shown is that the axioms are realizable. But Axiom 1 is trivially realized. Also Axiom 2 is proven in [4], and discussed in the introduction here. Axiom 3 is virtually the identity, given our interpretation of $\overset{k}{\forall}$ and $\overset{k}{\longrightarrow}$. The grading is quite straightforward.

2 Multicategories

In this section, we consider the structure of the propositional part of the calculus of polynomial-time realizers. In particular, we shall ignore variable declarations, and so the "primes" in the cut rule may be dropped. This gives us the structure of a "graded multicategory". (Due to deadline constraints, I must leave the strange structure of the quantifiers to a promised sequel. It is clear that we have something more complicated there than a straightforward notion of "weak adjoint", especially with the universal quantifiers $\overset{k}{\forall}$. The existential quantifier is a little more straightforward, and will be remarked upon briefly at the end of the paper.)

2.1 Definitions

Recall from [9,10] that

Definition 10 *A multicategory* **C** *consists of a set* $Ob(\mathbf{C})$ *of objects and a set* $M\varphi(\mathbf{C})$ *of morphisms, (also called arrows, multimorphisms, ...,) just like a category, except that the source of a morphism is a finite sequence of objects, rather than a single object. The target of a morphism is a single object as usual. So we have the two maps*

$$source : M\varphi(\mathbf{C}) \longrightarrow Ob(\mathbf{C})^*$$

$$target : M\varphi(\mathbf{C}) \longrightarrow Ob(\mathbf{C})$$

(where X^* *= the free monoid generated by* X*.)*

As with categories, we have identity morphisms $1_A : A \to A$*, and a notion of composition which is most simply given by the following "inference diagram":*

$$\frac{\Gamma, A, \Delta \xrightarrow{g} B \quad \Theta \xrightarrow{f} A}{\Gamma, \Theta, \Delta \xrightarrow{g(f)} B}$$

We have the following axioms:

1. $$\Gamma \xrightarrow{f} A \quad = \quad \frac{A \xrightarrow{1_A} A \quad \Gamma \xrightarrow{f} A}{\Gamma \xrightarrow{1_A(f)} A}$$

2. $$\Gamma, A, \Delta \xrightarrow{g} B \quad = \quad \frac{\Gamma, A, \Delta \xrightarrow{g} B \quad A \xrightarrow{1_A} A}{\Gamma, A, \Delta \xrightarrow{g(1_A)} B}$$

3. $$\frac{\Phi, B, \Psi \xrightarrow{h} C \qquad \dfrac{\Gamma, A, \Delta \xrightarrow{g} B \quad \Theta \xrightarrow{f} A}{\Gamma, \Theta, \Delta \xrightarrow{g(f)} B}}{\Phi, \Gamma, \Theta, \Delta, \Psi \xrightarrow{h(g(f))} C}$$

$$= \quad \frac{\dfrac{\Phi, B, \Psi \xrightarrow{h} C \quad \Gamma, A, \Delta \xrightarrow{g} B}{\Phi, \Gamma, A, \Delta, \Psi \xrightarrow{h(g)} C} \qquad \Theta \xrightarrow{f} A}{\Phi, \Gamma, \Theta, \Delta, \Psi \xrightarrow{h(g)(f)} C}$$

4.
$$\frac{\dfrac{\Phi, A, \Theta, B, \Psi \xrightarrow{h} C \quad \Delta \xrightarrow{g} B}{\Phi, A, \Theta, \Delta, \Psi \xrightarrow{h\langle g\rangle} C} \qquad \Theta \xrightarrow{f} A}{\Phi, \Gamma, \Theta, \Delta, \Psi \xrightarrow{h\langle g\rangle\langle f\rangle} C}$$

$$= \quad \frac{\dfrac{\Phi, A, \Theta, B, \Psi \xrightarrow{h} C \quad \Gamma \xrightarrow{f} A}{\Phi, \Gamma, \Theta, B, \Psi \xrightarrow{h\langle f\rangle} C} \qquad \Delta \xrightarrow{g} B}{\Phi, \Gamma, \Theta, \Delta, \Psi \xrightarrow{h\langle f\rangle\langle g\rangle} C}$$

(We have made slight changes to the notation of [10], mainly in reversing the order of presenting the composition, in order to make the comparison with Definition 6 more obvious. As Lambek pointed out, there is an ambiguity in the notation $g\langle f\rangle$, which we shall ignore, appealing to diagrams when necessary.)

As an example of a multicategory, we offer the following:

Example: Finite sequences of natural numbers form a multicategory $\mathbf{N}^{<\omega}$, which has one object (which we shall not bother to name), and whose morphisms are finite sequences of natural numbers. We shall denote such a morphism by the sequence concerned, and not refer to the objects giving the source and target, as they are obvious. The identity is the singleton $\langle 0\rangle$; composition is given by the diagram:

$$\frac{\langle \mathbf{k}_1, k, \mathbf{k}_2\rangle \qquad\qquad \mathbf{l}}{\langle \mathbf{k}_1, \mathbf{l}+k, \mathbf{k}_2\rangle}$$

Of course, this example is inspired by the grading of Definition 6. It is a simple exercise to verify the four axioms in this case.

Definition 11 *A graded multicategory* $\langle \mathbf{C}, G\rangle$ *consists of a multifunctor* G *between a multicategory* $\mathbf{C}$ *and* $\mathbf{N}^{<\omega}$.

This means that to every morphism $f : \Gamma \longrightarrow A$ of $\mathbf{C}$ there is associated a "grading" $G(f) = \mathbf{k}$, usually denoted by a superscript, as we have been doing for our graded sequent calculus:

$$f : \Gamma \xrightarrow{\mathbf{k}} A.$$

The point of this grading being a multifunctor is that it should be defined in such a way as to make the grading of a composite as suggested by the cut rule (of Definition 6), *viz.*:

$$\frac{g : \Gamma, A, \Delta \xrightarrow{\langle \mathbf{k}_1, k, \mathbf{k}_2\rangle} B \quad f : \Theta \xrightarrow{\mathbf{l}} A}{g\langle f\rangle : \Gamma, \Theta, \Delta \xrightarrow{\langle \mathbf{k}_1, \mathbf{l}+k, \mathbf{k}_2\rangle} B}$$

And finally, the grading ought to respect the axioms 1 to 4 of Definition 10.

I have presented this definition in such a way as to leave open the possibility of other types of grading than by sequences of natural numbers. However, in the absence of interesting examples, I prefer to leave this definition in this more restrictive form. A point, however: it will not be correct to merely change the target multicategory in order to capture the notion of dependent grading, since there must be a greater degree of interaction between the morphism $f : \Gamma \longrightarrow A$ being graded and the grade $\mathbf{k} = G(f)$. For instance, if $f : A, B \longrightarrow C$, then $G(f) = \langle k_1, k_2 \rangle$, where k_1 is a number, and k_2 is a number-valued function that may depend on both k_1 and $a \in A$, (where $\in$ is a suitable "membership" or "typing" relation—in our main example, $\in$ would be $\Vdash$.)

2.2 The main example

Now we come to the point of this note: it is clear from the notation that what is intended is to have the structure of realizers form a multicategory, with grading as given in Definition 6.

So we define the multicategory $\mathbf{C}$ as follows:

$Ob(\mathbf{C})$ consists of all formulae of $\mathcal{L}'$, the propositional part of our logical system. (In fact we shall really be dealing with a set $[A]$ of graded realizers of A, rather than with A itself, for any formula A—however, the identification is generally harmless.) $M\varphi(\mathbf{C})$ consists of all graded realizers of the corresponding sequents: this means that we are not allowing dependencies of the sort exemplified by Favorite Example 2. To have the axioms of Definition 10, we must factor out by those equations, in the usual manner. The grading G is of course then given by the "least grading" function, (*i.e.* take the least possible entry in each coordinate.) (Here is where we have included less structure in the definition of grading than our model shows, by ignoring the filtering.) We have then already shown that

Proposition 2 *The structure $\langle \mathbf{C}, G \rangle$ is in fact a graded multicategory.*

2.3 Structured graded multicategories

In [10], Lambek defines the notion of a *right closed multicategory, with Cartesian products and coproducts.* Here we sketch the corresponding notions for the graded case, and indicate how the main example has these properties.

Definition 12 *A right closed graded multicategory is a graded multicategory* $(\mathbf{C}, G)$ *with a family of (graded) internal homs* $\overset{k}{\supset}$, *(one for each natural number k), and graded morphisms*

$$ev_{AB} : A \overset{k}{\supset} B, A \overset{\langle 0,k \rangle}{\longrightarrow} B$$

inducing a bijection

$$\frac{f \quad : \quad \Gamma \overset{\mathbf{k}}{\rightarrow} (A \overset{k}{\supset} B)}{ev_{AB}\langle f \rangle \quad : \quad \Gamma, A \overset{\langle \mathbf{k},k \rangle}{\longrightarrow} B}$$

This means to each $f : \Gamma, A \overset{\langle \mathbf{k},k \rangle}{\longrightarrow} B$ there is a (unique) $f^\star : \Gamma \overset{\mathbf{k}}{\rightarrow} A \overset{k}{\supset} B$ such that $ev_{AB}\langle f^\star \rangle = f$. (Uniqueness amounts to $(ev_{AB}\langle f \rangle)^\star = f$.) (The equations for grading are automatic.)

In our situation, $f^\star$ is constructed by the ($\overset{k}{\supset}$ R) rule, and ev_{AB} is constructed from ($\overset{k}{\supset}$ L) as follows:

$$\frac{B \xrightarrow{0} B \quad A \xrightarrow{0} A}{A \overset{k}{\supset} B, A \xrightarrow{(0,k)} B}$$

(We might note that **C** is not left closed in any meaningful sense, as that would require some exchange:)

$$\frac{A, \Gamma \longrightarrow C}{\Gamma \longrightarrow A \not\subset C}$$

The structure of Cartesian products and coproducts is straightforward, and can be found in [10] without alteration. Of course, the product structure is not a tensor for $\overset{k}{\supset}$. However, **C** does have some extra structure given by the axiom scheme $A, B \xrightarrow{(0,1)} A \wedge B$.

3 Final remarks

3.1 Dependent grading

We have made reference throughout to allowing the grades of realizers of formulae that ocurr in sequents to depend on realizers of previous formulae in the sequent and on previous grades. Some final remarks are all that need be made here.

1. With dependent grading, we must now be more careful with the restrictions imposed on the deduction rules in Definition 6. Of course, the restriction from [5] must be imported: (we use the notation of Definition 6, but now understanding that the grades may be functions of appropriate arities.)

($\exists$ L) k must be a constant, (or at least, must be a bounded function.)

In addition to this we also have two further restrictions caused by the fact that $\overset{k}{\supset}$ and $\overset{k}{\forall}$ are graded by constants:

($\overset{k}{\supset}$ R) k must be a constant.

($\overset{k}{\forall}$ R) k must be a constant.

2. In the graded system described in Definition 6, the rule ($\overset{k}{\supset}$ R) in effect defines a bijection. In a system with dependent grades, this would no longer be possible, unless one replaced $\overset{k}{\supset}$ with $\supset$—but as pointed out by [11], there are problems there. It is hoped that this will be discussed in the final version of [5]. A similar remark holds for $\overset{k}{\forall}$.

3. From the preceeding remarks, it is clear that the multicategorical structure for the dependently graded situation is no longer as simple. (For example, the closed structure has been weakened by the absence of the bijections referred to above.) I have not had time to look into this properly, so that will have to await a sequel.

3.2 The quantifiers

The structure of the quantifiers is more complicated than I have time to give it at the time of writing, so as I mentioned earlier must await a sequel. The main complication is that with variable declarations, the definition of "graded multicategory" must change, in view of the implicit contraction allowed by the cut rule. A further complication with the universal quantifiers (not to mention that there are many of these) comes from the absence of a ($\forall$ L) rule—in place of that, we have a "counit" represented by Axiom 3 of Definition 6.

However, the existential quantifier does have some more recognizable structure—ignoring for the moment the problems with cut, we can see the possibility of a bijection

$$\frac{\Gamma, A, \pi\Delta \longrightarrow \pi B}{\Gamma, \Sigma A, \Delta \longrightarrow B}$$

where Σ is the multifunctor corresponding to the existential quantifier, and π the multifunctor corresponding to adding a dummy free variable. But we must leave the details of this for another day.

References

[1] S. Buss, *Bounded Arithmetic,* Doctoral dissertation, Princeton University, 1985.

[2] S. Buss, "The polynomial time hierarchy and intuitionistic bounded arithmetic", Proceedings of the First Symposium on Structures and Complexity, 1986, (IEEE Publications).

[3] S. Cook and A. Urquhart, "Functional interpretations of feasibly constructive arithmetic", Technical Report 210/88, Department of Computer Science, University of Toronto, 1988.

[4] J.N. Crossley, "Proofs, programs and run-times", Preprint, Monash University, 1989.

[5] J.N. Crossley, G.L. Mathai, and R.A.G. Seely, "A logical calculus for polynomial time realizability", Preprint, Monash University, 1989, (final version in preperation).

[6] G. Gentzen, "Investigations into logical deductions", in M.E. Szabo (ed.), *The Collected Papers of Gerhard Gentzen*, North-Holland, 1969.

[7] J.-Y. Girard, "Linear logic", J. Theoretical Computer Science 50 (1987), 1 – 102.

[8] J. Lambek, "Deductive systems and categories I", J. Math. Systems Theory 2 (1968), 278 – 318.

[9] J. Lambek, "Deductive systems and categories II", Springer LNM 86 (1969), 76 – 122.

[10] J. Lambek, "Multicategories revisited", Proceedings of the A.M.S. Summer Conference on Categories in Logic and Computer Science, Boulder 1987.

[11] A. Nerode, J.B. Remmel, and A. Scedrov, "Polynomially graded logic", Proceedings of the Fourth Symposium on Logic in Computer Science, Asilomar 1989 (IEEE Publications).

[12] R.A.G. Seely, "Hyperdoctrines, natural deduction, and the Beck condition", Zeitsch. f. math. Logik und Grundlagen d. Math. 29 (1983), 505 – 542.

On the semantics of second order lambda calculus: from Bruce-Meyer-Mitchell models to Hyperdoctrine models and vice-versa *

Bart Jacobs

Dep. Comp. Sc., Toernooiveld, 6525 ED Nijmegen, The Netherlands.

In the literature there are two notions of model for the second order polymorphic λ-calculus: one by Bruce, Meyer and Mitchell (the BMM-model, for short) in set-theoretical formulation and one category-theoretical by Seely, based on hyperdoctrines. Using notions from Hayashi [1985], we adapt Seely's definition in such a way that a model of (non-extensional) 2nd order λ-calculus is obtained that satisfies the ξ-rules for the interpretation with valuation functions. This hyperdoctrine model is called a (categorical) λ_2-model; it gives rise to a BMM-model. The other way around, we show that a BMM-model gives rise to a categorical λ_2-model. If that λ_2-model is transformed into a BMM-model again, the result is essentially the same as what we started with.

1. Introduction.

1.1 The 2nd order λ-calculus.

The syntax of the 2nd order λ-calculus is introduced in two steps.

TYPES. The statement $\Gamma \vdash \sigma{:}*$, expressing that σ is a type in context Γ is defined as follows. Let $\Gamma = \{\alpha_1{:}*,\dots,\alpha_n{:}*\}$ be a set of type variables; $\Gamma,\alpha{:}*$ denotes $\Gamma\cup\{\alpha{:}*\}$, where it is supposed that $\alpha{:}*\notin\Gamma$. The deduction rules for types are

$$\begin{aligned} \alpha{:}*\in\Gamma &\Rightarrow \Gamma\vdash\alpha{:}* \\ \Gamma\vdash\sigma{:}*,\ \Gamma\vdash\tau{:}* &\Rightarrow \Gamma\vdash\sigma\to\tau{:}* \\ \Gamma,\alpha{:}*\vdash\sigma{:}* &\Rightarrow \Gamma\vdash\forall\alpha{:}*.\sigma{:}*. \end{aligned}$$

TERMS. The statement $\Gamma+\Theta\vdash M:\sigma$, denoting that the term M has type σ in the type context Γ and the term context Θ is formed as follows. Let $\Theta = \{x_1{:}\sigma_1,\dots,x_m{:}\sigma_m\}$ be a set of statements about term variables such that $\Gamma\vdash\sigma_i{:}*$; the context $\Theta,y{:}\tau$ is $\Theta\cup\{y{:}\tau\}$, where $y{:}\tau\notin\Theta$. We write $\alpha\in FV(\Theta)$ iff there is a i with $\alpha\in FV(\sigma_i)$. The deduction rules for terms are

$$\begin{aligned} x{:}\sigma\in\Theta &\Rightarrow \Gamma+\Theta\vdash x{:}\sigma \\ \Gamma+\Theta\vdash M{:}\sigma\to\tau,\ \Gamma+\Theta\vdash N{:}\sigma &\Rightarrow \Gamma+\Theta\vdash MN{:}\tau \\ \Gamma+\Theta,x{:}\sigma\vdash M{:}\tau &\Rightarrow \Gamma+\Theta\vdash\lambda x{:}\sigma.M{:}\sigma\to\tau \\ \Gamma+\Theta\vdash M{:}\forall\alpha{:}*.\sigma,\ \Gamma\vdash\tau{:}* &\Rightarrow \Gamma+\Theta\vdash M\tau{:}\sigma[\alpha:=\tau] \\ \Gamma,\alpha{:}*+\Theta\vdash M{:}\sigma \text{ and } \alpha\notin FV(\Theta) &\Rightarrow \Gamma+\Theta\vdash\lambda\alpha{:}*.M{:}\forall\alpha{:}*.\sigma. \end{aligned}$$

The β-contraction rules are

$$(\lambda x{:}\sigma.M)N \to M[x:=N] \quad\text{and}\quad (\lambda\alpha{:}*.M)\tau \to M[\alpha:=\tau].$$

---

* Research partially performed at the University of Pisa, Italy, during the first half of 1989, supported by the "Jumelage" project ST2J–0374–C of the European Communitee.

Note that we do *not* require the η-rules $\lambda x{:}\sigma.Mx \to M$ and $\lambda\alpha{:}*.M\alpha \to M$.

1.2 BRUCE-MEYER-MITCHELL MODELS.

In Bruce & Meyer [1984] and Bruce, Meyer & Mitchell [1985] a notion of model for the second order polymorphic λ-calculus is introduced. The construction roughly consists of a hierarchy of sets (as for first order λ-calculus models, see Friedman [1975]) with collections of functions between them (as for reflexive domains). The domains of the abstraction functions are not completely specified but are required to be such that they contain the λ-polynomials (and analogously for the Δ-function used to interpret universal types). In this way the model definition is explicitly depending on the syntax. In principle, one might avoid this reference by using second order combinators; in Bruce, Meyer & Mitchell [1985] this is done for extensional BMM-models only. Such a description however is unmanageable and so we shall stick to the syntactical conditions.

In order to fix the notation, we briefly recall the definition of a BMM-model.

A *type domain* **T** is a tuple $\langle T, \leadsto, \Delta\rangle$, where T is a set of so-called type representatives, $\leadsto$ is a function $T\times T\to T$ and Δ a (partial) function $T^T\to T$. The domain of Δ is denoted by $[T\to T]$ and should be such that types can be interpreted in **T** as follows. Given a type valuation ξ : Typevar → T one has

$$\begin{aligned}[\![\alpha]\!]_\xi &= \xi(\alpha)\\ [\![\sigma\to\tau]\!]_\xi &= [\![\sigma]\!]_\xi \leadsto [\![\tau]\!]_\xi\\ [\![\forall\alpha{:}*.\sigma]\!]_\xi &= \Delta(\lambda a\in T.[\![\sigma]\!]_{\xi(\alpha:=a)}).\end{aligned}$$

Notice that one can immediately prove $[\![\sigma[\alpha:=\tau]]\!]_\xi = [\![\sigma]\!]_{\xi(\alpha:=[\![\tau]\!]_\xi)}$ and similarily for simultaneous substitution.

A *second order functional domain* is a tuple $\mathfrak{m} = \langle \mathbf{T}, \{D_a\}_{a\in T}, \{F_{ab}, G_{ab}\}_{a,b\in T}, \{F_f, G_f\}_{f\in[T\to T]}\rangle$ where the D's are sets and the F and G's are functions between them in the following way.

$$F_{ab} : D_{a\leadsto b} \to [D_a \to D_b] \qquad G_{ab} : [D_a \to D_b] \to D_{a\leadsto b},$$

satisfying $F_{ab}\circ G_{ab} = id_{[D_a\to D_b]}$, for a certain set $[D_a \to D_b] \subseteq D_b^{\,D_a}$.

$$F_f : D_{\Delta(f)} \to [\Pi a\in T.D_{f(a)}] \qquad G_f : [\Pi a\in T.D_{f(a)}] \to D_{\Delta(f)},$$

satisfying $F_f\circ G_f = id_{[\Pi a\in T.D_{f(a)}]}$, for a certain set $[\Pi a\in T.D_{f(a)}] \subseteq \Pi a\in T.D_{f(a)}$.

Such an $\mathfrak{m}$ is called a *second order model* or better a *BMM-model* if terms can be interpreted as follows. For a term $\Gamma+\Theta \vdash M{:}\sigma$ and a type valuation ξ : Typevar → T and a term valuation ρ : Termvar → $\bigcup_{a\in T}.D_a$ with $\xi,\rho\models\Gamma+\Theta$ — i.e. $\rho(y)\in D_{[\![\tau]\!]_\xi}$ for $y{:}\tau\in\Theta$ — one has

$$\begin{aligned}[\![x]\!]_{\xi,\rho} &= \rho(x)\\ [\![MN]\!]_{\xi,\rho} &= F_{[\![\sigma]\!]_\xi[\![\tau]\!]_\xi}([\![M]\!]_{\xi,\rho})([\![N]\!]_{\xi,\rho}) && \text{for } M{:}\sigma\to\tau\\ [\![\lambda x{:}\sigma.M]\!]_{\xi,\rho} &= G_{[\![\sigma]\!]_\xi[\![\tau]\!]_\xi}(\lambda z\in D_{[\![\sigma]\!]_\xi}.[\![M]\!]_{\xi,\rho(x:=z)}) && \text{for } M{:}\tau\\ [\![M\tau]\!]_{\xi,\rho} &= F_{\lambda a\in T.[\![\sigma]\!]_{\xi(\alpha:=a)}}([\![M]\!]_{\xi,\rho})([\![\tau]\!]_\xi) && \text{for } M{:}\forall\alpha{:}*.\sigma\\ [\![\lambda\alpha{:}*.M]\!]_{\xi,\rho} &= G_{\lambda a\in T.[\![\sigma]\!]_{\xi(\alpha:=a)}}(\lambda a\in T.[\![M]\!]_{\xi(\alpha:=a),\rho}) && \text{for } M{:}\sigma.\end{aligned}$$

As for type interpretation, a number of substitution lemmas can now be proved. In the sequel we always suppose that ξ and ρ validate the appropriate contexts whenever we write $[\![\sigma]\!]_\xi$ or $[\![M]\!]_{\xi,\rho}$.
Note that $\mathfrak{m}$ satisfies the ξ-rules.

$$\mathfrak{m} \models (\forall x{:}\sigma.\ M{=}N) \quad\rightarrow\quad \lambda x{:}\sigma.M = \lambda x{:}\sigma.N$$
$$\mathfrak{m} \models (\forall \alpha{:}*.\ M{=}N) \quad\rightarrow\quad \lambda \alpha{:}*.M = \lambda \alpha{:}*.N,$$

i.e. $\forall z\in D_{[\![\sigma]\!]_\xi}.\ [\![M]\!]_{\xi,\rho(x:=z)} = [\![N]\!]_{\xi,\rho(x:=z)} \ \Rightarrow\ [\![\lambda x{:}\sigma.M]\!]_{\xi,\rho} = [\![\lambda x{:}\sigma.N]\!]_{\xi,\rho}$,

$\forall a\in T.\ [\![M]\!]_{\xi(\alpha:=a),\rho} = [\![N]\!]_{\xi(\alpha:=a),\rho} \ \Rightarrow\ [\![\lambda\alpha{:}*.M]\!]_{\xi,\rho} = [\![\lambda\alpha{:}*.N]\!]_{\xi,\rho}$.

An example of a (non-extensional) BMM-model is the interval model from Martini [1987].

Let us explicitly state the conditions which have to be satisfied in order that a functional domain yields a BMM-model, see the last definition of §3 in Bruce & Meyer [1984] and the preceding alinea.

1. For every type σ and valuation ξ, $\lambda a\in T.[\![\sigma]\!]_{\xi(\alpha:=a)} \in [T\to T]$.

For every pair of corresponding valuation functions ξ,ρ and every term M,

2. if $M(x){:}\tau$ for $x{:}\sigma$, then $\lambda z\in D_{[\![\sigma]\!]_\xi}.[\![M]\!]_{\xi,\rho(x:=z)} \in [D_{[\![\sigma]\!]_\xi} \to D_{[\![\tau]\!]_\xi}]$;
3. if $M(\alpha){:}\sigma(\alpha)$ — with α not free in the types of free term variables in M — then $\lambda a\in T.[\![M]\!]_{\xi(\alpha:=a),\rho} \in [\prod a\in T.D_{[\![\sigma]\!]_{\xi(\alpha:=a)}}]$.

The notation F_{ab}, G_{ab}, F_f, and G_f is a bit cumbersome; we shall often write

$$x\cdot y \quad \text{for} \quad F_{ab}(x)(y) \text{ if } x\in D_{a\sim>b} \text{ and } y\in D_a$$
$$\lambda x{:}a.t(x) \quad \text{for} \quad G_{ab}(t) \text{ if } t\in [D_a \to D_b]$$
$$x\bullet a \quad \text{for} \quad F_f(x)(a) \text{ if } x\in D_{\Delta(f)}$$
$$\lambda a{:}T.s(a) \quad \text{for} \quad G_f(s) \text{ if } s\in [\textstyle\prod a\in T.D_{f(a)}].$$

Then $(\lambda x{:}a.t(x))\cdot y = t(y)$ and $(\lambda a{:}T.s(a))\bullet b = s(b)$. Hence if we know that the functions we use are syntactically definable, we can reason directly in the model. The following lemma is an application of this; it will be used later.

LEMMA. (pairing) Let $\mathfrak{m}$ be a BMM-model as described above. For $c,d\in T$ there is a type representative $c\&d\in T$ with associated pairing function $[\_,\_] : D_c \times D_d \to D_{c\&d}$ and projection functions $p\in [D_{c\&d}\to D_c]$ and $p'\in [D_{c\&d}\to D_d]$ satisfying $p([x,y]) = x$ and $p'([x,y]) = y$.

PROOF. Obviously we take the interpretation of the syntactically definable pairing from 2nd order λ-calculus, i.e.

$$c\&d \ = \ \Delta(\lambda a\in T.(c\sim>d\sim>a)\sim>a)$$
$$[x,y] \ = \ \lambda a{:}T\ \lambda z{:}c\sim>d\sim>a.\ z\cdot x\cdot y$$
$$p \ = \ \lambda z\in D_{c\&d}.\ z\bullet c\cdot(\lambda x{:}c\ \lambda y{:}d.x)$$
$$p' \ = \ \lambda z\in D_{c\&d}.\ z\bullet d\cdot(\lambda x{:}c\ \lambda y{:}d.y).\ \blacksquare$$

1.3. Semi-functors and semi-adjunctions.

In Hayashi [1985] the notions of semi-functor and semi-adjunction are introduced in order to describe non-extensional concepts categorically: semi-cartesian closed categories are defined and it is shown that they are essentially the same as first order (non-extensional) typed λ-calculi with (non-surjective) pairing. Here we shall use these notions to deal also with non-extensional second order abstraction. We recall the necessary elements from Hayashi [1985].

A *semi-functor* $F:\mathbf{C}\to\mathbf{D}$ is a 'functor' except that it need not preserve identies. Another semi-functor $G:\mathbf{D}\to\mathbf{C}$ is a right *semi-adjoint* of F — notation $F \dashv_s G$ — if there are collections of functions $\{\alpha_{X,Y}, \beta_{X,Y}\}_{X\in\mathbf{C},Y\in\mathbf{D}}$ such that the four squares in the following diagram commute.

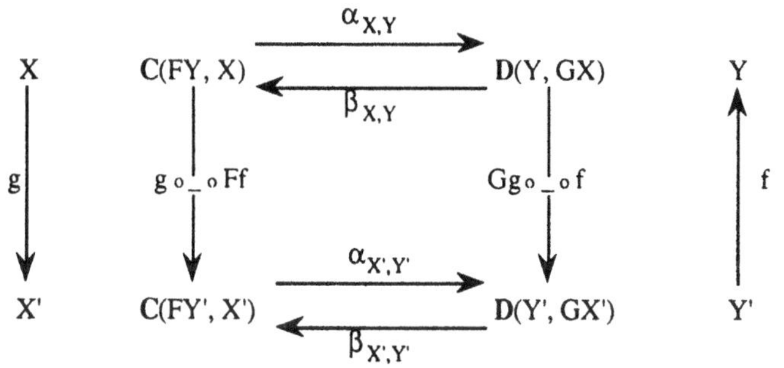

It is easy to see that if F is a (normal) functor, then — omitting indices — one has $(\beta \circ \alpha)(s) = 1 \circ \beta(\alpha(s)) \circ F1 = \beta(G1 \circ \alpha(s) \circ 1) = 1 \circ s \circ F1 = s$, i.e. $\mathbf{C}(FY, X)$ is a retract of $\mathbf{D}(Y, GX)$; in that case, one also has $\alpha(t \circ Fs) = \alpha(t) \circ s$, since $\alpha(t \circ Fs) = \alpha(1 \circ t \circ Fs) = G1 \circ \alpha(t) \circ s = G1 \circ \alpha(t) \circ 1 \circ s = \alpha(1 \circ t \circ F1) \circ s = \alpha(t) \circ s$.

A *semi-ccc* is defined as a category provided with the following three semi-adjunctions.

$$0 \dashv_s 1_{(\_)}$$
$$\Delta(\_) \dashv_s (\_)\times(\_)$$
$$(\_)\times Y \dashv_s (\_)^{Y}.$$

Algebraically it can be described as follows.

1. There is an object 1 and for every object X an arrow $!_X:X\to 1$ such that $!_Z \circ f = !_Y$ for every $f:Y\to Z$.
2. For all objects X,Y there is an object $X\times Y$ such that
   (a) for all $f:Z\to X$ and $g:Z\to Y$ there is an arrow $\langle f,g\rangle:Z\to X\times Y$ satisfying
   $$\langle f, g\rangle \circ h = \langle f \circ h, g \circ h\rangle.$$
   (b) there are morphisms $\pi:X\times Y\to X$ and $\pi':X\times Y\to Y$ with
   $$\pi \circ \langle f,g\rangle = f,$$
   $$\pi' \circ \langle f,g\rangle = g.$$
3. For all objects X,Y there is an object X^{Y} such that

(a) for all $h:Z\times Y\to X$ there is an arrow $\Lambda(h):Z\to X^Y$;

(b) there is an arrow $ev:X^Y\times Y\to X$ with

$$\begin{aligned} ev \circ \langle\Lambda(h)\circ f, g\rangle &= h\circ\langle f,g\rangle \\ \Lambda(h\circ\langle f\circ\pi, \pi'\rangle) &= \Lambda(h)\circ f \\ ev\circ\langle\pi,\pi'\rangle &= ev. \end{aligned}$$

2. HYPERDOCTRINE MODELS FOR 2nd ORDER λ-CALCULUS.

2.1 DEFINITION.

In Seely [1987] so-called PL-categories are introduced as models of extensional higher order polymorphic λ-calculus (with products and unit types). Below, we extract that part from it that corresponds (as will be shown) to BMM-models; we call those hyperdoctrine models (categorical) λ_2-models and show that the (non-extensional) 2nd order λ-calculus can be interpreted in them with valuation maps (as for BMM-models) in such a way that the ξ-rules are satisfied. In the definition of λ_2-models one can find requirements concerning "enough points", needed to validate these ξ-rules; they could be omitted, which amounts to defining categorical λ_2-algebras — extrapolating the notions from the untyped λ-calculus, see Barendregt [1984],ch.5. Such (non-extensional) "categorical λ_2-algebras" can be found in Martini [1988], 2, 5.9.

Remember that an object X (in a category with terminal object 1) has enough points if for all $f,g:X\to Y$ with $\forall x:1\to X.\ f\circ x = g\circ x$, one has $f=g$.

DEFINITION. A (categorical) λ_2-model consists of the following data.

(i) A base category **B** with finite products (including a terminal object 1) and a distinguished object Ω, which has enough points.

(ii) A functor $H:\mathbf{B}^{op}\to\mathbf{Cat}$ satisfying

(a) for $b\in\mathbf{B}$, there is an isomorphism $\phi_b:\mathbf{B}(b,\Omega)\cong \mathrm{Obj}\,(Hb)$, natural in the following way: for $h:b\to c$ in **B**, the functor $h^* = Hh : Hc\to Hb$ acts on objects $f\in Hc$ as

$$h^*(f) = \phi_b(\phi_c^{-1}(f)\circ h).$$

(b) for $b\in\mathbf{B}$, Hb is a semi-ccc; the functors h^* preserve the semi-ccc structure.

(c) all objects of H1 have enough points; moreover H *preserves enough points of Ω*, i.e. for arrows $t,s:f\to g$ in $H\Omega$ with $\forall h:1\to\Omega.\ h^*(t) = h^*(s)$ one has $t = s$.

(iii) For $b\in\mathbf{B}$, there is a semi-functor $\prod_b : H(b\times\Omega)\to Hb$, natural in b. Moreover, there is a semi-adjunction $\pi^*_{b,\Omega} \dashv_s \prod_b$, natural in b.

Notice that in comparison with Seely's PL-categories, we require in the definition of λ_2-models both less (only products in the base, semi-ccc and semi-adjunction for $\prod$) and more (enough points).

Following Seely [1987], 2.3 we take as an arrow between two λ_2-models $H : \mathbf{B}^{op} \to \mathbf{Cat}$ and $H' : \mathbf{B}'^{op} \to \mathbf{Cat}$, a pair $\langle F_0, F_1 \rangle$ where

$F_0 : \mathbf{B} \to \mathbf{B}'$ is a functor that preserves products and Ω

$F_1 : H \to F_0^* H'$ is a functor indexed over $\mathbf{B}$ — i.e. $(F_1)_b : Hb \to H'(F_0 b)$, commuting with h^* — preserving the semi-ccc structures and the semi-adjunctions $\pi^* \dashv_s \Pi$.

REMARKS. (i) The products in the base $\mathbf{B}$ are only used to arrange type variables. One could also require semi-products (as in 1.3) only, but in that case one has to demand that $1\times\Omega$ has enough points and that H preserves enough points of $1\times\Omega$. This is due to the fact that one no longer has $\Omega \cong 1\times\Omega$. We require a usual product (in fact, this is a minor point).

(ii) It is cleaner not to incorporate the requirement that $\Omega \in \mathbf{B}$ should have enough points in the definition of a categorical λ_2-model, since it is not used in the interpretation below, but only obtain a Δ-function in section 4. But to deal with it as an extra feature of λ_2-models is not of much interest here.

2.2. INTERPRETATION IN A λ_2-MODEL.

In this section we give an interpretation of types and terms in a λ_2-model as determined in the definition above, using valuations as in Koymans [1982] for the type free λ-calculus.

TYPES. Suppose a type statement $\Gamma \vdash \sigma:*$ is given for $\Gamma = \{\alpha_1:*,\dots,\alpha_n:*\}$. Let $\xi : \text{Typevar} \to \mathbf{B}(1, \Omega)$; we put $[\![\Gamma]\!] = \Omega^n$ $\quad \xi^\Gamma = \langle \xi(\alpha_1),\dots,\xi(\alpha_n)\rangle : 1 \to [\![\Gamma]\!]$.

First we define $[\![\sigma]\!]_\Gamma : [\![\Gamma]\!] \to \Omega$ in $\mathbf{B}$.

$$[\![\alpha_i]\!]_\Gamma = \pi_i$$
$$[\![\sigma\to\tau]\!]_\Gamma = \phi^{-1}_{[\![\Gamma]\!]}\left(\phi_{[\![\Gamma]\!]}([\![\tau]\!]_\Gamma)^{\phi_{[\![\Gamma]\!]}([\![\sigma]\!]_\Gamma)} \right)$$
$$[\![\forall\alpha:*.\sigma]\!]_\Gamma = \phi^{-1}_{[\![\Gamma]\!]}\left(\Pi_{[\![\Gamma]\!]}(\phi_{[\![\Gamma,\alpha:*]\!]}([\![\sigma]\!]_{\Gamma,\alpha:*}))\right).$$

Then we take $[\![\sigma]\!]_\xi = [\![\sigma]\!]_\Gamma \circ \xi^\Gamma : 1 \to \Omega$ in $\mathbf{B}$.

TERMS. Suppose we have $\Gamma+\Theta \vdash M:\sigma$ for $\Gamma = \{\alpha_1:*,\dots,\alpha_n:*\}$ again, and $\Theta = \{x_1:\tau_1,\dots,x_m:\tau_m\}$. Put

$$[\![\Theta]\!]_\Gamma = \phi_{[\![\Gamma]\!]}([\![\tau_1]\!]_\Gamma) \times \dots \times \phi_{[\![\Gamma]\!]}([\![\tau_m]\!]_\Gamma) \qquad [\![\Theta]\!]_\xi = \xi^{\Gamma *}([\![\Theta]\!]_\Gamma);$$

then one has $[\![\Theta]\!]_\xi = \phi_1([\![\tau_1]\!]_\xi) \times \dots \times \phi_1([\![\tau_m]\!]_\xi)$.

Let $\rho : \text{Termvar} \to \bigcup_{X\in \text{Obj}(H1)} H1(1, X)$ be such that $\xi,\rho \models \Gamma+\Theta$, i.e. $\rho(x_j) \in H1(1, \phi_1([\![\tau_j]\!]_\xi))$; analogously to ξ^Γ we take $\rho^\Theta = \langle \rho(x_1),\dots,\rho(x_m)\rangle : 1 \to [\![\Theta]\!]_\xi$.

We define $[\![M]\!]_{\Gamma+\Theta} : [\![\Theta]\!]_\Gamma \to \phi_{[\![\Gamma]\!]}([\![\sigma]\!]_\Gamma)$ in $H[\![\Gamma]\!]$, using the semi-adjunction

$$H(b\times\Omega)\left(\pi^*_{b,\Omega}(f),\ g\right) \quad \underset{\beta_b(f,g)}{\overset{\alpha_b(f,g)}{\rightleftarrows}} \quad H(b)\left(f,\ \Pi_b(g)\right)$$

In order to increase the readability, we shall omit the parameters f,g in $\alpha_b(f,g)$ and $\beta_b(f,g)$.

$$\begin{aligned}
[\![x_j]\!]_{\Gamma+\Theta} &= \pi_j \\
[\![MN]\!]_{\Gamma+\Theta} &= ev \circ \langle [\![M]\!]_{\Gamma+\Theta}, [\![N]\!]_{\Gamma+\Theta} \rangle \\
[\![\lambda x{:}\mu.M]\!]_{\Gamma+\Theta} &= \Lambda([\![M]\!]_{\Gamma+\Theta,x:\mu}) \\
[\![M\mu]\!]_{\Gamma+\Theta} &= \langle 1_{[\![\Gamma]\!]}, [\![\mu]\!]_{\Gamma} \rangle^{*} (\beta_{[\![\Gamma]\!]} ([\![M]\!]_{\Gamma,\alpha:*+\Theta})) \\
[\![\lambda\alpha{:}*.M]\!]_{\Gamma+\Theta} &= \alpha_{[\![\Gamma]\!]} ([\![M]\!]_{\Gamma,\alpha:*+\Theta}).
\end{aligned}$$

Finally, we take $[\![M]\!]_{\xi,\rho} = \xi^{\Gamma *}([\![M]\!]_{\Gamma+\Theta}) \circ \rho^{\Theta} : 1 \to \phi_1([\![\sigma]\!]_\xi)$ in H1.

As usual, one can now prove that $\Gamma+\Theta \vdash M=N \Rightarrow [\![M]\!]_{\Gamma+\Theta} = [\![N]\!]_{\Gamma+\Theta}$. The ξ-rules wrt. this interpretation (in a categorical λ_2-algebra) are satisfied because abstraction is interpreted by the Λ and α. As to the ξ-rules wrt. the interpretation with valuations, some more work has to be done, cf. Barendregt [1984], ch. 5, for the type free λ-calculus. The fact that all objects of H1 have enough points and that H preserves enough points of Ω will be used.

1. Suppose one has $\forall t{:}1\to\phi_1([\![\sigma]\!]_\xi).\ [\![M]\!]_{\xi,\rho(x:=t)} = [\![N]\!]_{\xi,\rho(x:=t)}$; we want to prove that $[\![\lambda x{:}\sigma.M]\!]_{\xi,\rho} = [\![\lambda x{:}\sigma.N]\!]_{\xi,\rho}$. Since

$$\begin{aligned}
[\![M]\!]_{\xi,\rho(x:=t)} &= \xi^{\Gamma *}([\![M]\!]_{\Gamma+\Theta,x:\sigma}) \circ \rho(x:=t)^{\Theta,x:\sigma} \\
&= \xi^{\Gamma *}([\![M]\!]_{\Gamma+\Theta,x:\sigma}) \circ \langle \rho^{\Theta}, t \rangle \\
&= \xi^{\Gamma *}([\![M]\!]_{\Gamma+\Theta,x:\sigma}) \circ \rho^{\Theta}\times 1_{\phi_1([\![\sigma]\!]_\xi)} \circ \langle 1_1, t \rangle,
\end{aligned}$$

and the fact that $1\times \phi_1([\![\sigma]\!]_\xi) \in$ H1 has enough points, one one may conclude

$$\xi^{\Gamma *}([\![M]\!]_{\Gamma+\Theta,x:\sigma}) \circ \rho^{\Theta}\times 1_{\phi_1([\![\sigma]\!]_\xi)} = \xi^{\Gamma *}([\![N]\!]_{\Gamma+\Theta,x:\sigma}) \circ \rho^{\Theta}\times 1_{\phi_1([\![\sigma]\!]_\xi)}.$$

By applying Λ on both sides we are done, since

$$\begin{aligned}
\Lambda(\xi^{\Gamma *}([\![M]\!]_{\Gamma+\Theta,x:\sigma}) \circ \rho^{\Theta}\times 1_{\phi_1([\![\sigma]\!]_\xi)}) &= \Lambda(\xi^{\Gamma *}([\![M]\!]_{\Gamma+\Theta,x:\sigma})) \circ \rho^{\Theta} \\
&= \xi^{\Gamma *}(\Lambda([\![M]\!]_{\Gamma+\Theta,x:\sigma})) \circ \rho^{\Theta}, \text{ since } \xi^{\Gamma *} \text{ preserves the semi-ccc structure} \\
&= \xi^{\Gamma *}([\![\lambda x{:}\sigma.M]\!]_{\Gamma+\Theta}) \circ \rho^{\Theta} \\
&= [\![\lambda x{:}\sigma.M]\!]_{\xi,\rho}.
\end{aligned}$$

2. Suppose $\forall h{:}1\to\Omega.\ [\![M]\!]_{\xi(\alpha:=h),\rho} = [\![M]\!]_{\xi(\alpha:=h),\rho}$; we now want $[\![\lambda\alpha{:}*.M]\!]_{\xi,\rho} = [\![\lambda\alpha{:}*.N]\!]_{\xi,\rho}$.

First,

$$\begin{aligned}
[\![M]\!]_{\xi(\alpha:=h),\rho} &= (\xi(\alpha:=h)^{\Gamma,\alpha:*})^{*} ([\![M]\!]_{\Gamma,\alpha:*+\Theta}) \circ \rho^{\Theta} \\
&= \langle \xi^{\Gamma}, h \rangle^{*} ([\![M]\!]_{\Gamma,\alpha:*+\Theta}) \circ \rho^{\Theta} \\
&= \langle 1_1, h \rangle^{*} \{ (\xi^{\Gamma}\times 1_{\Omega})^{*} ([\![M]\!]_{\Gamma,\alpha:*+\Theta}) \circ \pi_{1,\Omega}^{*}(\rho^{\Theta}) \}.
\end{aligned}$$

Using that H preserves enough points of $\Omega \cong 1\times\Omega$, we obtain

$$(\xi^{\Gamma}\times 1_{\Omega})^{*} ([\![M]\!]_{\Gamma,\alpha:*+\Theta}) \circ \pi_{1,\Omega}^{*}(\rho^{\Theta}) = (\xi^{\Gamma}\times 1_{\Omega})^{*} ([\![N]\!]_{\Gamma,\alpha:*+\Theta}) \circ \pi_{1,\Omega}^{*}(\rho^{\Theta}).$$

By applying α_1 on both sides, we get

$$\alpha_1((\xi^\Gamma\times 1_\Omega)^*([\![M]\!]_{\Gamma,\alpha:*+\Theta}) \circ \overset{*}{\pi}_{1,\Omega}(\rho^\Theta)) = \alpha_1((\xi^\Gamma\times 1_\Omega)^*([\![M]\!]_{\Gamma,\alpha:*+\Theta})) \circ \rho^\Theta, \text{ since } \overset{*}{\pi}_{1,\Omega} \text{ is a functor, cf. section 1.3}$$

$$= \xi^{\Gamma*}(\alpha_{[\![\Gamma]\!]}([\![M]\!]_{\Gamma,\alpha:*+\Theta})) \circ \rho^\Theta, \text{ by naturality of } \alpha$$

$$= \xi^{\Gamma*}([\![\lambda\alpha:*.M]\!]_{\Gamma+\Theta}) \circ \rho^\Theta$$

$$= [\![\lambda\alpha:*.M]\!]_{\xi,\rho}.$$

3. About the translations.

Before we start the technical translations from BMM-models to categorical λ_2-models and vice-versa, we take a global look. Since the notion of a BMM-model is depending on the syntax, the translations cannot take place at the semantical level only; instead a detour through syntax has to be made, roughly as in the following picture.

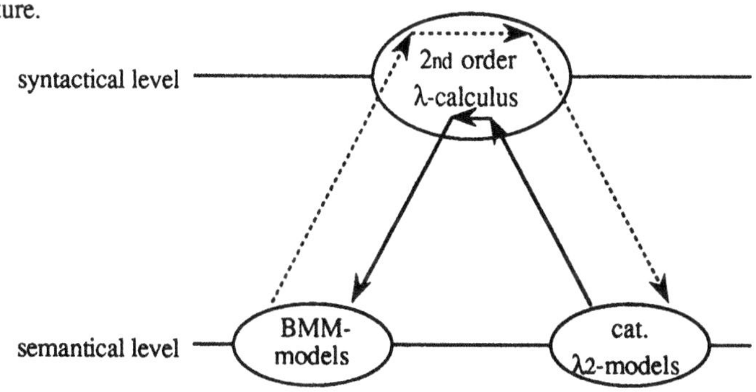

This detour makes the translations a bit complicated in their formulations; especially in going from BMM-models to λ_2-models the reference to syntax is quite often needed (in order to ensure well-definedness). It might be of use to ignore this reference at first reading to see the essentials of the construction; this amounts to projecting the dotted line on the semantical level.

(One might argue that the picture is not correct, in the sense that the BMM-models as described in paragraph 1.2 do not live completely at the semantical level. Our response is that the picture is simplifying in order to serve its purpose of conveying the intuition. On the other hand we are confident that a wholly semantical definition of BMM-models could be given with second order combinators.)

4. A BMM-model from a λ_2-model.

Having seen the interpretation in a λ_2-model in paragraph 2.2, it is not very hard to assemble a BMM-model from it. Take $T = \mathbf{B}(1, \Omega)$ and for $a \in T$, $D_a = H1(1, \phi_1(a))$.

A general difficulty with BMM-models is how to determine the domains $[T \to T]$, $[D_a \to D_b]$ and $[\Pi a\in T.D_{f(a)}]$ of the Δ-function and the abstraction functions, see paragraph 1.2. Here we follow our detour through syntax and restrict these domains to the type and term polynomials.

$$[T \to T] = \{ \lambda a\in T.[\![\sigma]\!]_{\xi(\alpha:=a)} \mid \sigma \text{ type, } \xi \text{ valuation} \}$$

$$[D_{[\![\sigma]\!]_\xi} \to D_{[\![\tau]\!]_\xi}] = \{ \lambda z\in D_{[\![\sigma]\!]_\xi}.[\![M]\!]_{\xi,\rho(x:=z)} \mid \text{term } M \text{ of type } \tau, \xi,\rho \text{ valuations} \}$$

$$[\Pi a\in T.D_{f(a)}] \;=\; \{\ \lambda a\in T.[\![M]\!]_{\xi(\alpha:=a),\rho} \mid \text{term } M \text{ of type } \sigma \text{ with } \alpha \text{ not free in the types of free term variables in } M,\ f=\lambda a\in T.[\![\sigma]\!]_{\xi(\alpha:=a)},\ \xi,\rho \text{ valuations}\}.$$

Notice that these interpretations are in the given categorical λ_2-model. Now it is straightforward how to define the rest of what is needed for a BMM-model.

For a,b$\in$T, put
$$a \rightsquigarrow b \;=\; \phi_1^{-1}(\phi_1(b)^{\phi_1(a)}) \in T,$$
and
$$\Delta(\lambda a\in T.[\![\sigma]\!]_{\xi(\alpha:=a)}) \;=\; \phi_1^{-1}(\Pi_1(\phi_{1\times\Omega}([\![\sigma]\!]_{\Gamma,\alpha:*} \circ \xi^{\Gamma}\times 1_{\Omega}))).$$
The fact that $\Omega\cong 1\times\Omega$ has enough points guarantees that Δ is well-defined: equal inputs yield equal outputs.

For $x\in D_{a\rightsquigarrow b}$ and $y\in D_a$, i.e. $x: 1 \rightarrow \phi_1(b)^{\phi_1(a)}$ and $y: 1 \rightarrow \phi_1(a)$ in H1, we define
$$F_{ab}(x)(y) \;=\; ev \circ \langle x, y\rangle \in D_b,$$
and
$$G_{[\![\sigma]\!]_\xi[\![\tau]\!]_\xi}(\lambda z\in D_{[\![\sigma]\!]_\xi}.[\![M]\!]_{\xi,\rho(x:=z)}) \;=\; \Lambda(\xi^{\Gamma*}([\![M]\!]_{\Gamma+\Theta,x:\sigma}) \circ \rho^{\Theta}\times 1_{\phi_1([\![\sigma]\!]_\xi)}),$$
which is well-defined because all objects of H1 have enough points (see under 1. in 2.2).

Let $f=\lambda a\in T.[\![\sigma]\!]_{\xi(\alpha:=a)}$ and $x\in D_{\Delta(f)}$; then $x: 1 \rightarrow \Pi_1(\phi_{1\times\Omega}([\![\sigma]\!]_{\Gamma,\alpha:*} \circ \xi^{\Gamma}\times 1_{\Omega}))$ in H1; hence we put
$$F_f(x)(b) \;=\; \langle 1_1, b\rangle^*(\beta_1(x)) \in D_{f(b)},$$
and
$$G_f(\lambda a\in T.[\![M]\!]_{\xi(\alpha:=a),\rho}) \;=\; \alpha_1((\xi^{\Gamma}\times 1_{\Omega})^*([\![M]\!]_{\Gamma,\alpha:*+\Theta}) \circ \pi^*_{1,\Omega}(\rho^{\Theta})),$$
which is well-defined because H preserves enough points of Ω (see under 2. in 2.2).

In a somewhat similar way the Δ and application functions are defined in Meseguer [1988],5.7 in order to obtain a (part of a) BMM-model from a universe model; the detour through syntax and the abstraction functions are not mentioned however.

5. A λ_2-MODEL FROM A BMM-MODEL.

Let us fix a BMM-model $\mathfrak{m}$ as described in paragraph 1.2. A categorical λ_2-model will be built from it step-by-step.

5.1 THE BASE CATEGORY B.

The category **B** is given by the following data.

Objects. $n\in\mathbb{N}$.

Arrows. $f:n\rightarrow m$ are set-theoretical functions $f:T^n\rightarrow T^m$ definable by types in the following way.

$f=(f_1,\dots,f_m): n\rightarrow m \;\Leftrightarrow\;$ there is a context $\Gamma=\{\alpha_1:*,\dots,\alpha_n:*\}$ together with types $\sigma_1,\dots,\sigma_m$ such that for $1\le i\le m$, $\Gamma \vdash \sigma_i:*$ & $f_i = \lambda\vec{a}\in T^n.[\![\sigma_i]\!]_{\xi(\vec{\alpha}:=\vec{a})}$.

(In fact the ξ is useless, but we shall write it nevertheless to avoid new notation.)

As a limit case, 0 can be understood as terminal object in **B**.

Composition. Given $f=(f_1,...,f_m):n\to m$ and $g=(g_1,...,g_k):m\to k$, put $g \circ f = (g_1 \circ f,... ,g_k \circ f)$. We show that $g\circ f:n\to k$. Suppose $f_i = \lambda\vec{a}\in T^n.[\![\sigma_i]\!]_{\xi(\vec{\alpha}:=\vec{a})}$ and $g_j = \lambda\vec{b}\in T^m.[\![\tau_j]\!]_{\xi(\vec{\beta}:=\vec{b})}$. Then

$$\begin{aligned} g_j \circ f &= \lambda\vec{a}\in T^n.[\![\tau_j]\!]_{\xi(\beta_1:=[\![\sigma_1]\!]_{\xi(\vec{\alpha}:=\vec{a})},...,\beta_m:=[\![\sigma_m]\!]_{\xi(\vec{\alpha}:=\vec{a})})} \\ &= \lambda\vec{a}\in T^n.[\![\tau_j[\beta_1:=\sigma_1,...,\beta_m:=\sigma_m]]\!]_{\xi(\vec{\alpha}:=\vec{a})}. \end{aligned}$$

Identity.

$$\begin{aligned} 1_m &= (\lambda\vec{a}\in T^n.a_1,...,\lambda\vec{a}\in T^n.a_m) \\ &= (\lambda\vec{a}\in T^n.[\![\alpha_1]\!]_{\xi(\vec{\alpha}:=\vec{a})},...,\lambda\vec{a}\in T^n.[\![\alpha_m]\!]_{\xi(\vec{\alpha}:=\vec{a})}). \end{aligned}$$

Product. The product of n and m in **B** is n+m. If $f=(f_1,...,f_n) : k\to n$ and $g=(g_1,...,g_m) : k\to m$, then $\langle f,g\rangle = (f_1,...,f_n,g_1,...,g_m) : k\to n+m$. As the associated projections one has

$$\begin{aligned} \pi &= (\lambda\vec{a}\in T^{n+m}.[\![\alpha_1]\!]_{\xi(\vec{\alpha}:=\vec{a})},...,\lambda\vec{a}\in T^{n+m}.[\![\alpha_n]\!]_{\xi(\vec{\alpha}:=\vec{a})}) \\ \pi' &= (\lambda\vec{a}\in T^{n+m}.[\![\alpha_{n+1}]\!]_{\xi(\vec{\alpha}:=\vec{a})},...,\lambda\vec{a}\in T^{n+m}.[\![\alpha_{n+m}]\!]_{\xi(\vec{\alpha}:=\vec{a})}). \end{aligned}$$

As the distinguished object Ω we take $1\in$ **B**. Obviously, it has enough points.

5.2 The functor H : $\mathbf{B}^{op} \to$ Cat.

For objects $n\in$ **B** a category Hn is defined.

Objects. $\bar{f} = \lambda\vec{a}\in T^n.\ D_{f(\vec{a})}$ for $f:n\to 1$ in **B**.

Arrows. $t : \bar{f} \to \bar{g}$ are functions $t\in \prod\vec{a}\in T^n.[D_{f(\vec{a})}\to D_{g(\vec{a})}]$ which are definable by terms: there must be a context $\Gamma = \{\alpha_1:*,...,\alpha_n:*\}$ with types σ,τ and a term $M(x)$ such that

$$\Gamma\vdash \sigma:*, \qquad \Gamma\vdash\tau:* \qquad \Gamma+x:\sigma\vdash M:\tau,$$

$$f = \lambda\vec{a}\in T^n.[\![\sigma]\!]_{\xi(\vec{\alpha}:=\vec{a})}, \qquad g = \lambda\vec{a}\in T^n.[\![\tau]\!]_{\xi(\vec{\alpha}:=\vec{a})} \qquad t = \lambda\vec{a}\in T^n\ \lambda z\in D_{f(\vec{a})}.[\![M]\!]_{\xi(\vec{\alpha}:=\vec{a}),\rho(x:=z)}.$$

Composition. Suppose $t : \bar{f} \to \bar{g}$ and $s : \bar{g} \to \bar{h}$; we take $s \circ t = \lambda\vec{a}\in T^n.\ s(\vec{a}) \circ t(\vec{a})$ in Hn. We have to check that $s \circ t$ is definable by a term again. Suppose

$$t = \lambda\vec{a}\in T^n\ \lambda z\in D_{f(\vec{a})}.[\![M]\!]_{\xi(\vec{\alpha}:=\vec{a}),\rho(x:=z)} \qquad s = \lambda\vec{a}\in T^n\ \lambda z\in D_{g(\vec{a})}.[\![N]\!]_{\xi(\vec{\alpha}:=\vec{a}),\rho(x:=z)},$$

then

$$s \circ t = \lambda\vec{a}\in T^n\ \lambda z\in D_{f(\vec{a})}.[\![N[y:=M]]\!]_{\xi(\vec{\alpha}:=\vec{a}),\rho(x:=z)}.$$

Identity.

$$\begin{aligned} 1_{\bar{f}} &= \lambda\vec{a}\in T^n.\ id_{D_{f(\vec{a})}} \\ &= \lambda\vec{a}\in T^n\ \lambda z\in D_{f(\vec{a})}.[\![x]\!]_{\xi(\vec{\alpha}:=\vec{a}),\rho(x:=z)}. \end{aligned}$$

Since we may assume that the D_a's are disjoint — otherwise, we can work with $D_a\times\{a\}$ — we have that $\phi_n : \mathbf{B}(n,1) \cong \mathrm{Obj}\ (Hn)$ is given by $\phi_n(f) = \bar{f}$ and $\phi_n^{-1}(\bar{f}) = f$.

For arrows $h:m\to n$ in **B** a functor $h^* = Hf : Hn \to Hm$ is defined by

$$h^*(\bar{f}) = \overline{f \circ h} \qquad \text{and} \qquad h^*(t) = t \circ h.$$

We have to check that $h^*(t)$ is definable by a term. Suppose that $t:\bar{f}\to\bar{g}$ and $\Gamma+x:\sigma\vdash M:\tau$, where σ,τ define f,g and M(x) defines t. If $(\rho_1,...,\rho_n)$ define h in the context Γ', then we also have $\Gamma'+x:\sigma[\vec{\alpha}:=\vec{\rho}] \vdash M[\vec{\alpha}:=\vec{\rho}]:\tau[\vec{\alpha}:=\vec{\rho}]$. Hence $M[\vec{\alpha}:=\vec{\rho}]$ defines $t \circ h = h^*(t)$ and guarantees that $h^*(t) : h^*(\bar{f}) \to h^*(\bar{g})$.

It is easy to check that h^* and H are functors and that H preserves enough points of 1; also that all objects of H0 have enough points.

5.3 THE SEMI-CCC STRUCTURE OF THE FIBRES Hn.

Terminal object. Since inhabited closed types behave like semi-terminal objects, we take $1_n = \lambda\vec{a}\in T^n.[\![\forall\alpha{:}*.\alpha\to\alpha]\!]_{\xi(\vec{\alpha}:=\vec{a})} : n\to 1$ in **B**, and for $\bar{f}\in$ Hn we take as $!_{\bar{f}} : \bar{f}\to \bar{1}_n$, $!_{\bar{f}} = \lambda\vec{a}\in T^n\,\lambda z\in D_{f(\vec{a})}.$ $[\![\lambda\alpha{:}*\ \lambda x{:}\alpha.x]\!]_{\xi(\vec{\alpha}:=\vec{a}),\rho(x:=z)}$. It is easy to see that for $t{:}\bar{g}\to\bar{f}$ one has $!_{\bar{f}}\circ t = !_{\bar{g}}$.

Product. For $\bar{f},\bar{g}\in$ Hn, put $\bar{f}\times\bar{g} = \lambda\vec{a}\in T^n.\,D_{f(\vec{a})\&g(\vec{a})}$, using the lemma from 1.2.
For $t{:}\bar{h}\to\bar{f}$ and $s{:}\bar{h}\to\bar{g}$, take $\langle t,s\rangle = \lambda\vec{a}\in T^n\,\lambda z\in D_{h(\vec{a})}.\ [t(\vec{a})(z),\ s(\vec{a})(z)]$, and

$$\pi = \lambda\vec{a}\in T^n\,\lambda z\in D_{f(\vec{a})\&g(\vec{a})}.\ p(z) \qquad \text{and} \qquad \pi' = \lambda\vec{a}\in T^n\,\lambda z\in D_{f(\vec{a})\&g(\vec{a})}.\ p'(z).$$

THEN (1) $\bar{f}\times\bar{g}$, $\langle t,s\rangle$, π and π' are well-defined.

(2) $\langle t, s\rangle\circ r = \langle t\circ r, s\circ r\rangle$

$\pi\circ\langle t, s\rangle = t,\quad \pi'\circ\langle t, s\rangle = s.$

PROOF (1) Suppose σ,τ define f,g; then the type $\sigma\&\tau = \forall\alpha{:}*.(\sigma\to\tau\to\alpha)\to\alpha$ defines a function $h{:}n\to 1$ with $\bar{h} = \bar{f}\times\bar{g}$; Hence $\bar{f}\times\bar{g}$ is well-defined.
Suppose furthermore that M(x),N(y) define t,s; the term $L(u) = [\ M[x:=u],\ N[y:=u]\ ] = \lambda\alpha{:}*\ \lambda v{:}\sigma\to\tau\to\alpha.\ v(M[x:=u])(N[y:=u])$ defines $\langle t,s\rangle$. The terms $P(z) = z\sigma(\lambda x{:}\sigma\,\lambda y{:}\tau.x)$ and $P'(z) = z\tau(\lambda x{:}\sigma\ \lambda y{:}\tau.y)$ define π and π'.

(2) Easy.

Exponent. For $\bar{f},\bar{g}\in$ Hn put $\bar{g}^{\bar{f}} = \lambda\vec{a}\in T^n.\,D_{f(\vec{a})\sim>g(\vec{a})}$.
For $t{:}\bar{h}\times\bar{f}\to\bar{g}$, take $\Lambda(t) = \lambda\vec{a}\in T^n\,\lambda z\in D_{h(\vec{a})}.\ \lambda y{:}f(\vec{a}).\ t(\vec{a})([z,\ y])$.
$ev = \lambda\vec{a}\in T^n\,\lambda z\in D_{(f(\vec{a})\sim>g(\vec{a}))\&f(\vec{a})}.\ p(z)\cdot p'(z).$

THEN (1) $\bar{g}^{\bar{f}}$, $\Lambda(t)$, and ev are well-defined.

(2)
$$\begin{aligned} ev\circ\langle\Lambda(t)\circ s,\ r\rangle &= t\circ\langle s,r\rangle \\ \Lambda(t\circ\langle s\circ\pi,\ \pi'\rangle) &= \Lambda(t)\circ s \\ ev\circ\langle\pi,\ \pi'\rangle &= ev. \end{aligned}$$

PROOF. (1) Suppose σ,τ,ρ define f,g,h and M(x) defines t. The type $\sigma\to\tau$ defines an arrow $k{:}n\to 1$ with $\bar{k} = \bar{g}^{\bar{f}}$. The term $L(u) = \lambda v{:}\sigma.\ M[x:=[u,\ v]]$, for $u{:}\rho$, defines $\Lambda(t)$ and ensures that $\Lambda(t){:}\bar{h}\to\bar{g}^{\bar{f}}$. Obviously ev is defined by P(u)P'(u), for $u{:}(\sigma\to\tau)\&\sigma$, where P and P' are from the previous proof (the types must be adapted of course).

(2) We shall take the first equation.

$$\begin{aligned} \{ev\circ\langle\Lambda(t)\circ s,\ r\rangle\}(\vec{a})(z) &= ev\,(\vec{a})\,([\ \Lambda(t)\,(\vec{a})\,(s(\vec{a})(z)),\ r(\vec{a})(z)\ ]) \\ &= \Lambda(t)\,(\vec{a})\,(s(\vec{a})(z))\cdot r(\vec{a})(z) \\ &= t\,(\vec{a})\,([\ s(\vec{a})(z),\ r(\vec{a})(z)\ ]) \\ &= \{t\circ\langle s,\ r\rangle\}(\vec{a})(z). \end{aligned}$$

Finally, it is easy to check that the semi-ccc structures defined above are preserved by functors h^*.

5.4 THE SEMI-FUNCTOR $\prod_n$.

We define $\prod_n : H(n+1) \to Hn$ as follows.

$$\prod_n(\bar{f}) = \hat{\bar{f}}, \qquad \text{where } \hat{f} = \lambda\vec{a}\in T^n.\ \Delta(\lambda b\in T.\ f(\vec{a},b)),$$
$$\prod_n(t:\bar{f}\to\bar{g}) = \lambda\vec{a}\in T^n\ \lambda z\in D_{\hat{f}(\vec{a})}.\ \lambda b{:}T.\ t(\vec{a},b)(z\bullet b).$$

THEN (1) $\prod_n$ is well-defined.

(2) $\prod_n$ is a semi-functor.

(3) $\prod_n$ is a functor if $G_h \circ F_h = id$ for all $h\in[T\to T]$.

(4) $\prod_n$ is natural in n.

PROOF. (1) Suppose $f = \lambda\vec{a},b\in T^{n+1}.[\![\sigma]\!]_{\xi(\vec{\alpha}:=\vec{a},\beta:=b)}$; then for all $\vec{a}\in T^n$, $\lambda b\in T.\ f(\vec{a},b)\in[T\to T]$ and so $\hat{f} = \lambda\vec{a}\in T^n.[\![\forall\beta{:}*.\sigma]\!]_{\xi(\vec{\alpha}:=\vec{a})}$.

Suppose we have $\Gamma,\beta{:}*+x{:}\sigma\vdash M{:}\tau$, where τ defines g and M(x) defines $t{:}\bar{f}\to\bar{g}$. Then $\Gamma,\beta{:}* + y{:}\forall\beta{:}*.\sigma\vdash M[x:=y\beta] : \tau$, and so $\Gamma + y{:}\forall\beta{:}*.\sigma \vdash \lambda\beta{:}*.M[x:=y\beta] : \forall\beta{:}*.\tau$. Hence $\lambda\beta{:}*.M[x:=y\beta]$ defines $\prod_n(t)$ and ensures that $\prod_n(t){:}\prod_n(\bar{f})\to\prod_n(\bar{g})$.

(2)
$$\begin{aligned}\prod_n(s\circ t)(\vec{a})(z) &= \lambda b{:}T.\ s(\vec{a},b)(\ t(\vec{a},b)(z\bullet b)\ )\\ &= \lambda b{:}T.\ s(\vec{a},b)(\ \prod_n(t)(\vec{a})(z)\bullet b)\ )\\ &= \prod_n(s)(\vec{a})(\ \prod_n(t)(\vec{a})(z)\ )\\ &= \{\prod_n(s)\circ\prod_n(t)\}(\vec{a})(z).\end{aligned}$$

(3)
$$\begin{aligned}\prod_n(1_{\bar{f}}) &= \lambda\vec{a}\in T^n\ \lambda z\in D_{\hat{f}(\vec{a})}.\ \lambda b{:}T.\ z\bullet b\\ &= \lambda\vec{a}\in T^n\ \lambda z\in D_{\hat{f}(\vec{a})}.\ (G_{f(\vec{a})}\circ F_{f(\vec{a})})(z)\\ &= \lambda\vec{a}\in T^n.\ id_{D_{\hat{f}(\vec{a})}}\\ &= 1_{\prod_n(\bar{f})}.\end{aligned}$$

(4) It is easy to check that for $h{:}n\to m$, $h^*\circ\prod_m = \prod_n\circ(h\times 1_1)^*$. ■

5.5 THE SEMI-ADJUNCTION $\overset{*}{\pi}_{n,1} \dashv_s \prod_n$.

We define suitable functions (see 1.3) between the hom-sets $H(n+1)\left(\overset{*}{\pi}_{n,1}(\bar{f}),\ \bar{g}\right)$ and $H(n)\left(\bar{f},\ \prod_n(\bar{g})\right)$:

$$\alpha_n(\bar{f},\bar{g})(u) = \lambda\vec{a}\in T^n\ \lambda z\in D_{f(\vec{a})}.\ \lambda b{:}T.\ u(\vec{a},b)(z)$$
$$\beta_n(\bar{f},\bar{g})(v) = \lambda\vec{a},b\in T^{n+1}\ \lambda z\in D_{f(\vec{a})}.\ v(\vec{a})(z)\bullet b.$$

THEN (1) $\alpha_n(\bar{f},\bar{g})$ and $\beta_n(\bar{f},\bar{g})$ are well-defined.

(2) $\{\alpha_n(\bar{f},\bar{g}),\ \beta_n(\bar{f},\bar{g})\}_{\bar{f},\bar{g}}$ determines a semi-adjunction $\overset{*}{\pi}_{n,1} \dashv_s \prod_n$.

(3) α_n and β_n are natural in n.

Since $\overset{*}{\pi}_{n,1}$ is a functor, we have that $H(n+1)\left(\overset{*}{\pi}_{n,1}(\bar{f}),\ \bar{g}\right)$ is a retract of $H(n)\left(\bar{f},\ \prod_n(\bar{g})\right)$.

PROOF (1) Suppose $f = \lambda\vec{a}\in T^n.[\![\sigma]\!]_{\xi(\vec{\alpha}:=\vec{a})}$ and $g = \lambda\vec{a},b\in T^{n+1}.[\![\tau]\!]_{\xi(\vec{\alpha}:=\vec{a},\beta:=b)}$. If $u : \overset{*}{\pi}_{n,1}(\bar{f})\rightarrow\bar{g}$ is defined by M(x), then $\Gamma,\beta{:}*{+}x{:}\sigma \vdash M{:}\tau$. Thus $\Gamma{+}x{:}\sigma \vdash \lambda\beta{:}*.M : \forall\beta{:}*.\tau$, since $\beta\notin FV(\sigma)$, and hence $\lambda\beta{:}*.M$ defines $\alpha_n(\bar{f},\bar{g})(u)$.

If $v : \bar{f}\rightarrow\prod_n(\bar{g})$ is defined by N(y), then $\Gamma{+}y{:}\sigma \vdash N{:}\forall\beta{:}*.\tau$. Thus $\Gamma,\beta{:}*{+}y{:}\sigma \vdash N\beta{:}\tau$, and so $N\beta$ defines $\beta_n(\bar{f},\bar{g})(v)$.

(2) Let $t : \bar{g}\rightarrow\bar{g}'$ and $s : \bar{f}'\rightarrow\bar{f}$; we have to prove the following equalities, cf. 1.3.

$$(\textstyle\prod_n(t)\circ\_\circ s)\circ\alpha_n(\bar{f},\bar{g}) = \alpha_n(\bar{f}',\bar{g}')\circ(t\circ\_\circ\overset{*}{\pi}_{n,1}(s))$$

$$(t\circ\_\circ\overset{*}{\pi}_{n,1}(s))\circ\beta_n(\bar{f},\bar{g}) = \beta_n(\bar{f}',\bar{g}')\circ(\textstyle\prod_n(t)\circ\_\circ s)$$

$$t\circ\_\circ\overset{*}{\pi}_{n,1}(s) = \beta_n(\bar{f}',\bar{g}')\circ(\textstyle\prod_n(t)\circ\_\circ s)\circ\alpha_n(\bar{f},\bar{g})$$

$$\textstyle\prod_n(t)\circ\_\circ s = \alpha_n(\bar{f}',\bar{g}')\circ(t\circ\_\circ\overset{*}{\pi}_{n,1}(s))\circ\beta_n(\bar{f},\bar{g}).$$

Let's do the first and the fourth.

$$\begin{aligned}
\{(\textstyle\prod_n(t)\circ_\circ s)\circ\alpha_n(\bar{f},\bar{g})\}(u)(\vec{a})(z) &= \textstyle\prod_n(t)(\vec{a})(\alpha_n(\bar{f},\bar{g})(u)(\vec{a})(s(\vec{a})(z)))\\
&= \lambda b{:}T.\ t(\vec{a},b)(\{\alpha_n(\bar{f},\bar{g})(u)(\vec{a})(s(\vec{a})(z))\}\bullet b)\\
&= \lambda b{:}T.\ t(\vec{a},b)(u(\vec{a},b)(s(\vec{a})(z)))\\
&= \lambda b{:}T.\ (t\circ u\circ\overset{*}{\pi}_{n,1}(s))(\vec{a},b)(z)\\
&= \{\alpha_n(\bar{f}',\bar{g}')\circ(t\circ_\circ\overset{*}{\pi}_{n,1}(s))\}(u)(\vec{a})(z).
\end{aligned}$$

$$\begin{aligned}
\{\alpha_n(\bar{f}',\bar{g}')\circ(t\circ_\circ\overset{*}{\pi}_{n,1}(s))\circ\beta_n(\bar{f},\bar{g})\}(v)(\vec{a})(z)&\\
&= \lambda b{:}T.\ t(\vec{a},b)(\beta_n(\bar{f},\bar{g})(v)(\vec{a},b)(\overset{*}{\pi}_{n,1}(s)(\vec{a},b)(z)))\\
&= \lambda b{:}T.\ t(\vec{a},b)(v(\vec{a})(s(\vec{a})(z))\bullet b)\\
&= \{\textstyle\prod_n(t)\circ_\circ s\}(v)(\vec{a})(z).
\end{aligned}$$

(3) It is not hard to check that for $h{:}n\rightarrow m$,

$$h^*\circ\alpha_m(\bar{f},\bar{g}) = \alpha_n(h^*(\bar{f}),(h\times 1_1)^*(\bar{g}))\circ(h\times 1_1)^*$$

$$(h\times 1_1)^*\circ\beta_m(\bar{f},\bar{g}) = \beta_n(h^*(\bar{f}),(h\times 1_1)^*(\bar{g}))\circ h^*.\ \blacksquare$$

Hence the construction is completed.

6. CONCLUDING REMARKS.

Given a BMM-model $\mathbf{m} = \langle\langle T, \leadsto, \Delta\rangle, \{D_a\}_{a\in T}, \{F_{ab}, G_{ab}\}_{a,b\in T}, \{F_f, G_f\}_{f\in[T\rightarrow T]}\rangle$, one may turn it into a λ_2-model and then again into a BMM-model; let the result be $\mathbf{m}' = \langle\langle T', \leadsto', \Delta'\rangle, \{D'_a\}_{a\in T'}, \{F'_{ab}, G'_{ab}\}_{a,b\in T'}, \{F'_f, G'_f\}_{f\in[T'\rightarrow T']}\rangle$.

The models $\mathbf{m}$ and $\mathbf{m}'$ are essentially the same: elements in $\mathbf{m}$ are turned into constant functions in $\mathbf{m}'$.

In a more abstract formulation:

There is an isomorphism $\psi : T \rightarrow T'$, such that $\psi(a\leadsto b) = \psi(a)\leadsto'\psi(b)$ and for $f = \lambda a\in T'.[\![\sigma]\!]_{\xi(\alpha:=a)}$, one has $\quad\Delta(\psi^{-1}\circ f\circ\psi) = \psi^{-1}(\Delta'(f))$.

Furthermore, for each $a\in T$, there is an isomorphism $\chi_a : D_a \rightarrow D_{\psi(a)}$ such that (omitting indices)

$$\chi(x\cdot y) = \chi(x)\cdot'\chi(y),$$

for $g = \lambda z \in D_{[\![\sigma]\!]_\xi}.[\![M]\!]_{\xi,\rho(x:=z)}$,

$$G(\chi^{-1} \circ g \circ \chi) = \chi^{-1}(G'(g)).$$

$$\chi(x \bullet a) = \chi(x) \bullet' \psi(a),$$

for $g = \lambda a \in T'.[\![M]\!]_{\xi(\alpha:=a),\rho}$,

$$G(\chi^{-1} \circ g \circ \psi) = \chi^{-1}(G'(g)).$$

To check all this requires a bit of patience.

Given a λ_2-model $H : \mathbf{B}^{op} \to \mathbf{Cat}$; let $H' : \mathbf{B}'^{op} \to \mathbf{Cat}$, be the result of transforming H into a BMM-model and then into a λ_2-model again. One might expect to find an arrow from H' to H, but in general one does not know if the defined semi-terminal objects and semi-products in the fibres of H' correspond with the given semi-terminal objects and semi-products in the fibres of H. This is due to the fact that semi-left and right adjoints need not be unique (up to natural isomorphisms) as for normal adjoints. Hence H and H' are incomparable.

One might wonder if it is really necessary to use Hayashi's non-extensional notions in a categorical λ_2-model: by requiring that a given BMM-model is extensional, i.e. satisfies $G_{ab_*} \circ F_{ab} = id$ and $G_h \circ F_h = id$, the abstraction in the fibres (defined in 5.3) is extensional and the adjunction $\pi_{n,1}^* \dashv_s \Pi_n$ is a normal one. But in that case, the defined terminal objects and products in the fibres are still semi-ones. Hence we have to use the semi-notions.

One could, of course, just extend the definition of an extensional BMM-model with products and terminal object. We do not favour that approach for the following reasons.

(i) The above results using the non-extensional concepts are more general.

(ii) A computer scientist — in contrast to e.g. a logician — looks at the terms of a typed λ-calculus as programs. The question of equality of programs is a very complex one and the extensional answer is rather limited: certainly not everyone (e.g. not in complexity theory) is willing to regard programs with the same input-output behaviour as equal.

Finally we mention two points for further research.

(i) The notions of categorical λ_2-algebra and λ_2-model deserve a separate treatment together with examples like open/closed term models, the interval model and the ideal model.

(ii) We expected that the definable "semi-weak" sum $\exists\alpha{:}*.\sigma \equiv \forall\beta{:}*.(\forall\alpha{:}*.\sigma\to\beta)\to\beta$ would yield a semi-left adjoint to the π^*'s in a categorical λ_2-model. But there seem to be problems with the semi-functoriality. The other way around — from a semi-left adjoint to semi-weak sums — is not clear either, since one cannot straightforwardly use the Frobenius reciprocity (cf. Seely [1987], p.981) in a non-extensional context. Hence it is open to what extend semi-adjunctions are useful in describing "semi" sums.

ACKNOWLEDGEMENT. I would like to thank Simone Martini, whose comments on an early draft resulted in considerable improvements. From Eugenio Moggi I learned that inhabited closed types behave like semi-terminal objects.

REFERENCES.

BARENDREGT, H.P.,

[1984] *The Lambda Calculus. Its syntax and semantics* (revised second edition), North Holland, Amsterdam.

BRUCE, K.B. and MEYER, A.R.,

[1984] The semantics of second-order lambda calculus, in: KAHN, G., MACQUEEN, D.B. and PLOTKIN, G., (eds.), *Proceedings of the Conference on Semantics of Data Types*, Sophia-Antipolis, June 1984, LNCS 173, p.131-144.

BRUCE, K.B., MEYER, A.R. and MITCHELL, J.C.,

[1985] The semantics of second-order lambda calculus, manuscript.

FRIEDMAN, H.,

[1975] Equality between functionals, in: PARIKH, R., (ed.) *Logic Colloquium, Symposium on logic held at Boston, 1972-1973*, LNM 453.

HAYASHI, S.,

[1985] Adjunction of semifunctors: categorical structures in non-extensional lambda-calculus, *Th. Comp. Sc.*, Vol 41-1, p.95-104.

KOYMANS, K.,

[1982] Models of the lambda calculus, *Inf. & Contr.* 52, p.306-332.

MARTINI, S.,

[1987] An interval model for second order lambda calculus, in: PITT, D.H., POIGNÉ, A. and RYDEHEARD, D. (eds.), *Category theory and Computer Science*, Edinburgh, sept '87, LNCS 283, p.219-237.

[1988] *Modelli non estensionali del polimorfismo in programmazione funzionale*, Ph.D thesis, University of Pisa.

MESEGUER, J.,

[1988] Relating models of polymorphism, Techn. Rep. 88-13, Comp. Sc. Lab., SRI International.

SEELY, R.A.G.,

[1987] Categorical semantics for higher order polymorphic lambda calculus, *J. Symb. Log.* 52, p.969-989.

Dictoses

Thomas Ehrhard
Ecole Polytechnique (CNRS URA 0169) and LIENS (CNRS URA 1327)

Introduction

This paper presents a notion of categories which are models of the calculus of constructions. These categories are in fact equivalent to the "locally cartesian closed categories with a internal small-complete category" that Hyland presented in [5]. However, our point of view is quite different, since we shall not make reference to the completeness of an internal category, but get it as a corollary of our axioms. In fact, the notion of internal completeness is very subtle.

So what we present here are locally cartesian closed categories with an object of propositions and an impredicative quantification principle, and furthermore a reflection principle of types over propositions. Our claim is that this notion is a natural generalisation of toposes, where proofs are first-class citizens, in contrast with toposes where proofs are not interpreted. That's why we called these categories "dictoses".

In this short paper, we just present the notion, give some very basic facts about dictoses (preservation by localisation, internal completeness of the internal category of propositions), and give the example of ω-sets (for reflection only, since the remainder is well known). We don't claim that this is a very original piece of work, but we only hope to provide another sight on objects which have been studied by category theorists for years.

1 Dictoses

The object of this section is to introduce a categorical notion intended to generalize elementary toposes, taking proofs into acount. Actually, it is well known that toposes are higher order intuitionistic theories of types (see [8,9]), with equality. But in these theories, the notion of truth is quite poor, with respect to the so-called "Heyting semantics" of intuitionistic logic. In a topos, a proposition is true if it is equal to "true" constant. So the proof used in order to establish the truth of this proposition isn't taken in any way into account in the semantics, and in a logical (and computer scientist, cf. Curry-Howard paradigm of programs-as-proofs) point of view, it seems to be a serious drawback.

1.1 Higher order locally cartesian closed categories

We recall first that a locally cartesian closed category is a category with finites limits (thus it has a terminal object) such that all slice categories are cartesian closed. If $\mathbf{C}$ is such a category, the last condition means that for any morphism $f : X \to Y$, the pull-back functor $f^* : \mathbf{C}/Y \to \mathbf{C}/X$ has a right adjoint $\Pi f : \mathbf{C}/X \to \mathbf{C}/Y$, and that the Beck-Chevalley condition is satisfied. Roughly speaking, this means that for any pull-back diagram in **C**

$$\begin{array}{ccc} X & \xrightarrow{f} & Y \\ {\scriptstyle g}\downarrow & & \downarrow{\scriptstyle h} \\ X' & \xrightarrow{f'} & Y' \end{array}$$

the following is a (pseudo-)commutative diagram of functors

$$\begin{array}{ccc} \mathbf{C}/X & \xrightarrow{\Pi f} & \mathbf{C}/Y \\ {\scriptstyle g^*}\uparrow & & \uparrow{\scriptstyle h^*} \\ \mathbf{C}/X' & \xrightarrow{\Pi f'} & \mathbf{C}/Y' \end{array}$$

(for more details, see for instance [6] where a perfectly clean treatment is done of this property).

About LCCC's, we shall need the following property:

Lemma 1 *Let* **C** *be a LCCC, and let* X *be an object of it. Then* $\mathbf{C}/X$ *is locally cartesian closed as well, and the functor*

$$\begin{array}{rcl} X^* : \mathbf{C} & \rightarrow & \mathbf{C}/X \\ Y & \mapsto & (\pi_2 : Y \times X \rightarrow X) \\ (f : Y \rightarrow Z) & \mapsto & f \times \mathrm{Id}_X \end{array}$$

is left exact and preserves local exponentials.

If we consider a LCCC as an intuitionistic theory of types in the spirit of Martin-Löf's first order system of [10], this result means essentially that we can do the same kind of reasonning, assuming a constant hypothesis X.

Now we introduce the intermediate notion of higher order locally cartesian closed category, which doesn't seem to be rich enough to provide interesting results. A LCCC $\mathcal{E}$ is a HLCCC if it has a special morphism $T : \Omega' \rightarrow \Omega$ such that, for any $\varphi : X \rightarrow \Omega$ and any $f : X \rightarrow Y$ there exists a morphism $\forall f\,(\varphi) : Y \rightarrow \Omega$ such that $\widetilde{\forall f\,(\varphi)} \cong \Pi f\,(\tilde{\varphi}) : Y \rightarrow \Omega$ where $\tilde{\varphi}$ is a shorthand for $\varphi^*(T)$. This last requirement will sometimes be called "axiom of impredicative product".

A category of this kind is a model of the Calculus of Constructions (see [3] for more details), with a strong sum and an equality type in the spirit of [10].

1.2 The internal full subcategory of propositions

We recall a classical construction which can be carried out in any locally cartesian closed category **C**. In such a category, let $u : U \rightarrow C_0$ be any morphism. We consider the projections $\pi_1, \pi_2 : C_0 \times C_0 \rightarrow C_0$ which define, by pulling back and local exponential transpose, a morphism

$$d = {\pi_0}^*(u) \Rightarrow {\pi_1}^*(u) : C_1 \rightarrow C_0 \times C_0$$

of which both components Cod $= \pi_2 \circ d$ and Dom $= \pi_1 \circ d$ define the codomain and domain morphisms respectively of an internal category in **C**. The "identity" morphism $C_0 \rightarrow C_1$ is obtained by internalizing the identity using exponential, and composition is defined in the same way, without any problems (see for instance [8]). This internal category will be noted $\mathrm{Full}_{\mathbf{C}}(u)$ and called the internal full subcategory generated by u.

We know furthermore that to any internal category C in the locally cartesian closed category **C**, it is possible to associate a split fibration

$$p_C : [C] \rightarrow \mathbf{C}$$

(see for instance [7]). Let us recall this construction, which internalizes the "category of families of a (small) category indexed by sets". The category $[C]$ has objects the morphisms $x : X \to C_0$ ("families of objects of C indexed by X"). A morphisme $(x : X \to C_0) \to (y : Y \to C_0)$ in $[C]$ is given by a morphism $f : X \to Y$ ("function between sets of indexes") and a $\varphi : X \to C_1$ ("family of morphisms indexed by X"), morphisms in $\mathbf{C}$ such that both following diagrams be commutative

$$\begin{array}{ccc} X & \xrightarrow{\varphi} & C_1 \\ {\scriptstyle x}\downarrow & & \downarrow{\scriptstyle d_0} \\ C_0 & = & C_0 \end{array} \qquad \begin{array}{ccc} X & \xrightarrow{\varphi} & C_1 \\ {\scriptstyle f}\downarrow & & \downarrow{\scriptstyle d_1} \\ Y & \xrightarrow{y} & C_0 \end{array}$$

Composition is defined as follows. Let $(f, \varphi) : (x : X \to C_0) \to (y : Y \to C_0)$ and $(g, \psi) : (y : Y \to C_0) \to (z : Z \to C_0)$ be morphisms in $\mathbf{C}$. We set

$$(g, \psi) \circ (f, \varphi) = (g \circ f, c(\psi \circ f, \varphi))$$

where c is the internal composition of the internal category C. The split fibration $p_C : [C] \to \mathbf{C}$ is then the functor

$$\begin{array}{rcl} p_C : [C] & \to & \mathbf{C} \\ (x : X \to C_0) & \mapsto & X \\ (f, \varphi) & \mapsto & f \end{array}$$

When furthermore C is an internal full subcategory in $\mathbf{C}$ built on a morphism $u : U \to C_0$, (with the notation previously introduced, $C = \mathrm{Full}_{\mathbf{C}}(u)$), we have a cartesian full and faithful functor

$$\tau_{u,\mathbf{B}} : p_{\mathrm{Full}_{\mathbf{C}}(u)} \to \mathrm{Cod}$$

where $\mathrm{Cod} : \mathbf{C}^{\mathbf{2}} \to \mathbf{C}$ is the codomain functor which is well known to be a fibration. This corresponds to the fact that C is a full subcategory of $\mathbf{C}$.

In a HLCCC $\mathcal{E}$, we shall consider this construction for the morphism T. The internal full subcategory $\mathrm{Full}_{\mathcal{E}}(T)$ will be called "the category of proposition". We shall note $\mathcal{P}$ the presheaf associated to the split fibration $p_{\mathrm{Full}_{\mathbf{C}}(u)}$. (Thus for any $X \in \mathcal{E}$, $\mathcal{P}(X)$ has as objects the morphisms $X \to \Omega$, and a morphism $\varphi \to \psi$, when $\varphi, \psi : X \to \Omega$, is simply a morphism $\tilde{\varphi} \to \tilde{\psi}$ in $\mathcal{E}/X$.)

1.3 A reflection principle

We know that in a topos there is a canonical factorization of any morphism into an epi followed by a mono (ie. the type associated with a predicate). We require a similar property in any dictos, and this will complete our axiomatization. This will allow to interpret "big sums" of propositions over types (which are not "strong sums", since this would be paradoxal, as it is well known), and the factorization we require here is deeply connected to the one described in [6], section 2.12. We use the same kind of reflection.

We give first the general definition of reflections:

Definition 1 *Let* $\mathbf{B}$ *and* $\mathbf{D}$ *be categories. We say that two functors* $I : \mathbf{B} \to \mathbf{D}$ *and* $R : \mathbf{D} \to \mathbf{B}$ *define a reflection if* I *is right adjoint to* R*, and if the counit of this adjunction is an iso.*

This notion must be understood as a kind of generalization of the "embedding-retractions" which are well-known in domain theory. I is a kind of embedding of $\mathbf{B}$ in $\mathbf{D}$ (actually, it is full and faithful since the co-unit is an iso). Sometimes, this functor will be called the "embedding part" of the reflection.

This notion admits an obvious generalization to the case where **B** and **D** are fibrations instead of simple categories. (Bénabou has shown that all what can be done with categories can as well be done with fibrations, see [1].)

Definition 2 *We say that a HLCCC $\mathcal{E}$ is reflexif (or that it is a dictos) if the cartesian functor $\tau_{u,\mathcal{E}}$ is the embedding part of a reflection between the fbrations $p_{\mathrm{Full}_\mathcal{E}(T)}$ and* $\mathrm{Cod}_\mathcal{E}$.

This means that for any object X in $\mathcal{E}$, the functor

$$\begin{array}{rcl} \mathrm{T}_X : \mathcal{P}(X) & \to & \mathcal{E}/X \\ \varphi & \mapsto & \tilde{\varphi} \\ (f : \varphi \to \psi) & \mapsto & f \end{array}$$

is the embedding part of a reflection. We shall note R_X the "retraction" part of this reflection. Furthermore, the requirement that the functor R be cartesian means that a kind of "Beck-Chevalley" condition be satisfied. This means that $f : X \to X'$ being a given morphism in $\mathcal{E}$, the following diagram must (pseudo-)commute:

$$\begin{array}{ccc} \mathcal{P}(X) & \xleftarrow{\mathcal{P}(f)} & \mathcal{P}(X') \\ {\scriptstyle \mathrm{R}_X}\uparrow & & \uparrow{\scriptstyle \mathrm{R}_{X'}} \\ \mathcal{E}/X & \xleftarrow{f^*} & \mathcal{E}/X' \end{array}$$

This reflection principle (assuming, to simplify a bit, that the counit is the identity, instead of being a more general iso) gives, for any $f : X \to Y$, morphism of $\mathcal{E}$, a unit morphism

$$\mu_f : X \to \widetilde{X}_{\mathrm{R}_X(f)}$$

such that

$$f = \widetilde{\mathrm{R}_X(f)} \circ \mu_f \quad \text{and} \quad \mathrm{R}_X(\mu_f) = \mathrm{Id}\ .$$

In that last equality, it is the morphism part of functor R_X which is applied. Furthermore, is $\varphi : X \to \Omega$ is a "predicate", we also have that $\mu_{\tilde{\varphi}} = \mathrm{Id}$.

We now explain why this reflection principle is a generalisation of the epi-mono factorization which holds in a elementary topos. For this, we need a categorical notion of orthogonality.

Definition 3 *Let $\mathcal{R}$ be a class of morphisms in a category* **C**. *Let $f : X \to Y$ be a morphism of* **C**. *We say that f if orthogonal to the class $\mathcal{R}$ if for any commutative square*

$$\begin{array}{ccc} X & \xrightarrow{f} & Y \\ {\scriptstyle a}\downarrow & & \downarrow{\scriptstyle b} \\ U & \xrightarrow{r} & V \end{array}$$

where $r \in \mathcal{R}$, there exists a unique morphisms $g : Y \to U$ such that

$$g \circ f = a \quad \text{and} \quad r \circ g = b$$

This definition implies readily that any decomposition of a morphism u of C in $r \circ f$ where $r \in \mathcal{R}$ and f orthogonal to $\mathcal{R}$ is unique (up to isomorphism). It is clear on the other hand that, in a topos, any epi is orthogonal to the class of monos (since monos are split). We show that the situation is similar in a dictos. First we make an obvious remark, which is nothing else but a rephrasement of the reflection condition:

Lemma 2 *For any $f : X \to Y$ in $\mathcal{E}$ and any decomposition of f:*

$$f = \tilde{\psi} \circ h$$

there exists a unique morphism $v : \tilde{Y}_{\mathrm{R}(f)} \to \tilde{Y}_{\psi}$ such that

$$\tilde{\psi} \circ v = \widetilde{\mathrm{R}(f)} \quad \text{and} \quad v \circ \mu_f = h$$

Now we can state that

Proposition 1 *For any $f : X \to Y$ in $\mathcal{E}$, μ_f is orthogonal to the morphisms of the form $\tilde{\varphi}$ where $\varphi : U \to \Omega$ is any morphism.*

Proof: Let $f : X \to Y$ and $\varphi : U \to \Omega$ be morphisms and assume given a commutative diagram

$$\begin{array}{ccc} X & \xrightarrow{\mu_f} & V \\ {\scriptstyle a}\downarrow & & \downarrow{\scriptstyle b} \\ \tilde{U}_{\varphi} & \xrightarrow{\tilde{\varphi}} & U \end{array}$$

where we have set $V = \tilde{Y}_{\mathrm{R}(f)}$. We consider then the pullback built on $\tilde{\varphi}$ and b:

$$\begin{array}{ccc} \tilde{V}_{\psi} & \xrightarrow{\tilde{\psi}} & V \\ {\scriptstyle b'}\downarrow & & \downarrow{\scriptstyle b} \\ \tilde{U}_{\varphi} & \xrightarrow{\tilde{\varphi}} & U \end{array}$$

where we have set $\psi = \varphi \circ b$. Let $h : X \to \tilde{V}_{\psi}$ be the unique morphism such that

$$b' \circ h = a \quad \text{et} \quad \tilde{\psi} \circ h = \mu_f \,.$$

Since $\mathrm{R}(\mu_f) = \mathrm{Id}$ (because the unit of the adjunction is the identity), there exists (cf. previous lemma) a unique morphism $v : V \to \tilde{V}_{\psi}$ such that

$$\tilde{\psi} \circ v = \mathrm{Id} \quad \text{and} \quad v \circ \mu_f = h$$

Then let $u = b' \circ v : V \to \tilde{U}_{\varphi}$. It is clear that this morphism renders both required triangles commutative. We conclude by verifying that it is alone satisfying these conditions. Actually, let u be such a morphism, and call v the unique morphism $V \to \tilde{V}_{\psi}$ such that

$$\tilde{\psi} \circ v = \mathrm{Id} \quad \text{and} \quad b' \circ v = u$$

Let $h' = v \circ \mu_f$. To conclude, it suffices to see that $h' = h$. But this results from the following equalities:

$$\tilde{\psi} \circ h' = \mu_f \quad \text{and} \quad b' \circ h' = a$$

∎

1.4 Logical functors between dictoses

Let us define a notion of morphism between dictoses.

Definition 4 *Let $\mathcal{E}$ and $\mathcal{F}$ be two dictoses. A logical functor between $\mathcal{E}$ and $\mathcal{F}$ is a functor $F : \mathcal{E} \to \mathcal{F}$ which is left exact, preserves the local exponentials, the generic morphisms T and the reflection.*

This is of course inspired by the homonym notion in topos theory.

Remark: It is not clear yet whether or not there exists a more relevant notion of morphisms between dictoses, like the geometric morphisms for toposes. This question should be clarified by the study of examples and by trying to generalize the construction of sheaves topose over a topological space.

We have in mind to give an important example of logical functor. In fact, we shall prove

Proposition 2 *Let $\mathcal{E}$ be a dictos. Let $X \in \mathcal{E}$ be an object in this dictos. Then the category $\mathcal{E}/X$ is a dictos. Furthermore, for any $u : X \to X'$, the functor $u^* : \mathcal{E}/X' \to \mathcal{E}/X$ is logical.*

Proof: First, let us prove that $\mathcal{E}/X$ is a dictos. We already know that it is locally cartesian closed by lemma 1. The object Ω_X of propositions in $\mathcal{E}/X$ is taken to be $\pi_2 : \Omega \times X \to X$, and in the same way $\Omega'_X = \pi_2 : \Omega \times X \to X$ defines the local object of proofs. As generic morphism $T_X : \Omega'_X \to \Omega_X$ we take of course $T \times X$ (where X stands for Id_X).

Now let $\varphi : f \to \Omega_X$ be a "predicate" in $\mathcal{E}/X$. Let $h : f \to g$ be a morphism, and let Y (resp. Z) be the codomain of h (resp. its domain). We define $\forall X\,(h)\,\varphi$ by

$$\forall X\,(h)\,\varphi = \langle \forall f\,(\pi_1\varphi)\,,\, g\rangle : g \to \Omega_X$$

It is easy to check that it satisfies the required condition. For this just remark that, for any $\varphi : f \to \Omega_X$, we have

$$\tilde{\varphi}^X = \widetilde{\pi_1\varphi}$$

where $\tilde{\varphi}^X$ is the type associated to φ in $\mathcal{E}/X$.

It remains to define a local version of the reflection which we shall note R^X. For any $h : f \to g$ morphism in $\mathcal{E}/X$ we set simply

$$\mathrm{R}^X_g(h) = \langle \mathrm{R}_Z(h)\,,\, g\rangle$$

where Z is the domain of g. And we let to the reader the straightforward verification that it defines the required reflection. Thus $\mathcal{E}/X$ is a dictos.

Let us proof the second part of the proposition. Let $u : X \to X'$ be a morphism. We already know by lemma 1 that the functor u^* is left exact and preserves local exponentials. The fact that it preserves $T : \Omega' \to \Omega$ results from the fact that, for any object $A \in \mathcal{E}$, the following diagram is a pullback:

$$\begin{array}{ccc} A \times X & \xrightarrow{\mathrm{Id}_A \times u} & A \times X' \\ \downarrow{\scriptstyle \pi_2} & & \downarrow{\scriptstyle \pi_2} \\ X & \xrightarrow{u} & X' \end{array}$$

It just remains to check that u^* preserves reflections. Let $h' : f' \to g'$ be an arrow in $\mathcal{E}/X'$ (where $f' : Y' \to X'$ and $g' : Z' \to X'$ are local objects). Then we have set

$$\mathrm{R}^{X'}_{g'}(h') = \langle \mathrm{R}_{Z'}(h')\,,\, g'\rangle$$

One easily checks that

$$u^*\left(\langle \mathrm{R}_{Z'}(h')\,,\, g'\rangle\right) = \langle \mathrm{R}_{Z'}(h') \circ u_{g'}\,,\, u^*g'\rangle\,.$$

On the other hand we get

$$\begin{aligned} \mathrm{R}^X_{g'}(u^*(h')) &= \langle \mathrm{R}_Z(u^*(h'))\,,\, u^*g'\rangle \\ &= \langle \mathrm{R}_Z(u_{g'}{}^*h')\,,\, u^*g'\rangle \\ &= \langle \mathrm{R}_{Z'}(h') \circ u_{g'}\,,\, u^*g'\rangle \end{aligned}$$

and this completes the proof. ∎

This proposition means that the notion of dictos is preserved by localization, and this has a clear logical significance: if we see a dictos as a logical theory, the theory we get by fixing a given hypothesis is again a theory of the same kind.

2 The internal category of propositions

This section is devoted to the announced study of the properties of the internal full subcategory of propositions $\mathrm{Full}_{\mathcal{E}}(T)$. We shall prove that it is an internal cartesian closed category which is internally small complete (more precisely, we shall only prove that it has all internal small products).

2.1 General results about reflections

Let $I : \mathbf{B} \to \mathbf{D}$ be the embedding part of a reflection between the categories $\mathbf{B}$ and $\mathbf{D}$, and let R be the retraction part of this reflection. Then

Lemma 3 *Let $F : \mathbf{J} \to \mathbf{C}$ be a functor such that $I \circ F$ admits a limit in $\mathbf{D}$. Then the functor F admits a limit in $\mathbf{B}$. More precisely we have, up to isomorphism*

$$\varprojlim F = R(\varprojlim IF) .$$

As a consequence, there exist in $\mathbf{B}$ all the limits that exist in $\mathbf{D}$.

We don't recall the proof of this well known result.

We shall also need the following

Lemma 4 *If $\mathbf{D}$ is cartesian closed (let $\Rightarrow_{\mathbf{D}}$ be its exponential), and if there exists a functor $\Rightarrow_{\mathbf{B}}$: $\mathbf{B}^{\mathrm{op}} \times \mathbf{B} \to \mathbf{B}$ such that the following commutes (up to natural isomorphism)*

$$\begin{array}{ccc} \mathbf{B}^{\mathrm{op}} \times \mathbf{B} & \stackrel{\Rightarrow_{\mathbf{B}}}{\longrightarrow} & \mathbf{B} \\ {\scriptstyle I^{\mathrm{op}} \times I}\downarrow & & \downarrow{\scriptstyle I} \\ \mathbf{D}^{\mathrm{op}} \times \mathbf{D} & \stackrel{\Rightarrow_{\mathbf{D}}}{\longrightarrow} & \mathbf{D} \end{array}$$

then $\mathbf{B}$ is cartesian closed.

Proof: We already know that it is cartesian by the preceeding lemma. We just have to show that $\Rightarrow_{\mathbf{B}}$ defines an exponential on $\mathbf{B}$. Let X, Y, Z be three objects of $\mathbf{B}$. Then

$$\begin{array}{rcl} \mathrm{Hom}_{\mathbf{B}}(X \times Y , Z) & = & \mathrm{Hom}_{\mathbf{B}}(R(I(X) \times I(Y)) , Z) \\ & \cong & \mathrm{Hom}_{\mathbf{D}}(I(X) \times I(Y) , I(Z)) \\ & \cong & \mathrm{Hom}_{\mathbf{D}}(I(X) , I(Y) \Rightarrow I(Z)) \\ & \cong & \mathrm{Hom}_{\mathbf{D}}(I(X) , I(Y \Rightarrow Z)) \\ & \cong & \mathrm{Hom}_{\mathbf{B}}(X , Y \Rightarrow Z) \end{array}$$

and we conclude, since all these isomorphisms are natural. ∎

As a corollary we get

Lemma 5 *For any $X \in \mathcal{E}$, the category $\mathcal{P}(X)$ is cartesian closed.*

Proof: In view of the two preceeding lemmas, we just have to build a functor $\Rightarrow_{\mathcal{P}(X)}$: $\mathcal{P}(X)^{\mathrm{op}} \times \mathcal{P}(X) \to \mathcal{P}(X)$ such that the following diagram commute:

$$\begin{array}{ccc} \mathcal{P}(X)^{\mathrm{op}} \times \mathcal{P}(X) & \stackrel{\Rightarrow_{\mathcal{P}(X)}}{\longrightarrow} & \mathcal{P}(X) \\ {\scriptstyle I^{\mathrm{op}} \times I}\downarrow & & \downarrow{\scriptstyle I} \\ \mathcal{E}/X^{\mathrm{op}} \times \mathcal{E}/X & \stackrel{\Rightarrow_{\mathcal{E}/X}}{\longrightarrow} & \mathcal{E}/X \end{array} \qquad (1)$$

where $I : \mathcal{P}(X) \to \mathcal{E}/X$ is the obvious full embedding. We just have to define this functor $\Rightarrow_{\mathcal{P}(X)}$ on objects, since on arrow it is completely defined by the previous diagram. If $\varphi, \psi : X \to \Omega$ are objects of $\mathcal{P}(X)$, we take

$$\varphi \Rightarrow_{\mathcal{P}(X)} \psi = \forall\varphi\,(\psi\widetilde{\varphi})$$

and the axiom of impredicative product insures that the diagram 1 commutes. ∎

2.2 The internal structure of $\mathrm{Full}_{\mathcal{E}}(T)$

We prove now

Proposition 3 *The internal full subcategory* $\mathrm{Full}_{\mathcal{E}}(T)$ *is internally cartesian closed and has all internal small products.*

Proof: We use the caracterisation of internal full subcategories with internal finite and small products, and with internal exponential given in [12]. We first show that it has internally a terminal object. For this, we just have to find an arrow $[1] : 1 \to \Omega$ such there is a pullback of the form

$$\begin{array}{ccc} 1 & \longrightarrow & \Omega' \\ {\scriptstyle 1}\big\downarrow & & \big\downarrow{\scriptstyle T} \\ 1 & \xrightarrow{[1]} & \Omega \end{array}$$

We take $[1] = \mathrm{R}_1(1)$. We just have to prove that the object $\widetilde{1}_1$ is terminal in $\mathcal{E}$. We know by lemma 3 that $\mathrm{R}_1(1)$ is terminal in $\mathcal{P}(1)$. But for any object $X \in \mathcal{E}$ we have

$$\mathrm{Hom}_{\mathcal{E}}\left(X\,,\,\widetilde{1}_1\right) \cong \mathrm{Hom}_{\mathcal{P}(1)}\left(\mathrm{R}_1(X)\,,\,\mathrm{R}_1(1)\right)$$

which is thus a set with a single element. This proves that $\widetilde{1}_1$ is terminal.

Let $\pi_1, \pi_2 : \Omega^2 \to \Omega$ be the two projections. To show that $\mathrm{Full}_{\mathcal{E}}(T)$ is cartesian we have to find an arrow $[\times] : \Omega^2 \to \Omega$ such that we have a pullback of the form

$$\begin{array}{ccc} U & \longrightarrow & \Omega' \\ {\scriptstyle \widetilde{\pi_1}\times\widetilde{\pi_2}}\big\downarrow & & \big\downarrow{\scriptstyle T} \\ \Omega^2 & \xrightarrow{[\times]} & \Omega \end{array}$$

We simply take $[\times] = \mathrm{R}_{\Omega^2}(\widetilde{\pi_1} \times \widetilde{\pi_2})$. We actually know by lemma 1 that the category $\mathcal{P}(\Omega^2)$ is cartesian. Hence, with respect to its product, we have just set

$$[\times] = \pi_1 \times \pi_2$$

Now it suffices to see that $\widetilde{\pi_1 \times \pi_2} \cong \widetilde{\pi_1} \times \widetilde{\pi_2}$. But this is obvious, since the functor $f \to \widetilde{f}$ preserves limits as a right adjoint.

To define internal exponential, we have to find an arrow $[\Rightarrow] : \Omega^2 \to \Omega$ such that we have a pullback of the form

$$\begin{array}{ccc} V & \longrightarrow & \Omega' \\ {\scriptstyle \widetilde{\pi_1}\Rightarrow\widetilde{\pi_2}}\big\downarrow & & \big\downarrow{\scriptstyle T} \\ \Omega^2 & \xrightarrow{[\Rightarrow]} & \Omega \end{array}$$

We take $[\Rightarrow] = \mathrm{R}_{\Omega^2}(\widetilde{\pi_1} \Rightarrow \widetilde{\pi_2})$. Again, we have to check that $\widetilde{[\Rightarrow]} \cong \widetilde{\pi_1} \Rightarrow \widetilde{\pi_2}$. But this results from the axiom of impredicative product. Actually, another equivalent definition for $[\Rightarrow]$ is

$$[\Rightarrow] = \forall\pi_1\,(\pi_2\widetilde{\pi_1})$$

Now we prove that $\mathrm{Full}_{\mathcal{E}}(T)$ has all internal small product (ie. products "indexed by objects of $\mathcal{E}$"). This amounts to find, for any object X of $\mathcal{E}$ an arrow $[\Pi_X] : \Omega^X \to \Omega$ such that we have a pullback of the form

$$\begin{array}{ccc} W & \longrightarrow & \Omega' \\ {\scriptstyle \Pi\pi_1(\widetilde{\mathrm{ev}})}\downarrow & & \downarrow{\scriptstyle T} \\ \Omega^X & \xrightarrow{[\Pi_X]} & \Omega \end{array}$$

where $\mathrm{ev} : \Omega^X \times X \to \Omega$ is the evaluation map and $\pi_1 : \Omega^X \times X \to \Omega^X$ is the first projection map. We simply take $[\Pi_X] = \forall\pi_1(\mathrm{ev})$ and the axiom of impredicative product gives the required pullback. ∎

As a corollary, we have another caracterisation of dictoses, coming from the work of M. Hyland :

Corollary 1 *A category is a dictos iff it is locally cartesian closed and has an internal full subcategory which is internally small complete.*

However we shall keep our caracterisation of dictoses which is more elementary, and closer to the idea of generalising topos theory. Furthermore, this notion may be easily weakened (cf. HLCCC's) and strengthened (for instance, it could seem natural to require all monos to be classified by T, since this phenomenon appears in any topos, and in $\omega-$**Set** as well, which is a non trivial dictos, as we shall see).

3 Reflection in $\omega-$Set

We won't recall how the category $\omega-$**Set** is a model of the theory of constructions (an HLCCC), since this is now a well known fact. We just want to express in this category the reflection principle, since it is very simple and nice. Remember just that an ω-set X is a pair $(|X|, \vdash_X)$ where $|X|$ is a set, called support of X and $\vdash_X$ is a relation over $\omega \times X$ such that for any $x \in |X|$ there exists a $n \in \omega$ such that $n \vdash_X x$. This means that n justifies x as an element of X. An ω-set is said to be modest if any integer justifies at most one point. For more details, see for instance [13], [2] or [3].

Let X be an ω-set. We associate to it a modest ω-set $\overline{X}$. Let $\sim$ be the transitive closure of the $\smile$ relation defined on $|X|$ by

$$x \smile x' \Leftrightarrow \exists n \in \omega \quad n \vdash_X x \text{ and } n \vdash_X x'$$

We set $|\overline{X}| = |X|/\sim$, and we define on this set of classes the following justification relation: if $\xi \in \overline{X}$ and $n \in \omega$, then

$$n \vdash_{\overline{X}} \xi \Leftrightarrow \exists x \in \xi \quad n \vdash_X x$$

With this justification relation, it is clear that $\overline{X}$ is a modest ω-set. Let P be any modest ω-set. Up to a trivial isomorphism, it is clear that $\overline{P} = P$. We check easily that

$$\mathrm{Hom}\,(X\ ,\ P) \cong \mathrm{Hom}\left(\overline{X}\ ,\ P\right)$$

actually, it suffices to remark that if $f : X \to P$ is a morphism of ω–**Set** (i.e. a justified function), then

$$\forall x, x' \in X \quad x \smile x' \Rightarrow f(x) = f(x')$$

since P is modest. From this construction, we deduce the reflection principle: let $f : X \to Y$ be a morphism of $\omega-$**Set**. We define $\mathrm{R}_Y(f) : Y \to \Omega$ by setting

$$\forall y \in |Y| \quad \mathrm{R}_Y(f)(y) = \overline{f^{-1}(y)}\ .$$

Conclusion

We have just presented a few facts about dictoses. We hope that some well known results of topos theory may be extended cleanly to this framework. For instance we may expect a connection between dictoses and PL-categories (see [16]) similar to the one which exists between toposes and triposes (see [11]). We hope also to show that, for any internal category in a dictos, the category of internal presheaves over this internal category in the dictos is a dictos.

This study may be seen as a semantical approach to powerful type theories like the calculus of constructions, and we may hope that such an approach could have some applications in the study of these type theories which take more and more importance in synthesis of proved programs.

References

[1] J. Bénabou "Fibered categories and the foundation of naive category theory" J. of Symbolic Logic 50(1985) 10-37.

[2] T. Ehrhard "A categorical semantics of constructions" Logic in Computer Science 1988.

[3] T. Ehrhard "Une sémantique catégorique des types dépendants. Application au Calcul des Constructions." Thèse, Paris VII, 1988.

[4] M. Hyland "The effective topos" in The L.E.J. Brouwer Centenary Symposium (ed. Troelstra et Van Dalen), North Holland (1982), 165-216.

[5] M. Hyland "A small complete category" Conference "Church's Thesis 50 years later" Zeiss (NL) 1986. Ann. Pure and Appl. Logic, à paraître.

[6] M. Hyland et A. Pitts "Theory of constructions: categorical semantics and topos-theoretic models" to appear in the proceedings of the conference Category in Computer Science and Logic, Contemporary Mathematics, AMS (1988).

[7] M. Hyland, E.P. Robinson et G. Rossolini "The discrete objects in the Effective Topos" To appear.

[8] P.T. Johnstone "Topos Theory" Academic Press, Inc.

[9] J. Lambek and P.J. Scott "Introduction to higher order categorical logic" Cambridge studies in advanced mathematics. Cambridge University Press (1986).

[10] P. Martin Löf "Intuisionistic type theory" Bibliopolis 1985

[11] A. Pitts "Tripos theory." Ph. D. Thesis.

[12] A. Pitts "Polymorphism is set-theoretic... constructively." proceedings de la conférence "Category theory and computer science" Edinburgh 1987. LNCS.

[13] A. Scedrov "Recursive realazibility semantics for calculus of constructions" Unpublished.

[14] R.A.G. Seely. "Hyperdoctrines, natural deduction and the Beck condition" Zeitschrift für Mat. Logik und Grundlagen d. Math. 29 505-542.

[15] R.A.G. Seely. "Locally Cartesian Closed Categories and Type Theory" Math. Proc. Camb. Phil. Soc. 95, (1984).

[16] R.A.G. Seely. "Categorical Semantics for Higher Order Polymorphic Lambda Calculus" Draft (1986).

[17] T. Streicher "Correctness and completeness of a semantics of the calculus of constructions with respect to interpretation in doctrines of constructions" Ph.D. Thesis, Passau (1988).

Declarative Continuations: an Investigation of Duality in Programming Language Semantics

Andrzej Filinski

DIKU - Computer Science Department, University of Copenhagen[1]
Universitetsparken 1, 2100 Copenhagen Ø, Denmark
uucp: andrzej@diku.dk

Abstract

This paper presents a formalism for including first-class continuations in a programming language as a *declarative* concept, rather than an *imperative* one. A symmetric extension of the typed λ-calculus is introduced, where values and continuations play dual roles, permitting mirror-image syntax for dual categorical concepts like products and coproducts. An implementable semantic description and a static type system for this calculus are presented. We also give a categorical description of the language, by presenting a correspondence with a system of combinatory logic, similar to a cartesian closed category, but with a completely symmetrical set of axioms.

Introduction

Category theory has proven a useful formalism for describing semantics of programming languages since many common language constructs have direct categorical counterparts. However, such descriptions concentrate very much on *data* flow, while many details of the *control* structure, including conditionals, evaluation strategies, *etc.* do not seem to be described adequately in existing presentations. In particular, applicative order reduction (with its termination problems) and imperative features such as "exceptions" and "non-local exits" seem somewhat incompatible with categorical concepts.

On the other hand, a continuation-based denotational semantics is very suitable for describing precisely those phenomena, and enables one to reason formally about programs and their termination properties regardless of the evaluation order of the description language. Also, it provides a clean foundation for handling run-time errors, and for concepts such as backtracking and coroutines. However, it introduces an additional level of abstraction in the semantic description, making it somewhat harder to automate reasoning about programs, for purposes of transformation or correctness proofs.

In this article, we will try to provide a bridge between these two methods of language description, and in particular to show that, in a suitable framework, first-class continuations can be given a sound declarative meaning, as opposed to being

[1] Address from Sept. '89: CS Dept., Carnegie Mellon Univ., Pittsburgh, PA 15213-3890

just a powerful but unstructured control primitive. The main idea is based on viewing values and continuations as dual concepts and on generalizing the usual value-based concepts to encompass continuations.

To permit formal reasoning about declarative continuations, the *symmetric λ-calculus (SLC)* is introduced. Here, dual concepts such as products and coproducts have mirror-image syntax with a 1-1 correspondence between value- and continuation-based operations. We hereby obtain a language with considerable expressive power, where many constructs usually presented in a higher-order framework have first-order descriptions.

However, we retain full compatibility with the ordinary λ-calculus. Every well-typed λ-term is also (modulo a slight syntactic difference) a valid term in the symmetric calculus and has the same meaning, which may – as usual – depend on the evaluation order. We also have the familiar conversion rules of λ-calculus, together with their mirror-image versions and a number of *bridge rules* relating values to continuations.

Finally, the SLC has a simple, easily parsable syntax, a small, directly implementable formal semantics, and a tractable, (optionally) polymorphic type system, permitting machine solution of problems too large to compute by hand.

There is a well-known correspondence between typed λ-calculi and cartesian closed categories (CCCs), which allows us to give an element-free description of the former. We do this by associating to every λ-term a morphism in the category and by replacing concepts such as abstraction and substitution with a set of combinators. We extend this equivalence by presenting a categorical framework, the *Symmetric Combinatory Logic, SCL*, for the SLC.

The SCL has a completely symmetrical set of core axioms, which correspond to the conversion rules of the SLC. This symmetry extends to the existence of *coexponential objects*, the natural dualization of CCC exponentials. While they do not seem to correspond to any simple set-theoretical concept, they have a natural explanation in terms of continuations.

Just as any λ-calculus is really a family of languages, parameterized by evaluation order, the SCL can be specialized with extra axioms to obtain a particular reduction strategy. One such specialization satisfies all the axioms of a CCC (and some more) and corresponds to normal-order reduction. However, other possibilities which do not even have the full product axioms exist, and one particular strategy, applicative-order reduction, will be treated in depth here.

We expect familiarity with the fundamentals of continuation semantics (*e.g.* [Stoy 77]). Some acquaintance with the ML (*e.g.*, [Rydeheard & Burstall 88]) and/or Scheme [Rees & Clinger 86] programming languages might facilitate understanding, but is not required.

The paper is organized as follows: In section 1, we give an informal presentation of the SLC. Section 2 presents its type system, and 3 gives a complete denotational description of its CBV version. In section 4, we treat the correspondence with SCL, and section 5 gives a formal semantics for SCL terms. In section 6, we briefly consider alternative evaluation strategies. Finally, the conclusion puts this work into perspective and suggests a number of directions for future research.

1 The symmetric λ-calculus

In this section, we give an informal overview of the symmetric λ-calculus, aiming to provide some intuition about first-class continuations. Thus, we will start with the central idea of the language and gradually add new concepts, such as structured types, higher-order constructs and recursion, trying at all times to point out how familiar, value-based concepts can be dualized in terms of continuations.

1.1 Functions, values and continuations

Functions are traditionally viewed as value transformers, and the common notation $x \mapsto \varphi(x)$ or syntactic variants thereof reinforce this view. However, a different perspective on functions is given by the data-flow paradigm of computation, namely as *request* transformers. For example, the function $odd : int \rightarrow bool$, which is usually thought of as transforming an integer value to a boolean value, could also be seen as transforming a request for a boolean value to a request for an integer.

We note that the roles of values and requests depend on the evaluation strategy. In an eager or *data-driven* evaluation, the flow of control is determined by values while the requests are implicit. On the other hand, in a lazy or *demand-driven* environment, the requests actively pull forward the evaluation while values have a passive existence.

We can extend this view to continuation semantics, where continuations in a sense act as requests. In a lazy language, they drive forward the computation as nothing is evaluated until someone asks for the result. In an eager language, on the other hand, continuations just accept already computed results. In both cases, however, a continuation can be thought of as the *lack* or *absence* of a value, just as having a negative amount of money represents a debt. This analogy even extends to 'debt transformation': one can repay a creditor by borrowing the money from someone else. In fact, some economic situations seem easier to understand in terms of changing debts and deficits rather than a genuine flow of money.

With this informal view of values and continuations as dual concepts, let us try to construct a language where the two are treated truly symmetrically. First, we must (temporarily) abandon the λ-calculus amalgamation of functions and values and distinguish sharply between three different syntactic classes: functions, values and continuations. Any function can be used either as a value transformer or a continuation transformer. Conversely, a function may be defined either in terms of its effect on the input value or on the output continuation.

We write a function application to a value using an explicit operator: $F \uparrow E$. This denotes the *result* of E transformed by F and corresponds exactly to juxtaposition in the λ-calculus. Symmetrically, we write a continuation C transformed by F as $C \downarrow F$. This denotes a new continuation which will first transform its *input* with F.

We can specify functions either as value abstractions $x \Rightarrow E$ or as continuation abstractions $y \Leftarrow C$. The former describes how to transform an input value into a result value, while the latter specifies how a request for the result is transformed into a request for the input. Since continuations are not treated as functions, we can never explicitly *apply* a continuation to a value, but only substitute the current continuation with another, possibly defined in terms of the original. Finally, in addition to explicit abstractions, we include a number of primitive functions (*e.g.*, arithmetic operators), corresponding to δ-rules.

As the language contains the λ-calculus as a proper subset, we call it the symmetric λ-calculus, or SLC for short. We extend the usual syntactic conventions, letting applications (of both kinds) bind stronger than abstractions. Abstractions associate to the right, as do continuation transformations, while ordinary (value-) applications associate to the left.

All the conversion rules of the λ-calculus still hold, together with their mirror images (with the usual restrictions to prevent capturing of free variables):

$$\begin{array}{rcl} a \Rightarrow E & \overset{\alpha}{\leftrightarrow} & b \Rightarrow E[b/a] \\ (x \Rightarrow E_1) \uparrow E_2 & \overset{\beta}{\leftrightarrow} & E_1[E_2/x] \\ (x \Rightarrow F \uparrow x) & \overset{\eta}{\leftrightarrow} & F, \text{ if } x \text{ not free in } F \end{array} \qquad \begin{array}{rcl} a \Leftarrow C & \overset{\bar{\alpha}}{\leftrightarrow} & b \Leftarrow C[b/a] \\ C_2 \downarrow (y \Leftarrow C_1) & \overset{\bar{\beta}}{\leftrightarrow} & C_1[C_2/y] \\ (y \Leftarrow y \downarrow F) & \overset{\bar{\eta}}{\leftrightarrow} & F, \text{ if } y \text{ not free in } F \end{array}$$

We also want the two possible definitions of identity and composition to mean the same. This gives rise to the following two rules, where x and y do not occur free in F and G:

$$\begin{array}{rcl} x \Rightarrow x & \overset{id}{\leftrightarrow} & y \Leftarrow y \\ x \Rightarrow F \uparrow (G \uparrow x) & \overset{\circ}{\leftrightarrow} & y \Leftarrow (y \downarrow F) \downarrow G \end{array}$$

Strictly speaking, a term like $x \Rightarrow x$ does not by itself denote a function, but rather a *function template*, from which we can make instances by supplying a type. In other words, there is a distinct identity morphism for every object. This will become significant when higher-order constructs are added to the language. We could also view $x \Rightarrow x$ as a single, polymorphically-typed function, but this adds an extra level of complication.

Unfortunately, having both the β and $\bar{\beta}$ rules in their most general form leads to an inconsistent system, where different reduction orders give widely differing results. For example, the function

$$(x \Rightarrow 2) \circ (y \Leftarrow k)$$

where k is a continuation from an outer scope may be reduced to either $x \Rightarrow 2$ or $y \Leftarrow k$. This is not a problem specific to the SLC, but a general manifestation of a weakness in every applied λ-calculus with partial functions as δ-rules, since runtime errors interfere with the current continuation. For example, the expression

```
let x=1/0 in 2
```

will evaluate to 2 in a call-by-name (CBN) language, but fail in a call-by-value (CBV) one.

Therefore, virtually all programming language definitions specify an evaluation order, *i.e.*, a set of restrictions on possible reductions. In the following we will assume a CBV strategy, so that β-conversion can only be performed on completely reduced arguments, but with no restrictions on the $\bar{\beta}$-rule. In section 6, alternative strategies will be considered.

The significance of the composition rule should now be apparent: it can be used to transform terms to a form where the restricted β-rule can be applied, by shifting nested function applications onto the continuation. This is exactly what happens in a continuation-based specification, such as the denotational semantics of the next section: during evaluation, an explicit continuation is built up, so that functions are only applied to simple values. It can be directly verified that all the given rules, except full β-reduction, hold in the CBV semantics, and that the identity and composition rules satisfy their associated axioms (neutrality and associativity).

Finally, let us point out a more categorical view of functions, values and continuations: we may consider the types as the objects in a category, with functions between them as the morphisms. A value of type T corresponds to a morphism $1 \rightarrow T$, where 1 is the (weak) terminal object of the category. We can see it as a degenerate function, with no input. Dually, a continuation accepting T-typed values corresponds to a morphism $T \rightarrow 0$, where 0 is the initial object, *i.e.*, a function with no output.

Applications (of both kinds) correspond to morphism composition, *e.g.*, if the function F from S to T denotes a morphism $f : S \rightarrow T$, and expression E of type S a morphism $e : 1 \rightarrow S$, then $F \uparrow E$ denotes the morphism $f \circ e : 1 \rightarrow T$. Similarly, if C is a T-accepting continuation, denoting the morphism $c : T \rightarrow 0$, then the S-accepting continuation given by $C \downarrow F$ denotes the morphism $c \circ f : S \rightarrow 0$.

Value and continuation abstractions are slightly more complicated, but are based on adjoining indeterminate arrows, denoted by variable symbols, to the category. Thus, an expression $E : T$, with morphism $e : 1 \rightarrow T$ in which the symbol x occurs as an S-typed value corresponds to a morphism $e^x : S \rightarrow T$, and a continuation $C : S$, *i.e.*, morphism $c : S \to 0$ expressed in terms of a continuation variable $y : T$ defines a morphism $c_y : S \to T$. This correspondence between SLC terms and categorical morphisms is pursued in section 4.

1.2 Structured types

Virtually all strongly-typed programming languages provide support for building new types out of existing ones. The goal of this subsection is to present the syntax for doing this in the SLC.

The two basic schemas for constructing new types out of old ones are the tuple (record, structure) and the disjoint sum (variant, union). The similarity of these with the categorical concepts of products and coproducts is obvious. Their associated syntax and semantics, however, seldom exhibit many similarities, perhaps mainly for historical reasons. In the SLC, we will directly adopt the simple, traditional way to write a projection:

$$\pi_1 \equiv (a, b) \Rightarrow a : A \times B \to A$$

Similarly, given functions $f : C \to A$ and $g : C \to B$, we write the mediating morphism:

$$\langle f, g \rangle \equiv c \Rightarrow (f \uparrow c, g \uparrow c) : C \to A \times B$$

The value-based reasoning behind this notation is clear: the function $\langle f, g \rangle$ constructs a pair of values, from which one can later be extracted using a projection. We note that the above syntax represents an *unlabelled* tuple, with *positional association* for the elements, but we could easily extend the notation to handle records with named fields.

In this setting, let us consider coproducts in terms of continuations: the injection $\iota_1 : A \to A + B$ transforms a request for either an A or a B into a request for an A. Now, a request for a coproduct type is really a pair of requests, of which either one can be satisfied, *i.e.*, a selection between two continuations. Perhaps the closest analogue to this is provided by the label-typed parameters of Algol 60, where a function may return at different points, depending on its result. In the SLC, the selection aspect is made explicit by the following syntax:

$$\iota_1 \equiv \{a, b\} \Leftarrow a : A \to A + B$$

Curly braces are used for clarity, to emphasize the fact that the definition is not expressed in terms of values but continuations. In fact, the direction of the arrow is sufficient to determine the meaning, so simple parentheses could be used instead.

Furthermore, given functions $f : A \to C$ and $g : B \to C$, the function $[f, g] : A + B \to C$ transforms a request for a C into a request for either an A or a B. In the first case, the function f is applied to obtain the result, otherwise, the function g:

$$[f,g] \equiv c \Leftarrow \{c\downarrow f, c\downarrow g\} : A + B \rightarrow C$$

Again, the above represents unlabelled coproducts, where the injection tags are implicit, and the order of components in a 'cotuple' is significant. We could easily add name association here as well, as in ML's **datatype** constructor.

1.3 Terminal and initial objects

Now, let us examine the border cases of products and coproducts: terminal and initial objects. Unfortunately, there seems to be no universally accepted way of writing the associated unique morphisms. We will therefore use a notation suggestive of the binary case, writing $\diamond : A \rightarrow 1$ and $\square : 0 \rightarrow A$, occasionally subscripting the symbols $\diamond$ and $\square$ with the type A.

The terminal object has an important role in programming. In some languages, it has an associated explicit type (usually called 'unit' or 'void'); in others it is represented by an empty (or absent) argument list on input, and the keyword 'procedure' instead of 'function' for output. Actually, 'void' is a misnomer, since the type does contain a value, which may just not be directly expressible.

Considering 1 as an empty product, we write the unique value of type *unit* as (). For every type A, we have a function

$$\diamond_A \equiv a \Rightarrow () : A \rightarrow \mathit{unit}$$

which ignores its parameter. This is important from a theoretical point of view, as it provides a framework for non-strict functions, but in practice functions of this type are only used for any side effects they might have. On the other hand, we have a bijection between T-typed expressions E, and functions $E' : \mathit{unit} \rightarrow T$, given by:

$$E' = () \Rightarrow E : \mathit{unit} \rightarrow T \qquad\qquad E = E' \uparrow () : T$$

Both denote the same morphism in the category, but are of different syntactic classes (function and expression, respectively). This equivalence will become important later, when higher-order constructs are added to the language. Often, such parameterless functions are called *suspensions* or *thunks*, and are used to force CBN-like behavior in a CBV language by delaying evaluation.

The initial object rarely manifests itself in programming, but has a natural interpretation as well. In the category of sets, it is simply the empty set, *i.e.*, the type *null* with *no* values. Again, the unique morphism $\square : 0 \rightarrow A$, naturally written like this:

$$\square_A \equiv a \Leftarrow \{\} : \mathit{null} \rightarrow A$$

is not very interesting by itself, but its significance is that it discards a *continuation*, and thus can be used to model imperative constructs, such as escapes, jumps, *etc.* Thus, both $\Diamond$ and $\Box$ play central roles in the categorical definition of SLC.

We also have a correspondence between S-accepting continuations C and functions $C' : S \to null$:

$$C' = \{\} \Leftarrow C : S \to null \qquad\qquad C = \{\} \downarrow C' : S$$

$S \to null$ is the type of a function which *never* returns to the point of call, as there is no possible value it can return with. Usually, no functions like that are written (on purpose), but some languages, notably 'C', present a number of control primitives as standard functions, *e.g.*, `exit` and `longjmp`. The return type of these could be conveniently declared as *null*.

Note that these constructs only present a syntactic formalism for products, coproducts and initial/terminal objects; while the associated morphisms are defined in all cases, whether they actually satisfy the axioms depends on the evaluation order of the language. For example, the SLC with CBV semantics does not have true products unless morphisms are restricted to total functions. In Lisp terms, the McCarthy axiom `car[cons[A,B]]` = `A` does not hold for *all* expressions `A` and `B`. In fact, not even empty products exist under CBV: even ignoring side effects such as assignments, I/O, *etc.*, there may be many different functions from a given type to *unit*, differing in termination behavior and escapes.

On the other hand, the CBV SLC does have true coproducts and a proper initial object, even in the presence of continuation abstractions or non-termination. The precise problems with existence and uniqueness are deferred to the section on translation to categorical terms.

1.4 Nesting and block structure

In the language presented so far, functions must be syntactical abstractions, and cannot be passed around. It may therefore seem that the variable-based notation is just syntactic sugar for the basic categorical constructs (product, coproduct, *etc.*), but this is not entirely true: by allowing nested function definitions, we have made it possible to define a number of morphisms which could not otherwise be expressed. This is because the body of an abstraction may refer to values and continuations abstracted at an outer lexical scope, *i.e.*, the language has a block structure.

Non-local values are usually considered a clear and efficient way of avoiding explicit function parameters that would otherwise just be carried around throughout a computation, but only used in a few places. On the other hand, non-local exits are commonly regarded as quick-and-dirty tricks, acceptable only in very special circumstances, *e.g.* for handling fatal errors. Even though many programming problems are naturally expressed in terms of continuations, there is a strong preference for value-based solutions.

Consider, for example, a (recursive) function that searches a tree for a given value, returning either 'found' or 'not found'. The 'clean' implementation will pass a 'found' answer back through all the nested recursive calls, which is clearly redundant, as it will not be modified on the way. If the searched tree is non-homogeneous, (*e.g.*, a syntax tree), and different parts must be searched with different functions, clarity suffers as well.

In such a case, a non-local exit with the answer 'found' is in fact more natural, just as a non-local value might be used to specify the element to be searched for. Unfortunately, non-local exits have strong connotations of 'error' in most languages, and are often performed using a powerful but unstructured construct like 'goto' in imperative languages, or 'call-with-current-continuation' in Scheme. In others, notably ML and Ada, 'exceptions' have an essentially dynamic scope, as control returns not to a statically-determined point, but to the last encountered 'handler'. A language where a non-local continuation would be considered just as natural and simple as a non-local value probably still remains to be seen, but the SLC at least provides a symmetrical, structured, and statically-scoped framework for such a facility.

1.5 Functions as values and as continuations

In the λ-calculus, there is an implicit equivalence between functions and values. In the SLC, this is also true, but symmetry dictates that there should be a correspondence between functions and continuations as well. While this is technically simple to add in a continuation semantics, its manifestation in the categorical interpretation of the SLC gives rise to a number of highly unusual and at first sight very counterintuitive properties. It is therefore important to remember that every result obtained can be verified in a simple and unambiguous way by reference to the semantic equations of section 3.

In a typed λ-calculus, the foundation of higher-order functions is provided by the fact that for all objects (types) A and B there exists an *exponential* or *function space* object, usually written B^A or $[A \rightarrow B]$, with a 1-1 correspondence between values of type $[A \rightarrow B]$ and functions $A \rightarrow B$. In the SLC, a dual relationship holds as well, namely as a *coexponential object* A_B or $[B \leftarrow A]$, and a correspondence between functions $A \rightarrow B$ and *continuations* of type $[B \leftarrow A]$.

In the SLC, explicit conversions between the three syntactic classes are not necessary in the concrete syntax, but are present as implied operators in the abstract one. In this sense, they are very much like the *coercions* found in many languages, permitting *e.g.*, an integer value to be used in a context where a real-typed one is required.

Computationally, a value of type $[A \rightarrow B]$ is known as a functional closure. In a continuation semantics, it denotes a function that will map an argument value and a result continuation to the final result of the program. We can therefore equivalently view a function $f : A \to B$ as a continuation accepting a pair consisting of an A-typed value and a B-accepting continuation. Such a pair will be called the *context* of a function application, and its type written as $[B \leftarrow A]$.

Perhaps the easiest way to think about a context-typed continuation is to view it as a Smalltalk object, (or, even better an *Actor* [Hewitt 79]). Such an object accepts requests ("messages") to compute a function. It does not know, however, where the request came from. Thus, if we want an answer, we must also supply a 'return address', *i.e.*, a continuation for it.

To handle free variables in a function body, we actually need more general versions of the correspondences mentioned above. For functional values we must have the properties of a cartesian closed category: a bijection between morphisms $f : A \times B \to C$ and their *curried* forms $f^* : A \to [B \to C]$. We can also express this as the existence of a right adjoint to the product functor. The special case $A = 1$ gives the original correspondence between morphisms $B \to C$ and $1 \to [B \to C]$.

We can write down all the relevant morphisms in the SLC: for functions $f : A \times B \to C$, and $g : A \to [B \to C]$, we define

$$\begin{aligned} f^* &\equiv a \Rightarrow b \Rightarrow f \uparrow (a, b) : A \to [B \to C] \\ ap &\equiv (p, b) \Rightarrow p \uparrow b : [B \to C] \times B \to C \end{aligned}$$

And we want the following equations, expressing existence and uniqueness, respectively, to hold:

$$ap \circ (f^* \times id_B) = f$$

$$(ap \circ (g \times id_B))^* = g$$

The general version of the dual case gives a bijection between morphisms $f : C \to A + B$ and $f_* : [B \leftarrow C] \to A$. Equivalently, we require the existence of a left adjoint to the coproduct functor. Again, for $A = 0$, the isomorphism between B and $0+B$ gives the desired bijection between morphisms $C \to B$ and $[B \leftarrow C] \to 0$. By reversing the arrows above, we immediately obtain the dual SLC definitions:

$$\begin{aligned} f_* &\equiv a \Leftarrow b \Leftarrow \{a, b\} \downarrow f : [B \leftarrow C] \to A \\ pa &= \{q, b\} \Leftarrow b \downarrow q : C \to [B \leftarrow C] + B \end{aligned}$$

and the existence and uniqueness equations:

$$(f_* + id_B) \circ pa = f$$

$$((g + id_B) \circ pa)_* = g$$

The first equation in particular is quite striking: The morphism $(f_* + id_B)$ does not change the injection tag, and in the case of the second inject not even the value. Nevertheless, a single morphism pa can be used for all f! This just shows that the usual element-based reasoning techniques must not be used indiscriminately in a categorical setting. Operationally, the equation is explained by the fact that the context-typed result of *pa* includes the continuation in effect just before the case construct, which permits the computation to backtrack slightly when f_* is applied. So, if $f = \iota_2$, *pa* will first return with the first inject, but make it possible for f_* to re-enter the case expression with the correct tag.

Let us now consider a simple practical application of "*cocurried*" functions, by analogy to partial (value) application: given a function $add : int \times int \to int$, we can define a specialized version $add3 \equiv add^* \uparrow 3 : int \to int$, which will add 3 to its argument. The dual case is analogous: let the function $elog : real \to unit + real$ be defined as returning the logarithm of its argument, if positive, and () on error. Let there also be given a continuation *fail*, accepting a *unit*-typed value. We can then define the function $log \equiv fail \downarrow elog_* : real \to real$ which will return the logarithm of a positive argument, but invoke the continuation *fail* on error.

Again, we must remember that, just as with the structured types, a particular semantics may not satisfy the above axioms in full generality, but only for "well-behaved" functions. In fact, it is one of the usual CCC axioms that does not quite hold in a CBV strategy: while a f^* with the required property exists for *every* f, it is not necessarily unique. Again this is true for any CBV-type language with runtime errors: the uncurried forms of the two following two ML functions

```
fn x=>let t=1/x in (fn y=>t)

fn x=>fn y=>1/x
```

both compute the inverse of the first argument, but behave differently (the first may fail) in curried form. To ensure uniqueness, we must put a suitable constraint of f.

On the other hand, both axioms for coexponentials hold in CBV. In fact, the underlying category could be called co-cartesian closed, as it has a proper initial object, binary coproducts and coexponentials, but not true products and exponentials.

Finally, it must be mentioned that the SLC with higher-order functions is somewhat more expressive than the category described above, as it is also possible to have references to non-local continuations in value abstractions and vice versa. To handle this, we need two special morphisms, similar in principle to the distributive law for products and coproducts. Details can be found in section 4.

1.6 Recursion and fixpoints

As the SLC is typed, we need a special operator to write recursive definitions. By doing so we will lose the strong normalization property. However, this presents surprisingly few new problems, as nontermination can be viewed as a special case of escaping. Thus, all the basic properties of the category still hold, and precisely because the axioms are weak enough, the inconsistency results presented in e.g. [Huwig & Poigné 86] do not apply. There are a number of technical problems, however, mainly due to the need of handling recursive types, so we will not treat recursion formally here, for lack of space.

Let us just note that the symmetry between values and continuations extends to recursive definitions as well. In lazy languages, we can form 'recursive values', which will be evaluated on need. For example, we can have a value "rec $x = (1, x)$", the type of which is a solution to the domain equation $D = int \times D$. We can also use the 'rec' operator to define recursive values from the function space, *i.e.*, recursive functions.

In CBV, recursive values are not possible, but instead we can have recursive continuations. In fact, recursive functions are defined as recursive, context-typed continuations. The implicit conversion between such continuations and functions permits recursive definitions using the obvious syntax, *e.g.*

$$\text{rec } f = n \Rightarrow \text{if } n = 0 \text{ then } 1 \text{ else } n \times f \uparrow (n-1)$$

However, recursive functions are just a special case of recursive continuations. In fact, we can write useful recursive continuations that are not functions, in the same way as we utilize recursive values such as lazy lists under CBN. For example, let us consider again the CBN domain of infinite lists of integers, $D = int \times D$. We can write a function which will map an integer to an infinite list of itself like this:

$$\text{rec } f = n \Rightarrow (n, f \uparrow n) : int \rightarrow D$$

Categorically, this represents the morphism $f = \langle id, f \rangle$.

However, we can also express the result directly as a recursive value, so we may write the same function as

$$n \Rightarrow \text{rec } d = (n, d) : int \rightarrow D$$

Now, let us consider the dual case. Our domain becomes the solution to $D' = int + D'$, *i.e.*, an integer embedded in a (finite) series of injection tags. (If we had used *unit* instead of *int* above, D' would be isomorphic to the natural numbers). We can write a function $g : D' \rightarrow int$, which will extract the integer from a D'-typed value:

$$\text{rec } g = n \Leftarrow \{n, n \downarrow g\} : D' \rightarrow int$$

We recall that this is the syntax for writing a 'case expression' in the SLC. In categorical terms, we have $g = [id, g]$. However, we can also do it directly, using a recursive continuation:

$$n \Leftarrow \text{rec } d = \{n, d\} : D' \rightarrow int$$

The continuation d will accept an element of the domain D', and pass it either to the result continuation n or to itself, stripping off the injection tag first.

Finally, mutual recursion also works in a dual way to CBN. While in CBN, we can define a pair of closure-typed values together, and project out the one we want, *e.g.*,

$$((f,g) \Rightarrow f) \uparrow (\text{rec } (f,g) = (\ldots,\ldots))$$

in CBV, we inject a computational context into the coproduct expected by the recursive continuation:

$$(\text{rec } \{f,g\} = \{\ldots,\ldots\}) \downarrow (\{f,g\} \Leftarrow f)$$

In the object-oriented view presented in subsection 1.5, we can think of recursion as sending messages to oneself. To distinguish between requests to compute f and g, we also supply an injection tag (the "message constructor").

1.7 Examples

Let us conclude this section by presenting a few more examples of SLC terms and their meaning.

First, we can define a function which takes a coproduct value and extracts its tag, according to the set-theoretical definition of coproducts as a disjoint sum. We can write such a function as the categorical term $[1 \circ \Diamond, 2 \circ \Diamond]$ or as a continuation abstraction in SLC:

$$tag \;\equiv\; n \Leftarrow \{n \downarrow (a \Rightarrow 1), n \downarrow (b \Rightarrow 2)\} : A + B \rightarrow int$$

The semantics does not include the usual conditional operator 'if'. But if we define the type *bool* as $unit + unit$, we can introduce it as a first-order syntactic extension:

$$\text{if } E_0 \text{ then } E_1 \text{ else } E_2 \;\equiv\; (r \Leftarrow \{r \downarrow (() \Rightarrow E_1), r \downarrow (() \Rightarrow E_2)\}) \uparrow E_0$$

Finally, let us consider Scheme's `call-with-current-continuation` or `call/cc`. This functional will pass an imperative functional abstraction, q, of the current continuation to its argument f. If f does not apply q, the result of `call/cc` will be f's result. But if f applies q to an argument a, evaluation of f's body will be abandoned, and a returned as the result of the `call/cc`. For example, we have:

`(+ 5 (call/cc (lambda (q) (+ 3 4))))` $\Longrightarrow 12$

`(+ 5 (call/cc (lambda (q) (+ 3 (q 4)))))` $\Longrightarrow 9$

`call/cc` is very similar to Landin's 'J'-operator and Reynolds's 'escape'. It is more general than an exception, since if f does not call q, but returns it as a functional value, the captured continuation may be re-invoked at any time, restoring control to the expression embedding the `call/cc`.

Clearly, we cannot define `call/cc` as a simple λ-abstraction, but in the SLC we can express it directly:

$$call/cc \;\equiv\; k \Leftarrow k\!\downarrow\!(f \Rightarrow f\!\uparrow\!(c \Leftarrow k)) : [[A \to B] \to A] \to A$$

This definition highlights the two key features of $call/cc$: the *duplication* of the continuation k at the point where $call/cc$ is called and the *discarding* of the current continuation c where the continuation-representing function (f's argument) is applied. The type also makes it clear that the result of $call/cc$ will be either f's result or q's argument, so that the two must have the same type A.

2 The SLC Type System

As mentioned earlier, the SLC has a simple, static type system that encompasses both values and continuations. We present it here in a concise notation, obtained by decorating the abstract syntax grammar with type symbols. In figure 1, subscripts represent output or result types, while superscripts denote input or entry types. For example, $C^{\alpha+\beta}$ denotes a continuation accepting a value of type $\alpha+\beta$. For symmetry reasons and to emphasize the distinction between functions and elements, we will also use the neutral notation F^{σ}_{τ} for a function with domain σ and codomain τ, and reserve the arrow notation for closure-typed expressions and context-typed continuations.

The usual scope rules of the λ-calculus apply, so that an abstraction body may also contain variables (of both kinds) bound at an outer lexical level. We also note that when variables occur on the left side of an abstraction, *i.e.*, as formal parameters, they behave differently, so that value variables denote inputs to the function, while continuation variables denote outputs. As an example, it is easily verified that the *expression* $\zeta \equiv (f,g) \Rightarrow c \Leftarrow \{c\!\downarrow\! f, c\!\downarrow\! g\}$, (expressing the coproduct morphism constructor as a constant), has type $[[A \to C] \times [B \to C] \to [A+B \to C]]$.

The rules can either describe a simple, monotyped system, where every type variable in the rules must be instantiated to a concrete type, or a polymorphic one, where the type of a function like identity can contain free type variables, as in ML. In the following, we will assume the monotyped version, as its categorical foundations are much simpler, but the semantic description of the SLC is also valid for polymorphic types.

3 A Denotational Semantics of the SLC

This section presents a full denotational semantics for the CBV version of the SLC. The abstract syntax is as in figure 1, and the semantic equations of figure 2 assume correctly-typed terms, so that no pattern decomposition will fail. In particular,

$$
\begin{array}{rcl}
E_\tau & ::= & cst_\tau \mid x_\tau \mid F^\sigma_\tau \uparrow E_\sigma \\
E_{unit} & ::= & () \\
E_{\alpha\times\beta} & ::= & (E_\alpha, E_\beta) \\
E_{[\sigma\to\tau]} & ::= & F^\sigma_\tau \\
\\
F^\sigma_\tau & ::= & X^\sigma \Rightarrow E_\tau \mid E_{[\sigma\to\tau]} \\
X^\tau & ::= & x_\tau \\
X^{unit} & ::= & () \\
X^{\alpha\times\beta} & ::= & (X^\alpha, X^\beta)
\end{array}
\qquad
\begin{array}{rcl}
C^\sigma & ::= & y^\sigma \mid C^\tau \downarrow F^\sigma_\tau \\
C^{null} & ::= & \{\} \\
C^{\alpha+\beta} & ::= & \{C^\alpha, C^\beta\} \\
C^{[\tau\leftarrow\sigma]} & ::= & F^\sigma_\tau \\
\\
F^\sigma_\tau & ::= & Y_\tau \Leftarrow C^\sigma \mid C^{[\tau\leftarrow\sigma]} \\
Y_\sigma & ::= & y^\sigma \\
Y_{null} & ::= & \{\} \\
Y_{\alpha+\beta} & ::= & \{Y_\alpha, Y_\beta\}
\end{array}
$$

Figure 1: Abstract Syntax and Type System of the SLC

for CBV, the continuation denoted by $\{\}$ can never be applied in a type-correct program, so the semantic equation gives its denotation as a 'case expression' with no choices.

By a slight abuse of notation, we identify syntactic and semantic constants. Similarly, the equation giving the denotation of primitive functions should be read as a family of equations, one for every p.

Even though the valuation function definitions are curried, we never apply a function to only a subset of its arguments. In fact, product types could have been used for the parameters, but they tend to clutter up the equations with tuple parentheses.

We must also remember that we are basically translating a symmetrical language into an asymmetrical one (the λ-calculus), so that the equations for expressions and continuations do not look symmetrical. In particular, the CBV strategy represents both products and coproducts as values. However, this is just one possible semantics respecting the basic axioms of the SLC. We can also specify a CBN-like semantics, where the continuation-selecting aspect of coproducts is expressed directly.

4 Symmetric Combinatory Logic

The denotational description of the last section gives an operational definition of the CBV version of SLC. In this section, we will consider a different, strategy-independent approach, which could be described as compiling to combinator code. We show that there is an equivalence between SLC terms and morphisms in a suitable category, analogously to the equivalence between the ordinary λ-calculus and the Categorical Combinatory Logic terms of [Curien 86]. By doing so, we replace the concept of value or continuation substitution by morphism composition, making explicit the flow of both data and control. We call this set of combinators

$$
\begin{aligned}
Val &= Basic + Unit() + Pair(Val \times Val) + In_1(Val) + In_2(Val) + \\
&\quad Closr(Val \to Cnt \to Ans) + Contx(Val \times Cnt) \\
Cnt &= Val \to Ans \\
Env &= Ide \to (Val + Cnt)
\end{aligned}
$$

$$
\begin{aligned}
\mathcal{E} : E &\to Env \to Cnt \to Ans \\
\mathcal{E}[\![cst]\!]\rho\kappa &= \kappa\; cst \\
\mathcal{E}[\![x]\!]\rho\kappa &= \text{let } val(v) = (\rho\; x) \text{ in } \kappa v \\
\mathcal{E}[\![(E_1, E_2)]\!]\rho\kappa &= \mathcal{E}[\![E_1]\!]\rho\; (\lambda v_1.\mathcal{E}[\![E_2]\!]\rho\; (\lambda v_2.\kappa\; pair(v_1, v_2))) \\
\mathcal{E}[\![()]\!]\rho\kappa &= \kappa\; unit() \\
\mathcal{E}[\![F \uparrow E]\!]\rho\kappa &= \mathcal{E}[\![E]\!]\rho\; (\lambda v.\mathcal{F}[\![F]\!]\rho v\kappa) \\
\mathcal{E}[\![F]\!]\rho\kappa &= \kappa\; closr(\lambda vc.\mathcal{F}[\![F]\!]\rho vc)
\end{aligned}
$$

$$
\begin{aligned}
\mathcal{C} : C &\to Env \to Val \to Ans \\
\mathcal{C}[\![y]\!]\rho v &= \text{let } cnt(\kappa) = (\rho\; y) \text{ in } \kappa v \\
\mathcal{C}[\![\{C_1, C_2\}]\!]\rho v &= \text{case } v \text{ of } in_1(t) : \mathcal{C}[\![C_1]\!]\rho t \;[\!]\; in_2(t) : \mathcal{C}[\![C_2]\!]\rho t \text{ esac} \\
\mathcal{C}[\![\{\}]\!]\rho v &= \text{case } v \text{ of esac} \\
\mathcal{C}[\![C \downarrow F]\!]\rho v &= \mathcal{F}[\![F]\!]\rho v\; (\lambda t.\mathcal{C}[\![C]\!]\rho t) \\
\mathcal{C}[\![F]\!]\rho v &= \text{let } contx(a, c) = v \text{ in } \mathcal{F}[\![F]\!]\rho ac
\end{aligned}
$$

$$
\begin{aligned}
\mathcal{F} : F &\to Env \to Val \to Cnt \to Ans \\
\mathcal{F}[\![p]\!]\rho v\kappa &= \kappa(pv) \\
\mathcal{F}[\![X \Rightarrow E]\!]\rho v\kappa &= \mathcal{E}[\![E]\!]\; ([\mathcal{X}[\![X]\!] \mapsto v]\rho)\kappa \\
\mathcal{F}[\![Y \Leftarrow C]\!]\rho v\kappa &= \mathcal{C}[\![C]\!]\; ([\mathcal{Y}[\![Y]\!] \mapsto \kappa]\rho)v \\
\mathcal{F}[\![E]\!]\rho v\kappa &= \mathcal{E}[\![E]\!]\rho\; (\lambda t.\text{let } closr(f) = t \text{ in } fv\kappa) \\
\mathcal{F}[\![C]\!]\rho v\kappa &= \mathcal{C}[\![C]\!]\rho\; contx(v, \kappa)
\end{aligned}
$$

$$
\begin{aligned}
\mathcal{X} : X &\to Val \to Env \to Env \\
[\mathcal{X}[\![x]\!] \mapsto v]\rho &= [x \mapsto val(v)]\rho \\
[\mathcal{X}[\![()]\!] \mapsto v]\rho &= \text{let } unit() = v \text{ in } \rho \\
[\mathcal{X}[\![(X_1, X_2)]\!] \mapsto v]\rho &= \text{let } pair(v_1, v_2) = v \text{ in } [\mathcal{X}[\![X_1]\!] \mapsto v_1, \mathcal{X}[\![X_2]\!] \mapsto v_2]\rho
\end{aligned}
$$

$$
\begin{aligned}
\mathcal{Y} : Y &\to Cnt \to Env \to Env \\
[\mathcal{Y}[\![y]\!] \mapsto \kappa]\rho &= [y \mapsto cnt(\kappa)]\rho \\
[\mathcal{Y}[\![\{\}]\!] \mapsto \kappa]\rho &= \rho \\
[\mathcal{Y}[\![\{Y_1, Y_2\}]\!] \mapsto \kappa]\rho &= [\mathcal{Y}[\![Y_1]\!] \mapsto (\lambda v.\kappa\; in_1(v)),\; \mathcal{Y}[\![Y_2]\!] \mapsto (\lambda v.\kappa\; in_2(v))]\rho
\end{aligned}
$$

Figure 2: A CBV semantics for the SLC

$$\frac{}{p_i : S \to T} \qquad \frac{}{id : A \to A} \qquad \frac{f : A \to B \quad g : B \to C}{g \circ f : A \to C}$$

$$\frac{}{\diamond : A \to 1} \qquad \frac{}{x_i : 1 \to T} \qquad \frac{}{y_i : S \to 0} \qquad \frac{}{\square : 0 \to A}$$

$$\frac{f : C \to A \quad g : C \to B}{\langle f, g \rangle : C \to A \times B} \qquad \frac{f : A \to C \quad g : B \to C}{[f, g] : A + B \to C}$$

$$\frac{}{\pi_1 : A \times B \to A} \qquad \frac{}{\pi_2 : A \times B \to B} \qquad \frac{}{\iota_1 : A \to A + B} \qquad \frac{}{\iota_2 : B \to A + B}$$

$$\frac{f : A \times B \to C}{f^* : A \to [B \to C]} \qquad \frac{f : C \to A + B}{f_* : [B \leftarrow C] \to A}$$

$$\frac{}{ap : [B \to C] \times B \to C} \qquad \frac{}{pa : C \to [B \leftarrow C] + B}$$

$$\frac{}{\phi : A \times [C \leftarrow B] \to [C \leftarrow A \times B]} \qquad \frac{}{\theta : [C \to A + B] \to A + [C \to B]}$$

Figure 3: The SCL morphisms

a *Symmetric Combinatory Logic, SCL.*

We need a category with a fair amount of structure, *i.e.*, a number of distinguished objects and morphisms, that will act as products, exponentials, projections, etc. While we will use the traditional names and notation for these, it must be remembered that their associated axioms are somewhat weaker than the usual ones, in that they sometimes only quantify over *well-behaved morphisms.* In fact, a category that obeyed all the axioms in full generality would collapse to a Boolean algebra.

The objects include all the basic types (integers, booleans etc). In addition to those, there must be an *initial object* 0 and a *terminal object* 1. Furthermore, for every pair of objects A and B, there must exist a specific *product* $A \times B$, *coproduct* $A + B$, *exponential* $[A \to B]$ and *coexponential* $[A \leftarrow B]$ object.

For each type, the category includes a denumerable set of value and continuation variables, corresponding to free variables in SLC terms. SLC constants can be thought of as either primitive functions $1 \to T$ or global (value) variables. Figure 3 gives a concise specification of all the morphisms in the category.

4.1 Axioms of SCL

Of course, the usual axioms of a category (neutrality of identity and associativity of composition) must be satisfied for *all* morphisms, even "badly-behaved" ones.

We now present the axioms specific to SCL. For this, we must first formalize

$$
\begin{array}{rclrcl}
\Diamond_1 & = & id_1 & \Box_0 & = & id_0 \\
\pi_1 \circ \langle f, t \rangle & = & f & [f, s] \circ \iota_1 & = & f \\
\pi_2 \circ \langle t, g \rangle & = & g & [s, g] \circ \iota_2 & = & g \\
\langle \pi_1, \pi_2 \rangle & = & id & [\iota_1, \iota_2] & = & id \\
\langle f, g \rangle \circ t & = & \langle f \circ t, g \circ t \rangle & s \circ [f, g] & = & [s \circ f, s \circ g] \\
(t_1 \times t_2) \circ \langle f, g \rangle & = & \langle t_1 \circ f, t_2 \circ g \rangle & [f, g] \circ (s_1 + s_2) & = & [f \circ s_1, g \circ s_2] \\
ap \circ (f^* \times id) & = & f & (f_* + id) \circ pa & = & f \\
(ap \circ (t \times id))^* & = & t & ((s + id) \circ pa)_* & = & s \\
\phi \circ \langle f \circ \Diamond, id \rangle & = & (pa \circ \langle f \circ \Diamond, id \rangle)_* & [\Box \circ f, id] \circ \theta & = & ([\Box \circ f, id] \circ ap)^* \\
(f \circ \pi_2)_* \circ \phi & = & f_* \circ \pi_2 & \theta \circ (\iota_2 \circ f)^* & = & \iota_2 \circ f^*
\end{array}
$$

Figure 4: The SCL axioms

the concept of well-behaved morphisms. Basically, they are precisely the ones which respect the (weak) initial and terminal objects of the category:

Definition A morphism $f : A \to B$ will be called *strict*, if $f \circ \Box_A = \Box_B$, and *total* or *costrict* if $\Diamond_B \circ f = \Diamond_A$.

The names are based on familiar terms, but have slightly different meanings. Under CBV, 0 is a (true) initial object, which means that *all* morphisms are strict. A total morphism is basically one without any side effects, including non-termination.

Under CBN, 1 is a (true) terminal object, so that all morphisms are said to be "total". A strict morphism in CBN is one which always evaluates its input. Note that this is slightly stronger than the usual definition of strictness ($f \uparrow \bot = \bot$), so that *e.g.*, the morphism denoted by $x \Rightarrow \bot$ is *not* strict according to our definition.

All basic morphisms are both strict and total, except that *pa* needs not be strict, and *ap* needs not be total. All morphisms constructed from strict (total) morphisms are strict (total), except that f^* is always total, but may not be strict, and f_* is always strict, but may not be total. In the SCL axioms of figure 4, t's denote total morphisms and s's strict ones.

4.2 Translating from SCL to SLC

Throughout section 1, it was demonstrated how categorical constructs such as products, exponentials, *etc.* could be naturally expressed with SLC terms. In figure 5, we formalize this idea, giving a translation from every SCL morphism to an equivalent SLC function.

$$
\begin{array}{rclrcl}
\mathcal{L}[\![p]\!] &=& p \;=\; x \Rightarrow p \uparrow x & \mathcal{L}[\![p]\!] &=& p \;=\; y \Leftarrow y \downarrow p \\
\mathcal{L}[\![id]\!] &=& x \Rightarrow x & \mathcal{L}[\![id]\!] &=& y \Leftarrow y \\
\mathcal{L}[\![f \circ g]\!] &=& x \Rightarrow \mathcal{L}[\![f]\!] \uparrow (\mathcal{L}[\![g]\!] \uparrow x) & \mathcal{L}[\![f \circ g]\!] &=& y \Leftarrow (y \downarrow \mathcal{L}[\![f]\!]) \downarrow \mathcal{L}[\![g]\!] \\
\mathcal{L}[\![x]\!] &=& () \Rightarrow x & \mathcal{L}[\![y]\!] &=& \{\} \Leftarrow y \\
\mathcal{L}[\![\Diamond]\!] &=& x \Rightarrow () & \mathcal{L}[\![\Box]\!] &=& y \Leftarrow \{\} \\
\mathcal{L}[\![\langle f,g \rangle]\!] &=& x \Rightarrow (\mathcal{L}[\![f]\!] \uparrow x, \mathcal{L}[\![g]\!] \uparrow x) & \mathcal{L}[\![[f,g]]\!] &=& y \Leftarrow \{y \downarrow \mathcal{L}[\![f]\!], y \downarrow \mathcal{L}[\![g]\!]\} \\
\mathcal{L}[\![\pi_1]\!] &=& (a,b) \Rightarrow a & \mathcal{L}[\![\iota_1]\!] &=& \{a,b\} \Leftarrow a \\
\mathcal{L}[\![\pi_2]\!] &=& (a,b) \Rightarrow b & \mathcal{L}[\![\iota_2]\!] &=& \{a,b\} \Leftarrow b \\
\mathcal{L}[\![f^*]\!] &=& a \Rightarrow b \Rightarrow \mathcal{L}[\![f]\!] \uparrow (a,b) & \mathcal{L}[\![f_*]\!] &=& a \Leftarrow b \Leftarrow \{a,b\} \downarrow \mathcal{L}[\![f]\!] \\
\mathcal{L}[\![ap]\!] &=& (g,b) \Rightarrow g \uparrow b & \mathcal{L}[\![pa]\!] &=& \{g,b\} \Leftarrow b \downarrow g \\
\mathcal{L}[\![\phi]\!] &=& & \mathcal{L}[\![\theta]\!] &=& \\
\multicolumn{3}{r}{(x,q) \Rightarrow (p \Leftarrow (c \Rightarrow p \uparrow (x,c))) \uparrow q} & \multicolumn{3}{r}{\{y,p\} \Leftarrow p \downarrow (q \Rightarrow (c \Leftarrow \{y,c\} \downarrow q))}
\end{array}
$$

Figure 5: Translation from SCL to SLC

Naturally, the variables introduced by the rules must not capture any free variables in the categorical term. Note that for identity and composition, we have two exactly equivalent translations, either as value or continuation abstractions. The translations for primitive functions demonstrate the η and $\bar{\eta}$ conversion rules of the SLC.

4.3 Translating from SLC to SCL

While the above result was reasonably straightforward, the inverse translation is somewhat more complicated. The main difference between SLC and SCL is that the latter has no variable bindings. We must therefore express all value and continuation abstractions in categorical terms: for every expression $E : T$, possibly containing a variable x of type S, we must have a function $F : S \to T$, s.t. $F \uparrow x = E$, and the dual result for continuations. That such functions always exist depends on the following property of the SCL category:

Theorem (Functional Completeness) Let $f : A \to B$ be a morphism expressed in terms of the S-typed (value) variable x. Then there exists a unique morphism $f^x : S \times A \to B$ not containing x, such that $f^x \circ \langle x \circ \Diamond, id \rangle = f$. Dually, for a variable $y : T \to 0$, there exists a unique morphism $f_y : A \to T + B$, such that $[\Box \circ y, id] \circ f_y = f$.

Proof For lack of space, we will only sketch the structure of the proof. The detailed verifications are rather verbose, but simple in principle. In fact, most of

$$
\begin{aligned}
x^x &= \pi_1 & y_y &= \iota_1\\
f^x &= f \circ \pi_2 \text{ (if } x \text{ not in } f) & f_y &= \iota_2 \circ f \text{ (if } y \text{ not in } f)\\
(f \circ g)^x &= f^x \circ \langle \pi_1, g^x \rangle & (f \circ g)_y &= [\iota_1, f_y] \circ g_y\\
\langle f, g \rangle^x &= \langle f^x, g^x \rangle & \langle f, g \rangle_y &= \bar{\delta} \circ \langle f_y, g_y \rangle\\
[f, g]^x &= [f^x, g^x] \circ \delta & [f, g]_y &= [f_y, g_y]\\
f^{*x} &= (f^x \circ \alpha)^* & f^*{}_y &= \theta \circ (f_y)^*\\
f_*{}^x &= (f^x)_* \circ \phi & f_{*y} &= (\bar{\alpha} \circ f_y)_*
\end{aligned}
$$

Figure 6: Abstraction rules

the CCC results from [Lambek & Scott 86] can be carried over directly, verifying that only the weaker forms of the axioms are used. Since the SCL axioms are symmetrical, the results dualize immediately to continuation abstractions.

First, we define a number of auxiliary morphisms, expressing associativity, commutativity and distributivity of products and coproducts:

$$
\begin{aligned}
\alpha &\equiv \langle \pi_1 \circ \pi_1, \langle \pi_2 \circ \pi_1, \pi_2 \rangle \rangle : (A \times B) \times C \to A \times (B \times C)\\
\bar{\alpha} &\equiv [\iota_1 \circ \iota_1, [\iota_1 \circ \iota_2, \iota_2]] : A + (B + C) \to (A + B) + C\\
\gamma &\equiv \langle \pi_2, \pi_1 \rangle : A \times B \to B \times A\\
\bar{\gamma} &\equiv [\iota_2, \iota_1] : A + B \to B + A\\
\delta &\equiv ap \circ ([(\iota_1 \circ \gamma)^*, (\iota_2 \circ \gamma)^*] \times id) \circ \gamma : A \times (B + C) \to (A \times B) + (A \times C)\\
\bar{\delta} &\equiv \bar{\gamma} \circ (\langle (\bar{\gamma} \circ \pi_1)_*, (\bar{\gamma} \circ \pi_2)_* \rangle + id) \circ pa : (A + B) \times (A + C) \to A + (B \times C)
\end{aligned}
$$

Of these, the first four are isomorphisms in any SCL category. δ and $\bar{\delta}$ are not, in general, but they do satisfy the following equations:

$$
\delta \circ \langle t \circ \Diamond, id \rangle = \langle t \circ \Diamond, id \rangle + \langle t \circ \Diamond, id \rangle \qquad [\Box \circ s, id] \circ \bar{\delta} = [\Box \circ s, id] \times [\Box \circ s, id]
$$

$$
(\pi_2 + \pi_2) \circ \delta = \pi_2 \qquad \bar{\delta} \circ (\iota_2 \times \iota_2) = \iota_2
$$

We can now define the *abstraction rules* of figure 6.

First, we show that this really is a definition, *i.e.*, that the result is not dependent on the order of application of the rules, *e.g.*, if x does not occur in $h = f \circ g$,

$$
(f \circ g)^x = f^x \circ \langle \pi_1, g^x \rangle = f \circ \pi_2 \circ \langle \pi_1, g \circ \pi_2 \rangle = f \circ g \circ \pi_2 = h^x
$$

We can then show that $f^x \circ \langle x \circ \Diamond, id \rangle = f$ by structural induction on f.

Finally, we prove uniqueness of f^x, *i.e.*, given an x-free morphism g, s.t. $g \circ \langle x \circ \Diamond, id \rangle = f$, we show that $g = f^x$. This is actually quite simple:

$$
\begin{aligned}
f^x = (g \circ \langle x \circ \Diamond, id \rangle)^x = g^x \circ \langle \pi_1, \langle x \circ \Diamond, id \rangle^x \rangle = g \circ \pi_2 \circ \langle \pi_1, \langle (x \circ \Diamond)^x, id^x \rangle \rangle =\\
g \circ \langle x^x \circ \langle \pi_1, \Diamond^x \rangle, \pi_2 \rangle = g \circ \langle \pi_1 \circ \langle \pi_1, \Diamond \circ \pi_2 \rangle, \pi_2 \rangle = g \circ \langle \pi_1, \pi_2 \rangle = g
\end{aligned}
$$

$$
\begin{array}{rcl@{\qquad}rcl}
\mathcal{E}[\![x]\!] &=& x & \mathcal{C}[\![y]\!] &=& y \\
\mathcal{E}[\![()]\!] &=& \Diamond & \mathcal{C}[\![\{\}]\!] &=& \Box \\
\mathcal{E}[\![(E_1,E_2)]\!] &=& \langle \mathcal{E}[\![E_1]\!], \mathcal{E}[\![E_2]\!] \rangle & \mathcal{C}[\![\{C_1,C_2\}]\!] &=& [\mathcal{C}[\![C_1]\!], \mathcal{C}[\![C_2]\!]] \\
\mathcal{E}[\![F \uparrow E]\!] &=& \mathcal{F}[\![F]\!] \circ \mathcal{E}[\![E]\!] & \mathcal{C}[\![C \downarrow F]\!] &=& \mathcal{C}[\![C]\!] \circ \mathcal{F}[\![F]\!] \\
\mathcal{E}[\![F]\!] &=& (\mathcal{F}[\![F]\!] \circ \pi_2)^* & \mathcal{C}[\![F]\!] &=& (\iota_2 \circ \mathcal{F}[\![F]\!])_* \\
\\
\mathcal{F}[\![p]\!] &=& p & \mathcal{F}[\![p]\!] &=& p \\
\mathcal{F}[\![X \Rightarrow E]\!] &=& \mathcal{E}[\![E]\!]^{\mathcal{X}[\![X]\!]} \circ \langle id, \Diamond \rangle & \mathcal{F}[\![Y \Leftarrow C]\!] &=& [id, \Box] \circ \mathcal{C}[\![C]\!]_{\mathcal{Y}[\![Y]\!]} \\
\mathcal{F}[\![E]\!] &=& ap \circ \langle \mathcal{E}[\![E]\!] \circ \Diamond, id \rangle & \mathcal{F}[\![C]\!] &=& [\Box \circ \mathcal{C}[\![C]\!], id] \circ pa \\
\\
f^{\mathcal{X}[\![x]\!]} &=& f^x & f_{\mathcal{Y}[\![y]\!]} &=& f_y \\
f^{\mathcal{X}[\![()]\!]} &=& f \circ \pi_2 & f_{\mathcal{Y}[\![\{\}]\!]} &=& \iota_2 \circ f \\
f^{\mathcal{X}[\![(X_1,X_2)]\!]} &=& (f^{\mathcal{X}[\![X_2]\!]})^{\mathcal{X}[\![X_1]\!]} \circ \alpha & f_{\mathcal{Y}[\![\{Y_1,Y_2\}]\!]} &=& \alpha' \circ (f_{\mathcal{Y}[\![Y_2]\!]})_{\mathcal{Y}[\![Y_1]\!]}
\end{array}
$$

Figure 7: Translation from SLC to SCL

Corollary For every morphism $e : 1 \to T$ containing a (value) variable $x : S$, there exists a unique morphism $f : S \to T$, s.t. $f \circ x = e$. Similarly, for every morphism $c : S \to 0$, in a (continuation) variable $y : T$, there exists a unique $g : S \to T$, s.t. $y \circ g = c$.

Proof Set $f = e^x \circ \langle id, \Diamond \rangle$ and $g = [id, \Box] \circ c_y$. ∥

This result makes it possible to give a full translation from SLC terms to morphisms in SCL, where every function $F : S \to T$, expression $E : T$ and continuation $C : S$ denotes a unique $f : S \to T$, $e : 1 \to T$ and $c : S \to 0$, respectively. The equations are given in figure 7.

The translation above is independent of the evaluation strategy. Consider for example the SLC function $(x \Rightarrow 2) \circ (y \Leftarrow k)$, where k is a continuation constant or variable from an outer scope. For both definitions of composition in SLC, its denotation is the morphism $2 \circ \Diamond \circ \Box \circ k$. In a CBN strategy, where 1 is a terminal object, this is equivalent to $2 \circ \Diamond$, while under CBV, where 0 is initial, it is the same as $\Box \circ k$.

Among other things, the translations $\mathcal{F}$ and $\mathcal{L}$ make it possible to mix the two styles freely, using categorical morphisms and morphism constructors as SLC functions, or value and continuation abstractions as SCL morphisms. For instance, the mixed term $f \Rightarrow f^* : [A \times B \to C] \to [A \to [B \to C]]$ can be seen as abbreviating either the SLC function $f \Rightarrow a \Rightarrow b \Rightarrow f \uparrow (a,b)$ or the SCL morphism $((ap \circ \alpha)^*)^*$.

This also shows that a (non-trivial) SCL category exists, since we can translate every SCL morphism to a SLC function which can be interpreted by the denotational semantics of section 3. It is more natural, however, to use an environment-free semantics interpreting SCL morphisms directly, and we will do so in the next section.

Let us also note that a SLC (and, as a special case, a λ-calculus) term that only uses the weak forms of the axioms will have the same meaning for every evaluation strategy. In particular, this restriction is satisfied by a language semantics written in continuation-passing style, so that the meaning of a construct does not depend on the evaluation order of the metalanguage.

5 A denotational semantics for categorical terms

In this section, a CBV denotational semantics for categorical terms is presented. It can be verified that this is equivalent to the SLC semantics of section 3, under the translations in section 4. The semantic domains and equations are given in figure 8. We note that there is only one valuation function $\mathcal{M}$, giving the denotation of SCL morphisms.

We can easily check that this semantics satisfies all of the axioms required for a SCL category. In fact, the continuation-based axioms (coproduct, *etc.*) hold in full generality, since *all* morphisms are strict in this semantics.

We also note that the above semantics is closely related to the Kleisli category [Lambek & Scott 86] of the triple (T, η, μ), where $T(A) = (A \rightarrow Ans) \rightarrow Ans$, $T(f) = \lambda tc.t(\lambda a.c(fa))$, $\eta(A) = \lambda ac.ca$, and $\mu(A) = \lambda pc.p(\lambda t.tc)$. In particular, the composition of two morphisms $f : A \rightarrow T(B)$ and $g : B \rightarrow T(C)$ is given by $g * f = \mu(C) \circ T(g) \circ f : A \rightarrow T(C)$, and $\eta(A) : A \rightarrow T(A)$ becomes the identity morphism in the Kleisli category.

6 Evaluation strategies

In the previous sections, we have mostly considered a CBV strategy. In the following, we will briefly (for lack of space) consider other possibilities.

In a CBV semantics, values are atomic, while continuations are functions mapping values to answers. We may consider the opposite concept, where continuations are atomic, and values are functions mapping continuations to answers. This is basically a CBN semantics, where function composition is defined as

$$\mathcal{M}[\![f \circ g]\!]\nu k = \mathcal{M}[\![f]\!](\lambda c.\mathcal{M}[\![g]\!]\nu c)k$$

In the CBN category, the usual axioms of a CCC hold in full generality, as all functions are costrict. However, we lose the general coproducts, etc. of CBV. In fact, we only have true coproducts for *strict* functions. Again, this is not specific to the SLC but is true for any CBN-like language, where, for example, $[f \circ \iota_1, f \circ \iota_2]$ is not in general semantically equal to f, if f is a non-strict function. For example,

$$
\begin{aligned}
Val &= Bas + Unit() + Pair(Val \times Val) + In_1(Val) + In_2(Val) + \\
&\quad Closr(Val \to Cnt \to Ans) + Contx(Val \times Cnt) \\
Cnt &= Val \to Ans
\end{aligned}
$$

$$
\begin{aligned}
\mathcal{M} : M &\to Val \to Cnt \to Ans \\
\mathcal{M}[\![x]\!]v\kappa &= \text{let } val(a) = \rho_{init}\, x \text{ in } \kappa\, a \\
\mathcal{M}[\![y]\!]v\kappa &= \text{let } cnt(c) = \rho_{init}\, y \text{ in } c\, v \\
\mathcal{M}[\![p]\!]v\kappa &= \kappa(pv) \\[1ex]
\mathcal{M}[\![id]\!]v\kappa &= \kappa\, v \\
\mathcal{M}[\![F \circ G]\!]v\kappa &= \mathcal{M}[\![G]\!]v(\lambda t.\mathcal{M}[\![F]\!]t\kappa) \\[1ex]
\mathcal{M}[\![\Diamond]\!]v\kappa &= \kappa\, unit() \\
\mathcal{M}[\![\langle f, g\rangle]\!]v\kappa &= \mathcal{M}[\![f]\!]v(\lambda s.\mathcal{M}[\![g]\!]v(\lambda t.\kappa\, pair(s,t))) \\
\mathcal{M}[\![\pi_1]\!]v\kappa &= \text{let } pair(v_1, v_2) = v \text{ in } \kappa\, v_1 \\
\mathcal{M}[\![\pi_2]\!]v\kappa &= \text{let } pair(v_1, v_2) = v \text{ in } \kappa\, v_2 \\[1ex]
\mathcal{M}[\![\Box]\!]v\kappa &= \text{case } v \text{ of esac} \\
\mathcal{M}[\![[f, g]]\!]v\kappa &= \text{case } v \text{ of } in_1(t) : \mathcal{M}[\![f]\!]t\kappa \mathbin{[\!]} in_2(t) : \mathcal{M}[\![g]\!]t\kappa \text{ esac} \\
\mathcal{M}[\![\iota_1]\!]v\kappa &= \kappa\, in_1(v) \\
\mathcal{M}[\![\iota_2]\!]v\kappa &= \kappa\, in_2(v) \\[1ex]
\mathcal{M}[\![f^*]\!]v\kappa &= \kappa\, closr(\lambda bc.\mathcal{M}[\![f]\!]pair(v, b)\; c) \\
\mathcal{M}[\![ap]\!]v\kappa &= \text{let } pair(g, a) = v \text{ in let } closr(f) = g \text{ in } fa\kappa \\
\mathcal{M}[\![\phi]\!]v\kappa &= \text{let } pair(x, q) = v \text{ in let } contx(a, c) = q \text{ in } \kappa\, contx(pair(x, a), c) \\[1ex]
\mathcal{M}[\![f_*]\!]v\kappa &= \text{let } contx(a, c) = v \text{ in} \\
&\qquad \mathcal{M}[\![f]\!]a(\lambda t.\text{case } t \text{ of } in_1(r) : \kappa r \mathbin{[\!]} in_2(s) : cs \text{ esac}) \\
\mathcal{M}[\![pa]\!]v\kappa &= \kappa\, in_1(contx(v, \lambda t.\kappa\, in_2(t))) \\
\mathcal{M}[\![\theta]\!]v\kappa &= \text{let } closr(f) = v \text{ in} \\
&\qquad \kappa\, in_2(closr(\lambda ac.fa(\lambda t.\text{case } t \text{ of } in_1(r) : \kappa t \mathbin{[\!]} in_2(s) : cs \text{ esac})))
\end{aligned}
$$

Figure 8: A CBV semantics of SCL

if $f \equiv x \Rightarrow 2$, the case expression above will evaluate its argument until at least the injection tag is known, but f itself will not.

A number of other characteristic properties of the CBV strategy are also pointed out by considering their duals. For example, we can use pattern-matching definitions dispatching on the required result of the function, in a style similar to [Hagino 87]. Also, CBN does not necessarily terminate more often than CBV: under CBV, we may discard a result continuation leading to an infinite loop, just as a CBN strategy may discard a non-terminating value. Again, for lack of space, we cannot go into details here; a full account will be given in [Filinski 89].

Conclusion and issues

We have presented a categorical framework for reasoning about continuations as a declarative concept. This suggests a number of possible practical applications and ideas for future research in both design and implementation of programming languages:

- *Symmetrical programming languages:* The SLC is not really a practical programming language, even though actual implementations exist, and have been used for experiments. Just like the λ-calculus, it requires a fair amount of preprocessing ("syntactic sugar") for realistic applications. It should be possible, however, to build a shell, in the style of, *e.g.*, ML, and create a useful language, which would have continuations as a basic descriptive primitive, rather than an imperative add-on like exceptions.

- *SLC and SCL as metalanguages:* Even though the SLC is quite terse, it operates at a somewhat higher level of abstraction than the simple λ-calculus. Therefore, it should be easier (with practice) to give denotations of traditional language constructs in SLC rather than as λ-calculus terms, at least partially bridging the "semantic gap". It is also conceivable that the SCL system could be used as a basis for an abstract machine, similar to the CAM, but with a firm categorical foundation for handling imperative aspects of control flow, such as run-time errors, exceptions, etc. In fact, for most current machine architectures, flexible handling of continuations for expressing flow of control is just as essential as value manipulation.

- *Exploring duals of existing ideas:* The symmetry provided by SLC descriptions makes it possible to reason clearly about duals of many programming language terms. For example, we may consider how the the class-based inheritance of Simula or Smalltalk could be dualized to coproduct types, and whether this would be a useful construct to integrate in a real programming language. Other ideas, such as the pattern-matching function definitions of Hope and ML also appear to have useful dualizations.

As another example, consider the program transformation technique of partial evaluation: given a program computing a function $f : A \times B \rightarrow C$, and a known value a, we may be able to construct a more efficient program, computing a function $f' : B \rightarrow C$, such that $f(a,b) = f'(b)$. There is a dual version of this problem: given a program for a function $g : C \rightarrow A + B$, and a known continuation a, we may be able to optimize the program by propagating the knowledge of what will happen to the A-typed result back through the definition of g.

- *Nondeterminism:* The SLC already contains the framework for nondeterministic programming, as it is possible to capture a continuation, and invoke it several times with different values. Thus, with a relatively simple translation, we can give SLC denotations of backtracking programs. It appears that the call-by-need optimization of lazy languages can be dualized to avoid re-traversing deterministic parts of the continuation.

- *Abstract Interpretation:* Finally, we may consider abstract interpretation of SLC terms, for constant propagation, strictness analysis etc. The SLC framework presents a unifying view of *forward* and *backward* analyses, as interpretations with abstract values and abstract continuations, respectively. The two equivalent definitions of function composition, in terms of either the value or the continuation exactly mirror these two uses.

  The categorical framework of SCL also suggests a dual to strictness analysis, namely *totality* analysis: the knowledge that a function will preserve its continuation can be used for optimizing, as a number of SCL axioms that do not necessarily hold in general can then be used for transformations.

- *Connections with Linear Logic:* [Girard 87] presents a deductive system called Linear Logic, which exhibits many similarities with SCL. It has a form of negation as a fundamental primitive, and this induces a question/answer symmetry similar to SCL's. Also, it recognizes two kinds of "multiplicatives" and "additives" which seem strongly connected with the eager and lazy products and coproducts of SCL, a topic not covered in this paper.

  However, this work (*classical* linear logic) appears to involve only proof theory, not general computation. In particular, it does not impose any structure on different morphisms between objects (although the *intuitionistic* subset of LL has been extended to this, *cf.* the Linear Abstract Machine [Lafont 88].). Another difference with SCL is that the latter does not consider a type and its "negation" as two different objects in the same category, but expresses the duality through the morphism structure instead. Finally, it is not clear how to introduce recursion in LL without losing consistency.

Acknowledgement

The author wishes to thank his thesis advisor, Olivier Danvy, for many long and fruitful discussions about the topics presented in this paper. Acknowledgements are also due to Hans Dybkjær, Neil D. Jones, Karoline Malmkjær, Austin Melton and Torben Mogensen for their patient rereading of various revisions of this paper and numerous helpful suggestions.

References

[Curien 86] P-L. Curien: *Categorical Combinators, Sequential Algorithms and Functional Programming*, Research Notes in Theoretical Computer Science, Vol. 1, Pitman, 1986

[Filinski 89] A. Filinski: *Master's thesis (forthcoming)*, DIKU, University of Copenhagen, June 1989

[Girard 87] J-Y. Girard: *Linear Logic*, Theoretical Computer Science, Vol. 50 (1987), pp 1–102

[Hagino 87] T. Hagino: *A Typed Lambda Calculus with Categorical Type Constructors*, Proc. Summer Conference on Category Theory and Computer Science, Edinburgh, pp 141-156, LNCS 283

[Hewitt 79] C. Hewitt: *Control Structure as Patterns of Passing Messages*, Artificial Intelligence: An MIT Perspective, Vol. 2, pp 434-465, MIT Press, 1979

[Huwig & Poigné 86] H. Huwig, A. Poigné: *A Note on Inconsistencies Caused by Fixpoints in a Cartesian Closed Category*, unpublished manuscript (1986)

[Lafont 88] Y. Lafont: *The Linear Abstract Machine*, Theoretical Computer Science, Vol. 59 (1988), pp 157–180

[Lambek & Scott 86] J. Lambek and P.J. Scott: *Introduction to Higher Order Categorical Logic*, Cambridge studies in advanced mathematics, Vol. 7, Cambridge University Press, 1986

[Rees & Clinger 86] J. Rees and W. Clinger (eds): *Revised^3 Report on the Algorithmic Language Scheme*, Sigplan Notices, Vol. 21, No 12 pp 37-79 (December 1986)

[Rydeheard & Burstall 88] D.E. Rydeheard and R.M. Burstall: *Computational Category Theory*, Prentice Hall International Series in Computer Science, Prentice Hall, 1988

[Stoy 77] J.E. Stoy: *Denotational Semantics: The Scott-Strachey Approach to Programming Language Theory*, MIT Press, 1977

Logic Representation in LF

Report on work in progress*

Robert Harper† Donald Sannella‡ Andrzej Tarlecki§

Abstract

The purpose of a logical framework such as LF is to provide a language for defining logical systems suitable for use in a logic-independent proof development environment. In previous work we have developed a theory of representation of logics in a logical framework and considered the behaviour of structured theory presentations under representation. That work was based on the simplifying assumption that logics are characterized as families of consequence relations on "closed" sentences. In this report we extend the notion of logical system to account for open formula, and study its basic properties. Following standard practice, we distinguish two types of logical system of open formulae that differ in the treatment of free variables, and show how they may be induced from a logical system of closed sentences. The technical notions of a logic presentation and a uniform encoding of a logical system in LF are generalized to the present setting.

1 Introduction

The Logical Framework (LF) [HHP87] is a language for defining formal systems. The language is a three-level typed λ-calculus with Π-types, closely related to the AUTOMATH type theories [dB80, vD80]. A formal system is specified by giving an LF signature, a finite list of constant declarations that specifies the syntax, judgement forms, and inference rules of the system. All of the syntactic apparatus of a formal system, including proofs, are represented as LF terms. The LF type system is sufficiently expressive to capture the uniformities of a large class of logical systems of interest to computer science, including notions of schematic rules and proofs, derived rules of inference, and higher-order judgement forms expressing consequence and generality. Throughout this paper we assume a reasonably good acquaintance with the concepts and formalism of LF as presented in [HHP87].

In [HST89] we have studied a notion of representation of a logical system in LF. A logical system (or *logic*) is formalized in [HST89] as a family of consequence relations between *sentences* of the logical system uniformly defined over *signatures* of the system. To represent such an object logic $\mathcal{L}$ in LF, a uniform presentation of $\mathcal{L}$-signatures as extensions of an LF signature is required. Then, for each signature, $\mathcal{L}$-sentences over this signature are mapped to closed LF types of a specified form in such a way that this yields a full and faithful embedding of the consequence relation $\vdash^{\mathcal{L}}$ in the consequence relation $\vdash^{\mathcal{LF}}$ of LF. (The consequence relation of LF is given by considering inhabitation assertions, as in NuPRL [Con86].) By focusing on the embedding of logical systems, LF may be viewed as a "universal metalogic" in which all inferential activity is to be conducted: object logics exists (for the purposes of implementation) only insofar as they are encodable in LF. In

*Most definitions in this paper are rather tentative and may change at further stages of our work on these problems.

†School of Computer Science, Carnegie Mellon University, Pittsburgh, PA 15213, USA.

‡Laboratory for Foundations of Computer Science, Department of Computer Science, Edinburgh University, Edinburgh, Scotland.

§Institute of Computer Science, Polish Academy of Sciences, Warsaw, Poland.

[HST89] we have studied in detail the issues concerned in lifting inferential activity in object logics to LF via their representations. In particular we studied the problems of inference in theories presented in a structured way, much as in [SB83].

In our earlier work on logic representation [HST89] we focused on the notion of a logical system as a family of *simple consequence relations* [Avr87] satisfying certain natural closure conditions. For the sake of simplicity we considered only logics of "closed" sentences (referred to in this paper as *ground logical systems*), taking no explicit account of the behaviour of variables in a logic. Although it is difficult to say in general what are "closed" sentences, this assumption is perhaps best explained by noting that in our notion of representation, the sentences of a logical system are naturally (compositionally) encoded in LF as closed types. The purpose of this work is to remove this simplifying assumption by considering a notion of logical system that includes an explicit treatment of open formulae, and to consider the representation of such systems in a logical framework. It should be stressed that we are still making the simplifying assumption that logics are presented as consequence relations. In future work we intend to consider not just consequence, but also proofs.

Open formulae are not just a fancy feature we wish to add to our framework for laughs. First of all, an important motivation for our work is to adequately model logical concepts as described and used in mathematical logic, and open formulae certainly occur there. More specifically and perhaps even more importantly, open formulae are necessary to adequately study standard finite presentations of some common mathematical theories. For example, the usual presentation of Peano arithmetic includes the following axiom schema (induction schema):

$$P(0) \wedge \forall n.(P(n) \supset P(succ(n))) \supset \forall n.P(n)$$

This is schematic in P, which should not be interpreted as ranging over all closed sentences "with holes for n" only. For instance, the associativity of $+$, $\forall k, n, m.\, k + (n + m) = (k + n) + m$, is *not derivable* from such closed instances of the induction schema. It may be proved, however, "by induction on m" using an instance of the schema with k and n *free*.

This paper is organized as follows. In Section 2 we recall from [HST89] the definition of a ground logical system as a family of consequence relations indexed by signatures that satisfies a certain uniformity condition with respect to change of signature. This resembles the formalization of a logical system as an *institution* from [GB84]; the crucial difference is that institutions present a model-theoretic view of logical systems while our formulation is centered directly on the notion of a consequence relation. (See also [FS88], which is based on the notion of a closure operation, and [Mes89], which encompasses both model-theoretic and proof-theoretic points of view.) The sorts of consequence relations that we consider are motivated by the strictures of encoding in LF, and thus are limited to one-sided consequence relations that are closed under weakening, permutation, contraction, and cut, and which satisfy compactness. Generalizing the methodology of [HHP87], we introduce the notion of a *representation* of one logical system in another, taking account of variability in signatures.

In Section 3 we generalize these ideas and introduce the concept of a logical system of open formulae. Roughly, to each signature of the logical system is associated a category of contexts and substitutions (*cf.* [Car86]), and then to each signature and context over this signature is associated a set of formulae over this signature and context. Every formula is always considered with an explicit indication of its context. Hence, the consequence relation associated with each signature of a logical system of open formulae is defined on pairs consisting of a context (over this signature) and a formula built in this context. Of course, this consequence relation is required to satisfy the uniformity condition induced by the morphism structure of the category of signatures. In mathematical logic two types of consequence relations on open formulae are usually considered. Validity-type consequence relation rely on an implicit universal quantification of the free variables in each open formula. Truth-type consequence relations universally quantify free variables "globally" in the whole consequence statement. We characterize the two types of logical systems of open formulae by imposing appropriate structural conditions on their consequence relations.

In Section 4 we show how open formulas may be introduced to ground logical systems. This follows the view in abstract model theory that variables are uninterpreted constants (*cf.* [Bar74]), explored in a similar way in the theory of institutions in [Tar86], [ST88]. We try to justify the formal construction of Section 4 using a model-theoretic view provided via the theory of institutions in Section 5.

In Section 6 we introduce the metalogic of interest, LF, and view it as a logical system of open formulae in two different ways, guided by the validity and truth interpretations of LF contexts. We then define the notion of a logic presentation. A logic presentation is essentially an LF signature equipped with an indication of which LF contexts encode contexts of the object logic and which LF types encode the judgements of the object logic. Such a presentation induces again two logical systems: one of validity type, the other of truth type. A uniform encoding of an object logic in LF is a representation of the object logic in a logic presented in LF satisfying certain additional conditions ensuring adequacy of the encoding of the syntax. We also indicate that all the methodological suggestions on constructing logical systems in a structured way via structuring their presentations as suggested in [HST89] for ground logical systems carry over to this more general framework as well.

Finally, in Section 7 we suggest directions for future research. We stress once more that the current paper is just a report on work very much in progress. Thus, most of these suggestions are in fact research obligations to round off the technical ideas presented here.

2 Consequence relations and ground logical systems

In this section we recall the basic definitions used in [HST89] to capture the concept of a ground logical system and of a representation of one ground logical system in another. We start with some categorial preliminaries.

By a *category with inclusions* we mean any category $\mathcal{K}$ with a "wide" preorder subcategory of morphisms, which will be referred to as *inclusions*, such that the identity map on each object $A \in |\mathcal{K}|$ is the (unique) inclusion of A into itself. Inclusions are designated by $\iota : A \hookrightarrow B$. When convenient we will write the target object B as B^{ι}, and sometimes even identify the inclusion with its target (when the source is clear from the context). For any two objects $A, B \in |\mathcal{K}|$, we say that A *is included in* B, written $A \hookrightarrow B$, if there is an inclusion $\iota : A \hookrightarrow B$. In many particular cases that we study, morphisms are functions of some kind; in such cases we will normally assume without explicit mention that the inclusions are inclusions in the usual sense.

We will also assume that each category with inclusions $\mathcal{K}$ *has canonical pushouts along inclusions*, *i.e.*, whenever $f : A \to A'$ and $\iota : A \hookrightarrow A''$ are morphisms of $\mathcal{K}$, the pushout of f and ι exists, and, moreover, the morphism opposite the inclusion in the pushout diagram is itself an inclusion:

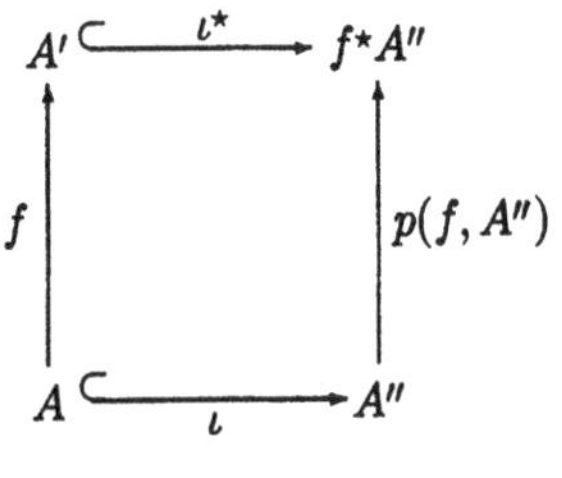

We require a canonical choice of $p(f, A'')$ (and $f^\star A''$) which is functorial in f, *i.e.*, $p(f;f', A'') = p(f, A''); p(f', f^\star A'')$ (dually to contextual categories, *cf.* [Car86]).

For any two morphisms $f : A \to B$ and $f' : A' \to B'$ in a category with inclusions $\mathcal{K}$, we say that f' *is an extension of* f if there are inclusions $\iota_A : A \hookrightarrow A'$ and $\iota_B : B \hookrightarrow B'$ such that $\iota_A;f' = f;\iota_B$. A family of morphisms $f_i : A_i \to B_i$, $i = 1, \ldots, n$, is *compatible* if for all $\{i_1, \ldots, i_k\} \subseteq \{1, \ldots, n\}$, for all objects A such that for $l = 1, \ldots, k$, $A \hookrightarrow A_{i_l}$, there is a morphism $f : A \to B$ (for some B) such that for $l = 1, \ldots, k$, f_{i_l} is an extension of f.

Throughout this paper, all functors between categories with inclusions will be assumed to preserve inclusions. The category of all categories with inclusions and inclusion-preserving functors will be denoted by **ICat**.

For any category with inclusions $\mathcal{K}$, *the category of functors into* $\mathcal{K}$, $\mathbf{Func}_{\mathbf{ICat}}(\mathcal{K})$, is defined as follows (*cf.* [TBG] where a similar category of diagrams in a given category is defined via an indexed category using the Grothendieck construction):

objects: are pairs $\langle \mathcal{I}, F \rangle$ consisting of a category with inclusions $\mathcal{I}$ and a functor $F : \mathcal{I} \to \mathcal{K}$.

morphisms: from $\langle \mathcal{I}, F \rangle$ to $\langle \mathcal{J}, G \rangle$ are pairs $\mu = \langle \mu^1, \mu^2 \rangle$ where $\mu^1 : \mathcal{I} \to \mathcal{J}$ is a functor and $\mu^2 : F \dot{\to} \mu^1;G$ is a natural transformation of functors in $\mathcal{I} \to \mathcal{K}$.

composition: is defined by[1]:

$$\langle \mu_1^1, \mu_1^2 \rangle;\langle \mu_2^1, \mu_2^2 \rangle = \langle \mu_1^1;\mu_2^1, \mu_1^2;(\mu_1^1;\mu_2^2) \rangle.$$

Our treatment of logical systems centers on consequence relations (see [Avr87] for a survey). We take a consequence relation to be a binary relation between finite subsets and elements of a set of "sentences" satisfying three conditions to be given below. We use φ and ψ to range over sentences, Φ to range over arbitrary sets of sentences, and Δ to range over finite sets of sentences. We write Δ, Δ' for union, and write φ, Δ for $\{\varphi\}, \Delta$. If $s : \Phi_1 \to \Phi_2$ is a function, then the extension of s to subsets of Φ_1 is denoted by s as well. Function application will often be denoted by juxtaposition, *e.g.*, $s\varphi$ stands for $s(\varphi)$.

Definition 2.1 *A* consequence relation (CR) *is a pair* $\langle S, \vdash \rangle$ *where* S *is a set of* sentences *and* $\vdash \subseteq Fin(S) \times S$ *is a binary relation such that*

1. *(Reflexivity)* $\varphi \vdash \varphi$;
2. *(Transitivity) If* $\Delta \vdash \varphi$ *and* $\varphi, \Delta' \vdash \psi$, *then* $\Delta, \Delta' \vdash \psi$.
3. *(Weakening) If* $\Delta \vdash \psi$, *then* $\varphi, \Delta \vdash \psi$.

If $S' \subseteq S$, *then* $\langle S, \vdash \rangle \restriction S'$ *is defined to be the consequence relation* $\langle S', \vdash \cap (Fin(S') \times S') \rangle$.

The choice of conditions on consequence relations is motivated by our intention to consider encodings of logical systems in LF (in a sense to be made precise below.)

Definition 2.2 *A* morphism of consequence relations (CR morphism) $s : \langle S_1, \vdash_1 \rangle \to \langle S_2, \vdash_2 \rangle$ *is a function* $s : S_1 \to S_2$ *(the* translation of sentences*) such that if* $\Delta \vdash_1 \varphi$, *then* $s\Delta \vdash_2 s\varphi$. *The CR morphism* s *is* conservative *if* $\Delta \vdash_1 \varphi$ *whenever* $s\Delta \vdash_2 s\varphi$. **CR** *is the category with inclusions whose objects are consequence relations and whose morphisms are CR morphisms. Identity, composition and inclusions are inherited from the category of sets. By* $|\_| : \mathbf{CR} \to \mathbf{Set}$ *we denote the functor which maps each consequence relation to its underlying set of sentences.*

CLEAR-like techniques for structuring theory presentations are based on the separation between the language of a theory and the set of axioms that generates it [BG80]. We therefore consider a logical system to be a family of consequence relations indexed by a collection of *signatures* which determine the language of a theory. Moreover, it is important for the development that consequence be preserved under variation in signature (for example, renaming constants or replacing constants by terms over another signature). This leads to the following definition:

[1] We use ";" to denote not only composition in a category (*e.g.*, the usual composition of functions and functors), written in diagrammatic order, but also both vertical composition of natural transformations and the composition of a natural transformation with a functor so that $(\mu_1^2;(\mu_1^1;\mu_2^2))_A = (\mu_1^2)_A;(\mu_2^2)_{\mu_1^1(A)}$.

Definition 2.3 *A* ground logical system, *or* ground logic, *is a functor* $\mathcal{G} : \mathbf{Sig}^{\mathcal{G}} \to \mathbf{CR}$ *where* $\mathbf{Sig}^{\mathcal{G}}$ *is a category with inclusions and* $\mathcal{G}$ *is an inclusion-preserving functor.*[2].

The category $\mathbf{Sig}^{\mathcal{G}}$ is called the *category of signatures* of $\mathcal{G}$, with objects denoted by Σ and morphisms by $\sigma : \Sigma_1 \to \Sigma_2$. A signature morphism $\sigma : \Sigma_1 \to \Sigma_2$ is to be thought of as specifying a "relative interpretation" of the language defined by Σ_1 into the language defined by Σ_2. Writing $\mathcal{G}(\Sigma) = (|\mathcal{G}|_\Sigma, \vdash^{\mathcal{G}}_\Sigma)$, the definition of logical system implies that if $\sigma : \Sigma_1 \to \Sigma_2$ and $\Delta \vdash^{\mathcal{G}}_{\Sigma_1} \varphi$, then $\mathcal{G}(\sigma)(\Delta) \vdash^{\mathcal{G}}_{\Sigma_2} \mathcal{G}(\sigma)(\varphi)$. The function $|\mathcal{G}|(\sigma)$ underlying the CR morphism is called the *translation function* induced by σ (we write $|\mathcal{G}|$ for the composition $\mathcal{G};|\_| : \mathbf{Sig}^{\mathcal{G}} \to \mathbf{Set}$). To simplify notation, we write $\sigma(\varphi)$ for $\mathcal{G}(\sigma)(\varphi)$ and $\sigma(\Delta)$ for $\mathcal{G}(\sigma)(\Delta)$ when no confusion is likely.

Definition 2.4 *The category of ground logical systems,* **GLog**, *is defined as* $\mathbf{Func}_{\mathbf{ICat}}(\mathbf{CR})$. *Hence, a morphism of ground logics* $\gamma : \mathcal{G} \to \mathcal{G}'$ *is a pair* $(\gamma^{Sig}, \gamma^{CR})$ *where* $\gamma^{Sig} : \mathbf{Sig}^{\mathcal{G}} \to \mathbf{Sig}^{\mathcal{G}'}$ *is a functor and* $\gamma^{CR} : \mathcal{G} \to \gamma^{Sig};\mathcal{G}' : \mathbf{Sig}^{\mathcal{G}} \to \mathbf{CR}$ *is a natural transformation.*

A morphism of ground logics is to be thought of as an "encoding" of one logical system in another in such a way that consequence is preserved. Let $\gamma : \mathcal{G} \to \mathcal{G}'$ be a morphism of ground logics. To simplify notation, we write $\gamma(\Sigma)$ for $\gamma^{Sig}(\Sigma)$, and $\gamma(\varphi)$ for $\gamma^{CR}_\Sigma(\varphi)$ (for appropriate choice of Σ).

Definition 2.5 *A ground logic morphism* $\gamma : \mathcal{G} \to \mathcal{G}'$ *is a* representation *if* γ^{Sig} *is an embedding and each* γ^{CR}_Σ *is conservative. A representation is* surjective *if each* γ^{CR}_Σ *is surjective as a function on the underlying sets.*

We refer to [HST89] for simple examples of ground logics and their representations.

3 Logical systems of open formulae

In the previous section we studied ground logical systems, where the logical sentences considered are closed (intuitively, built entirely out of the symbols given in the signature). These may perhaps be best characterized by referring to model theory: the truth of a closed sentence is unambigously determined by an interpretation of the symbols in the signature (*i.e.*, a model over this signature). In many logical systems studied in mathematical logic, however, logical formulae may additionally contain "free variables". To determine the truth of such a formula, an interpretation must be provided not only for the symbols in the signature, but also for the free variables (*i.e.*, not only a model, but also a "valuation" of the free variables in the model must be given). The free variables of an open formula, usually together with information on their typing, form the "context" in which the formula is built. We will avoid the sloppiness of leaving implicit the context in which an open formula is built, and always consider formulae over a given signature together with an explicitly indicated context. Moreover, where for ground logical systems we have assumed that signatures form a category and that signature morphisms induce translation of sentences, here we deal as well with a category of signatures, where signature morphisms induce translations of both contexts and formulae. Furthermore, the contexts over each signature are assumed to form a category (with morphisms which may be thought of as substitutions of terms for variables) and context morphisms then induce a translation of formulae. As usual, the translations induced are assumed to be mutually consistent.

Thus, the formulae of a logical system of open formulae are given by a functor

$$\mathcal{F} : \mathbf{Sig} \to \mathbf{Func}_{\mathbf{ICat}}(\mathbf{Set}).$$

The functor $\mathcal{F}$, for each signature $\Sigma \in |\mathbf{Sig}|$, yields a functor $\mathcal{F}(\Sigma) : \mathbf{Ctxt}_\Sigma \to \mathbf{Set}$, where $\mathbf{Ctxt}_\Sigma$ is a category of Σ-*contexts* with inclusions. Then, for any Σ-context $\Gamma \in |\mathbf{Ctxt}_\Sigma|$, the set $\mathcal{F}(\Sigma)(\Gamma)$ is a

[2] Of course this definition captures only one aspect of what is usually meant by the informal notion of "logical system."

set of Σ-*formulae in context* Γ. As mentioned before, we will always consider open formulae together with the context they are built in, and so we in fact will be dealing with the following set of open Σ-formulae:

$$Form_{\mathcal{F}}(\Sigma) = \{ \langle \Gamma, \varphi \rangle \mid \Gamma \in |\mathbf{Ctxt}_\Sigma|, \varphi \in \mathcal{F}(\Sigma)(\Gamma) \}.$$

The functor $\mathcal{F}$ also determines translations of contexts and formulae as mentioned above: for any signature morphism $\sigma : \Sigma \to \Sigma'$, $\mathcal{F}(\sigma)$ is a morphism in $\mathbf{Func}_{\mathbf{ICat}}(\mathbf{Set})$ from $\mathcal{F}(\Sigma) : \mathbf{Ctxt}_\Sigma \to \mathbf{Set}$ to $\mathcal{F}(\Sigma') : \mathbf{Ctxt}_{\Sigma'} \to \mathbf{Set}$. By definition (*cf.* Sec. 2), we thus have a functor $\mathcal{F}(\sigma)^1 : \mathbf{Ctxt}_\Sigma \to \mathbf{Ctxt}_{\Sigma'}$ and a natural transformation $\mathcal{F}(\sigma)^2 : \mathcal{F}(\Sigma) \dot{\to} \mathcal{F}(\sigma)^1;\mathcal{F}(\Sigma')$, and hence for each Σ-context Γ, a function $\mathcal{F}(\sigma)^2_\Gamma : \mathcal{F}(\Sigma)(\Gamma) \to \mathcal{F}(\Sigma')(\mathcal{F}(\sigma)^1(\Gamma))$. When no confusion is likely, we will write $\sigma(\Gamma)$ for $\mathcal{F}(\sigma)^1(\Gamma)$ (for any $\Gamma \in |\mathbf{Ctxt}_\Sigma|$), $\sigma_\Gamma(\varphi)$ for $\mathcal{F}(\sigma)^2_\Gamma(\varphi)$ (for any $\Gamma \in |\mathbf{Ctxt}_\Sigma|$ and $\varphi \in \mathcal{F}(\Sigma)(\Gamma)$) and $\gamma(\varphi)$ for $\mathcal{F}(\Sigma)(\gamma)(\varphi)$ (for any Σ-context morphism $\gamma : \Gamma \to \Gamma'$ and $\varphi \in \mathcal{F}(\Sigma)(\Gamma)$).

This in turn induces a natural extension of the map $\Sigma \mapsto Form_{\mathcal{F}}(\Sigma)$ to a functor

$$Form_{\mathcal{F}} : \mathbf{Sig} \to \mathbf{Set}$$

where for any signature morphism $\sigma : \Sigma \to \Sigma'$, for any $\langle \Gamma, \varphi \rangle \in Form_{\mathcal{F}}(\Sigma)$, $Form_{\mathcal{F}}(\sigma)(\langle \Gamma, \varphi \rangle) = \langle \sigma(\Gamma), \sigma_\Gamma(\varphi) \rangle$.

Definition 3.1 *A* logical system of open formulae *(or, a* logic of open formulae*, or simply a* logic*)* $\mathcal{L}$ *consists of*

- *a functor* $\mathcal{F}_\mathcal{L} : \mathbf{Sig}^\mathcal{L} \to \mathbf{Func}_{\mathbf{ICat}}(\mathbf{Set})$*, called the* formula functor of $\mathcal{L}$*, and*
- *a ground logical system* $C_\mathcal{L} : \mathbf{Sig}^\mathcal{L} \to \mathbf{CR}$*, called the* consequence functor of $\mathcal{L}$*,*

such that the underlying sentence functor $|C_\mathcal{L}| : \mathbf{Sig}^\mathcal{L} \to \mathbf{Set}$ *of the consequence functor of* $\mathcal{L}$ *coincides with the functor* $Form_{\mathcal{F}_\mathcal{L}} : \mathbf{Sig}^\mathcal{L} \to \mathbf{Set}$ *as determined by the formula functor of* $\mathcal{L}$*.*

Example 3.2 As a simple example of a logic of open formulae we present first-order logic. Since this is very standard and well known, we will omit many standard definitions and refer to the reader's intuition.

A *first-order signature* is a set of operation and predicate names with indicated arities (≥ 0). A first-order signature morphism maps operation names to terms of the same arity and renames predicate names preserving their arities. This defines the category $\mathbf{Sig}^{\mathcal{FO}}$ of first-order logic signatures.

For any first-order signature Σ, Σ-*contexts* are finite sets of variables, and so the category $\mathbf{Ctxt}^{\mathcal{FO}}_\Sigma$ of Σ-contexts has finite sets (of variables) as contexts and substitutions of terms with variables Y for variables X as morphisms from X to Y. Any first-order signature morphism $\sigma : \Sigma \to \Sigma'$ determines an obvious functor from $\mathbf{Ctxt}^{\mathcal{FO}}_\Sigma$ to $\mathbf{Ctxt}^{\mathcal{FO}}_{\Sigma'}$, which is the identity on objects (sets of variables).

Finally, for any first-order signature Σ and Σ-context X, the set of *first-order formulae* is defined in the usual way; these are first-order formulae over Σ with all free variables in X. Any context morphism (substitution) determines the usual translation of first-order formulae, as does any signature morphism.

All this defines the formula functor $\mathcal{F}_{\mathcal{FO}} : \mathbf{Sig}^{\mathcal{FO}} \to \mathbf{Func}_{\mathbf{ICat}}(\mathbf{Set})$. The consequence relation of first-order logic may be defined model-theoretically as follows.

For any first-order signature Σ, a *first-order* Σ-*structure* A consists of a non-empty carrier set $|A|$ and an interpretation of operation names in Σ as functions on $|A|$, and of the predicate names in Σ as relations on $|A|$, of the arity indicated in Σ. Let $\mathbf{Str}^{\mathcal{FO}}(\Sigma)$ be the collection of all first-order Σ-structures (this forms a category with Σ-homomorphisms as morphisms).

Consider now a Σ-structure $A \in \mathbf{Str}^{\mathcal{FO}}(\Sigma)$, a set of variables X and a *valuation* $v : X \to |A|$. For any Σ-formula with free variables in X, $\varphi \in \mathcal{F}_{\mathcal{FO}}(\Sigma)(X)$, *the satisfaction of* φ *in* A *under* v, written $A[v] \models^{\mathcal{FO}}_\Sigma \varphi$, is defined in the usual way.

This standard notion of satisfaction may be used to determine consequence relations over the set of first-order formulae in two different ways. One possibility is to consider free variables as always implicitly universally quantified. This leads to the following family of *validity* consequence relations: for each first-order signature Σ, sets X_i of variables and open first-order formulae $\varphi_i \in \mathcal{F}_{\mathcal{FO}}(\Sigma)(X_i)$, $i = 0, \ldots, n$,

$\{ \langle X_i, \varphi_i \rangle \}_{i=1}^{n} \vdash_{\Sigma}^{\mathcal{FO}^v} \langle X_0, \varphi_0 \rangle$ if and only if for all $A \in \mathbf{Str}^{\mathcal{FO}}(\Sigma)$, $A[v_0] \models_{\Sigma}^{\mathcal{FO}} \varphi_0$ for all valuations $v_0 : X_0 \to |A|$ whenever $A[v_i] \models_{\Sigma}^{\mathcal{FO}} \varphi_i$ for $i = 1, \ldots, n$, for all valuations $v_i : X_i \to |A|$.

This yields the logical system $\mathcal{FO}^v$ of open formulae of first-order logic under the validity interpretation. From the point of view of proof theory, the usual Hilbert-type presentations of first-order logic present the same validity consequence relation.

Another possible view is to consider open formulae as truly open, and hence to identify occurrences of the same variable in different formulae. This leads to the following family of *truth* consequence relations: for each first-order signature Σ, sets X_i of variables and open first-order formulae $\varphi_i \in \mathcal{F}_{\mathcal{FO}}(\Sigma)(X_i)$, $i = 0, \ldots, n$,

$\{ \langle X_i, \varphi_i \rangle \}_{i=1}^{n} \vdash_{\Sigma}^{\mathcal{FO}^t} \langle X_0, \varphi_0 \rangle$ if and only if for all $A \in \mathbf{Str}^{\mathcal{FO}}(\Sigma)$, for all valuations $v_i : X_i \to |A|$, $i = 0, \ldots, n$, such that for all $\{ i_1, \ldots, i_k \} \subseteq \{ 0, \ldots, n \}$ the v_{i_l} coincide on $X = \bigcap_{l=1}^{k} X_{i_l}$, $A[v_0] \models_{\Sigma}^{\mathcal{FO}} \varphi_0$ whenever $A[v_i] \models_{\Sigma}^{\mathcal{FO}} \varphi_i$ for $i = 1, \ldots, n$.

This yields the logical system $\mathcal{FO}^t$ of open formulae of first-order logic under the truth interpretation. Natural-deduction-style presentations of first-order logic present the truth consequence relation via the notion of derivation under hypotheses. □

The above example illustrates two different views of the rôle of free variables in open formulae. The first option is to assume that free variables in an open formula are "local" to the formula, and so open formulae are always implicitly universally closed. This corresponds to so-called *validity* consequence relations, determined by a model-theoretic satisfaction relation according to the scheme:

$\Phi \vdash^v \varphi$ if and only if in every model, if Φ holds under every valuation then φ holds under every valuation as well.

A characteristic structural property of such consequence relations is that its conclusion (the formula on the right) may be instantiated, and its premises (formulae on the left) may be generalized.

Definition 3.3 *A logic of open formulae $\mathcal{L}$ is* of validity type *if its consequence relations admit instantiation on the right, i.e., for any signature $\Sigma \in |\mathbf{Sig}^{\mathcal{L}}|$ and open Σ-formulae $\Delta \subseteq Form_{\mathcal{F}_{\mathcal{L}}}(\Sigma)$ and $\langle \Gamma, \varphi \rangle \in Form_{\mathcal{F}_{\mathcal{L}}}(\Sigma)$, whenever*

$$\Delta \vdash_{\Sigma}^{\mathcal{L}} \langle \Gamma, \varphi \rangle$$

then for any Σ-context morphism $\gamma : \Gamma \to \Gamma'$,

$$\Delta \vdash_{\Sigma}^{\mathcal{L}} \langle \Gamma', \gamma(\varphi) \rangle.$$

Proposition 3.4 *If $\mathcal{L}$ is a logical system of validity type then its consequence relations admit generalisation on the left, i.e., for any signature $\Sigma \in |\mathbf{Sig}^{\mathcal{L}}|$ and open formulae $\langle \Gamma_i, \varphi_i \rangle \in Form_{\mathcal{F}_{\mathcal{L}}}(\Sigma)$, $i = 0, \ldots, n$, whenever*

$$\{ \langle \Gamma_i, \varphi_i \rangle \}_{i=1}^{n} \vdash_{\Sigma}^{\mathcal{L}} \langle \Gamma_0, \varphi_0 \rangle$$

then for any Σ-context morphisms $\gamma_i : \Gamma_i' \to \Gamma_i$ and formulae $\varphi_i' \in \mathcal{F}_{\mathcal{L}}(\Sigma)(\Gamma_i')$ such that $\varphi_i = \gamma_i(\varphi_i')$, $i = 1, \ldots, n$,

$$\{ \langle \Gamma_i', \varphi_i' \rangle \}_{i=1}^{n} \vdash_{\Sigma}^{\mathcal{L}} \langle \Gamma_0, \varphi_0 \rangle.$$

Proof Using instantiation on the right, we get $\langle \Gamma_i', \varphi_i' \rangle \vdash_{\Sigma}^{\mathcal{L}} \langle \Gamma_i, \varphi_i \rangle$ for $i = 1, \ldots, n$. The conclusion then follows by the transitivity of the consequence relation $\vdash_{\Sigma}^{\mathcal{L}}$. □

Proposition 3.5 *If $\mathcal{L}$ is a logical system of validity type then its consequence relations admit renamings, i.e., for any signature $\Sigma \in |\mathbf{Sig}^{\mathcal{L}}|$ and open formulae $\langle \Gamma_i, \varphi_i \rangle \in Form_{\mathcal{F}_{\mathcal{L}}}(\Sigma)$, $i = 0, \ldots, n$, whenever*

$$\{ \langle \Gamma_i, \varphi_i \rangle \}_{i=1}^{n} \vdash_{\Sigma}^{\mathcal{L}} \langle \Gamma_0, \varphi_0 \rangle$$

then for any Σ-context isomorphisms $r_i : \Gamma_i \to \Gamma'_i$, $i = 0, \ldots, n$,

$$\{ \langle \Gamma'_i, r_i(\varphi_i) \rangle \}_{i=1}^{n} \vdash_{\Sigma}^{\mathcal{L}} \langle \Gamma'_0, r_0(\varphi_0) \rangle.$$

Proof Use instantiation on the right w.r.t. $r_0 : \Gamma_0 \to \Gamma'_0$ and generalisation on the left w.r.t. $r_i^{-1} : \Gamma'_i \to \Gamma_i$. □

Proposition 3.6 *If $\mathcal{L}$ is a logical system of validity type then its consequence relations admit elimination of dummy variables on the left and introduction of dummy variables on the right, i.e., for any signature $\Sigma \in |\mathbf{Sig}^{\mathcal{L}}|$ and open formulae $\langle \Gamma_i, \varphi_i \rangle \in Form_{\mathcal{F}_{\mathcal{L}}}(\Sigma)$, $i = 0, \ldots, n$, whenever*

$$\{ \langle \Gamma_i, \varphi_i \rangle \}_{i=1}^{n} \vdash_{\Sigma}^{\mathcal{L}} \langle \Gamma_0, \varphi_0 \rangle$$

then for any Σ-contexts $\Gamma'_i \hookrightarrow \Gamma_i$ such that $\varphi_i \in \mathcal{F}_{\mathcal{L}}(\Sigma)(\Gamma'_i)$, $i = 1, \ldots, n$, and $\Gamma_0 \hookrightarrow \Gamma'_0$,

$$\{ \langle \Gamma'_i, \varphi_i \rangle \}_{i=1}^{n} \vdash_{\Sigma}^{\mathcal{L}} \langle \Gamma'_0, \varphi_0 \rangle.$$

Proof Use instantiation on the right and generalisation on the left w.r.t. the context inclusions. □

It is worth emphasizing that validity consequence relations need not admit introduction of dummy variables on the left, nor elimination of dummy variables on the right.

A different view of free variables in open formulae is that they denote an arbitrary but (in some sense) fixed value throughout all the formulae they occur in. Consequently, occurrences of the same variable on both sides of the consequence relation are assumed to denote the same value. This corresponds to so-called *truth* consequence relations, determined by a model-theoretic satisfaction relation according to the scheme:

$\Phi \vdash^{v} \varphi$ if and only if in every model under every valuation, if Φ holds then φ holds as well.

A characteristic structural property of such consequence relations is that any variable may be instantiated at the same time on both sides of the consequence relation.

Definition 3.7 *A logic of open formulae $\mathcal{L}$ is* of truth-type *if its consequence relations admit global instantiation, i.e., for any signature $\Sigma \in |\mathbf{Sig}^{\mathcal{L}}|$ and open formulae $\langle \Gamma_i, \varphi_i \rangle \in Form_{\mathcal{F}_{\mathcal{L}}}(\Sigma)$, $i = 0, \ldots, n$, whenever*

$$\{ \langle \Gamma_i, \varphi_i \rangle \}_{i=1}^{n} \vdash_{\Sigma}^{\mathcal{L}} \langle \Gamma_0, \varphi_0 \rangle$$

then for any compatible family of Σ-context morphisms $\gamma_i : \Gamma_i \to \Gamma'_i$, $i = 0, \ldots, n$,

$$\{ \langle \Gamma'_i, \gamma_i(\varphi_i) \rangle \}_{i=1}^{n} \vdash_{\Sigma}^{\mathcal{L}} \langle \Gamma'_0, \gamma_i(\varphi_0) \rangle.$$

Proposition 3.8 *If $\mathcal{L}$ is a logical system of the truth type then its consequence relations admit introduction of dummy variables on the left and on the right, i.e., for any signature $\Sigma \in |\mathbf{Sig}^{\mathcal{L}}|$ and open formulae $\langle \Gamma_i, \varphi_i \rangle \in Form_{\mathcal{F}_{\mathcal{L}}}(\Sigma)$, $i = 0, \ldots, n$, whenever*

$$\{ \langle \Gamma_i, \varphi_i \rangle \}_{i=1}^{n} \vdash_{\Sigma}^{\mathcal{L}} \langle \Gamma_0, \varphi_0 \rangle$$

then for any Σ-contexts Γ'_i such that $\Gamma_i \hookrightarrow \Gamma'_i$ for $i = 0, \ldots, n$,

$$\{ \langle \Gamma'_i, \varphi_i \rangle \}_{i=1}^{n} \vdash_{\Sigma}^{\mathcal{L}} \langle \Gamma'_0, \varphi_0 \rangle.$$

Proof This is a particular instance of global instantiation; just notice that any family of inclusions is compatible. □

It is important to realize that truth consequence relations in general do not admit elimination of dummy variables, neither on the left nor on the right.

Turning now to morphisms of logical systems, let us note that formula functors, as functors into the category $\mathbf{Func_{ICat}}(\mathbf{Set})$, come naturally equipped with a notion of morphism. Given two functors $\mathcal{F} : \mathbf{Sig} \to \mathbf{Func_{ICat}}(\mathbf{Set})$ and $\mathcal{F}' : \mathbf{Sig}' \to \mathbf{Func_{ICat}}(\mathbf{Set})$, a morphism $\mu : \mathcal{F} \to \mathcal{F}'$ consists of a functor $\mu^1 : \mathbf{Sig} \to \mathbf{Sig}'$ and a natural transformation $\mu^2 : \mathcal{F} \dot{\to} \mu^1;\mathcal{F}'$. The latter, in turn, is a family of morphisms in $\mathbf{Func_{ICat}}(\mathbf{Set})$, consisting for each $\Sigma \in |\mathbf{Sig}|$ of a functor $(\mu^2_\Sigma)^1 : \mathbf{Ctxt}_\Sigma \to \mathbf{Ctxt}'_{\mu^1(\Sigma)}$ and a natural transformation $(\mu^2_\Sigma)^2 : \mathcal{F}(\Sigma) \dot{\to} (\mu^2_\Sigma)^1;\mathcal{F}'(\mu^1(\Sigma))$. Finally, $(\mu^2_\Sigma)^2$ is a family indexed by Σ-contexts Γ of functions $(\mu^2_\Sigma)^2_\Gamma : \mathcal{F}(\Sigma)(\Gamma) \to \mathcal{F}'(\mu^1(\Sigma))((\mu^2_\Sigma)^1(\Gamma))$. When no confusion is likely, for any $\Sigma \in |\mathbf{Sig}|$, $\Gamma \in |\mathbf{Ctxt}_\Sigma|$ and $\varphi \in \mathcal{F}(\Sigma)(\Gamma)$, we write $\mu(\Sigma)$, $\mu_\Sigma(\Gamma)$ and $\mu_{\Sigma,\Gamma}(\varphi)$ for $\mu^1(\Sigma)$, $(\mu^2_\Sigma)^1(\Gamma)$ and $(\mu^2_\Sigma)^2_\Gamma(\varphi)$, respectively. Notice also that any such morphism $\mu : \mathcal{F} \to \mathcal{F}'$ defines a morphism $\hat{\mu} : Form_\mathcal{F} \to Form_{\mathcal{F}'}$, where $\hat{\mu}^1 : \mathbf{Sig} \to \mathbf{Sig}'$ is simply μ^1 and the natural transformation $\hat{\mu}^2 : Form_\mathcal{F} \dot{\to} \mu^1;Form_{\mathcal{F}'}$ between functors in $\mathbf{Sig} \to \mathbf{Set}$ is given by: for each $\Sigma \in |\mathbf{Sig}|$ and $\langle \Gamma, \varphi \rangle \in Form_\mathcal{F}(\Sigma)$, $\hat{\mu}^2_\Sigma(\langle \Gamma, \varphi \rangle) = \langle \mu_\Sigma(\Gamma), \mu_{\Sigma,\Gamma}(\varphi) \rangle$.

Definition 3.9 *Consider two logical systems of open formulae $\mathcal{L}$ and $\mathcal{L}'$. A* logic morphism $\mu : \mathcal{L} \to \mathcal{L}'$ *is a morphism between their formula functors $\mu : \mathcal{F}_\mathcal{L} \to \mathcal{F}_{\mathcal{L}'}$ such that $\hat{\mu} : Form_{\mathcal{F}_\mathcal{L}} \to Form_{\mathcal{F}_{\mathcal{L}'}}$ is a ground logic morphism $\hat{\mu} : C_\mathcal{L} \to C_{\mathcal{L}'}$.* **Log** *is the category of logical systems of open formulae and their morphism (it is easy to see that a composition of logic morphisms is a logic morphism).*

Just as for ground logical system, logic morphisms are a bit too crude to model the informal notion of logic representation.

Definition 3.10 *A logic morphism $\mu : \mathcal{L} \to \mathcal{L}'$ is a* representation *if $\hat{\mu} : C_\mathcal{L} \to C_{\mathcal{L}'}$ is a ground logic representation, i.e., if for each $\Sigma \in |\mathbf{Sig}^\mathcal{L}|$, $\hat{\mu}^2_\Sigma : Form_{\mathcal{F}_\mathcal{L}}(\Sigma) \to Form_{\mathcal{F}_{\mathcal{L}'}}(\mu^1(\Sigma))$ is a conservative morphism of consequence relations $\hat{\mu}^2_\Sigma : C_\mathcal{L}(\Sigma) \to C_\mathcal{L}(\mu^1(\Sigma))$, and all the functors $\mu^2_\Sigma : \mathbf{Ctxt}_\Sigma \to \mathbf{Ctxt}'_{\mu^1(\Sigma)}$ for $\Sigma \in |\mathbf{Sig}^\mathcal{L}|$ are injective. A logic representation $\mu : \mathcal{L} \to \mathcal{L}'$ is* exact *if in addition all the functors $\mu^2_\Sigma : \mathbf{Ctxt}_\Sigma \to \mathbf{Ctxt}'_{\mu^1(\Sigma)}$ for $\Sigma \in |\mathbf{Sig}^\mathcal{L}|$ are surjective and $\hat{\mu} : C_\mathcal{L} \to C_{\mathcal{L}'}$ is a* surjective *ground logic representation.*

4 From closed sentences to open formulae

The treatment of variables in abstract model theory [Bar74] suggests that at least in many typical cases the complete structure of a logical system of open formulae is in a sense redundant, since it may be recovered from a corresponding ground logical system of closed sentences. We have used the ideas of [Bar74] to develop a concept simulating open formulae in an arbitrary institution in [Tar86], [ST88]. Here we apply the same technique to construct a logical system of open formulae out of a ground logical system.

In the example of first-order logic sketched in the previous section, for any first-order signature Σ and Σ-context (set of variables X), the open Σ-formulae built in context X are exactly the closed sentences over the signature $\Sigma(X)$ defined as the extension of Σ by the variables in X as new constants (0-ary operation names). Moreover, given any Σ-structure A, a valuation of X in $|A|$ corresponds to an expansion of A to a $\Sigma(X)$-structure which indeed additionally determines an interpretation of the new constants, *i.e.*, a valuation of variables. Then, an open first-order Σ-formula φ with free variables in X holds in a structure A under a valuation $v : X \to |A|$ if and only if φ viewed as a closed $\Sigma(X)$-sentence holds in the expansion of A to a $\Sigma(X)$-structure determined by v. Hence, a consequence relation on open Σ-formulae with variables X may be recovered from the satisfaction relation on closed $\Sigma(X)$-sentences, and (much less directly) from the consequence relation on closed sentences. This idea readily generalizes to an arbitrary ground logical system. Notice that the "global" view of logical systems as uniformly-defined families indexed by a category of signatures is crucial here.

Consider an arbitrary ground logical system $\mathcal{G} : \mathbf{Sig}^{\mathcal{G}} \to \mathbf{CR}$, fixed throughout this section.

Given the category $\mathbf{Sig}^{\mathcal{G}}$ of signatures (with inclusions), we define contexts as signature extensions. More formally, for each $\Sigma \in |\mathbf{Sig}^{\mathcal{G}}|$, the category $\mathbf{Ctxt}^{\mathcal{G}}_{\Sigma}$ of Σ-contexts is the full subcategory of the slice category $\Sigma{\downarrow}\mathbf{Sig}^{\mathcal{G}}$ determined by inclusions $\iota : \Sigma \hookrightarrow \Sigma'$. That is, $\mathbf{Ctxt}^{\mathcal{G}}_{\Sigma}$ has as objects pairs consisting of a signature Σ' and a signature inclusion $\iota : \Sigma \hookrightarrow \Sigma'$; and a Σ-context morphism from $\iota_1 : \Sigma \to \Sigma_1$ to $\iota_2 : \Sigma \to \Sigma_2$ is a signature morphism $\gamma : \Sigma_1 \to \Sigma_2$ such that $\iota_1;\gamma = \iota_2$. As before, we will denote contexts by Γ; the target signature of the extension is then written as Σ^{Γ}, and the inclusion $\Sigma \hookrightarrow \Sigma^{\Gamma}$ is then determined unambigously. This allows us to identify contexts with signature inclusions or even with their target signatures when no confusion is likely. Notice that for any signature Σ, the identity on Σ, which is a signature inclusion, is the "empty context" $\langle\rangle \in |\mathbf{Ctxt}^{\mathcal{G}}_{\Sigma}|$. This is the least object in $|\mathbf{Ctxt}^{\mathcal{G}}_{\Sigma}|$ w.r.t. the inclusion ordering.

The map $\Sigma \mapsto \mathbf{Ctxt}^{\mathcal{G}}_{\Sigma}$ extends to a functor $\mathbf{Ctxt}^{\mathcal{G}} : \mathbf{Sig}^{\mathcal{G}} \to \mathbf{ICat}$ using the canonical pushout construction in $\mathbf{Sig}^{\mathcal{G}}$. Namely, for any signature morphism $\sigma : \Sigma \to \Sigma'$, $\mathbf{Ctxt}^{\mathcal{G}}(\sigma) : \mathbf{Ctxt}^{\mathcal{G}}_{\Sigma} \to \mathbf{Ctxt}^{\mathcal{G}}_{\Sigma'}$ is defined on objects as follows: for any $\Gamma \in |\mathbf{Ctxt}^{\mathcal{G}}_{\Sigma}|$, $\mathbf{Ctxt}^{\mathcal{G}}(\sigma)(\Gamma) = \Gamma^{\star}$.

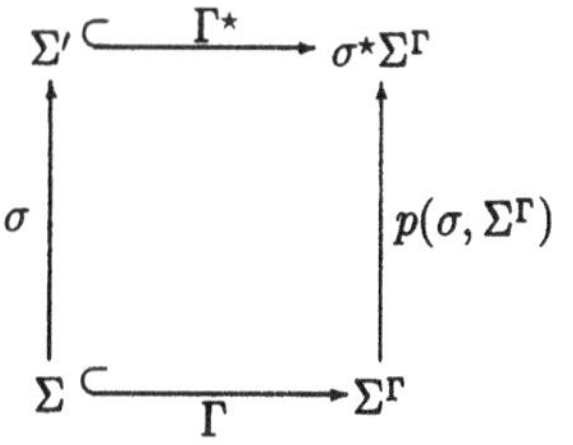

This extends to context morphisms using the pushout property.

The open formulae determined by $\mathcal{G}$ are given by the functor

$$\mathcal{F}^{\mathcal{G}} : \mathbf{Sig}^{\mathcal{G}} \to \mathbf{Func}_{\mathbf{ICat}}(\mathbf{Set})$$

defined as follows:

- for each $\Sigma \in |\mathbf{Sig}^{\mathcal{G}}|$, $\mathcal{F}^{\mathcal{G}}(\Sigma) : \mathbf{Ctxt}^{\mathcal{G}}_{\Sigma} \to \mathbf{Set}$ is defined by:
  - for each $\Gamma \in \mathbf{Ctxt}^{\mathcal{G}}_{\Sigma}$, $\mathcal{F}^{\mathcal{G}}(\Sigma)(\Gamma) = |\mathcal{G}|(\Sigma^{\Gamma})$
  - for each $\gamma : \Gamma \to \Gamma'$ in $\mathbf{Ctxt}^{\mathcal{G}}_{\Sigma}$, *i.e.*, $\gamma : \Sigma^{\Gamma} \to \Sigma^{\Gamma'}$ in $\mathbf{Sig}^{\mathcal{G}}$, $\mathcal{F}^{\mathcal{G}}(\Sigma)(\gamma) = |\mathcal{G}|(\gamma)$
- for each $\sigma : \Sigma \to \Sigma'$ in $\mathbf{Sig}^{\mathcal{G}}$
  - $\mathcal{F}^{\mathcal{G}}(\sigma)^1 : \mathbf{Ctxt}^{\mathcal{G}}_{\Sigma} \to \mathbf{Ctxt}^{\mathcal{G}}_{\Sigma'}$ is the functor $\mathbf{Ctxt}^{\mathcal{G}}(\sigma)$ as defined above
  - for each $\Gamma \in |\mathbf{Ctxt}^{\mathcal{G}}_{\Sigma}|$, $\mathcal{F}^{\mathcal{G}}(\sigma)^2_{\Gamma} : |\mathcal{G}|(\Sigma^{\Gamma}) \to |\mathcal{G}|(\mathbf{Ctxt}^{\mathcal{G}}(\sigma)(\Gamma))$ is the function $|\mathcal{G}|(p(\sigma, \Sigma^{\Gamma})) : |\mathcal{G}|(\Sigma^{\Gamma}) \to |\mathcal{G}|(\sigma^{\star}\Sigma^{\Gamma})$.

It is easy to see that everything is well-defined here; in particular that all the functoriality and naturality requirements follow.

As before, $\mathcal{F}^{\mathcal{G}} : \mathbf{Sig}^{\mathcal{G}} \to \mathbf{Func}_{\mathbf{ICat}}(\mathbf{Set})$ determines open formulae of the logics we derive from $\mathcal{G}$. The functor $Form_{\mathcal{F}^{\mathcal{G}}} : \mathbf{Sig}^{\mathcal{G}} \to \mathbf{Set}$ is defined as in Sec. 3. We have yet to define the consequence relations of the system. This may be done in two different ways, depending on whether we consider the validity or the truth interpretation of consequence relations on open formulae.

Definition 4.1 *The* validity logic $\mathcal{L}^v(\mathcal{G})$ of open formulae determined by the ground logic $\mathcal{G} : \mathbf{Sig}^{\mathcal{G}} \to \mathbf{CR}$ *consists of the formula functor*

$$\mathcal{F}^{\mathcal{G}} : \mathbf{Sig}^{\mathcal{G}} \to \mathbf{Func}_{\mathbf{ICat}}(\mathbf{Set})$$

and the consequence functor

$$C_{\mathcal{L}^v(\mathcal{G})} : \mathbf{Sig}^{\mathcal{G}} \to \mathbf{CR}$$

where for each $\Sigma \in |\mathbf{Sig}^{\mathcal{G}}|$, *the consequence relation* $\vdash_{\Sigma}^{\mathcal{L}^v(\mathcal{G})}$ *on* $Form_{\mathcal{F}^{\mathcal{G}}}(\Sigma)$ *is defined as follows: for any* $\Gamma_i \in |\mathbf{Ctxt}_{\Sigma}^{\mathcal{G}}|$ *and* $\varphi_i \in \mathcal{F}^{\mathcal{G}}(\Sigma)(\Gamma_i)$ $(i = 0, \ldots, n)$,

$\{ \langle \Gamma_i, \varphi_i \rangle \}_{i=1}^{n} \vdash_{\Sigma}^{\mathcal{L}^v(\mathcal{G})} \langle \Gamma_0, \varphi_0 \rangle$ *if and only if for some* $\{ i_1, \ldots, i_k \} \subseteq \{ 1, \ldots, n \}$ *there exist* Σ*-context morphisms* $\gamma_l : \Gamma_{i_l} \to \Gamma_0$, $l = 1, \ldots, k$, *such that*[3] $\{ \gamma_l(\varphi_{i_l}) \}_{l=1}^{k} \vdash_{\Sigma^{\Gamma_0}}^{\mathcal{G}} \varphi_0$.

In the above definition, $i_1, \ldots, i_k$ are not assumed to be distinct: we may use premises in many different ways.

Proposition 4.2 *$\mathcal{L}^v(\mathcal{G})$ as defined in Def. 4.1 is indeed a logical system of open formulae of validity type.*

Proof

- For each $\Sigma \in |\mathbf{Sig}^{\mathcal{G}}|$, $\vdash_{\Sigma}^{\mathcal{L}^v(\mathcal{G})}$ is indeed a consequence relation:

  Reflexivity and weakening follow directly from the definition. Transitivity follows from the fact that the consequence relations of $\mathcal{G}$ are preserved under translations induced by signature morphisms, which applies to context morphisms in $\mathcal{L}^v(\mathcal{G})$ as well since they are in fact signature morphisms in $\mathcal{G}$, and from the transitivity of the consequence relations of $\mathcal{G}$.

- For each $\sigma : \Sigma \to \Sigma'$, $\vdash_{\Sigma}^{\mathcal{L}^v(\mathcal{G})}$ is preserved under the translation of formulae induced by σ.

  This again follows from the fact that the consequence relations of $\mathcal{G}$ are preserved under translations induced by signature morphisms. Here are some details: let $\Gamma_i \in |\mathbf{Ctxt}_{\Sigma}^{\mathcal{G}}|$, $\varphi_i \in \mathcal{F}^{\mathcal{G}}(\Sigma)(\Gamma_i)$, $i = 0, \ldots, n$, be such that $\{ \langle \Gamma_i, \varphi_i \rangle \}_{i=1}^{n} \vdash_{\Sigma}^{\mathcal{L}^v(\mathcal{G})} \langle \Gamma_0, \varphi_0 \rangle$. We have to prove that

  $$\{ \langle \sigma^\star \Sigma^{\Gamma_i}, \mathcal{G}(p(\sigma, \Sigma^{\Gamma_i}))(\varphi_i) \rangle \}_{i=1}^{n} \vdash_{\Sigma'}^{\mathcal{L}^v(\mathcal{G})} \langle \sigma^\star \Sigma^{\Gamma_0}, \mathcal{G}(p(\sigma, \Sigma^{\Gamma_0}))(\varphi_0) \rangle.$$

  By definition, there are $\{ i_1, \ldots, i_k \} \subseteq \{ 1, \ldots, n \}$ and Σ-context morphisms $\gamma_l : \Gamma_{i_l} \to \Gamma_0$, $l = 1, \ldots, k$, such that $\{ \mathcal{F}^{\mathcal{G}}(\Sigma)(\gamma_l)(\varphi_{i_l}) \}_{l=1}^{k} \vdash_{\Sigma^{\Gamma_0}}^{\mathcal{G}} \varphi_0$.

  Since $\vdash_{\Sigma^{\Gamma_0}}^{\mathcal{G}}$ is preserved by translation of formulae induced by $p(\sigma, \Sigma^{\Gamma_0}) : \Sigma^{\Gamma_0} \to \sigma^\star \Sigma^{\Gamma_0}$, it follows that

  $$\{ \mathcal{G}(p(\sigma, \Sigma^{\Gamma_0}))(\mathcal{F}^{\mathcal{G}}(\Sigma)(\gamma_l)(\varphi_{i_l})) \}_{l=1}^{k} \vdash_{\sigma^\star \Sigma^{\Gamma_0}}^{\mathcal{G}} \mathcal{G}(p(\sigma, \Sigma^{\Gamma_0}))(\varphi_0).$$

  For $l = 1, \ldots, k$, by the pushout property of $\sigma^\star \Sigma^{\Gamma_{i_l}}$, there is a Σ'-context morphism γ'_l : $\sigma^\star \Sigma^{\Gamma_{i_l}} \to \sigma^\star \Sigma^{\Gamma_0}$ such that

  $$p(\sigma, \Sigma^{\Gamma_{i_l}}); \gamma'_l = \gamma_l; p(\sigma, \Sigma^{\Gamma_0}).$$

  Then, for $l = 1, \ldots, k$,

  $$\begin{aligned} \mathcal{G}(p(\sigma, \Sigma^{\Gamma_0}))(\mathcal{F}^{\mathcal{G}}(\Sigma)(\gamma_l)(\varphi_{i_l})) &= \mathcal{G}(\gamma_l; p(\sigma, \Sigma^{\Gamma_0}))(\varphi_{i_l}) \\ &= \mathcal{G}(p(\sigma, \Sigma^{\Gamma_{i_l}}); \gamma'_l)(\varphi_{i_l}) \\ &= \mathcal{F}^{\mathcal{G}}(\Sigma')(\gamma'_l)(\mathcal{G}(p(\sigma, \Sigma^{\Gamma_{i_l}})(\varphi_{i_l})). \end{aligned}$$

  Thus:

  $$\{ \mathcal{F}^{\mathcal{G}}(\Sigma')(\gamma'_l)(\mathcal{G}(p(\sigma, \Sigma^{\Gamma_{i_l}}))(\varphi_{i_l})) \}_{l=1}^{k} \vdash_{\sigma^\star \Sigma^{\Gamma_0}}^{\mathcal{G}} \mathcal{G}(p(\sigma, \Sigma^{\Gamma_0}))(\varphi_0)$$

  which completes the proof.

[3] We are using freely the notational conventions introduced in Sec. 3. In particular, $\gamma_l(\varphi_{i_l})$ is really $\mathcal{F}^{\mathcal{G}}(\Sigma)(\gamma_l)(\varphi_{i_l})$.

- For each $\Sigma \in |\mathbf{Sig}^{\mathcal{G}}|$, $\vdash_{\Sigma}^{\mathcal{L}^v(\mathcal{G})}$ admits instantiation on the right:

  This again follows from the fact that the consequence relations of $\mathcal{G}$ are preserved under translations induced by signature morphisms.

  Let $\Gamma_i \in |\mathbf{Ctxt}_{\Sigma}^{\mathcal{G}}|$, $\varphi_i \in \mathcal{F}^{\mathcal{G}}(\Sigma)(\Gamma_i)$, $i = 0, \ldots, n$, be such that $\{ \langle \Gamma_i, \varphi_i \rangle \}_{i=1}^{n} \vdash_{\Sigma}^{\mathcal{L}^v(\mathcal{G})} \langle \Gamma_0, \varphi_0 \rangle$. Consider a Σ-context morphism $\gamma : \Gamma_0 \to \Gamma'_0$. By definition, there are $\{ i_1, \ldots, i_k \} \subseteq \{ 1, \ldots, n \}$ and Σ-context morphisms $\gamma_l : \Gamma_{i_l} \to \Gamma_0$, $l = 1, \ldots, k$, such that

$$\{ \mathcal{F}^{\mathcal{G}}(\Sigma)(\gamma_l)(\varphi_{i_l}) \}_{l=1}^{k} \vdash_{\Sigma^{\Gamma_0}}^{\mathcal{G}} \varphi_0.$$

  Then also

$$\{ \mathcal{F}^{\mathcal{G}}(\Sigma)(\gamma_l;\gamma)(\varphi_{i_l}) \}_{l=1}^{k} \vdash_{\Sigma^{\Gamma'_0}}^{\mathcal{G}} \mathcal{G}(\gamma)(\varphi_0)$$

  from which it follows that

$$\{ \langle \Gamma_i, \varphi_i \rangle \}_{i=1}^{n} \vdash_{\Sigma}^{\mathcal{L}^v(\mathcal{G})} \langle \Gamma'_0, \mathcal{F}^{\mathcal{G}}(\Sigma)(\gamma)(\varphi_0) \rangle.$$

□

As a special case of the closure condition embodied in the above definition we have the following "rule of universal closure" and "rule of universal elimination."

Corollary 4.3 *Let $\mathcal{G} : \mathbf{Sig}^{\mathcal{G}} \to \mathbf{CR}$ be any ground logic, let $\Sigma \in |\mathbf{Sig}^{\mathcal{G}}|$ be any signature of $\mathcal{G}$, and let $\Gamma \in |\mathbf{Ctxt}_{\Sigma}^{\mathcal{G}}|$ be any Σ-context of $\mathcal{G}$.*

1. *For any formulae $\varphi_i \in \mathcal{F}^{\mathcal{G}}(\Sigma)(\langle\rangle)$, $i = 1, \ldots, n$, and $\psi \in \mathcal{F}^{\mathcal{G}}(\Sigma)(\Gamma)$, if $\{ \varphi_i \}_{i=1}^{n} \vdash_{\Sigma^{\Gamma}}^{\mathcal{G}} \psi$ then $\{ \langle \langle\rangle, \varphi_i \rangle \}_{i=1}^{n} \vdash_{\Sigma}^{\mathcal{L}^v(\mathcal{G})} \langle \Gamma, \psi \rangle$.*
2. *For any formula $\varphi \in \mathcal{F}^{\mathcal{G}}(\Sigma)(\Gamma)$ and any context morphism $\gamma : \Gamma \to \langle\rangle$, $\langle \Gamma, \varphi \rangle \vdash_{\Sigma}^{\mathcal{L}^v(\mathcal{G})} \langle \langle\rangle, \gamma(\varphi) \rangle$.*

Definition 4.4 *The* truth logic $\mathcal{L}^t(\mathcal{G})$ of open formulae determined by the ground logic $\mathcal{G} : \mathbf{Sig}^{\mathcal{G}} \to \mathbf{CR}$ *consists of the formula functor*

$$\mathcal{F}^{\mathcal{G}} : \mathbf{Sig}^{\mathcal{G}} \to \mathbf{Func}_{\mathbf{ICat}}(\mathbf{Set})$$

and the consequence functor

$$C_{\mathcal{L}^t(\mathcal{G})} : \mathbf{Sig}^{\mathcal{G}} \to \mathbf{CR}$$

where for each $\Sigma \in |\mathbf{Sig}^{\mathcal{G}}|$, the consequence relation $\vdash_{\Sigma}^{\mathcal{L}^t(\mathcal{G})}$ on $Form_{\mathcal{F}^{\mathcal{G}}}(\Sigma)$ is defined as the transitive closure[4] *of the relation defined as follows: for any $\Gamma_i \in |\mathbf{Ctxt}_{\Sigma}^{\mathcal{G}}|$ and $\varphi_i \in \mathcal{F}^{\mathcal{G}}(\Sigma)(\Gamma_i)$ $(i = 0, \ldots, n)$,*

$\{ \langle \Gamma_i, \varphi_i \rangle \}_{i=1}^{n} \vdash_{\Sigma}^{\mathcal{L}^t(\mathcal{G})} \langle \Gamma_0, \varphi_0 \rangle$ *if and only if for some $\{ i_1, \ldots, i_k \} \subseteq \{ 1, \ldots, n \}$ there is a Σ-context $\Gamma \in |\mathbf{Ctxt}_{\Sigma}^{\mathcal{G}}|$ such that $\Gamma \hookrightarrow \Gamma_{i_l}$ for $l = 1, \ldots, k$ and $\Gamma \hookrightarrow \Gamma_0$, and for some formulae $\psi_l \in \mathcal{F}^{\mathcal{G}}(\Sigma)(\Gamma)$, $l = 0, \ldots, k$, such that*

- $\varphi_{i_l} \vdash_{\Sigma^{\Gamma_{i_l}}}^{\mathcal{G}} \psi_l$ *for* $l = 1, \ldots, k$
- $\{ \psi_l \}_{l=1}^{k} \vdash_{\Sigma^{\Gamma}}^{\mathcal{G}} \psi_0$
- $\psi_0 \vdash_{\Sigma^{\Gamma_0}}^{\mathcal{G}} \varphi_0$

If all the contexts are the same, say $\Gamma_i = \Gamma$ for $i = 0, \ldots, n$, then the condition stated in the definition amounts to the requirement that

$$\{ \varphi_i \}_{i=1}^{n} \vdash_{\Sigma^{\Gamma}}^{\mathcal{G}} \varphi_0.$$

Moreover, the relation defined in such a way is a consequence relation between formulas with a fixed context.

[4] *i.e.*, the least relation containing the relation defined below and satisfying the transitivity condition of Def. 2.1.

Proposition 4.5 *$\mathcal{L}^t(\mathcal{G})$ as defined in Def. 4.4 is indeed a logical system of open formulae of truth type.*

Proof

- For each $\Sigma \in |\mathbf{Sig}^{\mathcal{G}}|$, $\vdash_{\Sigma}^{\mathcal{L}^t(\mathcal{G})}$ is indeed a consequence relation:

  Transitivity follows directly from the definition. Reflexivity and weakening obviously hold for the generating relation, and it is easy to see that they are preserved by the transitive closure.

- For each $\sigma : \Sigma \to \Sigma'$, $\vdash_{\Sigma}^{\mathcal{L}^t(\mathcal{G})}$ is preserved under the translation of formulae induced by σ.

  It is enough to show that the generating relation is preserved. Consider: $\Gamma_i \in |\mathbf{Ctxt}_{\Sigma}^{\mathcal{G}}|$, $\varphi_i \in \mathcal{F}^{\mathcal{G}}(\Sigma)(\Gamma_i)$, $i = 0, \ldots, n$, such that $\{\langle \Gamma_i, \varphi_i \rangle\}_{i=1}^{n} \vdash_{\Sigma}^{\mathcal{L}^t(\mathcal{G})} \langle \Gamma_0, \varphi_0 \rangle$ and such that for some $\{i_1, \ldots, i_k\} \subseteq \{1, \ldots, n\}$ there is a Σ-context $\Gamma \in |\mathbf{Ctxt}_{\Sigma}^{\mathcal{G}}|$ such that $\Gamma \hookrightarrow \Gamma_{i_l}$ for $l = 1, \ldots, k$ and $\Gamma \hookrightarrow \Gamma_0$, and there exist formulae $\psi_l \in \mathcal{F}^{\mathcal{G}}(\Sigma)(\Gamma)$, $l = 0, \ldots, k$, such that

  - $\varphi_{i_l} \vdash_{\Sigma^{\Gamma_{i_l}}}^{\mathcal{G}} \psi_l$ for $l = 1, \ldots, k$
  - $\{\psi_l\}_{l=1}^{k} \vdash_{\Sigma^{\Gamma}}^{\mathcal{G}} \psi_0$
  - $\psi_0 \vdash_{\Sigma^{\Gamma_0}}^{\mathcal{G}} \varphi_0$

  Then $\sigma^\star\Gamma$ is a Σ'-context such that $\sigma^\star\Gamma \hookrightarrow \sigma^\star\Gamma_{i_l}$ for $l = 1, \ldots, k$ and $\sigma^\star\Gamma \hookrightarrow \sigma^\star\Gamma_0$. Moreover,

  - $\mathcal{G}(p(\sigma, \Sigma^{\Gamma_{i_l}}))(\varphi_{i_l}) \vdash_{\sigma^\star\Sigma^{\Gamma_{i_l}}}^{\mathcal{G}} \mathcal{G}(p(\sigma, \Sigma^{\Gamma}))(\psi_l)$ for $l = 1, \ldots, k$
  - $\{\mathcal{G}(p(\sigma, \Sigma^{\Gamma}))(\psi_l)\}_{l=1}^{k} \vdash_{\sigma^\star\Sigma^{\Gamma}}^{\mathcal{G}} \mathcal{G}(p(\sigma, \Sigma^{\Gamma}))(\psi_0)$
  - $\mathcal{G}(p(\sigma, \Sigma^{\Gamma}))(\psi_0) \vdash_{\sigma^\star\Sigma^{\Gamma_0}}^{\mathcal{G}} \mathcal{G}(p(\sigma, \Sigma^{\Gamma_0}))(\varphi_0)$

  Thus indeed:

  $$\{\langle \sigma^\star\Sigma^{\Gamma_i}, \mathcal{G}(p(\sigma, \Sigma^{\Gamma_i}))(\varphi_i)\rangle\}_{i=1}^{n} \vdash_{\Sigma'}^{\mathcal{L}^t(\mathcal{G})} \langle \sigma^\star\Sigma^{\Gamma_0}, \mathcal{G}(p(\sigma, \Sigma^{\Gamma_0}))(\varphi_0)\rangle$$

- For each $\Sigma \in |\mathbf{Sig}^{\mathcal{G}}|$, $\vdash_{\Sigma}^{\mathcal{L}^t(\mathcal{G})}$ admits global instantiation.

  Again, it is enough to prove this for the generating relation. Consider: $\Gamma_i \in |\mathbf{Ctxt}_{\Sigma}^{\mathcal{G}}|$, $\varphi_i \in \mathcal{F}^{\mathcal{G}}(\Sigma)(\Gamma_i)$, $i = 0, \ldots, n$, such that $\{\langle \Gamma_i, \varphi_i \rangle\}_{i=1}^{n} \vdash_{\Sigma}^{\mathcal{L}^t(\mathcal{G})} \langle \Gamma_0, \varphi_0 \rangle$ and such that for some set $\{i_1, \ldots, i_k\} \subseteq \{1, \ldots, n\}$ there is a Σ-context $\Gamma \in |\mathbf{Ctxt}_{\Sigma}^{\mathcal{G}}|$ such that $\Gamma \hookrightarrow \Gamma_{i_l}$ for $l = 1, \ldots, k$ and $\Gamma \hookrightarrow \Gamma_0$, and for some formulae $\psi_l \in \mathcal{F}^{\mathcal{G}}(\Sigma)(\Gamma)$, $l = 0, \ldots, k$, such that

  - $\varphi_{i_l} \vdash_{\Sigma^{\Gamma_{i_l}}}^{\mathcal{G}} \psi_l$ for $l = 1, \ldots, k$
  - $\{\psi_l\}_{l=1}^{k} \vdash_{\Sigma^{\Gamma}}^{\mathcal{G}} \psi_0$
  - $\psi_0 \vdash_{\Sigma^{\Gamma_0}}^{\mathcal{G}} \varphi_0$

  Let then $\gamma_i : \Gamma_i \to \Gamma'_i$, $i = 0, \ldots, n$ be a compatible family of Σ-context morphisms. By definition, there is a Σ-context morphism $\gamma : \Gamma \to \Gamma'$ such that for $i = 0, \ldots, n$, γ_i is an extension of γ. In particular this implies that for $i = 0, \ldots, n$, $\Gamma' \hookrightarrow \Gamma'_i$. Since the consequence relations of $\mathcal{G}$ are preserved under translations induced by signature (and hence Σ-context) morphisms, we have:

  - $\mathcal{F}^{\mathcal{G}}(\Sigma)(\gamma_{i_l})(\varphi_{i_l}) \vdash_{\Sigma^{\Gamma'_{i_l}}}^{\mathcal{G}} \mathcal{F}^{\mathcal{G}}(\Sigma)(\gamma)(\psi_{i_l})$ for $l = 1, \ldots, k$
  - $\{\mathcal{F}^{\mathcal{G}}(\Sigma)(\dot{\gamma})(\psi_{i_l})\}_{l=1}^{k} \vdash_{\Sigma^{\Gamma'}}^{\mathcal{G}} \mathcal{F}^{\mathcal{G}}(\Sigma)(\gamma)(\psi_0)$
  - $\mathcal{F}^{\mathcal{G}}(\Sigma)(\gamma)(\psi_0) \vdash_{\Sigma^{\Gamma'_0}}^{\mathcal{G}} \mathcal{F}^{\mathcal{G}}(\Sigma)(\gamma_0)(\varphi_0)$

Thus, as required

$$\{\langle\Gamma'_i, \mathcal{F}^{\mathcal{G}}(\Sigma)(\gamma_i)(\varphi_i)\rangle\}_{i=1}^n \vdash_\Sigma^{\mathcal{L}^t(\mathcal{G})} \langle\Gamma'_0, \mathcal{F}^{\mathcal{G}}(\Sigma)(\gamma)(\varphi_0)\rangle.$$

□

The simplest case (and the most typical for considerations on truth consequence relations) is when the context of open formulae used in a deduction sequence is fixed. The closure property embodied in the condition in Def. 4.4 then amounts to the following:

Proposition 4.6 *For any ground logic* $\mathcal{G} : \mathbf{Sig}^{\mathcal{G}} \to \mathbf{CR}$, *any signature* $\Sigma \in |\mathbf{Sig}^{\mathcal{G}}|$, Σ*-context* $\Gamma \in |\mathbf{Ctxt}^{\mathcal{G}}_\Sigma|$ *and formulae* $\varphi_i \in \mathcal{F}^{\mathcal{G}}(\Sigma)(\Gamma)$, $i = 0, \ldots, n$, *if* $\{\varphi_i\}_{i=1}^n \vdash^{\mathcal{G}}_{\Sigma\Gamma} \varphi_0$ *then* $\{\langle\Gamma, \varphi_i\rangle\}_{i=1}^n \vdash_\Sigma^{\mathcal{L}^t(\mathcal{G})} \langle\Gamma, \varphi_0\rangle$.

5 Logical systems and institutions

One justification for the definitions we gave in Section 4 may be sought in model theory. The theory of *institutions* (*cf.* [GB84]) provides a framework to study this issue at a sufficiently abstract level.

Definition 5.1 *An* institution $\mathcal{I}$ *consists of:*

- *a category* $\mathbf{Sig}^{\mathcal{I}}$ *(of signatures);*
- *a functor* $\mathbf{Sen}^{\mathcal{I}}:\mathbf{Sig}^{\mathcal{I}} \to \mathbf{Set}$ *(*$\mathbf{Sen}^{\mathcal{I}}$ *gives for any signature* Σ *the set* $\mathbf{Sen}^{\mathcal{I}}(\Sigma)$ *of* Σ*-sentences and for any signature morphism* $\sigma:\Sigma \to \Sigma'$ *the function* $\mathbf{Sen}^{\mathcal{I}}(\sigma):\mathbf{Sen}^{\mathcal{I}}(\Sigma) \to \mathbf{Sen}^{\mathcal{I}}(\Sigma')$ *translating* Σ*-sentences to* Σ'*-sentences);*
- *a functor* $\mathbf{Mod}^{\mathcal{I}}:\mathbf{Sig}^{\mathcal{I}} \to \mathbf{Cat}^{op}$ *(where* $\mathbf{Cat}$ *is the category of all categories;* $\mathbf{Mod}^{\mathcal{I}}$ *gives for any signature* Σ *the category* $\mathbf{Mod}^{\mathcal{I}}(\Sigma)$ *of* Σ*-models and for any signature morphism* $\sigma:\Sigma \to \Sigma'$ *the* σ*-reduct functor* $\mathbf{Mod}^{\mathcal{I}}(\sigma):\mathbf{Mod}^{\mathcal{I}}(\Sigma') \to \mathbf{Mod}^{\mathcal{I}}(\Sigma)$ *translating* Σ'*-models to* Σ*-models); and*
- *a satisfaction relation* $\models_{\mathcal{I},\Sigma} \subseteq |\mathbf{Mod}^{\mathcal{I}}(\Sigma)| \times \mathbf{Sen}^{\mathcal{I}}(\Sigma)$ *for each signature* Σ.

such that for any signature morphism $\sigma:\Sigma \to \Sigma'$ *the translations* $\mathbf{Mod}^{\mathcal{I}}(\sigma)$ *of models and* $\mathbf{Sen}^{\mathcal{I}}(\sigma)$ *of sentences preserve the satisfaction relation, i.e. for any* $\varphi \in \mathbf{Sen}^{\mathcal{I}}(\Sigma)$ *and* $M' \in |\mathbf{Mod}^{\mathcal{I}}(\Sigma')|$,

$$M' \models_{\mathcal{I},\Sigma'} \mathbf{Sen}^{\mathcal{I}}(\sigma)(\varphi) \iff \mathbf{Mod}^{\mathcal{I}}(\sigma)(M') \models_{\mathcal{I},\Sigma} \varphi \qquad \textit{(Satisfaction condition)}$$

In the following we will assume in addition that the institutions we consider have categories of signatures *with inclusions* and that the sentence functors $\mathbf{Sen}^{\mathcal{I}}$ preserve inclusions. For any signature morphism $\sigma : \Sigma \to \Sigma'$, the function $\mathbf{Sen}^{\mathcal{I}}(\sigma)$ will be written simply as σ and the reduct functor $\mathbf{Mod}^{\mathcal{I}}(\sigma)$ as $\_|_\sigma$. Moreover, for any signature inclusion $\iota : \Sigma \hookrightarrow \Sigma'$, the reduct functor $\_|_\iota$ will be written as $\_|_\Sigma$.

Definition 5.2 *An institution* $\mathcal{I}$ *determines a ground logical system*

$$\mathcal{G}(\mathcal{I}) : \mathbf{Sig}^{\mathcal{I}} \to \mathbf{CR}$$

where $|\mathcal{G}(\mathcal{I})| : \mathbf{Sig}^{\mathcal{I}} \to \mathbf{Set}$ *is just* $\mathbf{Sen}^{\mathcal{I}} : \mathbf{Sig}^{\mathcal{I}} \to \mathbf{Set}$ *and for each* $\Sigma \in |\mathbf{Sig}^{\mathcal{I}}|$, *for* $\varphi_i \in \mathbf{Sen}^{\mathcal{I}}(\Sigma)$, $i = 0, \ldots, n$,

$\{\varphi_i\}_{i=1}^n \vdash_\Sigma^{\mathcal{G}(\mathcal{I})} \varphi_0$ *if and only if for all models* $M \in |\mathbf{Mod}^{\mathcal{I}}(\Sigma)|$, $M \models_{\mathcal{I},\Sigma} \varphi_0$ *whenever* $M \models_{\mathcal{I},\Sigma} \varphi_i$ *for* $i = 1, \ldots, n$.

Proposition 5.3 *For any insitution $\mathcal{I}$, the logical system $\mathcal{G}(\mathcal{I})$ given by Definition 5.2 is indeed a ground logical system.*

Definition 5.4 *Let $\mathcal{I}$ be an institution. By a ground logic* sound for $\mathcal{I}$ *we mean any ground logic $\mathcal{G} : \mathbf{Sig}^{\mathcal{I}} \to \mathbf{CR}$ such that $|\mathcal{G}| = \mathbf{Sen}^{\mathcal{I}}$ and for each $\Sigma \in |\mathbf{Sig}^{\mathcal{I}}|$ and $\varphi_i \in \mathbf{Sen}^{\mathcal{I}}(\Sigma)$, $i = 0, \ldots, n$, if $\{\varphi_i\}_{i=1}^n \vdash_\Sigma^{\mathcal{G}} \varphi_0$ then $\{\varphi_i\}_{i=1}^n \vdash_\Sigma^{\mathcal{G}(\mathcal{I})} \varphi_0$. If the opposite implication holds as well, we say that $\mathcal{G}$ is* complete for $\mathcal{I}$.

The sentence part of any institution $\mathcal{I}$ may be used to determine open formulae in this institution in exactly the same way as in Section 4 for ground logics, giving a functor $\mathcal{F}_{\mathbf{Sen}^{\mathcal{I}}} : \mathbf{Sig}^{\mathcal{I}} \to \mathbf{Func}_{\mathbf{ICat}}(\mathbf{Set})$. To use the model-theoretic satisfaction relation of $\mathcal{I}$ to determine consequence relations on open formulae, we need an "institutional" version of the notion of a valuation. This may be introduced in a rather straightforward way: for any signature $\Sigma \in |\mathbf{Sig}^{\mathcal{I}}|$, Σ-context Γ (a signature extension), and model $M \in |\mathbf{Mod}^{\mathcal{I}}(\Sigma)|$, a *valuation* of context Γ in the model M is any expansion of M to a Σ^Γ-model, *i.e.*, a model $M' \in \mathbf{Mod}^{\mathcal{I}}(\Sigma^\Gamma)$ such that $M'|_\Sigma = M$.

Definition 5.5 *Let $\mathcal{I}$ be an arbitrary institution. The* validity logic *$\mathcal{L}^v(\mathcal{I})$* of open formulae determined by *$\mathcal{I}$ consists of the formula functor $\mathcal{F}_{\mathbf{Sen}^{\mathcal{I}}} : \mathbf{Sig}^{\mathcal{I}} \to \mathbf{Func}_{\mathbf{ICat}}(\mathbf{Set})$ with the consequence relations on open formulas defined as follows: for each signature $\Sigma \in |\mathbf{Sig}^{\mathcal{I}}|$, Σ-contexts Γ_i and formulae $\varphi_i \in \mathcal{F}_{\mathbf{Sen}^{\mathcal{I}}}(\Sigma)(\Gamma_i)$ for $i = 0, \ldots, n$*

$\{\langle\Gamma_i, \varphi_i\rangle\}_{i=1}^n \vdash_\Sigma^{\mathcal{L}^v(\mathcal{I})} \langle\Gamma_0, \varphi_0\rangle$ *if and only if for each model $M \in |\mathbf{Mod}^{\mathcal{I}}(\Sigma)|$, $M_0 \models_{\Sigma^{\Gamma_0}} \varphi_0$ for all $M_0 \in |\mathbf{Mod}^{\mathcal{I}}(\Sigma^\Gamma)|$ such that $M_0|_\Sigma = M$ whenever for all $i = 1, \ldots, n$, $M_i \models_{\Sigma^{\Gamma_0}} \varphi_i$ for all $M_i \in |\mathbf{Mod}^{\mathcal{I}}(\Sigma^\Gamma)|$ such that $M_i|_\Sigma = M$.*

Proposition 5.6 *For any institution $\mathcal{I}$, the logical system $\mathcal{L}^v(\mathcal{I})$ as defined in Def. 5.5 is indeed a logical system of open formulae of validity type.*

Definition 5.7 *Let $\mathcal{I}$ be an institution and $\mathcal{L}$ be a logic of open formulae. We say that $\mathcal{L}$* is sound for $\mathcal{I}$ under the validity interpretation *if*

- *$\mathcal{L}$ is of validity type*
- *$\mathcal{F}_{\mathcal{L}}$ is $\mathcal{F}_{\mathbf{Sen}^{\mathcal{I}}}$*
- *For all signatures $\Sigma \in |\mathbf{Sig}^{\mathcal{I}}|$, Σ-contexts Γ_i and formulae $\varphi_i \in \mathcal{F}_{\mathbf{Sen}^{\mathcal{I}}}(\Sigma)(\Gamma_i)$, $i = 0, \ldots, n$, if $\{\langle\Gamma_i, \varphi_i\rangle\}_{i=1}^n \vdash_\Sigma^{\mathcal{L}} \langle\Gamma_0, \varphi_0\rangle$ then $\{\langle\Gamma_i, \varphi_i\rangle\}_{i=1}^n \vdash_\Sigma^{\mathcal{L}^v(\mathcal{I})} \langle\Gamma_0, \varphi_0\rangle$.*

If the implication opposite to the one in the last condition holds as well, we say that $\mathcal{L}$ is complete for $\mathcal{I}$ under the validity interpretation.

Proposition 5.8 *If a ground logical system $\mathcal{G}$ is sound for an institution $\mathcal{I}$, then $\mathcal{L}^v(\mathcal{G})$ is sound for $\mathcal{I}$ under the validity interpretation.*

Proof Consider any $\Sigma \in |\mathbf{Sig}^{\mathcal{I}}|$, Σ-contexts Γ_i and formulae $\varphi_i \in \mathcal{F}_{\mathbf{Sen}^{\mathcal{I}}}(\Sigma)(\Gamma_i)$, $i = 0, \ldots, n$, such that $\{\langle\Gamma_i, \varphi_i\rangle\}_{i=1}^n \vdash_\Sigma^{\mathcal{L}^v(\mathcal{G})} \langle\Gamma_0, \varphi_0\rangle$. Then, for some $\{i_1, \ldots, i_k\} \subseteq \{1, \ldots, n\}$ there exist Σ-context morphisms $\gamma_l : \Gamma_{i_l} \to \Gamma_0$, $l = 1, \ldots, k$, such that $\{\gamma_l(\varphi_{i_l})\}_{l=1}^k \vdash_{\Sigma^{\Gamma_0}}^{\mathcal{G}} \varphi_0$. The soundness of $\mathcal{G}$ implies that $\{\mathcal{G}(\gamma_l)(\varphi_{i_l})\}_{l=1}^k \vdash_{\Sigma^{\Gamma_0}}^{\mathcal{G}(\mathcal{I})} \varphi_0$, *i.e.*, for every $M_0 \in |\mathbf{Mod}^{\mathcal{I}}(\Sigma^{\Gamma_0})|$, $M_0 \models_{\mathcal{I},\Sigma^{\Gamma_0}} \varphi_0$ whenever $M_0 \models_{\mathcal{I},\Sigma^{\Gamma_0}} \mathbf{Sen}^{\mathcal{I}}(\gamma_l)(\varphi_{i_l})$ for all $l = 1, \ldots, k$.

Consider now $M \in |\mathbf{Mod}^{\mathcal{I}}(\Sigma)|$ such that for all $i = 1, \ldots, n$, for all $M_i \in |\mathbf{Mod}^{\mathcal{I}}(\Sigma^{\Gamma_i})|$ such that $M_i|_\Sigma = M$, $M_i \models_{\mathcal{I},\Sigma^{\Gamma_i}} \varphi_i$. Let then $M_0 \in |\mathbf{Mod}^{\mathcal{I}}(\Sigma^{\Gamma_0})|$ be such that $M_0|_\Sigma = M$. Then for $l = 1, \ldots, k$, since $\mathbf{Mod}^{\mathcal{I}}(\gamma_l)(M_0)|_\Sigma = M_0|_\Sigma = M$, the satisfaction condition implies that $M_0 \models_{\mathcal{I},\Sigma^{\Gamma_0}} \mathbf{Sen}^{\mathcal{I}}(\gamma_l)(\varphi_{i_l})$. Hence, $M_0 \models_{\mathcal{I},\Sigma^{\Gamma_0}} \varphi_0$ as well, and we conclude

$$\{\langle\Gamma_i, \varphi_i\rangle\}_{i=1}^n \vdash_\Sigma^{\mathcal{L}^v(\mathcal{I})} \langle\Gamma_0, \varphi_0\rangle$$

□

Completeness, as usual, is much more difficult. In general, $\mathcal{L}^v(\mathcal{G})$ need not be complete for $\mathcal{I}$ under the validity interpretation even if $\mathcal{G}$ is complete for $\mathcal{I}$. We are working on natural conditions on the institution $\mathcal{I}$ that would ensure this to be the case.

We can, however, ensure so-called *weak completeness*: if $\mathcal{G}$ is weakly complete for $\mathcal{I}$, *i.e.* all theorems (consequences of the empty set of premises) of $\mathcal{I}$ are theorems of $\mathcal{G}$, then $\mathcal{L}^v(\mathcal{G})$ is weakly complete for $\mathcal{I}$ under the validity interpretation.

To introduce the logic of open formulae of an institution with the truth interpretation, we need one more technical concept.

Definition 5.9 *Let $\mathcal{I}$ be an institution. Consider $\Sigma \in |\mathbf{Sig}^{\mathcal{I}}|$ and Σ-contexts Γ_i, $i = 0,\dots,n$. We say that a family of models $\{M_i\}_{i=0}^n$, $M_i \in |\mathbf{Mod}^{\mathcal{I}}(\Sigma^{\Gamma_i})|$ for $i = 0,\dots,n$, is* compatible *if for all $j,k \in \{0,\dots,n\}$, for all Σ-contexts Γ such that $\Gamma \hookrightarrow \Gamma_j$ and $\Gamma \hookrightarrow \Gamma_k$, $M_j|_{\Sigma^\Gamma} = M_k|_{\Sigma^\Gamma}$.*

Definition 5.10 *Let $\mathcal{I}$ be an arbitrary institution. The* truth logic $\mathcal{L}^t(\mathcal{I})$ of open formulae *determined by $\mathcal{I}$ consists of the formula functor $\mathcal{F}_{\mathbf{Sen}^{\mathcal{I}}} : \mathbf{Sig}^{\mathcal{I}} \to \mathbf{Func}_{\mathbf{ICat}}(\mathbf{Set})$ with the consequence relations on open formulas defined as follows: for each signature $\Sigma \in |\mathbf{Sig}^{\mathcal{I}}|$, Σ-contexts Γ_i and formulae $\varphi_i \in \mathcal{F}_{\mathbf{Sen}^{\mathcal{I}}}(\Sigma)(\Gamma_i)$ for $i = 0,\dots,n$*

$\{\langle\Gamma_i,\varphi_i\rangle\}_{i=1}^n \vdash_\Sigma^{\mathcal{L}^t(\mathcal{I})} \langle\Gamma_0,\varphi_0\rangle$ *if and only if (for each model $M \in |\mathbf{Mod}^{\mathcal{I}}(\Sigma)|$) for every compatible family of models $\{M_i \in |\mathbf{Mod}^{\mathcal{I}}(\Sigma^{\Gamma_i})|\}_{i=0}^n$ (such that $M_i|_\Sigma = M$ for any $i = 0,\dots,n$) $M_0 \models_{\Sigma^{\Gamma_0}} \varphi_0$ whenever $M_i \models_{\Sigma^{\Gamma_0}} \varphi_i$ for all $i = 1,\dots,n$.*

OOPS! Unfortunately, in general this is not well-defined, since the relations $\vdash_\Sigma^{\mathcal{L}^t(\mathcal{I})}$ as specified above need not be transitive. The (lack of) existence of valuations for intermediate contexts causes the problem.

Definition 5.11 *An institution $\mathcal{I}$ is* regular *if for any signature Σ, family of Σ-contexts Γ_i, $i = 0,\dots,n$, and compatible family $\{M_i \in |\mathbf{Mod}^{\mathcal{I}}(\Sigma^{\Gamma_i})|\}_{i=1}^n$ there exists a model $M_0 \in |\mathbf{Mod}^{\mathcal{I}}(\Sigma^{\Gamma_0})|$ such that the family $\{M_i \in |\mathbf{Mod}^{\mathcal{I}}(\Sigma^{\Gamma_i})|\}_{i=0}^n$ is compatible.*

In the most typical situations (or more abstractly, under some additional assumptions on the inclusion ordering of the collection of signatures) the requirement of regularity is equivalent to the condition that all reduct functors induced by signature inclusions are surjective. For example, the usual institution of first-order logic *where the structures are assumed to have non-empty carriers* satisfies this requirement. There are, however, numerous natural institutions which are not regular.

Proposition 5.12 *If $\mathcal{I}$ is a regular institution then $\mathcal{L}^t(\mathcal{I})$ as defined in Def. 5.10 is indeed a logical system of open formulae of truth type.*

Proof The only problem is to prove the transitivity of $\vdash_\Sigma^{\mathcal{L}^t(\mathcal{I})}$, which follows in a rather straightforward way when regularity is assumed. □

Definition 5.13 *Let $\mathcal{I}$ be an institution and $\mathcal{L}$ be a logic of open formulae. We say that $\mathcal{L}$ is* sound *for $\mathcal{I}$* under the truth interpretation *if*

- *$\mathcal{L}$ is of truth type*
- *$\mathcal{F}_{\mathcal{L}}$ is $\mathcal{F}_{\mathbf{Sen}^{\mathcal{I}}}$*
- *For all signatures $\Sigma \in |\mathbf{Sig}^{\mathcal{I}}|$, Σ-contexts Γ_i and formulae $\varphi_i \in \mathcal{F}_{\mathbf{Sen}^{\mathcal{I}}}(\Sigma)(\Gamma_i)$, $i = 0,\dots,n$, if $\{\langle\Gamma_i,\varphi_i\rangle\}_{i=1}^n \vdash_\Sigma^{\mathcal{L}} \langle\Gamma_0,\varphi_0\rangle$ then $\{\langle\Gamma_i,\varphi_i\rangle\}_{i=1}^n \vdash_\Sigma^{\mathcal{L}^t(\mathcal{I})} \langle\Gamma_0,\varphi_0\rangle$.*

If the implication opposite to the one in the last condition holds as well, we say that $\mathcal{L}$ is complete *for $\mathcal{I}$* under the truth interpretation.

Proposition 5.14 *If a ground logical system $\mathcal{G}$ is sound for a regular institution $\mathcal{I}$ then $\mathcal{L}^t(\mathcal{G})$ is sound for $\mathcal{I}$ under the truth interpretation.*

Proof Consider any $\Sigma \in |\mathbf{Sig}^{\mathcal{I}}|$, Σ-contexts Γ_i and formulae $\varphi_i \in \mathcal{F}_{\mathbf{Sen}^{\mathcal{I}}}(\Sigma)(\Gamma_i)$, $i = 0, \ldots, n$, such that $\{ \langle \Gamma_i, \varphi_i \rangle \}_{i=1}^{n} \vdash_{\Sigma}^{\mathcal{L}^t(\mathcal{G})} \langle \Gamma_0, \varphi_0 \rangle$. It is enough to prove the soundness for the generating relations of $\vdash_{\Sigma}^{\mathcal{L}^t(\mathcal{G})}$. So, we can assume that for some $\{ i_1, \ldots, i_k \} \subseteq \{ 1, \ldots, n \}$ there is a Σ-context Γ such that $\Gamma \hookrightarrow \Gamma_{i_l}$ for $l = 1, \ldots, k$ and $\Gamma \hookrightarrow \Gamma_0$, and there exist formulae $\psi_l \in \mathbf{Sen}^{\mathcal{I}}(\Sigma^{\Gamma})$, $l = 0, \ldots, k$, such that

- $\varphi_{i_l} \vdash^{\mathcal{G}}_{\Sigma^{\Gamma_{i_l}}} \psi_l$ for $l = 1, \ldots, k$
- $\{ \psi_l \}_{l=1}^{k} \vdash^{\mathcal{G}}_{\Sigma^{\Gamma}} \psi_0$
- $\psi_0 \vdash^{\mathcal{G}}_{\Sigma^{\Gamma_0}} \varphi_0$

Since $\mathcal{G}$ is sound for $\mathcal{I}$, we have

- $\varphi_{i_l} \vdash^{\mathcal{G}(\mathcal{I})}_{\Sigma^{\Gamma_{i_l}}} \psi_l$ for $l = 1, \ldots, k$
- $\{ \psi_l \}_{l=1}^{k} \vdash^{\mathcal{G}(\mathcal{I})}_{\Sigma^{\Gamma}} \psi_0$
- $\psi_0 \vdash^{\mathcal{G}(\mathcal{I})}_{\Sigma^{\Gamma_0}} \varphi_0$

Consider now any compatible family of models $\{ M_i \in |\mathbf{Mod}^{\mathcal{I}}(\Sigma^{\Gamma_i})| \}_{i=0}^{n}$ such that for $i = 1, \ldots, n$, $M_i \models_{\mathcal{I},\Sigma^{\Gamma_i}} \varphi_i$. Then for $l = 1, \ldots, k$, $M_{i_l} \models_{\mathcal{I},\Sigma^{\Gamma_{i_l}}} \psi_l$ and hence by the satisfaction condition $M_{i_l}|_{\Sigma^{\Gamma}} \models_{\mathcal{I},\Sigma^{\Gamma}} \psi_l$. This implies (by the compatibility of the family $\{ M_i \}_{i=0}^{n}$) that $M_0|_{\Sigma^{\Gamma}} \models_{\mathcal{I},\Sigma^{\Gamma}} \psi_0$ and hence, again by the satisfaction condition, $M_0 \models_{\mathcal{I},\Sigma^{\Gamma_0}} \psi_0$. Thus, $M_0 \models_{\mathcal{I},\Sigma^{\Gamma_0}} \varphi_0$, which proves $\{ \langle \Gamma_i, \varphi_i \rangle \}_{i=1}^{n} \vdash_{\Sigma}^{\mathcal{L}^t(\mathcal{I})} \langle \Gamma_0, \varphi_0 \rangle$ □

In general, completeness is not preserved in the case of the truth interpretation either. However, if $\mathcal{G}$ is complete for $\mathcal{I}$, then the logic $\mathcal{L}^t(\mathcal{G})$ is complete for $\mathcal{I}$ under the truth interpretation relative to the restriction of the consequence relations to formulae with the same context.

6 Logical systems and LF

In order to discuss representations of logical systems in LF, we first recall from [HST89] the logical system associated with the LF type theory. The basic form of assertion in this logic is that a closed type is inhabited. We then remove the restriction to closed types and define two logical systems of open formulae determined by the LF type theory, one of validity type, the other of truth type.

Definition 6.1 *An* LF signature morphism $\sigma : \Sigma \to \Sigma'$ *is a function σ mapping constants to closed terms such that if $c{:}A$ ($c{:}K$) occurs in Σ, then $\vdash_{\Sigma'} \sigma(c) : \sigma^{\natural}A$ ($\vdash_{\Sigma'} \sigma(c) : \sigma^{\natural}K$). (The function $\sigma^{\natural}$ is the natural extension of σ to LF terms.)* $\mathbf{Sig}^{\mathcal{LF}}$ *is the category with inclusions of LF signatures and LF-signature morphisms, with composition defined in the obvious way.*

Proposition 6.2 *If $\sigma : \Sigma \to \Sigma'$ and $\vdash_{\Sigma} \alpha$, then $\vdash_{\Sigma'} \sigma^{\natural}\alpha$ for each assertion α of the LF type system.*

Definition 6.3 *Let Σ be an LF signature. $\mathcal{GLF}(\Sigma)$ is the pair $\langle \mathrm{Types}_{\Sigma}, \vdash_{\Sigma}^{\mathcal{LF}} \rangle$ where* $\mathrm{Types}_{\Sigma} = \{ A \mid \vdash_{\Sigma} A : \mathsf{Type} \}$ *and*

$$A_1, \ldots, A_n \vdash_{\Sigma}^{\mathcal{LF}} A \quad \textit{iff} \quad x_1{:}A_1, \ldots, x_n{:}A_n \vdash_{\Sigma} M : A$$

for some M and any pairwise distinct variables $x_1, \ldots, x_n$.

This consequence relation has a straightforward Gentzen-style axiomatization similar to that used in NuPRL [Con86].

Definition 6.4 *The* ground logic of LF *is a functor* $\mathcal{GLF} : \mathbf{Sig}^{\mathcal{LF}} \to \mathbf{CR}$ *which is the extension of the map* $\Sigma \mapsto \mathcal{GLF}(\Sigma)$ *defined by taking* $\mathcal{GLF}(\sigma)$, *for* $\sigma : \Sigma \to \Sigma'$, *to be* $\sigma^{!}\upharpoonright \mathrm{Types}_\Sigma$, *the restriction of* $\sigma^{!}$ *to closed* Σ*-types.*

Logical systems of open formulae are determined by the LF type theory in much the same style. We have already defined the category of signatures. Contexts are just LF contexts with substitutions of terms for variables as morphisms:

Definition 6.5 *For any LF signature* $\Sigma \in |\mathbf{Sig}^{\mathcal{LF}}|$ *and any* Σ*-contexts* Γ *and* Γ', *an* LF context morphism $\gamma : \Gamma \to \Gamma'$ *is a function* γ *mapping variables to terms such that for each declaration* $x : A$ *in* Γ, $\Gamma' \vdash_\Sigma \gamma(x){:}\gamma^{!}A$. *(The function* $\gamma^{!}$ *is the obvious extension of* γ *to LF terms.) The category* $\mathbf{Ctxt}_\Sigma^{\mathcal{LF}}$ *of LF* Σ*-contexts and their morphisms is defined in a straightforward way.*

Open formulae of the systems we define are just LF types:

Definition 6.6 *For any LF signature* $\Sigma \in |\mathbf{Sig}^{\mathcal{LF}}|$ *and LF context* $\Gamma \in |\mathbf{Ctxt}_\Sigma^{\mathcal{LF}}|$, Σ*-formulae in context* Γ *are LF types formed over signature* Σ *in context* Γ,

$$\mathcal{F}_{\mathcal{LF}}(\Sigma)(\Gamma) = \{ A \mid \Gamma \vdash_\Sigma A{:}\mathsf{Type} \}$$

Definition 6.7 *The* $\mathcal{LF}$ *formula functor* $\mathcal{F}_{\mathcal{LF}} : \mathbf{Sig}^{\mathcal{LF}} \to \mathbf{Func}_{\mathbf{ICat}}(\mathbf{Set})$ *is the obvious extension of the mappings defined in Definitions 6.5 and 6.6 to a functor: for any signature morphisms* σ, $\mathcal{F}_{\mathcal{LF}}(\sigma)$ *is essentially given by the natural translation of LF terms induced by* σ*; for any signature* Σ *and a* Σ*-context morphism* γ, $\mathcal{F}_{\mathcal{LF}}(\Sigma)(\gamma)$ *is again the natural translation of LF terms induced by* γ.

Proposition 6.8 *There is an obvious, componentwise inclusion morphism*

$$S_{\mathcal{LF}} : \mathcal{F}_{\mathcal{LF}} \to \mathcal{F}_{\mathcal{GLF}}$$

where $\mathcal{F}_{\mathcal{GLF}}$ *is the formula functor determined by the ground logical system of LF,* $\mathcal{GLF}$, *as in Section 4. The inclusions are proper at the context level: LF contexts used in* $\mathcal{F}_{\mathcal{LF}}$ *are extensions of signatures by object constants (constants of a type), which excludes extensions by type constants (constants of a kind).*

Definition 6.9 *The* validity LF logic $\mathcal{LF}^v$ *is the logic of open formulae with the formula functor* $\mathcal{F}_{\mathcal{LF}} : \mathbf{Sig}^{\mathcal{LF}} \to \mathbf{Func}_{\mathbf{ICat}}(\mathbf{Set})$ *and, for each LF signature* $\Sigma \in |\mathbf{Sig}^{\mathcal{LF}}|$, *with the consequnce relation on* $Form_{\mathcal{F}_{\mathcal{LF}}}(\Sigma)$ *given as follows: for any* Σ *contexts* Γ_i *and* A_i *such that* $\Gamma_i \vdash_\Sigma A_i{:}\mathsf{Type}$, *for* $i = 0, \ldots, n$,

$$\{ \langle \Gamma_i, A_i \rangle \}_{i=1}^n \vdash_\Sigma^{\mathcal{LF}^v} \langle \Gamma_0, A_0 \rangle$$

if and only if

$$x_1{:}\Pi\Gamma_1.A_1, \ldots, x_n{:}\Pi\Gamma_n.A_n \vdash_\Sigma M{:}\Pi\Gamma_0.A_0$$

for some mutually distinct variables $x_1, \ldots, x_n$ *and term* M. *(We use the informal notation* $\Pi\Gamma_i.A_i$ *for the type obtained by* Π*-closure of* A_i *w.r.t. the variables in the context* Γ_i.*)*

Proposition 6.10 $\mathcal{LF}^v$ *as defined in Def. 6.9 is indeed a logical system of open formulae of the validity type.*

Definition 6.11 *The* truth LF logic $\mathcal{LF}^t$ *is the logic of open formulae with the formula functor* $\mathcal{F}_{\mathcal{LF}} : \mathbf{Sig}^{\mathcal{LF}} \to \mathbf{Func}_{\mathbf{ICat}}(\mathbf{Set})$ *and, for each LF signature* $\Sigma \in |\mathbf{Sig}^{\mathcal{LF}}|$, *with the consequnce relation on* $Form_{\mathcal{F}_{\mathcal{LF}}}(\Sigma)$ *defined as the transitive closure of the relation given as follows: for any* Σ*-contexts* Γ_i *and types* A_i *such that* $\Gamma_i \vdash_\Sigma A_i{:}\mathsf{Type}$ *for* $i = 0, \ldots, n$,

$\{ \langle \Gamma_i, A_i \rangle \}_{i=1}^n \vdash_\Sigma^{\mathcal{LF}^t} \langle \Gamma_0, A_0 \rangle$ *if and only if for some* $\{ i_1, \ldots, i_k \} \subseteq \{ 1, \ldots, n \}$, *there exists a* Σ*-context* Γ *and types* A'_l, $l = 1, \ldots, k$, *such that for* $l = 1, \ldots, k$, $\Gamma \vdash_\Sigma A'_l{:}\mathsf{Type}$ *and moreover*

- $\Gamma_{i_l}, x{:}A_{i_l} \vdash_\Sigma M_l{:}A'_l$ *for* $l = 1, \ldots, k$, *for some variable* x *and term* M_l
- $\Gamma, x_1{:}A'_1, \ldots, x_k{:}A'_k \vdash_\Sigma M{:}A'_0$ *for some mutually distinct variables* $x_1, \ldots, x_k$ *and term* M
- $\Gamma_0, x{:}A'_0 \vdash_\Sigma M_0{:}A_0$

Remark 6.12 *If all contexts in Def. 6.11 are the same, say* $\Gamma_i = \Gamma$ *for* $i = 0, \ldots, n$, *then the condition on the generating relation given in the definition is equivalent to the requirement that for some mutually distinct variables* $x_1, \ldots, x_n$ *and term* N,

$$\Gamma, x_1{:}A_1, \ldots, x_n{:}A_n \vdash_\Sigma N{:}A_0.$$

Moreover, the relation defined in this way is already transitive on the formulae built in the same context.

Proposition 6.13 $\mathcal{LF}^t$ *as defined in Def. 6.11 is indeed a logical system of open formulae of truth type.*

Proposition 6.14 *The inclusion morphism* $S_{\mathcal{LF}} : \mathcal{F}_{\mathcal{LF}} \to \mathcal{F}_{\mathcal{GLF}}$ *is a logic morphism*

$$S_{\mathcal{LF}} : \mathcal{LF}^t \to \mathcal{L}^t(\mathcal{GLF})$$

A reasonable question at this point is whether $S_{\mathcal{LF}}$ is a logic representation between $\mathcal{LF}^t$ and $\mathcal{L}^t(\mathcal{GLF})$ (and/or between $\mathcal{LF}^v$ and $\mathcal{L}^v(\mathcal{GLF})$). Unfortunately, this is not the case in general; in most cases it seems that $\mathcal{L}^t(\mathcal{GLF})$ (resp. $\mathcal{L}^v(\mathcal{GLF})$) is somewhat too weak. This topic needs further study.

For the purposes of encoding a logical system $\mathcal{L}$, we are interested in "specializations" of the logical system determined by the LF type theory obtained by fixing a "base" signature $\Sigma_\mathcal{L}$ specifying the syntax, assertions, and rules of $\mathcal{L}$ [HHP87]. The signatures of $\mathcal{L}$ are then represented as extensions to $\Sigma_\mathcal{L}$, and signature morphisms are represented as LF signature morphisms on these extensions leaving $\Sigma_\mathcal{L}$ fixed.

Definition 6.15 *Let* Σ *be an LF signature. The* category of extensions *of* Σ, *written* $\mathbf{Sig}^{\mathcal{LF}}_\Sigma$, *is the full subcategory of the slice category* $\Sigma\downarrow\mathbf{Sig}^{\mathcal{LF}}$ *determined by the inclusions* $\iota : \Sigma \hookrightarrow \Sigma'$.

Every LF signature induces logical systems based on that signature as follows:

Definition 6.16 *Let* Σ *be an LF signature.*

- *The* validity logical system presented by Σ, $\mathcal{LF}^v_\Sigma$, *is the restriction of* $\mathcal{LF}^v$ *to* $\mathbf{Sig}^{\mathcal{LF}}_\Sigma$.
- *The* truth logical system presented by Σ, $\mathcal{LF}^t_\Sigma$, *is the restriction of* $\mathcal{LF}^t$ *to* $\mathbf{Sig}^{\mathcal{LF}}_\Sigma$.

An encoding of a logical system $\mathcal{L}$ in LF consists not only of an LF signature $\Sigma_\mathcal{L}$, but also of an "internal type family" distinguishing the basic judgements of $\mathcal{L}$ in the encoding [HST89]. For example, in the encoding of first-order logic given in [HHP87], the constant *true* of kind $o \to \mathsf{Type}$ represents the basic judgement form of first-order logic. The significance of *true* for the encoding becomes apparent in the statement of the adequacy theorem: terms of type *true*$(\hat{\varphi})$ in a context with variables x_i of type *true*$(\hat{\varphi}_i)$ represent proofs of φ from the φ_i's. Similarly, we also indicate which LF contexts are used to represent $\mathcal{L}$-contexts. In the encoding of first-order logic given in [HHP87], first-order contexts are represented by LF contexts with variables of type ι (a distinguished type of individuals). This methodology is formalized in our setting as follows.

Definition 6.17 *An* internal type family *of* Σ *is a term* F *such that* $\vdash_\Sigma F : K$ *for some kind* K. *(Note that if* $\vdash_\Sigma K$*, then* K *has normal form* $\Pi x_1{:}A_1.\ldots.\Pi x_k{:}A_k.\mathsf{Type}$ *for some* $x_1,\ldots,x_k$ *and* $A_1,\ldots,A_k$.*)* The range *of an internal type family* F *of* Σ *in a* Σ*-context* Γ *is defined to be the set*

$$\mathrm{Rng}^\Gamma_\Sigma(F) = \{\, F\,M_1\,\ldots\,M_k \mid \Gamma \vdash_\Sigma F\,M_1\,\ldots\,M_k : \mathsf{Type}\,\},$$

(where terms are identified up to $\beta\eta$*-conversion.) If* $\mathcal{J}$ *is a set of internal type families of* Σ*, then*

$$\mathrm{Rng}^\Gamma_\Sigma(\mathcal{J}) = \bigcup_{F\in\mathcal{J}} \mathrm{Rng}^\Gamma_\Sigma(F).$$

Definition 6.18 *A* logic presentation *is a triple* $\langle\Sigma,\mathcal{T},\mathcal{J}\rangle$ *where* Σ *is an LF signature and* $\mathcal{T}$ *and* $\mathcal{J}$ *are finite sets of internal type families of* Σ.

Definition 6.19 *Let* $\langle\Sigma,\mathcal{T},\mathcal{J}\rangle$ *be a logic presentation.*
The validity (truth, respectively) logical system presented by $\langle\Sigma,\mathcal{T},\mathcal{J}\rangle$, $\mathcal{P}^v(\Sigma,\mathcal{T},\mathcal{J})$ *(*$\mathcal{P}^t(\Sigma,\mathcal{T},\mathcal{J})$*, respectively) is the restriction of* $\mathcal{LF}^v_\Sigma$ *(*$\mathcal{LF}^t_\Sigma$*, respectively):*

- *to signatures and signature morphisms in* $\mathbf{Sig}^{\mathcal{LF}}_\Sigma$,
- *for each signature* $\Sigma' \in |\mathbf{Sig}^{\mathcal{LF}}_\Sigma|$*, to* Σ'*-contexts of the form* $\langle x_1{:}A_1,\ldots,x_n{:}A_n\rangle$ *where* $A_i \in \mathrm{Rng}^{\langle x_1:A_1,\ldots,x_{i-1}:A_{i-1}\rangle}_{\Sigma'}(\mathcal{T})$ *for* $i = 1,\ldots,n$,
- *for each signature* $\Sigma' \in |\mathbf{Sig}^{\mathcal{LF}}_\Sigma|$ *and* Σ'*-context* Γ *(satisfying the above requirement), to the formulae that are types in* $\mathrm{Rng}^\Gamma_\Sigma(\mathcal{J})$.

Definition 6.20 *A logical system is* uniformly validity-encodable *(*uniformly truth-encodable, *respectively) in LF if there exists a logic presentation* $\langle\Sigma_\mathcal{L},\mathcal{T}_\mathcal{L},\mathcal{J}_\mathcal{L}\rangle$ *and a surjective exact representation* $\rho_\mathcal{L} : \mathcal{L} \to \mathcal{P}^v(\Sigma_\mathcal{L},\mathcal{J}_\mathcal{L})$ *(*$\rho_\mathcal{L} : \mathcal{L} \to \mathcal{P}^t(\Sigma_\mathcal{L},\mathcal{J}_\mathcal{L})$*, respectively). The tuple* $\langle\Sigma_\mathcal{L},\mathcal{J}_\mathcal{L},\rho_\mathcal{L}\rangle$ *with an indication of the type of the system* $\mathcal{L}$ *is called a* uniform encoding *of* $\mathcal{L}$ *in LF.*

The word "uniform" reflects the fact that we require a "natural" (or "compositional") encoding of the entire family of consequence relations of $\mathcal{L}$ in LF, rather than a signature-by-signature encoding as is suggested by the account in [HHP87]. The requirement of exactness ensures that $\mathcal{T}$ accurately describes the images of $\mathcal{L}$-contexts in LF. The requirement of surjectivity ensures that $\mathcal{J}$ accurately describes the images of $\mathcal{L}$-sentences in LF. For example, in the encoding of first-order logic in [HHP87], only proofs of true(M) in contexts with variables of type ι (in addition to those labelling assumptions) are considered, for otherwise a complete correspondence with first-order logic cannot in general be expected.

All the methodological consequences of the notions presented above, as described in [HST89] for ground logical systems, carry over to the present framework of logical systems of open formulae, their presentations and encodings in LF. We refer to that paper, as well as to [HHP87] and [AHM87] for examples of logic encodings in LF that may be readily adapted to the framework we have introduced here.

The following technicalities indicate that the ideas on presenting logics in a structured way using the pushout construction suggested in [HST89] carry over as well.

Definition 6.21 *A* logic presentation morphism $\sigma : \langle\Sigma,\mathcal{T},\mathcal{J}\rangle \to \langle\Sigma',\mathcal{T}'\mathcal{J}'\rangle$ *is a signature morphism* $\sigma : \Sigma \to \Sigma'$ *in* $\mathbf{Sig}^{\mathcal{LF}}$ *such that for every* $F \in \mathcal{T}$ *(*$F \in \mathcal{J}$*, respectively) with*

$$\vdash_\Sigma F : \Pi x_1{:}A_1.\ldots.x_k{:}A_k.\mathsf{Type},$$

there exists $F' \in \mathcal{T}'$ *(*$F' \in \mathcal{J}'$*, respectively) such that*

$$\sigma^\sharp F =_{\beta\eta} \lambda x_1{:}\sigma^\sharp A_1.\ldots.x_k{:}\sigma^\sharp A_k.F'(M_1,\ldots,M_n)$$

for some M_1*, …,* M_n *of suitable type.* **LogPres** *is the category of logic presentations and logic presentation morphisms.*

Proposition 6.22 *The assignments* $\langle\Sigma,\mathcal{T},\mathcal{J}\rangle \mapsto \mathcal{P}^v(\Sigma,\mathcal{T},\mathcal{J})$ *and* $\langle\Sigma,\mathcal{T},\mathcal{J}\rangle \mapsto \mathcal{P}^t(\Sigma,\mathcal{T},\mathcal{J})$ *extends to functors* $\mathcal{P}^v : \mathbf{LogPres} \to \mathbf{Log}$ *and* $\mathcal{P}^t : \mathbf{LogPres} \to \mathbf{Log}$, *respectively.*

Sketch of construction Consider a presentation morphism $\sigma : \langle\Sigma_1,\mathcal{T}_1,\mathcal{J}_1\rangle \to \langle\Sigma_2,\mathcal{T}_2,\mathcal{J}_2\rangle$. The logic morphism $\mathcal{P}^{v(t)}(\sigma) : \mathcal{P}^{v(t)}(\Sigma_1,\mathcal{T}_2,\mathcal{J}_1) \to \mathcal{P}^{v(t)}(\Sigma_2,\mathcal{T}_2,\mathcal{J}_2)$ may be defined as follows:

- $\mathcal{P}^{v(t)}(\sigma)^{Sig} : \mathbf{Sig}^{\mathcal{LF}}_{\Sigma_1} \to \mathbf{Sig}^{\mathcal{LF}}_{\Sigma_2}$ is defined on objects using the canonical pushout construction: $\mathcal{P}^{v(t)}(\sigma)^{Sig}(\iota_1 : \Sigma_1 \hookrightarrow \Sigma_1') = (\iota_2 : \Sigma_2 \hookrightarrow \Sigma_2')$ where $\Sigma_2' = \sigma^\star\Sigma_1'$ and $\iota_2 : \Sigma_2 \hookrightarrow \sigma^\star\Sigma_1'$ is the inclusion morphism to the pushout of σ and ι_1 in $\mathbf{Sig}^{\mathcal{LF}}$. This extends to morphisms using the co-universal property of pushouts.

- $\sigma' : \Sigma_1' \to \Sigma_2'$ in the construction above induces the translation $(\sigma')^\natural$ of LF terms over Σ_1' to LF terms over Σ_2' and of Σ_1'-contexts to Σ_2'-contexts. Moreover, for any Σ_1'-context Γ_1', $(\sigma')^\natural : \mathrm{Rng}^{\Gamma_1'}_{\Sigma_1'}(\mathcal{T}_1) \to \mathrm{Rng}^{(\sigma')^\natural(\Gamma_1')}_{\Sigma_2'}(\mathcal{T}_2)$ and similarly $(\sigma')^\natural : \mathrm{Rng}^{\Gamma_1'}_{\Sigma_1'}(\mathcal{J}_1) \to \mathrm{Rng}^{(\sigma')^\natural(\Gamma_1')}_{\Sigma_2'}(\mathcal{J}_2)$. (This uses the fact that σ is a logic presentation morphism.) It is easy to see that this translation preserves consequence relations as required.

We propose to use colimits in the category of logic presentations to build logics in a structured fashion. Although the category of logic presentations is not finitely co-complete, it may be shown that a diagram in **LogPres** has a colimit iff its projection to $\mathbf{Sig}^{\mathcal{LF}}$ has a colimit. The most pertinent case is that of pushouts along inclusions:

Proposition 6.23 LogPres *is a category with inclusions, where a logic presentation morphism* $\iota : \langle\Sigma,\mathcal{T},\mathcal{J}\rangle \hookrightarrow \langle\Sigma',\mathcal{T}',\mathcal{J}'\rangle$ *is an* inclusion *if* $\iota : \Sigma \hookrightarrow \Sigma'$ *is an inclusion,* $\mathcal{T} \subseteq \mathcal{T}'$ *and* $\mathcal{J} \subseteq \mathcal{J}'$. *In particular,* **LogPres** *has pushouts along inclusions.*

A **LogPres** inclusion can be seen as a parameterized logic presentation where the pushout of this morphism with a "fitting" morphism amounts to instantiation, by analogy with parameterized structured theory presentations. Small examples of this are given in [HST89] for ground logical systems, and may be generalized to the framework of logics of open formulae we present here. Let us just stress here once again that the category **LogPres** of logic presentations, not the category **Log** of logics, seems appropriate for "putting logics together"

7 Directions for further research

The paper presents only a sketch of some of the technicalities necessary to adequately grasp the notion of a logical system of open formulae and of a representation of such systems in a universal logical framework like LF.

An obvious technical gap in the presentation flow of this paper may be found in Section 5 where we try to connect a formal construction on logics with model theory as given by the theory of institutions. Clearly, the issue of (in)completeness of the construction needs more study. A less evident but equally important problem is how to understand (introduce?) a notion of logic encoding in LF via model theory of the encoded logic on one hand and of LF on the other. It seems to us at the moment that there may be some intrinsic difficulties there, as the model theory and proof theory offer inherently different views of logical systems.

A closely related point is to study situations in which a validity-type logical system $\mathcal{L}$ may be viewed as $\mathcal{L}^v(\mathcal{G})$ for the associated ground logical system $\mathcal{G}$ obtained by restricting $\mathcal{L}$ to closed sentences (and similarly for truth-type logical systems). It seems that in most cases the two constructions given in given in Section 4 do not yield the original logical system, but rather a somewhat weaker logic of the appropriate type. In particular, the validity logic of LF cannot be characterized as the validity-type extension of the ground LF logic to open types (due to the presence of binding operators).

Problems with general truth-type logical systems as presented here (the natural truth-type "consequence relations" are not transitive, *cf.* Sections 4 and 5) indicate that perhaps we should adopt a different formalisation, where a "truth context" is fixed throughout a deduction process, rather than being attached to individual formulae (as in this paper, and in the most straightforward view of first-order logic with open formulae, where a reasonable effect like the transitivity of the consequence relation is only due to the implicit assumption that structure carriers are never empty). Consequently, in truth-type logics consequence relations would be defined separately not only for each signature but also for each context over any signature as well. This would also allow us to combine the two types of logical systems (validity and truth) by considering a notion of a logic where for each signature Σ and for each context Γ over Σ, we would have a validity-type consequence relation on formulae built in the (truth) context Γ extended by a (validity) context which is explicitly indicated in the formula.

Part of the motivation for studying open formulae was to enable an adequate treatment of axiom schemes. We believe that the framework presented (or its alternative version mentioned in the previous paragraph) provides an appropriate basis for such treatment — but this remains to be investigated in detail. Finally, the issues of structured logic presentations need further study. Although the definitions in this paper provide the possibility of presenting a greater variety of logics than those in [HST89], concrete examples which exploit this increased flexibility have not yet been worked out.

Acknowledgements: Thanks to Rod Burstall for several useful discussions on the relation between institutions and LF. This research has been partially supported by the U.K. Science and Engineering Research Council and Edinburgh University (DS, AT), Carnegie Mellon University (RH), the Polish Academy of Sciences and Linköping University (AT).

References

[AHM87] A. Avron, F. Honsell, and I. Mason. *Using typed lambda calculus to implement formal systems on a machine.* Technical Report ECS–LFCS–87–31, Laboratory for the Foundations of Computer Science, Edinburgh University, June 1987.

[Avr87] A. Avron. *Simple consequence relations.* Technical Report ECS–LFCS–87–30, Laboratory for the Foundations of Computer Science, Edinburgh University, June 1987.

[Bar74] K.J. Barwise. Axioms for abstract model theory. *Ann. of Mathematical Logic* 7:221–265, 1974.

[BG80] R. Burstall and J. Goguen. The semantics of CLEAR, a specification language. In *Proc. of Advanced Course on Abstract Software Specifications*, Copenhagen, pages 292–332, Springer LNCS 86, 1980.

[Car86] J. Cartmell. Generalised algebraic theories and contextual categories. *Annals of Pure and Applied Logic*, 32:209–243, 1986.

[Con86] R.L. Constable, *et. al. Implementing Mathematics with the NuPRL Proof Development System.* Prentice–Hall, 1986.

[dB80] N.G. de Bruijn. A survey of the project AUTOMATH. In J.P. Seldin and J.R. Hindley, editors, *To H.B. Curry: Essays in Combinatory Logic, Lambda Calculus, and Formalism*, pages 589–606, Academic Press, 1980.

[FS88] J. Fiadeiro and A. Sernadas. Structuring theories on consequence. In D. Sannella and A. Tarlecki, editors, *Selected Papers from the Fifth Workshop on Specification of Abstract Data Types*, Gullane, Scotland, pages 44–72, Springer LNCS 332, 1988.

[GB84] J. Goguen and R. Burstall. Introducing institutions. In E. Clarke and D. Kozen, editors, *Logics of Programs*, pages 221–256, Springer LNCS 164, 1984.

[HHP87] R. Harper, F. Honsell, and G. Plotkin. A framework for defining logics. In *Proc. of the 2nd IEEE Symp. on Logic in Computer Science*, Ithaca, New York, pages 194–204, 1987.

[HST89] R. Harper, D. Sannella, A. Tarlecki. Structure and representation in LF. In *Proc. of the 4th IEEE Symp. on Logic in Computer Science*, Asilomar, California, pages 226–237, 1989.

[Mes89] J. Meseguer. General logics. In *Proc. Logic Colloquium '87*, North-Holland, 1989.

[SB83] D. Sannella and R. Burstall. Structured theories in LCF. In *Proc. of the 8th Colloq. on Algebra and Trees in Programming*, L'Aquila, Italy, pages 377–391, Springer LNCS 159, 1983.

[ST88] D. Sannella and A. Tarlecki. Specifications in an arbitrary institution. *Information and Computation*, 76:165–210, 1988.

[Tar86] A. Tarlecki. Quasi-varieties in abstract algebraic institutions. *Journal of Computer and System Sciences*, 33:333–360, 1986.

[TBG] A. Tarlecki, R. Burstall, and J. Goguen. Some fundamental algebraic tools for the semantics of computation, Part III: Indexed categories. To appear, *Theoretical Computer Science.*

[vD80] D.T. van Daalen. *The Language Theory of AUTOMATH.* Ph.D. thesis, Technical University of Eindhoven, Eindhoven, Netherlands, 1980.

Unification Properties of Commutative Theories: A Categorical Treatment

Franz Baader

Institut für Mathematische Maschinen und Datenverarbeitung (Informatik) 1, Universität Erlangen-Nürnberg, Martensstraße 3, 8520 Erlangen, F.R.G.

Abstract

A general framework for unification in "commutative" theories is investigated, which is based on a categorical reformulation of theory unification. This yields algebraic characterizations of unification type unitary (resp. finitary for unification with constants). We thus obtain the well-known results for abelian groups, abelian monoids and idempotent abelian monoids as well as some new results as corollaries to a general theorem. In addition, it is shown that constant-free unification problems in "commutative" theories are either unitary or of unification type zero and we give an example of a "commutative" theory of type zero.

1. Introduction

E-unification is concerned with solving term equations modulo an equational theory E. The theory E is finitary, if the solutions of an equation can always be represented by finitely many "most general" solutions (see Section 2 for formal definitions). Equational theories which are of unification type finitary play an important rôle in automated theorem provers with built-in equational theories (see e.g. Plotkin (1972), Nevins (1974), Slage (1974) or Stickel (1985)), in generalizations of the Knuth-Bendix algorithm (see e.g. Huet (1980), Peterson-Stickel (1981), Jouannaud (1983) and Jouannaud-Kirchner (1986)) and in logic programming with equality (see e.g. Jaffar-Lassez-Maher (1984)). Examples of finitary theories are the theory of abelian groups (Lankford-Butler-Brady (1984)), the theory of abelian monoids (Livesey-Siekmann (1978), Stickel (1981), Fages (1984), Fortenbacher (1985), Büttner (1986), Herold (1987)) and the theory of idempotent abelian monoids (Livesey-Siekmann (1978), Baader-Büttner (1988)). The proofs of these finitary-results make use of the following property, which the three theories have in common: The finitely generated free objects are direct products of the free objects in one generator.

This paper is concerned with equational theories which satisfy this and some additional properties. In Section 5 we give a characterization of these theories which justifies the name "commutative theories". A categorical reformulation of E-unification (Rydeheard-Burstall (1985)) shows that commutative theories correspond to semiadditive categories, i.e. categories which allow an associative, commutative binary operation on morphisms distributing with the composition of morphisms (Section 4). Using this fact we get algebraic characterizations of unification type unitary for constant-free unification and type finitary for unification with constants: The set of unifiers for a constant-free unification problem must correspond to a finitely generated ideal (see Condition 6.6), while for a unification problem with constants we need a finite union of cosets of a finitely generated ideal (see Condition 7.1). The above mentioned results for abelian groups etc. and some new results (for abelian monoids with an involution, idempotent abelian monoids with an involution, abelian groups with an involution, abelian groups with a homomorphism and abelian groups of exponent m) can thus be obtained as corollaries to a general theorem. This shows, which parts of the proofs are common for all these theories and which parts are specific for the theory in question. Furthermore we shall show that constant-free unification problems in commutative theories are either unitary or of unification type zero and we give an example of a commutative theory of type zero. The fact that this theory has type zero will also be proved in an algebraic way.

Before starting with the details, I would like to point out two advantages of a categorical setting for the description of unification problems. First, unification theory is not only interested in specific unification algorithms, but also in general results for whole classes of theories. Therefore an appropriate level of abstraction has to be found which allows to exhibit common structures. This paper shows that – at least for "commutative" theories – categories yield such a level of abstraction. Second, well-known results about certain categories – here semiadditive categories – can be exploited to obtain unification theoretic results. Werner Nutt (Nutt (1988)) has investigated commutative theories, which he calls monoidal, in a more algebraic way. The result for abelian groups with a homomorphism is due to him.

In the following we assume that the reader is familiar with the basic notions of universal algebra (see e.g. Cohn (1965), Grätzer (1968)). For more information about unification theory see Siekmann (1986). The composition of mappings and morphisms will be written from left to right, i.e. f ∘ g or simply fg means first f and then g. Consequently we use suffix notation for mappings.

2. E-unification

Let E be an equational theory and $=_E$ be the equality of terms, induced by E. We assume that terms are Ω-terms (with variables) for a given signature Ω. For a function symbol $\underline{f}$ in Ω we shall write f for its realization in any Ω-algebra. An *E-unification problem* is a finite set of equations denoted by $\Gamma = < s_i = t_i;\ 1 \le i \le n >_E$, where s_i and t_i are terms. A substitution θ is called an *E-unifier* of Γ iff $s_i\theta =_E t_i\theta$ for each i, i = 1, ..., n. The set of all E-unifiers of Γ is denoted by $U_E(\Gamma)$. We are mostly interested in complete sets of E-unifiers, i.e. sets of E-unifiers from which $U_E(\Gamma)$ may be generated by instantiation. More formally, we extend $=_E$ to $U_E(\Gamma)$ and define a quasi-ordering $\le_E$ on $U_E(\Gamma)$ by

$\sigma =_E \theta$ iff $x\sigma =_E x\theta$ for all variables x occurring in s_i or t_i for some i, i = 1, ..., n,

$\sigma \le_E \theta$ iff there exists a substitution λ such that $\sigma =_E \theta \circ \lambda$.

In this case σ is called an E-instance of θ. As usual the quasi-ordering $\le_E$ induces an equivalence relation $\equiv_E$ on $U_E(\Gamma)$, namely $\sigma \equiv_E \theta$ iff $\sigma \le_E \theta$ and $\theta \le_E \sigma$. A *complete set* $cU_E(\Gamma)$ *of E-unifiers* of Γ is defined as

(1) $cU_E(\Gamma) \subseteq U_E(\Gamma)$,

(2) For all $\theta \in U_E(\Gamma)$ there exists $\sigma \in cU_E(\Gamma)$ such that $\theta \le_E \sigma$.

A *minimal complete set* $\mu U_E(\Gamma)$ is a complete set of E-unifiers of Γ satisfying the minimality condition

(3) For all $\sigma, \theta \in \mu U_E(\Gamma)$ the relation $\sigma \le_E \theta$ implies $\sigma = \theta$.

A set $\mu U_E(\Gamma)$ may not always exist, but if it does it is unique up to $\equiv_E$-equivalence (Fages-Huet (1986)). Consequently equational theories may be classified according to the cardinality or existence of μU_E as follows:

(1) If $\mu U_E(\Gamma)$ exists for all E-unification problems Γ and has at most one element, then E is called *unitary*.

(2) If $\mu U_E(\Gamma)$ exists for all E-unification problems Γ and has finite cardinality, then E is called *finitary*.

(3) If $\mu U_E(\Gamma)$ exists for all E-unification problems Γ and for some E-unification problem is denumerable, then E is called *infinitary*.

(4) If for some E-unification problem Γ the minimal complete set $\mu U_E(\Gamma)$ does not exist, then E is said to be of *unification type zero*.

3. A Categorical Reformulation of E-unification

An equational theory E defines a *variety V(E)*, i.e. the class of all algebras (of the given signature Ω), which satisfy each identity of E. For any set X of generators, V(E) contains a *free algebra over V(E) with generators X*, which will be denoted by $F_E(X)$. Thus any mapping of X into an algebra $B \in V(E)$ can be uniquely extended to a homomorphism of $F_E(X)$ into B.

Let $\Gamma = < s_i = t_i;\ 1 \le i \le n >_E$ be an E-unification problem and X be the (finite) set of variables x occurring in some s_i or t_i. Evidently we can consider the s_i and t_i as elements of $F_E(X)$. Since we do not distinguish between $=_E$-equivalent unifiers, any E-unifier of Γ can be regarded as a homomorphism of $F_E(X)$ into $F_E(Y)$ for some finite set Y (of variables). Let $I = \{ x_1, ..., x_n \}$ be a set of cardinality n. We define homomorphisms

$$\sigma, \tau: F_E(I) \to F_E(X) \text{ by } x_i\sigma := s_i \text{ and } x_i\tau := t_i \quad (i = 1, ..., n).$$

Now $\delta: F_E(X) \to F_E(Y)$ is an E-unifier of Γ iff $x_i\sigma\delta = s_i\delta = t_i\delta = x_i\tau\delta$ for $i = 1, ..., n$, i.e. iff $\sigma\delta = \tau\delta$. Thus an E-unification problem can be written as a pair $< \sigma = \tau >_E$ of morphisms $\sigma, \tau: F_E(I) \to F_E(X)$ in the following category:

DEFINITION 3.1. Let E be an equational theory and V be a denumerable set. Then the *category C(E)* is defined as follows:

(1) The objects of C(E) are the algebras $F_E(X)$ for finite subsets X of V. We denote the class of these objects by F(E).
(2) The morphisms of C(E) are the homomorphisms between these objects.
(3) The composition of morphisms is the usual composition of mappings.

Note that in C(E) epimorphisms need not be surjective. But the isomorphisms (in the categorical sense) are just the bijective homomorphisms. Two objects $F_E(X)$, $F_E(Y)$ of C(E) are isomorphic iff $|X| = |Y|$. An E-unifiers of the unification problem $< \sigma = \tau >_E$ is a morphism δ such that $\sigma\delta = \tau\delta$. For morphisms $\sigma: F_E(X) \to F_E(Y)$, $\gamma: F_E(X) \to F_E(Z)$ we have $\sigma \le_E \gamma$ iff there is a morphism $\lambda: F_E(Z) \to F_E(Y)$ such that $\sigma = \gamma\lambda$.

Now the notions complete and minimal complete set of E-unifiers and unification type of a theory E are defined as in Section 2.

In this paper we shall consider equational theories E such that C(E) is a semiadditive category. Thus the well-known structure of these categories (see Freyd (1964) and Herrlich-

Strecker (1973)) can be exploited to obtain results about unification properties of these theories.

4. Semiadditive Categories

Before defining semiadditive categories we recall some basic concepts of category theory. Let C be a category and A, B be objects of C. We denote by hom(A,B) the set of morphisms with domain A and codomain B. The identity morphism in hom(A,A) is denoted by 1_A or just 1. We say that the object P is a *product* of A, B iff there exist morphisms p_1: P → A, p_2: P → B such that for every pair of morphisms f: X → A, g: X → B there is a unique morphism h: X → P such that the product diagram of Figure 4.1 commutes.

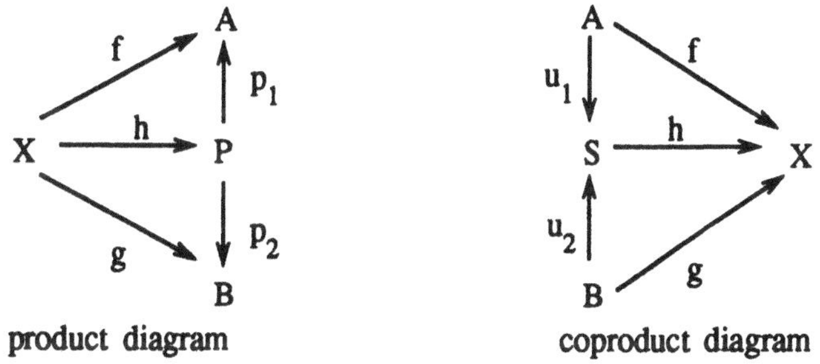

FIGURE 4.1

A product of two objects may not exist, but if it exists it is unique up to isomorphism. We denote the product of A and B by A×B and call the corresponding morphisms projections. The dual of the product is the coproduct. An object S is a *coproduct* of A, B iff there exist morphisms u_1: A → S, u_2: B → S such that for every pair of morphisms f: A → X, g: B → X there is a unique morphism h: S → X such that the coproduct diagram of Figure 4.1 commutes. We denote the coproduct of A and B (if it exists) by A+B and call the corresponding morphisms injections. Products and coproducts of more than two objects are defined in an analogous way. Given a coproduct S of the objects A_1, ..., A_n and a product P of the objects B_1, ..., B_n, every morphism f: S → P is uniquely determined by the matrix M_f = (f_{ij}), where $f_{ij} := u_i f p_j \in$ hom(A_i,B_j) for i = 1, ..., n and j = 1, ..., m. For n = 1 (resp. m = 1) we take u_1 = 1 (resp. p_1 = 1) in this definition.

An object A is called *initial* (*terminal*) iff for every object B, hom(A,B) (hom(B,A)) is a singleton. An object which is both initial and terminal is called *zero object*. If C has a zero object 0 we define the *zero morphism* $0_{A,B}$: A → B to be the composite of the unique morphism in hom(A,0) and the unique morphism in hom(0,B). It is easy to see that in this definition it does not matter which zero object of C is used. Let f: C $\rightarrow$ A, g: B $\rightarrow$ C be morphisms. Then we have $f \circ 0_{A,B} = 0_{C,B}$ and $0_{A,B} \circ g = 0_{A,C}$. In the following we shall omit the index and write 0 for any zero morphism.

Now we can define semiadditive categories:

DEFINITION 4.2. A category C is *semiadditive* iff

(1) C has a zero object.

(2) For every pair of objects there is a coproduct.

(3) For any pair of objects A, B there is a binary operation "+" on hom(A,B) such that

 (3.1) $0_{A,B}$ is a neutral element for "+" on hom(A,B).

 (3.2) For any objects A, B, C, D and any morphisms a, b $\in$ hom(A,B), c $\in$ hom(C,A) and d $\in$ hom(B,D) we have c(a + b) = ca + cb and (a + b)d = ad + bd.

The following theorem yields an alternative characterization of semiadditive categories.

THEOREM 4.3. *Let C be a category satisfying (1) and (2) of Definition 4.2 and let A+B with the injections u_1, u_2 be a coproduct of A and B. The morphisms p_1: A+B $\rightarrow$ A, p_2: A+B $\rightarrow$ B are defined by the commuting diagrams of Figure 4.4.*

FIGURE 4.4

Then the following statements are equivalent:

(1) For any pair of objects A, B the coproduct A+B is also the product of A, B relative to the morphisms p_1, p_2 defined above.

(2) C is semiadditive.

PROOF. The following sketch of the proof is included to give an idea of how these two properties are linked. Complete proofs can be found in Freyd (1964) and Herrlich-Strecker (1973).

If C satisfies (1), "+" may be defined as follows:

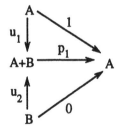

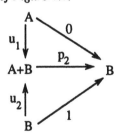

For a, b ∈ hom(A,B) define $a + b := (a,b) \circ \binom{1}{1}$.

Recall that (a,b) and $\binom{1}{1}$ are the unique morphisms such that the diagrams of Figure 4.5 commute.

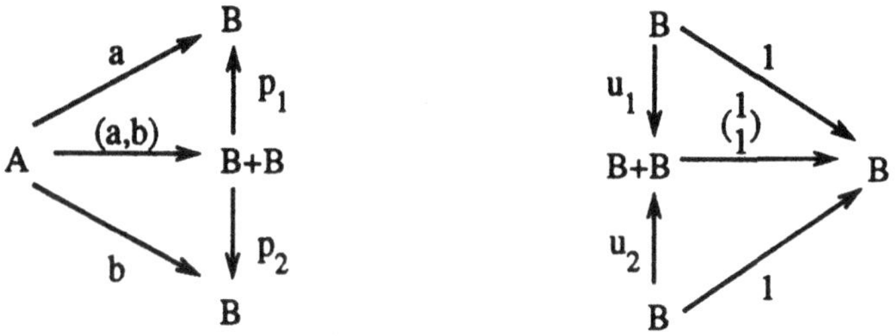

FIGURE 4.5

It can be shown that "+" satisfies (3.1) and (3.2) of Definition 4.2. Moreover "+" is associative and commutative (see e.g. Freyd (1964) pp. 47-49).

Conversely let C satisfy (2) and let a : X → A, b: X → B be morphisms. Then $x = au_1 + bu_2$ is the unique morphism such that $xp_1 = a$ and $xp_2 = b$ (see Freyd (1964) p. 50). ❑

COROLLARY 4.6. *The operation "+" of Definition 4.2 is unique and thus associative and commutative.*

PROOF. See Freyd (1964) p. 60 or Herrlich-Strecker (1973), Theorem 40.13. ❑

Let C be a semiadditive category and let A (resp. B, C) be coproduct and product of the objects A_i (resp. B_j, C_k). The morphisms a: A → B and b: B → C are uniquely determined by the matrices $M_a = (a_{i,j})$ and $M_b = (b_{j,k})$. It can be shown (see Freyd (1964) p. 49) that the usual rules of matrix multiplication hold, i.e. we have

COROLLARY 4.7. *Let* $c = ab$ *and* $M_c = (c_{i,k})$. *Then* $c_{i,k} = \Sigma_j\, a_{i,j} b_{j,k}$. ❑

5. Commutative Theories

We now characterize the equational theories, for which C(E) is a semiadditive category. A constant symbol (i.e. a nullary function symbol) $\underline{e} \in \Omega$ is called *idempotent in E* iff for any $\underline{f} \in \Omega$ we have $\underline{f}(\underline{e},...,\underline{e}) =_E \underline{e}$, i.e. in any algebra A ∈ V(E), f(e,...,e) = e holds. Note that for nullary $\underline{f}$ this means $\underline{f} =_E \underline{e}$.

Let **K** be a class of algebras (of signature Ω). An n-ary *implicit operation* in **K** is a family $f = \{ f_A;\ A \in \mathbf{K} \}$ of mappings $f_A\colon A^n \to A$ which is compatible with all homomorphisms, i.e. for any homomorphism $h\colon A \to B$ with $A, B \in \mathbf{K}$ and all $a_1, \ldots, a_n \in A$, $f_A(a_1,\ldots,a_n)h = f_B(a_1h,\ldots,a_nh)$ holds. In the following we omit the index and just write f for any f_A. Obviously an Ω-term induces an implicit operation on any class of Ω-algebras.

EXAMPLE 5.1. Let Ω consist of a nullary function symbol $\underline{e}$ and a binary function symbol $*$. $E := \{ x * x = e, x * e = x, (x * (e * y)) * (e * z) = (z * (e * y)) * (e * x) \}$.

The Ω-terms $x * (e * y)$ and $e * x$ define a binary implicit operation $\bullet$ and a unary implicit operation $^{-1}$ in V(E) and F(E) as follows:

Let $A \in V(E)$ and $a, b \in A$. Then $a \bullet b := a * (e * b)$ and $a^{-1} := e * a$.

It is easy to see, that $\bullet$, $^{-1}$ satisfy the abelian group axioms in any $A \in V(E)$. The constant symbol $\underline{e}$ is idempotent in E.

Since V(E) contains (F(E) consists of) all E-free algebras with finite set of generators, any implicit operation in V(E) (resp. F(E)) is given by an Ω-term (see Lawvere (1963)).

In the following we assume that E is not trivial, i.e. that V(E) contains algebras of cardinality greater 1. The next proposition characterizes the theories E for which C(E) has a zero object.

PROPOSITION 5.2. *The following three conditions are equivalent:*

(1) C(E) contains a zero object.

(2) $|F_E(\emptyset)| = 1$.

(3) Ω *contains a constant symbol* $\underline{e}$*, which is idempotent in E.*

PROOF. The equivalence of (2) and (3) is obvious. The definition of E-free objects shows that $F_E(\emptyset)$ is initial in C(E). If in addition $|F_E(\emptyset)| = 1$, it is also terminal. This proves "2→1". Conversely, since $F_E(\emptyset)$ is initial, any zero object of C(E) is isomorphic to $F_E(\emptyset)$. It is easy to see that for $|F_E(\emptyset)| > 1$ we have at least two morphisms of $F_E(x)$ into $F_E(\emptyset)$ and for $|F_E(\emptyset)| = 0$ we have no morphism of $F_E(x)$ into $F_E(\emptyset)$. Hence "1→2" holds. ❑

Evidently the zero morphism of $F_E(X)$ into $F_E(Y)$ is defined by $x \mapsto e$ for $x \in X$.

Condition (2) of Definition 4.2 holds in C(E) for any equational theory E. In fact, the coproduct of $F_E(X)$ and $F_E(Y)$ is given by $F_E(X \dot{\cup} Y)$ where $\dot{\cup}$ means disjoint union. Note that $F_E(X)$ is the coproduct of the objects $F_E(x)$ for $x \in X$ and the $F_E(x)$ are isomorphic to each other. We now consider Condition (3) of Definition 4.2.

PROPOSITION 5.3. *Let C(E) contain a zero object and let $\underline{e}$ be the constant symbol, which is idempotent in E. Then the following assertions are equivalent:*

(1) C(E) is semiadditive.

(2) There is a binary implicit operation $*$ in F(E) such that

(2.1) The constant e is a neutral element for $*$ in any algebra $A \in F(E)$.

(2.2) For any n-ary function symbol $\underline{f} \in \Omega$, any algebra $A \in F(E)$ and any $s_1, \ldots, s_n$, $t_1, \ldots, t_n \in A$ we have $f(s_1 * t_1, \ldots, s_n * t_n) = f(s_1, \ldots, s_n) * f(t_1, \ldots, t_n)$.

PROOF. We first prove (1) implies (2). The operation + on the morphisms of C(E) induces an operation $*$ in the $F_E(X)$ as follows: Given $s, t \in F_E(X)$ we define morphisms σ, τ: $F_E(x) \to F_E(X)$ by $\sigma: x \mapsto s$, $\tau: x \mapsto t$. Now define $s * t := x(\sigma + \tau)$.

The operation $*$ is implicit, since for $\lambda: F_E(X) \to F_E(Y)$ we have $(s * t)\lambda = x(\sigma + \tau)\lambda = x(\sigma\lambda + \tau\lambda) = (s\lambda) * (t\lambda)$. Assertion (2.1) holds, since the zero morphism is neutral for the operation +. To show that (2.2) holds, we consider morphisms σ, τ: $F_E(x_1,\ldots,x_n) \to F_E(X)$, defined by $x_i\sigma = s_i$ ($i = 1, \ldots, n$) and $x_i\tau = t_i$ ($i = 1, \ldots, n$). Now $x_i(\sigma + \tau) = s_i * t_i$ and since $\sigma + \tau$ is a homomorphism we have

$$f(x_1,\ldots,x_n)(\sigma + \tau) = f(x_1(\sigma + \tau),\ldots,x_n(\sigma + \tau)) = f(s_1 * t_1,\ldots,s_n * t_n).$$

On the other hand we may consider γ: $F_E(x) \to F_E(x_1,\ldots,x_n)$ defined by $x\gamma = f(x_1,\ldots,x_n)$.

This yields $f(x_1,\ldots,x_n)(\sigma + \tau) = x\gamma(\sigma + \tau) = x(\gamma\sigma + \gamma\tau) = (x\gamma\sigma) * (x\gamma\tau) =$

$$= f(s_1,\ldots,s_n) * f(t_1,\ldots,t_n).$$

Conversely, let E satisfy (2). The implicit operation $*$ induces an operation + on the morphisms of C(E). Let σ, τ: $F_E(X) \to F_E(Y)$ and $s \in F_E(X)$. Then $s(\sigma + \tau) := (s\sigma) * (s\tau)$. Using (2.2) it is easy to show that $\sigma + \tau$ is really a morphism of C(E), i.e. a homomorphism of $F_E(X)$ into $F_E(Y)$. Obviously (2.1) yields (3.1) of Definition 4.2. The fact that $*$ is an implicit operation yields (3.2) of Definition 4.2. ❑

The implicit operation • of Example 5.1 satisfies (2.1) and (2.2) of the proposition. Thus, for the theory E of the example, C(E) is semiadditive.

COROLLARY 5.4.

The implicit operation of Theorem 5.3 is associative and commutative.

PROOF. This is an immediate consequence of Corollary 4.6. ❑

Note that the explicit operation $*$ of Example 5.1 is neither commutative nor associative. Corollary 5.4 justifies the following definition.

DEFINITION 5.5.

An equational theory E is called *commutative* iff C(E) is semiadditive.

We now consider examples of commutative theories. In all these examples the implicit operation is given by a function symbol, which is associative and commutative in the corresponding theory.

EXAMPLES 5.6. We consider the following signatures:

$\Omega_1 := \{ \cdot, 1 \}$ where $\cdot$ is binary and 1 is nullary.

$\Omega_2 := \Omega_1 \cup \{ ^{-1} \}$ and $\Omega_3 := \Omega_1 \cup \{ h \}$ where $^{-1}$ and h are unary.

$\Omega_4 := \Omega_2 \cup \Omega_3$.

(1) The theory of abelian monoids. The signature is Ω_1 and

$AM := \{ x \cdot 1 = x, x \cdot (y \cdot z) = (x \cdot y) \cdot z, x \cdot y = y \cdot x \}$.

(2) The theory AIM of idempotent abelian monoids. The signature is Ω_1 and

$AIM := AM \cup \{ x \cdot x = x \}$.

(3) The theory AMH of abelian monoids with a homomorphism.

The signature is Ω_3 and $AMH := AM \cup \{ h(x \cdot y) = h(x) \cdot h(y), h(1) = 1 \}$.

(4) The theory AIMH of idempotent abelian monoids with a homomorphism.

The signature is Ω_3 and $AIMH := AIM \cup \{ h(x \cdot y) = h(x) \cdot h(y), h(1) = 1 \}$.

(5) The theory AMI of abelian monoids with an involution. The signature is Ω_3 and

$AMI := AM \cup \{ h(x \cdot y) = h(x) \cdot h(y), h(h(x)) = x \}$.

(6) The theory AIMI of idempotent abelian monoids with an involution.

We have signature Ω_3 and $AIMI := AMI \cup \{ x \cdot x = x \}$.

(7) The theory AG_m of abelian groups of exponent m ($m \in \mathbb{N}$) is given by signature Ω_2 and $AG_m := AM \cup \{ x \cdot x^{-1} = 1, x^m = 1 \}$. $AG = AG_0$ is the theory of abelian groups. Any variety of abelian groups is defined in this way.

(8) The theory AGI of abelian groups with an involution.
We take signature Ω_4 and define AGI := AG $\cup$ AMI.

(9) The theory AGH of abelian groups with a homomorphism.
We take signature Ω_4 and define AGH := AG $\cup$ $\{ h(x \cdot y) = h(x) \cdot h(y) \}$.

It is easy to see that these theories are commutative. Note that the implicit operation induced by the term $x \cdot y$ (for a binary function symbol $\cdot$) satisfies (2.2) of Proposition 5.3 for $f = \cdot$ iff $(a \cdot b) \cdot (c \cdot d) = (a \cdot c) \cdot (b \cdot d)$ holds in any algebra $A \in F(E)$.
We shall now consider unification in commutative theories and, in the end, determine the unification types of the theories defined above.

6. Commutative Theories and Unification

First we show the following theorem:

THEOREM 6.1. *Commutative theories are either unitary or of unification type zero.*

This is an easy consequence of the following two lemmata. Now let E be a commutative theory.

LEMMA 6.2. *Let* $\Gamma = < \sigma = \tau >_E$ *be an E-unification problem and let* $\{ \gamma_1, ..., \gamma_n \}$ *be a finite complete set of E-unifiers of* Γ. *Then there exists an E-unifier* γ *of* Γ *such that the singleton* $\{ \gamma \}$ *is a complete set of E-unifiers of* Γ.

PROOF. We have $\sigma, \tau: F_E(I) \to F_E(X)$ and $\gamma_i: F_E(X) \to F_E(Y_i)$.

With $Y = Y_1 \dot\cup ... \dot\cup Y_n$, $F_E(Y)$ is coproduct and product of the $F_E(Y_i)$. Let $p_1, ..., p_n$ be the corresponding projections. Then there exists a unique morphism $\gamma: F_E(X) \to F_E(Y)$ such that $\gamma p_i = \gamma_i$ for $i = 1, ..., n$. The morphism γ is an E-unifier of Γ, since $\sigma\gamma = \tau\gamma$ iff $\sigma\gamma p_i = \tau\gamma p_i$ for $i = 1, ..., n$ (by the definition of product).

Let δ be an E-unifier of Γ. Since $\{ \gamma_1, ..., \gamma_n \}$ is complete there is an index i and a morphism λ such that $\delta = \gamma_i\lambda$. But now $\delta = (\gamma p_i)\lambda = \gamma(p_i\lambda)$. Hence $\{ \gamma \}$ is complete. ❑

LEMMA 6.3. *Let* $\Gamma = < \sigma = \tau >_E$ *be an E-unification problem and let* $U = \{ \gamma_1, \gamma_2, \gamma_3, ... \}$ *be an infinite set of E-unifiers of* Γ *such that the* γ_i *do not lie (w.r.t.* $\leq_E$ *) below a single E-unifier of* Γ*. Then there does not exist a minimal complete set* $\mu U_E(\Gamma)$.

PROOF. We have $\sigma, \tau: F_E(I) \to F_E(X)$ and $\gamma_n: F_E(X) \to F_E(Y_n)$.

The morphisms δ_n are defined as follows: δ_1 is just γ_1.

Let $\delta_n: F_E(X) \to F_E(Z_n)$ be already defined and let $F_E(Z_{n+1})$ with the projections p_1, p_2 be a product of $F_E(Y_{n+1})$ and $F_E(Z_n)$. Then $\delta_{n+1}: F_E(X) \to F_E(Z_{n+1})$ is defined to be the unique morphism such that $\delta_{n+1}p_1 = \gamma_{n+1}$ and $\delta_{n+1}p_2 = \delta_n$ (see Figure 6.4).

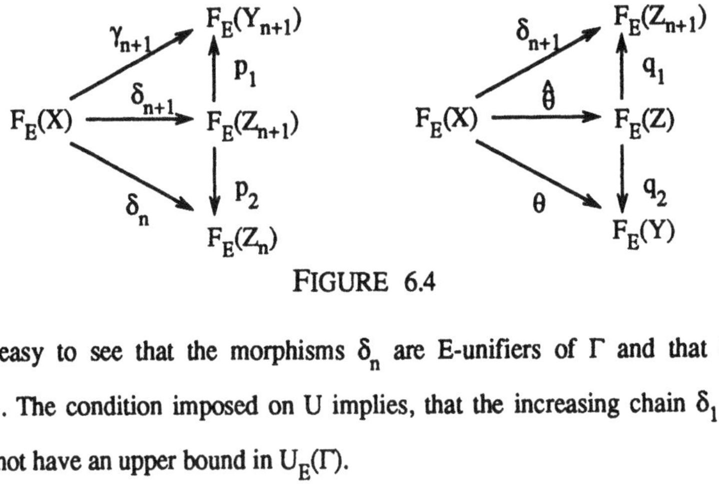

FIGURE 6.4

It is easy to see that the morphisms δ_n are E-unifiers of Γ and that $\delta_n \leq_E \delta_{n+1}$ for all $n \geq 1$. The condition imposed on U implies, that the increasing chain $\delta_1 \leq_E \delta_2 \leq_E \delta_3 \leq_E ...$ does not have an upper bound in $U_E(\Gamma)$.

Now we assume that $\mu U_E(\Gamma)$ exists and derive a contradiction. Since $\mu U_E(\Gamma)$ is complete, there is $\theta \in \mu U_E(\Gamma)$ such that $\delta_1 \leq_E \theta$. The fact that $\delta_1 \leq_E \delta_2 \leq_E \delta_3 \leq_E ...$ has no upper bound in $U_E(\Gamma)$ yields an $n \geq 1$ satisfying $\delta_n \leq_E \theta$, but not $\delta_{n+1} \leq_E \theta$. Let $\theta: F_E(X) \to F_E(Y)$ and let $F_E(Z)$ with the the projections q_1, q_2 be the product of $F_E(Z_{n+1})$ and $F_E(Y)$. The morphism $\hat{\theta}: F_E(X) \to F_E(Z)$ is defined to be the unique morphism such that $\hat{\theta}q_1 = \delta_{n+1}$ and $\hat{\theta}q_2 = \theta$ (see the right diagram of Figure 6.4). Obviously $\hat{\theta}$ is an E-unifier of Γ, $\delta_{n+1} \leq_E \hat{\theta}$ and $\theta \leq_E \hat{\theta}$. Since $\mu U_E(\Gamma)$ is complete there is $\theta' \in \mu U_E(\Gamma)$ such that $\hat{\theta} \leq_E \theta'$.

Now $\theta \leq_E \theta'$ for $\theta, \theta' \in \mu U_E(\Gamma)$ yields $\theta = \theta'$ by minimality of $\mu U_E(\Gamma)$. But then $\delta_{n+1} \leq_E \hat{\theta} \leq_E \theta' = \theta$ is a contradiction. ❑

A similar argument was used in Baader (1987) to show that most varieties of idempotent semigroups are defined by type zero theories. For more information on unification type zero consult Baader (1989a). In Section 9 it will be shown that the theory AMH of Example 5.6 is of type zero. In the remaining part of this section we establish a necessary and sufficient condition for a commutative theory to be unitary.
Let $\Gamma = <\sigma = \tau>_E$ (where σ, τ: $F_E(I) \rightarrow F_E(X)$) be an E-unification problem.

LEMMA 6.5. *Let δ: $F_E(X) \rightarrow F_E(Z)$ be an E-unifier of Γ and let y be an arbitrary variable. Then $\{ \delta \}$ is a complete set of E-unifiers of Γ, if for any E-unifier γ: $F_E(X) \rightarrow F_E(y)$ of Γ there is a morphism λ: $F_E(Z) \rightarrow F_E(y)$ such that $\gamma = \delta\lambda$.*

PROOF. Let α: $F_E(X) \rightarrow F_E(Y)$ be an E-unifier of Γ and let $|Y| = n$. Hence $F_E(Y)$ is the n-th power of $F_E(y)$. We call the corresponding projections $p_1, \ldots, p_n$. Since αp_i: $F_E(X) \rightarrow F_E(y)$ are E-unifier of Γ, the assumption yields morphisms λ_i: $F_E(Z) \rightarrow F_E(y)$ such that $\alpha p_i = \delta\lambda_i$.
Let λ: $F_E(Z) \rightarrow F_E(Y)$ be the unique morphism such that $\lambda p_i = \lambda_i$ for $i = 1, \ldots, n$. Now $\alpha p_i = \delta\lambda_i = (\delta\lambda)p_i$ for $i = 1, \ldots, n$ and thus $\alpha = \delta\lambda$. ❑

Hence if one wants to check completeness of a set of E-unifiers, it is sufficient to examine unifiers, which introduce a single variable. Therefore the following condition only considers unifiers of $F_E(X)$ into $F_E(y)$.

CONDITION 6.6. Let y be an arbitrary variable. For any given E-unification problem $<\sigma = \tau>_E$ (where σ, τ: $F_E(I) \rightarrow F_E(X)$) there are finitely many E-unifiers $\alpha_1, \ldots, \alpha_r$: $F_E(X) \rightarrow F_E(y)$ such that any E-unifier δ: $F_E(X) \rightarrow F_E(y)$ is representable as

$$\delta = \Sigma_{i=1}^{i=r} \alpha_i \lambda_i ,$$

where λ_i: $F_E(y) \rightarrow F_E(y)$ are morphisms.

THEOREM 6.7. *A commutative theory E is unitary iff it satisfies Condition 6.6.*

PROOF. We assume that E is a commutative theory satisfying 6.6. Let $\Gamma = < \sigma = \tau >_E$ be an E-unification problem and let $\alpha_1, ..., \alpha_r$: $F_E(X) \to F_E(y)$ be as in the condition. For Z = { $z_1, ..., z_r$ }, $F_E(Z)$ is the r-th power and the r-th co-power of $F_E(y)$. Let $p_1, ..., p_r$ ($u_1, ..., u_r$) be the corresponding projections (injections) such that $u_i\ p_i = 1$ and $u_i\ p_j = 0$ (for $i \neq j$).

We define an E-unifier α: $F_E(X) \to F_E(Z)$ as follows: α is the unique morphism such that $\alpha p_i = \alpha_i$ for i = 1, .., n. Thus $\alpha = (\alpha_1 \dots \alpha_r)$.

Obviously α is an E-unifier of Γ. It remains to be shown that { α } is complete. Because of Lemma 6.5 we only have to consider E-unifiers β: $F_E(X) \to F_E(y)$. Condition 6.6 yields morphisms $\lambda_1, ..., \lambda_r$: $F_E(y) \to F_E(y)$ such that

$$\beta = \Sigma_{i=1}^{i=r} \alpha_i\, \lambda_i \,.$$

Let λ be the unique morphism such that $u_i\lambda = \lambda_i$ for i = 1, ..., r.

Thus $\lambda = \begin{pmatrix} \lambda_1 \\ \vdots \\ \lambda_r \end{pmatrix}$ in matrix notation. Now $\alpha\lambda = (\alpha_1 \dots \alpha_r) \begin{pmatrix} \lambda_1 \\ \vdots \\ \lambda_r \end{pmatrix} = \Sigma_{i=1}^{i=r} \alpha_i\, \lambda_i = \beta$,

which establishes the if-part of the theorem.

Conversely, let E be unitary and let y be an arbitrary variable. Any E-unification problem $\Gamma = < \sigma = \tau >_E$ has a complete set { α } of E-unifiers. Let α: $F_E(X) \to F_E(Y)$ for $|Y| = r$. Then $F_E(Y)$ is r-th power and the r-th co-power of $F_E(y)$. Let $p_1, ..., p_r$ ($u_1, ..., u_r$) be the corresponding projections (injections). We define $\alpha_i := \alpha p_i$, which means $\alpha = (\alpha_1 \dots \alpha_r)$.

For any E-unifier δ: $F_E(X) \to F_E(y)$, there is a morphism λ: $F_E(Y) \to F_E(y)$ with $\alpha\lambda = \delta$. Now $\lambda_i := u_i\lambda$ yields $\delta = \alpha\lambda = \Sigma_{i=1}^{i=r} \alpha_i\, \lambda_i$ ❑

If $F_E(X)$ and $F_E(y)$ are finite, the set $hom(F_E(X),F_E(y))$ is also finite and Condition 6.6 trivially holds. This yields

COROLLARY 6.8. *Let E be a commutative theory for which the finitely generated free objects are finite. Then E is unitary.*

The theories AIM, AIMI and AG_m for $m \geq 1$ are examples of commutative theories satisfying the assumption of Corollary 6.8. Thus we have

PROPOSITION 6.9. *The theories AIM, AIMI and AG_m for $m \geq 1$ are unitary.*

Direct proofs for AIM can be found in Livesey-Siekmann (1978) and Baader-Büttner (1988).

7. Unification with Constants in Commutative Theories

In some applications of E-unifications it is advantageous to consider – in addition to the variables – syntactic constants. These constants have no meaning in the theory E, i.e. they behave like variables w.r.t. $=_E$. But they differ from variables in that they must not be replaced by substitutions. In the categorical context this can be formulated as follows: Let V (vid. Definition 3.1) be the disjoint union of denumerable sets U and C. A c-morphism is a morphism of C(E), which is the identity on elements of C. The subcategory $C_c(E)$ of C(E) is obtained from C(E) by restricting the morphisms to *c-morphisms*. Now *E_c-unification* is defined as E-unification with $C_c(E)$ in place of C(E).

In the following let I, W, X, Y, Z (resp. A, B) denote finite subsets of U (resp. C) and let E be a commutative theory. A unitary theory E need not be unitary w.r.t. E_c-unification. In this section we characterize the unitary commutative theory, which are finitary w.r.t. E_c-unification. Let $\Gamma = < \sigma = \tau >_E$, where $\sigma, \tau: F_E(I) \rightarrow F_E(X \cup A)$, be an E_c-unification problem. First note that we may confine ourselves to E_c-unifiers which do not introduce new constants, i.e. to c-morphisms $\gamma: F_E(X \cup A) \rightarrow F_E(Y \cup A)$. Otherwise the additional constants can be replaced by new variables, which yields an E_c-unifier δ such that $\gamma \leq_E \delta$ (see Baader-Büttner (1988), Lemma 3.1).

CONDITION 7.1. For any morphism (of C(E)) $\delta: F_E(A) \rightarrow F_E(Z)$ there exist finite sets M, N such that:

(1) The elements of M are morphisms $\mu: F_E(Z) \rightarrow F_E(A)$ satisfying $\delta\mu = 1$.

(2) The elements of $N = \{ \nu_1, ..., \nu_r \}$ are morphisms $\nu_i: F_E(Z) \rightarrow F_E(Z_i)$ with $\delta\nu_i = 0$.

(3) For any λ: $F_E(Z) \to F_E(A)$ with $\delta\lambda = 1$ there are $\mu \in M$ and morphisms $\lambda_1, ..., \lambda_r$ (where λ_i: $F_E(Z_i) \to F_E(A)$) satisfying:

$$\lambda = \mu + \Sigma_{i=1}^{i=r} v_i \lambda_i$$

THEOREM 7.2.

Let E be a unitary commutative theory. Then the following assertions are equivalent:

(1) E satisfies Condition 7.1.

(2) E is finitary w.r.t. E_c-unification.

PROOF. First we assume that E is unitary and satisfies Condition 7.1. Let $\Gamma = < \sigma = \tau >_E$, where σ, τ: $F_E(I) \to F_E(X \cup A)$, be an E_c-unification problem. We begin with considering Γ as E-unification problem, i.e. the elements of A are treated as variables. Let δ: $F_E(X \cup A) \to F_E(Z)$ be an E-unifier of Γ such that $\{ \delta \}$ is a complete set of E-unifiers. Since any E_c-unifier γ: $F_E(X \cup A) \to F_E(Y \cup A)$ is also an E-unifier, there is a morphism λ (which need not be a c-morphism) satisfying $\gamma = \delta\lambda$. Since E is commutative, $F_E(X \cup A)$ (resp. $F_E(Y \cup A)$) is coproduct and product of $F_E(X)$ and $F_E(A)$ (resp. $F_E(Y)$ and $F_E(A)$). Thus γ is uniquely determined by a matrix

$$\begin{pmatrix} \gamma_1 & \gamma_2 \\ \gamma_3 & \gamma_4 \end{pmatrix}, \quad \text{where} \quad \begin{array}{l} \gamma_1: F_E(X) \to F_E(Y),\ \gamma_2: F_E(X) \to F_E(A), \\ \gamma_3: F_E(A) \to F_E(Y),\ \gamma_4: F_E(A) \to F_E(A). \end{array}$$

Since γ is an E_c-unifier, we have $\gamma_3 = 0$ and $\gamma_4 = 1$. Accordingly, δ is determined by a matrix

$$\begin{pmatrix} \delta_1 \\ \delta_2 \end{pmatrix}, \quad \text{where} \quad \begin{array}{l} \delta_1: F_E(X) \to F_E(Z), \\ \delta_2: F_E(A) \to F_E(Z), \end{array}$$

and λ is determined by a matrix ($\lambda_1 \quad \lambda_2$), where λ_1: $F_E(Z) \to F_E(Y)$ and λ_2: $F_E(Z) \to F_E(A)$. Now $\gamma = \delta\lambda$ yields $\delta_1\lambda_1 = \gamma_1$, $\delta_1\lambda_2 = \gamma_2$, $\delta_2\lambda_1 = 0$ and $\delta_2\lambda_2 = 1$.

Applying Condition 7.1 for δ_2: $F_E(A) \to F_E(Z)$, we get finite sets M, N of morphisms satisfying (1), (2), (3) of 7.1. Thus there are $\mu \in M$ and morphisms $\lambda_1, ..., \lambda_r$ such that

$$\lambda_2 = \mu + \Sigma_{i=1}^{i=r} v_i \lambda_i.$$

Since $\delta_2\lambda_1 = 0 = 0\lambda_1$, λ_1 is an E-unifier of the E-unification problem $\Delta = < \delta_2 = 0 >_E$, where 0: $F_E(A) \to F_E(Z)$ is the zero morphism. Let $\{ \kappa \}$ be a complete set of E-unifiers of Δ, where κ: $F_E(Z) \to F_E(W)$. Then there is a morphism ρ: $F_E(W) \to F_E(Y)$ with the property $\lambda_1 = \kappa\rho$. The matrix $(\ \kappa\ \nu_1\ \dots\ \nu_r\ \mu\ )$ defines a morphism of $F_E(Z)$ into $F_E(W)\times F_E(Z_1)\times\dots\times F_E(Z_r)\times F_E(A)$. Note that this morphism only depends on δ_2, but not on λ. Furthermore we have

$$(\ \kappa\ \nu_1\ \dots\ \nu_r\ \mu\ )\begin{pmatrix} \rho & 0 \\ 0 & \lambda_1 \\ \vdots & \\ 0 & \lambda_r \\ 0 & 1 \end{pmatrix} = (\ \kappa\rho\quad \mu + \textstyle\sum_{i=1}^{i=r} \nu_i\,\lambda_i\ ) = (\ \lambda_1\quad \lambda_2\ ) = \lambda \text{ and the morphism}$$

$$\begin{pmatrix} \rho & 0 \\ 0 & \lambda_1 \\ \vdots & \\ 0 & \lambda_r \\ 0 & 1 \end{pmatrix} : F_E(W \dot{\cup} Z_1 \dot{\cup} \dots \dot{\cup} Z_r \dot{\cup} A) \to F_E(Y \dot{\cup} A) \text{ is obviously a c-morphism.}$$

We define $K = \{\ \delta \circ (\ \kappa\ \nu_1\ \dots\ \nu_r\ \mu\ );\ \mu \in M\ \}$, which is a finite set of morphisms of $F_E(X \cup A)$ into $F_E(W \dot{\cup} Z_1 \dot{\cup} \dots \dot{\cup} Z_r \dot{\cup} A)$. It remains to be shown that K is a complete set of E_c-unifiers of Γ. Since δ is an E-unifier of Γ, the elements of K are also E-unifiers of Γ. They are E_c-unifiers, because

$$\begin{pmatrix} \delta_1 \\ \delta_2 \end{pmatrix}(\ \kappa\ \nu_1\ \dots\ \nu_r\ \mu\ ) = \begin{pmatrix} \delta_1\kappa & \delta_1\nu_1 & \dots & \delta_1\nu_r & \delta_1\mu \\ \delta_2\kappa & \delta_2\nu_1 & \dots & \delta_2\nu_r & \delta_2\mu \end{pmatrix}$$

and Condition 7.1 asserts $\delta_2\nu_1 = \dots = \delta_2\nu_r = 0$, $\delta_2\mu = 1$, and we have $\delta_2\kappa = 0$ by definition of κ.

In the preceeding we have seen that an arbitrary E_c-unifier γ is of the form $\gamma = \delta\lambda$ where

$$\lambda = (\ \kappa\ \nu_1\ \dots\ \nu_r\ \mu\ )\begin{pmatrix} \rho & 0 \\ 0 & \lambda_1 \\ \vdots & \\ 0 & \lambda_r \\ 0 & 1 \end{pmatrix} \quad \text{for some } \mu \in M \text{ and morphisms } \lambda_1, \dots, \lambda_r. \text{ Thus}$$

$$\gamma = (\ \delta \circ\ (\ \kappa\ \nu_1\ \dots\ \nu_r\ \mu\ )\ )\begin{pmatrix} \rho & 0 \\ 0 & \lambda_1 \\ \vdots & \\ 0 & \lambda_r \\ 0 & 1 \end{pmatrix}, \text{ which shows completeness of K.}$$

But then a minimal complete set of E_c-unifiers exists and is also finite. This completes the proof of "$1\to 2$".

Conversely, let E be finitary w.r.t. unification with constants and let $\delta: F_E(A) \to F_E(Z)$ be a morphism. If X, Y $\subseteq$ U are two disjoint copies of A then, for I := X $\cup$ Y, $F_E(I)$ is product and coproduct of $F_E(A)$ with itself. We consider morphisms δ_1, δ_2: $F_E(I) \to F_E(Z \cup A)$, defined by the matrices

$$\delta_1 = \begin{pmatrix} \delta & 0 \\ 0 & 1 \end{pmatrix} \text{ and } \delta_2 = \begin{pmatrix} 0 & 1 \\ \delta & 0 \end{pmatrix}.$$

Let $\beta = \begin{pmatrix} \beta_1 & \beta_2 \\ 0 & 1 \end{pmatrix}$: $F_E(Z \cup A) \to F_E(Z' \cup A)$ be an E_c-unifier of $\Gamma := < \delta_1 = \delta_2 >_E$.

That means $\begin{pmatrix} \delta\beta_1 & \delta\beta_2 \\ 0 & 1 \end{pmatrix} = \begin{pmatrix} 0 & 1 \\ \delta\beta_1 & \delta\beta_2 \end{pmatrix}$ and thus $\delta\beta_1 = 0$ and $\delta\beta_2 = 1$.

Since E is finitary w.r.t. unification with constants, there exists a finite complete set $\{ \alpha_1, ..., \alpha_s \}$ of E_c-unifiers of Γ, where the morphisms α_i are of the form

$$\alpha_i = \begin{pmatrix} \nu_i & \mu_i \\ 0 & 1 \end{pmatrix} : F_E(Z \cup A) \to F_E(Z_i \cup A) .$$

We define M := $\{ \mu_1, ..., \mu_s \}$ and N := $\{ \nu_1, ..., \nu_s \}$. Since the morphisms α_i are E_c-unifier of Γ, we have $\delta\nu_i = 0$ and $\delta\mu_i = 1$.

Let $\lambda: F_E(Z) \to F_E(A)$ be a morphism with $\delta\lambda = 1$. The morphism β, defined by the matrix $\beta := \begin{pmatrix} 0 & \lambda \\ 0 & 1 \end{pmatrix}$: $F_E(Z \cup A) \to F_E(Z' \cup A)$, is an E_c-unifier of Γ, since

$$\begin{pmatrix} \delta & 0 \\ 0 & 1 \end{pmatrix}\begin{pmatrix} 0 & \lambda \\ 0 & 1 \end{pmatrix} = \begin{pmatrix} 0 & 1 \\ 0 & 1 \end{pmatrix} = \begin{pmatrix} 0 & 1 \\ \delta & 0 \end{pmatrix}\begin{pmatrix} 0 & \lambda \\ 0 & 1 \end{pmatrix}.$$

Because $\{ \alpha_1, ..., \alpha_s \}$ is complete, there exist i, $1 \le i \le s$, and a morphism

$$\begin{pmatrix} \lambda_1 & \lambda_2 \\ 0 & 1 \end{pmatrix} \text{ such that } \begin{pmatrix} 0 & \lambda \\ 0 & 1 \end{pmatrix} = \begin{pmatrix} \nu_i & \mu_i \\ 0 & 1 \end{pmatrix}\begin{pmatrix} \lambda_1 & \lambda_2 \\ 0 & 1 \end{pmatrix} = \begin{pmatrix} \nu_i\lambda_1 & \mu_i + \nu_i\lambda_2 \\ 0 & 1 \end{pmatrix}$$

Thus $\lambda = \mu_i + \sum_{j=1}^{j=r} \nu_j \gamma_j$, if we define $\gamma_i := \lambda_2$ and $\gamma_j := 0$ for $j \neq i$. This yields the direction "2→1" of the theorem. ❑

The proof of "1→2" shows that the cardinality of the minimal complete set is bounded by the cardinality of M.

COROLLARY 7.3. *If the set M of Condition 7.1 is a singleton for all E_c-unification problems Γ, then E is unitary w.r.t. E_c-unification.*

As in the case of E-unification we have

COROLLARY 7.4. *Let E be a commutative theory for which the finitely generated free objects are finite. Then E is finitary w.r.t. E_c-unification.*

And thus

PROPOSITION 7.5. *The theories AIM, AIMI and AG_m for $m \geq 1$ are finitary w.r.t. unification with constants.*

Note that AG_m for $m \geq 1$ is even unitary. This can be shown analogously to the proof of Proposition 8.3 below. In the following sections we shall consider the theories of Example 5.6, which do not satisfy the finiteness condition for the finitely generated free objects.

8. The Theories AM, AMI, AG and AGI

In this section it will be shown that the theories AM, AMI, AG and AGI (see Example 5.6) are unitary (resp. finitary w.r.t. unification with constants).

Let $X = \{ x_1, ..., x_n \}$, $Y = \{ y_1, ..., y_m \}$ and let $\delta: F_{AG}(X) \to F_{AG}(Y)$ be a morphism of C(AG). Then δ is uniquely determined by an n×m-matrix $M_\delta \in M_{n,m}(\mathbb{Z})$ where $(M_\delta)_{i,j} \in \mathbb{Z}$ is the exponent of y_j in $x_i\delta$. Sum and composition of morphisms correspond to sum and product of matrices, i.e. $M_{\sigma\delta} = M_\sigma \cdot M_\delta$ and $M_{\sigma+\delta} = M_\sigma + M_\delta$. The morphism δ is the zero morphism iff all entries of M_δ are zero and for X = Y, δ is the identity morphism iff M_δ is the unit matrix E.

Accordingly, any morphism $\delta: F_{AM}(X) \to F_{AM}(Y)$ of C(AM) corresponds to a matrix $M_\delta \in M_{n,m}(\mathbb{N})$.

For a morphism $\delta\colon F_{AGI}(X) \to F_{AGI}(Y)$ of C(AGI) we define matrices A_δ, B_δ as follows: $A_\delta = (a_{i,j})$, where $a_{i,j} \in \mathbb{Z}$ is the exponent of y_j in $x_i\delta$.

$B_\delta = (b_{i,j})$, where $b_{i,j} \in \mathbb{Z}$ is the exponent of $h(y_j)$ in $x_i\delta$.

We associate with the morphism δ the matrix

$$M_\delta = \begin{pmatrix} A_\delta & B_\delta \\ B_\delta & A_\delta \end{pmatrix} \in M_{2n,2m}(\mathbb{Z})$$

It is easy to see that $M_{\sigma\delta} = M_\sigma \cdot M_\delta$ and $M_{\sigma+\delta} = M_\sigma + M_\delta$. The set

$$D_{n,m}(\mathbb{Z}) = \left\{ \begin{pmatrix} A & B \\ B & A \end{pmatrix} ; A, B \in M_{n,m}(\mathbb{Z}) \right\}$$

with addition of matrices is a subgroup of the abelian group $M_{2n,2m}(\mathbb{Z})$.

Accordingly, any morphism $\delta\colon F_{AMI}(X) \to F_{AMI}(Y)$ of C(AMI) corresponds to a matrix $M_\delta \in D_{n,m}(\mathbb{N})$. The set

$$D_{n,m}(\mathbb{N}) = \left\{ \begin{pmatrix} A & B \\ B & A \end{pmatrix} ; A, B \in M_{n,m}(\mathbb{N}) \right\}$$

with addition of matrices is a stable subsemigroup of $M_{2n,2m}(\mathbb{N})$.

A subsemigroup T of a commutative semigroup S is called *stable* iff for all $a, b \in S$ the fact that $a, a+b \in T$ implies $b \in T$. Note that S is a stable subsemigroup of S.

For morphisms $\sigma, \tau, \delta, \lambda$ of C(AG) (resp. C(AM), C(AGI), C(AMI)) we have

$\sigma\delta = \tau\delta$ iff $M_{\sigma\delta} = M_{\tau\delta}$ iff $M_\sigma \cdot M_\delta = M_\tau \cdot M_\delta$ iff $(M_\sigma - M_\tau)M_\delta = 0$ and

$\sigma\lambda = 1$ iff $M_{\sigma\lambda} = M_1 = E$ iff $M_\sigma \cdot M_\lambda = E$.

Thus Condition 6.6 and Condition 7.1 for AG, AM, AGI and AMI translate into statements about certain solutions of systems of linear diophantine equations.

LEMMA 8.1. *Let G be a subgroup of $M_{n,m}(\mathbb{Z})$ and let A be an element of $M_{k,n}(\mathbb{Z})$. The solutions of the equation $AX = 0$ in G are $\mathbb{Z}$-linear combinations of $r \le n \cdot m$ basic solutions in G.*

PROOF. Let $U = \{ Y \in M_{n,m}(\mathbb{Z}); AY = 0 \}$ be the set of solutions in $M_{n,m}(\mathbb{Z})$. U and hence $U \cap G$ are subgroups of the free abelian group of rank $n \cdot m$ $M_{n,m}(\mathbb{Z})$.

Hence $U \cap G$ is a free abelian group of rank $r \le n \cdot m$ (see e.g. Kurosh (1960), p. 145). ❑

If we take these basic solutions, we get any solution in G as $\mathbb{Z}$-linear combination of $s \leq n \cdot m$ solutions in G. This yields Condition 6.6 for AG and AGI. The solutions of the inhomogeneous equation are obtained in the usual way.

LEMMA 8.2. *Let G be a subgroup of $M_{n,m}(\mathbb{Z})$ and let A, B be elements of $M_{k,n}(\mathbb{Z})$, $M_{k,m}(\mathbb{Z})$. Let $Y_0 \in G$ be an arbitrary solution of $AX = B$. Any solution $Y \in G$ is of the form $Y = Y_0 + Z$ where Z is a solution of $AX = 0$.*

This yields Condition 7.1 for AG and AGI. Note that we need only one special solution Y_0 of $AX = B$. Thus we have

PROPOSITION 8.3. *The theories AG and AGI are unitary. In addition, they are unitary w.r.t. unification with constants.*

See Lankford-Butler-Brady (1984) for a direct proof of the result for AG.

LEMMA 8.4. *Let S be a stable subsemigroup of $M_{n,m}(\mathbb{N})$ and let A be an element of $M_{k,n}(\mathbb{Z})$. The solutions of the equation $AX = 0$ in S are $\mathbb{N}$-linear combinations of finitely many basic solutions.*

PROOF. The proof is similar to that of Corollary 9.19 of Clifford-Preston (1967).
On the elements of $M_{n,m}(\mathbb{N})$ we define the partial ordering $\leq$ component wise. Since $\leq$ on $\mathbb{N}$ is a well partial ordering (wpo), its $n \cdot m$-fold cartesian product on $M_{n,m}(\mathbb{N})$ is also a wpo (see Nash-Williams (1963)). The set $H = \{ Y \in S; AY = 0 \}$ is a stable subsemigroup of $M_{n,m}(\mathbb{N})$. Let H_0 be the set of minimal elements of H. H_0 is finite (finite antichains) and any element of H lies above a minimal element (finite chains). We now show that H_0 is the set of basic solutions we are searching for. For any solution $Y \in H \setminus H_0$, there is a solution $Y_0 \in H_0$ such that $Y < Y_0$, i.e. there exists $Z \in M_{n,m}(\mathbb{N})$ with $Y = Y_0 + Z$. Since H is stable, we have $Z \in H$ and obviously $Z < Y$. By noetherian induction, we thus have proved the lemma. ❑

LEMMA 8.5. *Let S be a stable subsemigroup of $M_{n,m}(\mathbb{N})$ and let A, B be elements of $M_{k,n}(\mathbb{Z})$, $M_{k,m}(\mathbb{Z})$. There exists a finite set T_0 of solutions of $AX = B$ in S such that any* *solution Y of AX = B in S is of the form $Y = Y_0 + Z$, where $Y_0 \in T_0$ and $Z \in S$ is a solution of AX = 0.*

PROOF. Let T_0 be the finite set of minimal elements in $T = \{ Y \in S;\ AY = B \}$. For $Y \in T_0$ we have $Y = Y + 0$. Otherwise $Y_0 < Y$ for some $Y_0 \in T_0$, i.e. $Y = Y_0 + Z$ for some $Z \in M_{n,m}(\mathbb{N})$. Since S is stable and Y, $Y_0 \in S$ we have $Z \in S$. Now $B = AY = AY_0 + AZ = B + AZ$ yields $AZ = 0$. ❑

Lemma 8.4 and 8.5 establish Condition 6.6 and 7.1 for AM and AMI. Thus we have

PROPOSITION 8.6. *The theories AM and AMI are unitary and they are finitary w.r.t. unification with constants.*

For alternative proofs of this result for AM see the references in the Introduction.

Effective methods to solve systems of linear diophantine equations in $\mathbb{Z}$ can be found in Niven-Zuckerman (1972) and Knuth (1973). For solutions in $\mathbb{N}$ see e.g. Makanin (1977) (Lemma 1.1), Huet (1978), Fortenbacher (1985), Lambert (1987) and Clausen-Fortenbacher (1988). Efficient unification algorithms for the theories AG, AM, AGI and AMI depend upon efficient implementations of these methods.

9. The Theories AMH, AIMH and AGH

In an earlier version of this paper (Baader (1988)) it was shown that the unification problem $< h(x_1)h(x_2) = x_2h^2(x_3) >_{AIMH}$ has no minimal complete set of AIMH-unifiers. The same proof can be used for AMH in place of AIMH. In this section we give a more algebraic proof of the fact that AMH has type zero. We shall also show that AGH is unitary.

Let σ: $F_{AGH}(x) \to F_{AGH}(y)$ be a morphism of C(AGH). Then there are $k \geq 0$ and $a_0, \ldots, a_k \in \mathbb{Z}$ such that $x\sigma = y^{a_0}h(y^{a_1})\ldots h^k(y^{a_k})$. We associate with the morphism σ the polynomial $a_0 + a_1X + \ldots + a_kX^k$. Thus any morphism δ: $F_{AGH}(Y) \to F_{AGH}(Z)$ for $|Y| = m$ and $|Z| = n$ corresponds to a matrix $M_\delta \in M_{m,n}(\mathbb{Z}[X])$. This correspondence was first mentioned by Werner Nutt (Nutt (1988)). It is easy to see that $M_{\sigma\delta} = M_\sigma \cdot M_\delta$ and $M_{\sigma+\delta} = M_\sigma + M_\delta$.

Accordingly, any morphism $\delta: F_{AMH}(Y) \to F_{AMH}(Z)$ of C(AMH) corresponds to a matrix $M_\delta \in M_{m,n}(\mathbb{N}[X])$.

We shall now show that AMH has type zero, utilizing the fact that the semiring $\mathbb{N}[X]$ (see Kuich-Salomaa (1986) for the definition of "semiring") has ideals, which are not finitely generated (i.e. $\mathbb{N}[X]$ is not a noetherian semiring).

Consider the unification problem $\Gamma = < h(x_1)h(x_2) = x_2h^2(x_3) >_{AMH}$. If we translate the morphisms into polynomials we get the linear equation

$$(*) \quad Xx_1 + Xx_2 = x_2 + X^2x_3,$$

which has to be solved by a vector $\underline{p} = (p_1, p_2, p_3)$ in $(\mathbb{N}[X])^3$. Obviously, for any $n \geq 0$ the vector $\underline{p}^{(n)} = (p_1^{(n)}, p_2^{(n)}, p_3^{(n)}) = (1, X + X^2 + ... + X^{n+1}, X^n)$ is a solution of (*).

LEMMA 9.1. *There does not exist a solution $\underline{p}$ of (*) in $(\mathbb{N}[X])^3$ such that $p_1 + p_3 = 1$.*

PROOF. For $p_1 = 0$ and $p_3 = 1$ we get $Xp_2 = p_2 + X^2$, which yields $(X - 1)p_2 = X^2$ in $\mathbb{Z}[X]$. But $X - 1$ is not a divisor of X^2. The case $p_1 = 1$ and $p_3 = 0$ leads to a similar contradiction. ❑

It is easy to see that $I_{1+3} := \{ p_1 + p_3$; There exists p_2 such that (p_1, p_2, p_3) solves (*) $\}$ is an ideal in $\mathbb{N}[X]$. We know that $1 + X^n \in I_{1+3}$ for any $n \geq 0$ and $1 \notin I_{1+3}$.

LEMMA 9.2. *An ideal $I \subseteq \mathbb{N}[X]$ such that $1 + X^n \in I$ for any $n \geq 0$ and $1 \notin I$ is not finitely generated.*

PROOF. Evidently $1 + X^n = f \cdot g$ for $f, g \in \mathbb{N}[X]$ or $1 + X^n = f + g$ for $f, g \in \mathbb{N}[X] \setminus \{ 0 \}$ implies $f = 1$ or $g = 1$. But $1 \notin I$. ❑

PROPOSITION 9.3. *The theory AMH is of unification type zero.*

PROOF. Assume that AMH has not type zero. Then AMH is unitary by Theorem 6.1 and satisfies Condition 6.6 by Proposition 6.7. If we translate the morphisms into polynomials, Condition 6.6 means that $\underline{I} := \{ \underline{p} \in (\mathbb{N}[X])^3; \underline{p}$ is a solution of (*) $\}$ is a finitely generated $\mathbb{N}[X]$-semimodule, i.e. there exist $\underline{g}^{(1)}, ..., \underline{g}^{(s)} \in \underline{I}$ such that $\underline{I} = \{ \underline{g}^{(1)}f_1 + ... + \underline{g}^{(s)}f_s; f_1, ..., f_s \in \mathbb{N}[X] \}$. But then $I_{1+3} = \{ p_1 + p_3$; There exists p_2 such that $(p_1, p_2, p_3) \in \underline{I} \}$ would also be finitely generated, which contradicts Lemma 9.2. ❑

To establish Condition 6.6 for AGH, we have to consider systems of homogeneous linear equations in $\mathbb{Z}[X]$, i.e. systems $f_{1i}x_1 + \ldots + f_{ki}x_k = 0$ ($i = 1, \ldots, s$), where the coefficients f_{ij} and the desired solutions are elements of $\mathbb{Z}[X]$.

The set of solutions $\underline{I} \subseteq (\mathbb{Z}[X])^k$ is a $\mathbb{Z}[X]$-module, which is finitely generated by Hilbert's Basis Theorem (see e.g. Jacobson (1980)). This yields Condition 6.6. Any solution of an inhomogeneous system is of the form $\underline{p} = \underline{p}^{(0)} + \underline{q}$ where $\underline{p}^{(0)}$ is a special solution of this system and $\underline{q}$ is a solution of the corresponding homogeneous system. Thus AGH satisfies Condition 7.1 with $|M| = 1$.

PROPOSITION 9.4. (Nutt (1988))
The theory AGH is unitary and it is also unitary w.r.t. unification with constants. ❑

Since the proof of Hilbert's Basis Theorem is not effective, the above argument does not yield an AGH-unification algorithm. An effective method to solve linear equations in $\mathbb{Z}[X]$ using Gröbner Bases is described in Buchberger (1985). If we apply this method to the equation (*) $Xx_1 + Xx_2 = x_2 + X^2x_3$, we obtain ($X - 1, -X, 0$) and ($-X^2, X^2, -1$) as generators for the solutions in $\mathbb{Z}[X]$ (see also Baader (1989b) for a description of the method and more examples). Because Hilbert's Basis Theorem holds for $\mathbb{Z}[X_1,\ldots,X_r]$, Proposition 9.4 is also valid for the theory of abelian groups with finitely many commuting homomorphisms.
If we consider the theory of abelian groups with two (non-commuting) homomorphisms, a morphism δ translates into a matrix $M_\delta \in M_{m,n}(\mathbb{Z}\langle X,Y\rangle)$. The elements of $\mathbb{Z}\langle X,Y\rangle$ are polynomials in two non-commuting indeterminates X, Y. The set of solutions of a system $f_{1i}x_1 + \ldots + f_{ki}x_k = 0$ ($i = 1, \ldots, s$) is a $\mathbb{Z}\langle X,Y\rangle$-right module. In spite of the fact that $\mathbb{Z}\langle X,Y\rangle$ is not right noetherian, it can be shown that those $\mathbb{Z}\langle X,Y\rangle$-right modules, which are obtained as sets of solution of linear equations, are finitely generated (Baader (1989b), see also Mora (1986) for Gröbner Bases for non-commutative polynomial rings).

10. Conclusion

In this paper, we were less interested in deriving efficient unification algorithms for a specific theory. Instead, we gave a general framework for unification in the whole class of commutative theories. An important result is the fact that commutative theories, where the finitely generated objects are finite, are always unitary (finitary w.r.t. unification with constants). But even in this case, the construction of an efficient unification algorithm, which computes the most general unifier, is yet another problem. This algorithm should produce unifiers, which introduce a minimal number of variables (i.e. the number r of E-unifiers in Condition 6.6 should be as small as possible). In the case of unification with constants, we want to obtain a minimal complete set (i.e. the set M of Condition 7.1 has to be as small as possible) rather than just a complete set (see e.g. Baader-Büttner (1988) where this problem is solved for AIM).

The categorical reformulation of E-unification allows to characterize the class of commutative theories by properties of the category C(E) of finitely generated E-free objects: C(E) has to be a semiadditive category. The definition of semiadditive categories provides an algebraic structure on the morphism sets, which can be used to obtain algebraic characterizations of the unification types. Hence unification algorithms for commutative theories can be derived with the help of well-known algebraic methods (e.g. Gröbner Base algorithms).

Because coproducts in semiadditive categories are also products, finitely many morphisms of C(E) may be coded into one morphism. The prove of Theorem 6.1 – which states that commutative theories are either unitary or of type zero – is mainly based upon this fact. Thus well-known results of category theory (Theorem 4.3) can be used to get results in unification theory.

References

Baader, F. (1987). Unification in Varieties of Idempotent Semigroups. *Semigroup Forum* **36**.

Baader, F. (1988). Unification in Commutative Theories. To appear in *J. Symbolic Computation*.

Baader, F. (1989a). Characterizations of Unification Type Zero. Proceedings of the RTA '89 Chapel Hill (USA). *Springer Lec. Notes Comp. Sci.* **355**.

Baader, F. (1989b). Unification in Commutative Theories, Hilbert's Basis Theorem and Gröbner Bases. Preprint.

Baader, F., Büttner, W. (1988). Unification in Commutative Idempotent Monoids. *TCS* **56**.

Buchberger, B. (1985). Gröbner Bases: An Algorithmic Method in Polynomial Ideal Theory. In Bose, N. K. (Ed.). *Recent Trends in Multidimensional System Theory*.

Büttner, W. (1986). Unification in the Data Structure Multiset. *J. Automated Reasoning* **2**.

Clausen, M., Fortenbacher, A. (1988). Efficient Solution of Linear Diophantine Equations. To appear in *J. Symbolic Computation*, Special Issue on Unification.

Clifford, A.H., Preston, G.B. (1967). *The Algebraic Theory of Semigroups, Vol. 2*. AMS Mathematical Surveys **7**.

Cohn, P.M. (1965). *Universal Algebra*. New York: Harper and Row.

Fages, F. (1984). Associative-Commutative Unification. Proceedings of the CADE '84 Napa (USA). *Springer Lec. Notes Comp. Sci.* **170**.

Fages, F., Huet, G. (1986). Complete Sets of Unifiers and Matchers in Equational Theories. *TCS* **43**.

Fortenbacher, A. (1985). An Algebraic Approach to Unification under Associativity and Commutativity. Proceedings of the RTA '85 Dijon (France). *Springer Lec. Notes Comp. Sci.* **202**.

Freyd, P. (1964). *Abelian Categories*. New York: Harper and Row.

Grätzer, G. (1968). *Universal Algebra*. Princeton: Van Nostrand Company.

Herold, A. (1987). Combination of Unification Algorithms in Equational Theories. Ph.D. Dissertation, Universität Kaiserslautern.

Herrlich, H., Strecker, G.E. (1973). *Category Theory*. Boston: Allyn and Bacon Inc.

Huet, G. (1978). An Algorithm to Generate the Basis of Solutions to Homogeneous Diophantine Equations. *Information Processing Letters* **7**.

Huet, G. (1980). Confluent Reductions: Abstract Properties and Applications to Term Rewriting Systems. *J. ACM* **27**.

Jacobson, N. (1980). *Basic Algebra II*. San Francisco: Freeman and Company.

Jaffar, J., Lassez, J.L., Maher, M.J. (1984). A Theory of Complete Logic Programs with Equality. *J. Logic Programming* **1**.

Jouannaud, J.P. (1983). Confluent and Coherent Equational Term Rewriting Systems, Applications to Proofs in Abstract Data Types. Proceedings of the CAAP '83 Pisa (Italy). *Springer Lec. Notes Comp. Sci.* **159**.

Jouannaud, J.P., Kirchner, H. (1986). Completion of a Set of Rules Modulo a Set of Equations. *SIAM J. Comp.* **15**.

Kuich, W., Salomaa, A. (1986). *Semirings, Automata, Languages*. Berlin: Springer Verlag.

Knuth, D.E. (1973). *The Art of Computer Programming, Vol 2*. Reading: Addison-Wesley.

Kurosh, A.G. (1960). *The Theory of Groups, Vol 1*. New York: Chelsea Publishing Company.

Lambert, J-L. (1987). Une borne pour les générateurs des solutions entières positives d' une équation diophantienne linéaire. *CRASP* **305**.

Lankford, D., Butler, G., Brady, B. (1984). Abelian Group Unification Algorithms for Elementary Terms. *Contemporary Mathematics* **29**.

Lawvere, F.W. (1963). Functional Semantics of Algebraic Theories. Ph.D. Dissertation, Columbia University.

Livesey, M., Siekmann, J. (1978). Unification in Sets and Multisets. SEKI Technical Report, Universität Kaiserslautern.

Makanin, G.S. (1977). The Problem of Solvability of Equations in a Free Semigroup. *Mat. Sbornik* **103**. English transl. in *Math USSR Sbornik* **32** (1977).

Mora, F. (1986). Gröbner Bases for Non-Commutative Polynomial Rings. Proceedings of the AAECC3. *Springer Lec. Notes Comp. Sci.* **228**.

Nash-Williams, C.St.J.A. (1963). On Well-Quasi-Ordering Finite Trees. *Proc. Cambridge Philos. Soc.* **59**.

Nevins, A.J. (1974). A Human Oriented Logic for Automated Theorem Proving. *J. ACM* **21**.

Niven, I., Zuckerman, H.S. (1972). *An Introduction to the Theory of Numbers*. New York: John Wiley and Sons.

Nutt, W. (1988). Talk at the Second Workshop on Unification, Val d' Ajol (France).

Peterson, G., Stickel, M. (1981). Complete Sets of Reductions for Some Equational Theories. *J. ACM* **28**.

Plotkin, G. (1972). Building-in Equational Theories. *Machine Intelligence* **7**.

Rydeheard, D.E., Burstall, R.M. (1985). A Categorical Unification Algorithm. Proceedings of the Workshop on Category Theory and Computer Programming. *Springer Lec. Notes Comp. Sci.* **240**.

Siekmann, J. (1986). Universal Unification. Proceedings of the 7th ECAI.

Slage, J.R. (1974). Automated Theorem Proving for Theories with Simplifiers, Commutativity and Associativity. *J. ACM* **21**.

Stickel, M. (1981). A Unification Algorithm for Associative-Commutative Functions. *J. ACM* **28**.

Stickel, M. (1985). Automated Deduction by Theory Resolution. *J. Automated Reasoning* **1**.

AN ABSTRACT FORMULATION FOR REWRITE SYSTEMS

A.J. Power

Department of Mathematics and Statistics
Case Western Reserve University
Cleveland, Ohio 44106
U.S.A.

ABSTRACT. Herein, we describe an abstract algebraic object which can be used to discuss rewrite systems. We define then use the notion of 2–category to encapsulate some fundamental properties of rewrites. Recent developments in the general theory of 2–categories are then used to give a theorem stating conditions under which rewrite derivations exist, and to give further conditions under which there is a normal form for a derivation, hence a deterministic method to express a derivation as a composite of rewrites in a particular order. Several examples are pursued through the course of the paper in order to make clear precisely how the general theory may be applied.

1. INTRODUCTION

The concept of 2–category arose in Mathematics for reasons essentially irrelevant to rewrite systems. Ross Street, as suggested by his introduction of the term "computad" in [11], expected a relationship between 2–categories and computing to develop, but little came of it until recently. However, the concept of 2–category does seem to be remarkably well suited to an abstract algebraic formulation of a large class of rewriting problems. That has already been realized explicitly in some papers such as Rydeheard and Stell's [9], and is implicit in others such as Hoare's [3].

Recently, there have been two advances in the general theory of 2–categories that have specific bearing on rewrite systems. One is the formulation and proof of a "pasting" theorem along the lines adumbrated in Kelly and Street's paper [6]; the other is the development of a normal form for pasting composition, hence a deterministic method by which to evaluate a composite.

Herein, we first attempt to make explicit the relationship between 2–categories and rewrite systems, generalising the important class of examples offered by Rydeheard and Stell. We define the concept of 2–category and exhibit three fundamental classes of rewrite systems including that of Rydeheard and Stell, as 2–categories. That constitutes Section 2.

We then explain the first of the recent developments in the general theory of 2–categories outlined above, and we illustrate its specific relevance to rewrite systems by pursuing our three fundamental examples. One of the strengths of the general theory of 2–categories is that the pasting composition, which is the most general form of composition in a 2–category, has a particularly elegant planar depiction. Only recently has there emerged a theorem making precise the relationship between pasting and pictures in the plane. That serves to replace rather awkward formal expressions subject to a list of equations by a single, simple diagram in the plane. It means that entire rewrite derivations can be depicted elegantly in the plane, with the depiction making it clear to the eye precisely where and how each step is used. That forms the content of Section 3.

Finally, Section 4 outlines another relevant recent development in the theory of 2–categories, the development of a notion of normal form for a pasting composite, and a theorem stating conditions under which a normal form exists. That is important because as illustrated in Section 4, it is possible for different pasting diagrams to represent the same composite.

The notion of 2–category has long been in Mathematical literature such as [6]. Rydeheard and Stell explicitly use and develop the ideas for equational proofs and unification algorithms in [9]. The central theorem of Section 3 is to appear as [8], which contains no reference to rewrite systems; so Section 3 is essentially new. The main result of Section 4, by Eilenberg and Street, is part of [2]. I should like to thank Eilenberg and Street for their kind permission to report their result here.

Several people, including Huet [4], and especially Michael Johnson [5], and indubitably others, have also advocated a relationship between coherence problems and rewriting, but to the best of my knowledge, none has yet invoked the main theorem of Section 3. However, Curien, Friere and Golgo have begun work upon implementing the geometry relevant to Section 3 to display 2–categorical proofs on the computer. Other examples of the application of 2–categories to rewriting appear in Seely's [10] and Wells' [12].

2. 2–CATEGORIES

DEFINITION 2.1. A *2–category* $\mathscr{A}$ consists of a set ob $\mathscr{A}$ of objects together with, for each $A,B \in$ ob $\mathscr{A}$, a category $\mathscr{A}(A,B)$, for each $A,B,C \in$ ob $\mathscr{A}$, a functor $\circ\colon \mathscr{A}(A,B) \times \mathscr{A}(B,C) \longrightarrow \mathscr{A}(A,C)$, and for each $A \in$ ob $\mathscr{A}$, a functor $j\colon 1 \longrightarrow \mathscr{A}(A,A)$ where 1 is the one object, one arrow category, such that

$$\begin{array}{ccc}
\mathscr{A}(A,B) \times \mathscr{A}(B,C) \times \mathscr{A}(C,D) & \xrightarrow{\ \circ\times 1\ } & \mathscr{A}(A,C) \times \mathscr{A}(C,D) \\
{\scriptstyle 1\times\circ}\big\downarrow & & \big\downarrow{\scriptstyle \circ} \\
\mathscr{A}(A,B) \times \mathscr{A}(B,D) & \xrightarrow[\ \circ\ ]{} & \mathscr{A}(A,D)
\end{array}$$

and

$$\begin{array}{ccc}
\mathscr{A}(A,B) & \xrightarrow{\ j\times 1\ } & \mathscr{A}(A,A) \times \mathscr{A}(A,B) \\
{\scriptstyle 1\times j}\big\downarrow & \searrow^{1} & \big\downarrow{\scriptstyle \circ} \\
\mathscr{A}(A,B) \times \mathscr{A}(B,B) & \xrightarrow[\ \circ\ ]{} & \mathscr{A}(A,B)
\end{array}$$

commute.

The objects of any $\mathcal{A}(A,B)$ are called *1–cells* of $\mathcal{A}$; the arrows of any $\mathcal{A}(A,B)$ are called *2–cells*; $\circ$ is called *horizontal composition*; the composition within any $\mathcal{A}(A,B)$ is called *vertical composition*. If α is a 2–cell from f to g, we may write α: f => g: A → B, where dom f = dom g = A and cod f = cod g = B; and we express this information pictorially by

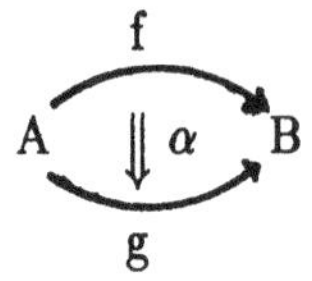

The main axiom is the first of the two commutativities demanded, and the main point of that is called the *interchange law*: that for 2–cells α, α', β and β' with appropriately matching domains and codomains,

$$(\alpha\cdot\alpha')\circ(\beta\cdot\beta') = (\alpha\circ\beta)\cdot(\alpha'\circ\beta').$$

CONSTRUCTION 2.2 Each category X yields a 2–category X_d whose objects are the objects of X, and whose hom–categories $X_d(A,B)$ are the sets X(A,B) regarded as discrete categories.

Conversely, for any 2–category $\mathcal{A}$, we may define the *underlying* category $\mathcal{A}_0$ of $\mathcal{A}$ to have objects precisely the objects of $\mathcal{A}$, and to have arrows precisely the 1–cells of $\mathcal{A}$. Composition in $\mathcal{A}_0$ is given by the object function of $\circ$, similarly for identities.

It is often useful, by the above, to regard a 2–category as an ordinary category together with 2–cells between arrows f and g such that dom f = dom g and cod f = cod g.

CONSTRUCTION 2.3 Given a category X together with a set X_2 and two functions ∂_0,∂_1: $X_2 \to \mathrm{Arr}\,X$ such that dom ∂_0 = dom ∂_1 and cod ∂_0 = cod ∂_1, we may add the elements of X_2 freely to X as 2–cells to yield a 2–category $\mathcal{A}$. Explicitly, the underlying category of $\mathcal{A}$ is X and the 2–cells in $\mathcal{A}$ are given as follows:

·the arrows in X are to be the identity 2–cells in $\mathcal{A}$ with respect to vertical composition

·given an element x of X_2 and arrows h and k such that dom $\partial_0(x)$ = cod h and cod $\partial_0(x)$ = dom k, then hoxok: $h\circ\partial_0 x\circ k$ => $h\circ\partial_1 x\circ k$ is to be a 2–cell in $\mathcal{A}$, sometimes called a "whisker" in $\mathcal{A}$.

·given non–identity 2–cells α:f => g and β: g => h in $\mathcal{A}$, then $\alpha\cdot\beta$: f => h is to be a 2–cell in $\mathcal{A}$,

subject to the identifications given by the commutativities in the definition of 2–category. Observe that the horizontal composite of 2–cells is always given by a vertical composite of two whiskers.

EXAMPLE 2.4 Given an alphabet L and a collection of rewrite rules on strings of symbols in L, we define a 2-category $\mathscr{L}$ as follows:

$\mathscr{L}$ has one object.

An arrow in $\mathscr{L}$ is a string of symbols in L.

$\mathscr{L}$ is generated as in Construction 2.3 by the rewrite rules.

It is trivial to see that the object and arrows of $\mathscr{L}$ form a category, with composition given by concatenation of strings; the identity arrow is the empty sting. Construction 2.3 ensures that $\mathscr{L}$ is thus a 2-category.

This example of a 2-category is special because it has only one object. However, even in this special case, recognition of the 2-categorical structure allows substantial simplification of the presentation of a derivation, cf. [7].

EXAMPLE 2.5. Given a strictly typed programming language L, let ob $\mathscr{P}$ be the set of all data types of the language. An arrow in $\mathscr{P}$ is a program; its domain is the input data type of the program; its codomain is its output data type. Composition is given by composition of programs; identity arrows are trivial programs from a data type to itself that takes an input to itself as output. Given programs P and Q from A to B, introduce a 2-cell α: P => Q if Q is a refinement of P. Then, $\mathscr{P}$ is to be the 2-category given by Construction 2.3. This definition of the 2-category $\mathscr{P}$ is implicit in Hoare's paper [3]: Hoare made the construction but did not state that it is a 2-categorical construction.

EXAMPLE 2.6 (See [9]) Let Ω be an operator domain and let $T_\Omega(X)$ be the set of Ω-terms with variables in X. The 2-category $\mathscr{T}$ has all sets as objects. An arrow f: X → Y is a function f: X → $T_\Omega(Y)$. Given f: X → Y and g: Y → Z, the composite fog is given by observing that the function g: Y → $T_\Omega(Z)$ extends to a function $T_\Omega(g)$: $T_\Omega(Y) \to T_\Omega(Z)$, and defining fog to be the composite $T_\Omega(g)\cdot f$ of functions. It follows from a general categorical construction called the Kleisli construction that this yields a category $\mathscr{T}_0$; identities are given by the obvious inclusions. $\mathscr{T}$ is given as follows: first, Construction 2.3 is applied to $\mathscr{T}_0$ together with the rewrite rules in Ω. For instance, given an operator domain Ω containing a binary operation ×, a possible rewrite rule is "commute": x × y => y × x in {x,y}. This is to correspond to the 2-cell

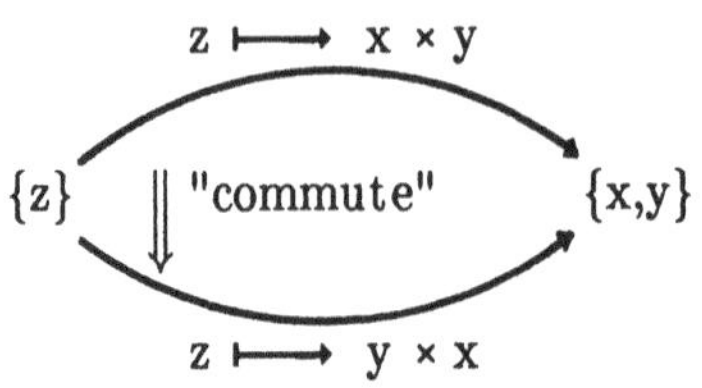

The reason for introducing z is to allow left composition with the "commute" rule, which amounts to substitution of z by more complex terms with commutativity applied to subterms. To complete the definition of $\mathcal{T}$, observe that $\mathcal{T}_0$ has binary coproducts (given as in Set). Then, by a process analogous to Construction 2.3, we can construct the 2-category $\mathcal{T}$ as follows: given α:f => f': a → c and β: g => g': b → c, add a 2-cell (α,β): (f,g) => (f',g'): a + b → c subject to the evident coherence conditions. The appropriate categorical setting for this sort of construction is another very recent development in the general theory of 2-categories; some early results appear in Street's [11], but the general treatment is only now being formulated. More details of this particular example appear in [9].

Observe that this example generalises Example 2.4, but also requires somewhat more sophistication. Observe also that although Rydeheard and Stell restrict themselves to a single-sorted theory, precisely the same categorical constructions hold for multi-sorted theories.

3. PASTING COMPOSITION IN 2-CATEGORIES

In the definition of 2-category, the only sorts of composition of 2-cells defined were horizontal and vertical composition. These generate a more general form of composition, called pasting, first introduced by Benabou in [1]. Pasting is the fundamental operation in a 2-category.

The idea is as follows: we draw a picture such as

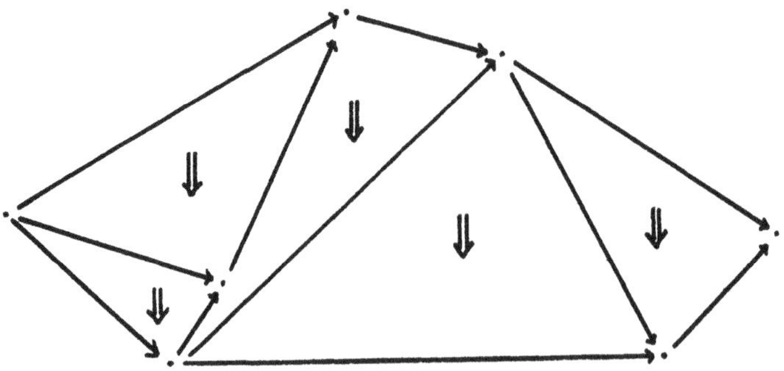

Given a 2-category $\mathcal{A}$, each vertex is labelled by a 0-cell (or object) in $\mathcal{A}$, each arc i labelled by a 1-cell in $\mathcal{A}$, and each face is labelled by a 2-cell, such that domains an codomains are preserved. Then, the above "pasting diagram" is meant to indicate a vertical composite of horizontal composites of 2–cells of the form

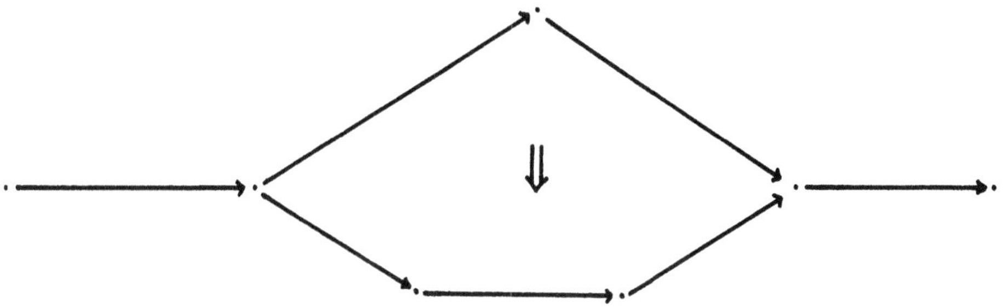

There is usually a choice in the order in which the composites are taken. Only recently has there emerged a theorem stating precisely which pictures may be drawn so that such a composite exists and is uniquely determined.

EXAMPLE 3.1 Pursuing Example 2.4, a group is a set G with an associative binary operation $\cdot$, an identity element e, and such that for each $g \in G$, there is an unique element $g^{-1} \in G$ such that $g^{-1} \cdot g = e$. We apply directions $g \cdot e \Rightarrow g$ and $g^{-1} \cdot g \Rightarrow e$ to each equality for each element of G. The proof that $(g \cdot h)^{-1} = h^{-1} \cdot g^{-1}$ may be expressed by

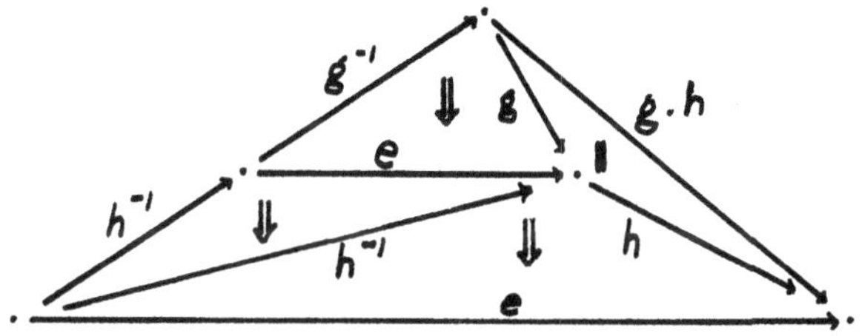

which shows that $h^{-1} \cdot g^{-1}$ satisfies the defining property of $(g \cdot h)^{-1}$.

Of course, this example has long been handled satisfactorily by displaying the information along a line. However, we find the 2–dimensional display often easier for the human eye to see precisely how the individual rewrites are composed. This example illustrates how the conventional 1–dimensional displays may be conveniently regarded as 2–dimensional displays.

EXAMPLE 3.2 Pursuing Example 2.5, in data refinement, typically one refines only part of a program at any time, so a typical picture of a succession of refinements is

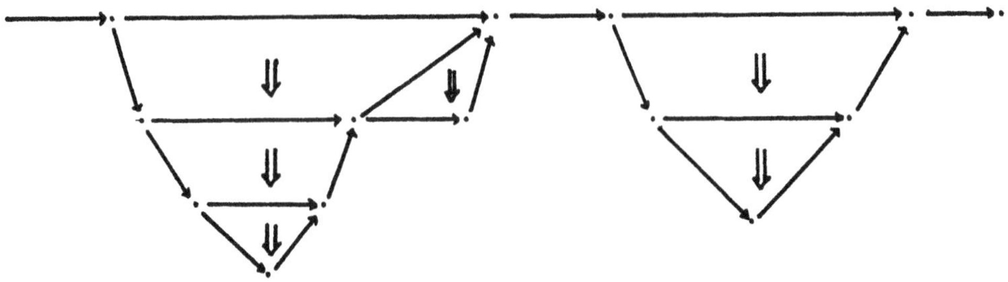

EXAMPLE 3.3 Pursuing Example 2.6, Rydeheard and Stell define a particular 2–cell in [7] by

(("commute"∘k)·("right ident"))∘(("commute"∘k')· "right inverse") (1)

This is the pasting composite

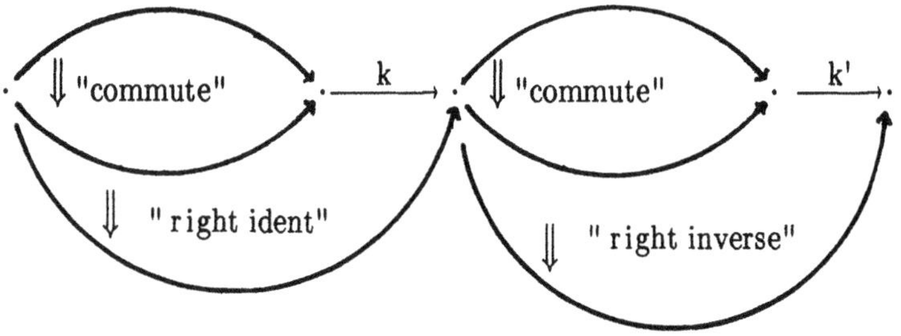

They mention that (1) must be identified with another expression of similar length: that fact follows immediately from the pasting theorem, because both expressions are the same labelling of the same pasting diagram. In fact, the above diagram can be described as a vertical composite of whiskers in a total of 6 possible ways, all of which must be identified.

Motivated by the above examples, we now formulate the pasting theorem.

DEFINITIONS 3.4 (i) A *plane graph* consists of finite set of points in the plane, called *vertices*, together with a finite set of directed arcs between any two vertices, such that any two arcs touch only at their endpoints. Herein, we restrict our attention to connected plane graphs: those plane graphs with a non–empty set of vertices and such that for any two distinct vertices, there is a non–directed path from one to the other, i.e. an alternating sequence of vertices and arcs $v_0 a_1 v_1 \ldots v_n$ such that a_i has endpoints v_{i-1} and v_i, and such that all of the vertices v_i are distinct.

(ii) A *directed path* is a path for which each a_i has head v_i.

(iii) A plane graph G partitions the rest of the plane into a finite number of connected regions, which are called the *faces* of G. Each plane graph has exactly one unbounded face, called the *exterior* face; the other faces are called *interior* faces. We regard the boundary of a face as an alternating sequence of vertices and arcs $v_0 a_1 \ldots v_n$ such that $v_0 = v_n$, the endpoints of a_i are v_{i-1} and v_i, and if one moves from v_0 via a_1 to v_1, etcetera, one always has F on one's right hand side.

(iv) Given a path γ from u to v, the corresponding path from v to u is denoted by γ^*.

(v) A *plane graph* G *with source* s *and sink* t is a plane graph G with distinct vertices s and t in the exterior face of G such that for each vertex v, there are directed paths from s to v and from v to t.

DEFINITIONS 3.5 (i) A *pasting diagram* is a plane graph G with source and sink such that for every interior face F, there exist distinct vertices s(F) and t(F) and directed paths $\sigma(F)$ and $\tau(F)$ from s(F) to t(F) such that the boundary of F is $\sigma(F)(\tau(F))^*$. Then, $\sigma(F)$ and $\tau(F)$ are called the *domain* and the *codomain* of F respectively.

(ii) A *labelling* of a pasting diagram G in a 2-category $\mathcal{A}$ is a computad morphism from G to $\mathcal{A}$, i.e. for each vertex, arc and face in G other than the exterior face, the assignment of a k-cell in $\mathcal{A}$, for k = 0,1,2 respectively, preserving domains and codomains.

3.6 PASTING THEOREM (Power [8]) Every labelling of a pasting diagram has an unique composite.

The more difficult part of the theorem is to prove the existence of a composite. The proof appears in [8].

Observe that the pasting theorem has as a corollary a partial answer to the question of confluence: given pasting diagrams such that the superimposition of one upon the another yields a matching, the union of the pasting diagrams exhibits confluence. For instance,

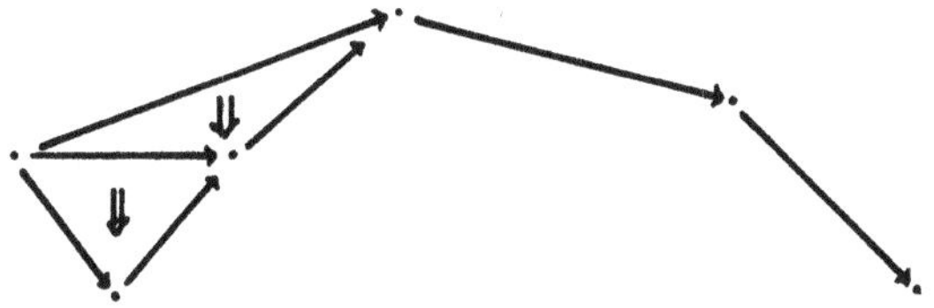

and

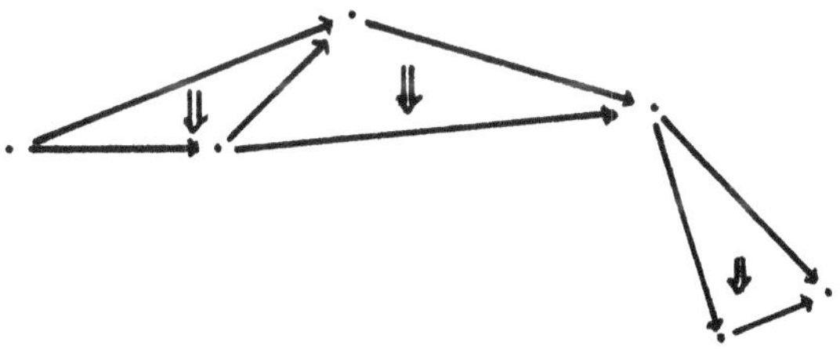

yield

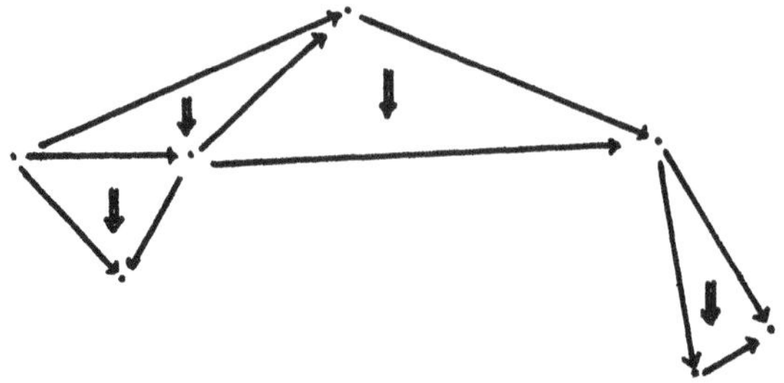

So, the Church–Rosser property becomes elementary to verify in many cases: if the two diagrams superimpose, the Church–Rosser property holds.

At this point, it seems important to say something about implementation of this work. Unfortunately, as a novice in Computer Science, I can only adumbrate some ideas. However, Curien, Friere and Golgo are currently working upon an implementation of pasting proofs.

A few points seem clear. First, one may assume that pasting diagrams are defined so that all arcs are given by a finite set of line segments. Just as computers can recognize the end of data on a tape, it should be possible for them to recognize a face of a plane graph, and also to recognize whether it is the exterior face, since we only consider finite graphs. It is part of the study of Algebraic Topology to recognize when two plane graphs are equivalent: when one restricts to edges given by finitely many line segments, it is a relatively easy task.

A source in a pasting diagram is necessarily unique; so one can identify the source. The first arc from the source is that on the exterior face such that, if there is more than one arc from the source, then moving clockwise around the source from the first segment of the first arc, one enters an interior face. Proceeding inductively, one can determine the topmost path from the source to the sink. Two pasting diagrams are equivalent only if they have the same number of arcs in their topmost paths.

It follows from the definition of pasting diagram that if a diagram has an interior face, it must have one, F, such that $\sigma(F)$ lies entirely on the topmost path. Take the leftmost. Two equivalent diagrams must have the same number of arcs in the codomain $\tau(F)$ of the leftmost face whose domain lies entirely on the topmost path. Moreover, of course, that leftmost face must occur after the same number of arcs in the topmost path. Continuing inductively, having removed F, this gives an algorithm to check whether two pasting diagrams are equivalent. (Pasting diagrams never have loops.) It is then straightforward to check when two labellings of equivalent pasting diagrams are equivalent.

Having done this, we now have an explicit description of the composites given by pasting rather than a list of composites all of which are to be identified, as in Rydeheard and Stell's manuscript.

4. NORMAL FORMS FOR PASTING

The notion of pasting is central to the theory of 2–categories, and for many Mathematical purposes, Theorem 3.6 gives all of the information that is desired. However, it does not give an unique description of every possible composite given by a set of 2–cells. We have already shown that every composite generated by a set of 2–cells can be depicted by a labelled pasting diagram; however, it is simply not true in general that any two such depictions are necessarily equivalent. For a large class of examples, those in which every arc is labelled by a non–identity 1–cell, it is true. However, in full generality, the situation is somewhat more delicate, hence the following discussion of work by Eilenberg and Street [2].

The problem that arises with identity 1–cells is that the data for a composite may be described as a labelled pasting diagram in several different ways. For instance,

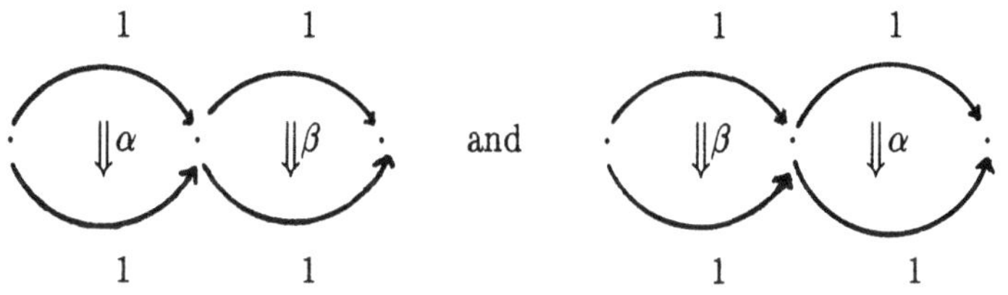

are non–equivalent labelled pasting diagrams, but they represent the same composite 2–cell, i.e. $\alpha \circ \beta = \beta \circ \alpha$. The same sort of difficulty arises if only the domain (or only the codomain) of a 2–cell is an identity 1–cell. To the best of my knowledge, this problem is not yet fully resolved; but Eilenberg and Street have given a substantial partial answer to it.

DEFINITIONS 4.1 (i) Given a finite directed graph G, let Dir Path (G) be the set of all directed paths in G, including the trivial path from a vertex to itself. The *length* of a directed path p is denoted by $|p|$ and is the number n, where $p = v_0 a_1 \ldots v_n$.

(ii) A *finite computad* C consists of a finite directed graph G, together with a finite set G_2 of 2–cells, and functions $\partial_0, \partial_1 \colon G_2 \to$ Dir Path (G) such that dom ∂_0 = dom ∂_1 and cod ∂_0 = cod ∂_1.

(iii) An *elementary derivation* in C is a triple (p,α,q), where α is a 2–cell and p and q are directed paths such cod p = dom $\partial_0(\alpha)$ and dom q = cod $\partial_0(\alpha)$.

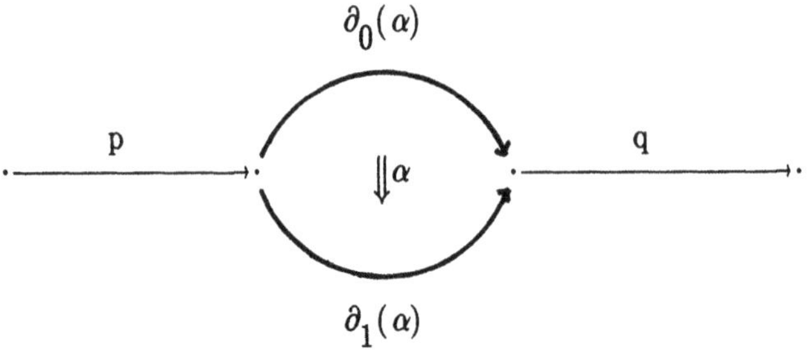

(iv) A *derivation* D in C is a finite sequence $<(p_0,\alpha_0,q_0),\ldots,(p_n,\alpha_n q_n)>$ of elementary derivations such that for each $0 \leq i < n$, $p_i\partial_1(\alpha_i)q_i = p_{i+1}\partial_0(\alpha_{i+1})q_{i+1}$.

It is evident how any derivation may be expressed as a pasting diagram together with a labelling in the free 2–category on C. It is also evident what constitutes a *labelling* of a derivation in a 2–category $\mathcal{A}$ and how that determines a labelling of the associated pasting diagram. However, as illustrated above, we can still have different labelled derivations determine different labelled pasting diagrams but with the same composite. So we seek a "normal form" condition so that every derivation is equivalent to a derivation in normal form, and any two labelled derivations in normal form have the same composite if and only if they are identical labelled derivations. Then, we would have an explicit description of all possible composites generated by C.

DEFINITIONS 4.2 (i) A 2–cell α in a finite computad C is called *central* if $\partial_0\alpha = \partial_1\alpha = v$ for some vertex v, i.e., α is of the form

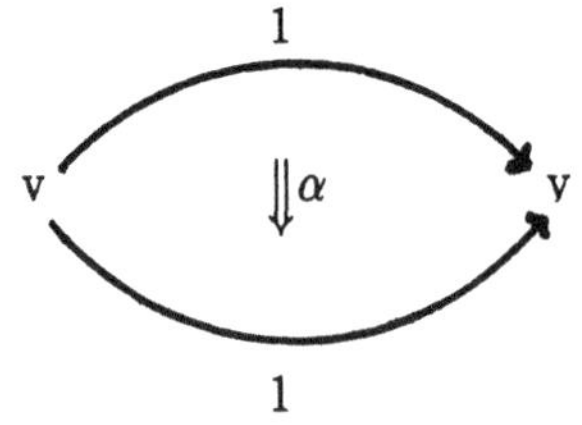

We henceforth assume that the set of central 2–cells in C is ordered.

(ii) A 2–cell α in C is called *penetrating* if $\partial_0\alpha = v$ for some vertex v; α is *strictly penetrating* if it is penetrating but not central. *Copenetrating* is the dual of penetrating.

(iii) A derivation is *pure* if the following conditions hold:

1) if there exists a strictly penetrating element $\alpha: 1 \Rightarrow y: v \to v$, then there is no copenetrating element β with $\partial_1\beta = v$, and

2) the dual of 1).

Observe that a derivation in which all 2–cells are central is pure, as is any derivation that contains no penetrating elements (but which may contain many strictly copenetrating elements). We now give a procedure to associate with a derivation D an equivalent derivation D' satisfying the conditions outlined above, i.e., an equivalent derivation D' in "normal form". We then give a theorem stating conditions under which the procedure halts and therefore succeeds. Under those conditions, we have a complete solution, i.e. an explicit description of all composites that arise from the given set of 2–cells.

PROCEDURE 4.3 Given a derivation D and successive elements $d_i = (p_i, \alpha_i, q_i)$, $d_{i+1} = (p_{i+1}, \alpha_{i+1}, q_{i+1})$ in D, if

(a) $|p_i| < |p_{i+1}\partial_1(\alpha_{i+1})|$

or (b) $|p_i| = |p_{i+1}|$, α_i and α_{i+1} are central, and $\alpha_i \leq \alpha_{i+1}$,

do not change D. Otherwise, we have a derivation of length 2 of the form

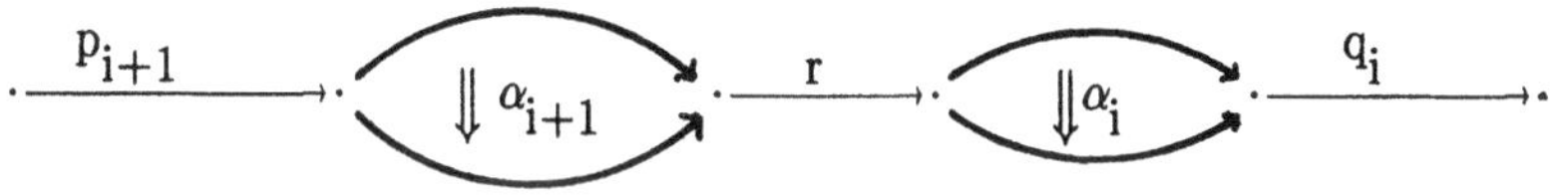

where $|r| \geq 0$; in this case, swap d_i and d_{i+1} and make the evident adjustments to p_i, p_{i+1}, q_i and q_{i+1}.

Iterate this procedure.

A derivation is said to be in "normal form" if it occurs as an output of the above procedure. Not every derivation has a normal form, but any two derivations in normal form are equivalent if and only if they generate the same composite.

THEOREM 4.4 (Eilenberg and Street [2]) If D is pure, Procedure 4.3 halts (and therefore gives a derivation equivalent to D and in a "normal form").

5. CONCLUSION

There have been several recent deep developments in the general theory of 2–categories. Moreover, several of those are of specific relevance to the Theory of Computing. However, it seemed to me that the above two are the ones that are most directly relevant to rewrite systems, hence my choice of those to present here.

ACKNOWLEDGEMENTS

I should like to thank Charles Wells for intelligent advice and valuable suggestions.

BIBLIOGRAPHY

[1] J. BENABOU, Introduction to bicategories, Lecture Notes in Math. 47 (1967), 1–77.

[2] S. EILENBERG and R.H. STREET, Seminar notes by R.H. Street delivered at Sydney Category Seminars, 1987–88.

[3] C.A.R. HOARE, Data refinement in a categorical setting, Preprint, June 1987.

[4] G. HUET, Equational systems for Category Theory and Intuitionistic Logic, draft, 1983.

[5] M. JOHNSON, Pasting diagrams in n–categories with applications to coherence theorems and categories of paths, Ph.D. thesis, University of Sydney, 1987.

[6] G.M. KELLY and R.H. STREET, Review of the elements of 2–categories, Lecture Notes in Math. 420 (1974), 75–103.

[7] DONALD E. KNUTH and PETER B. BENDIX, Simple Word Problems in Universal Algebras, Computational Problems in Abstract Algebra, Permagon Press, Oxford, 1970, 263–297.

[8] A.J. POWER, A 2–categorical pasting theorem, J. Algebra (to appear).

[9] D.E. RYDEHEARD and J.G. STELL, Foundations of Equational Deduction: A Categorical Treatment of Equational Proofs and Unification Algorithms, Preprint.

[10] R.A.G. SEELY, Modelling computations – a 2–categorical approach, Proc. 2nd Symp. on Logic in Comp. Science, Ithaca, New York, IEEE Publications 1987.

[11] R.H. STREET, Limits indexed by category–valued 2–functors, J. Pure Appl. Algebra 8, 149–181.

[12] C. WELLS, Path grammars, Conf. on Algebraic Methods and Software Technology, University of Iowa, 24th May 1989.

From Petri Nets to Linear Logic*

Narciso Martí-Oliet and José Meseguer
SRI International, Menlo Park, CA 94025, and
Center for the Study of Language and Information
Stanford University, Stanford, CA 94305

Abstract

Linear logic has been recently introduced by Girard as a logic of actions that seems well suited for concurrent computation. In this paper, we establish a systematic correspondence between Petri nets, linear logic theories, and linear categories. Such a correspondence sheds new light on the relationships between linear logic and concurrency, and on how both areas are related to category theory. Categories are here viewed as concurrent systems whose objects are states, and whose morphisms are transitions. This is an instance of the Lambek-Lawvere correspondence between logic and category theory that cannot be expressed within the more restricted framework of the Curry-Howard correspondence.

1 Introduction

During the last twenty years, the Curry-Howard [10] correspondence has proved very fruitful as a methodological tool for exploiting the deep connections between variants of intuitionistic logic and different typed lambda calculi. A direct payoff of this correspondence has been the design of functional programming languages with powerful type systems. This correspondence or isomorphism can be summarized as follows:

$$\begin{array}{lcl} \textit{Formulas} & \longleftrightarrow & \textit{Types} \\ \textit{Proofs} & \longleftrightarrow & \textit{Functions} \end{array}$$

At approximately the same time that Howard circulated his original note, a different correspondence, the Lambek-Lawvere correspondence, was found by these renowned mathematicians [14,15,16,17]. This new correspondence reads as follows:

$$\begin{array}{lcl} \textit{Formulas} & \longleftrightarrow & \textit{Objects} \\ \textit{Proofs} & \longleftrightarrow & \textit{Morphisms} \end{array}$$

Since for many categories the morphisms are functions, and for the particular case of typed lambda calculi the associated categories were cartesian closed categories, this less well-known correspondence has to a large extent passed unnoticed as yet another variant of the Curry-Howard correspondence in categorical garb. The crucial point being ignored, though, is that category theory is an *abstract* theory of mathematical structures and that, therefore, *morphisms* in a category *are not functions* in general, although they can happen to be functions in particular categories. An inspection of the original papers by Lambek and Lawvere makes absolutely clear

*Supported by Office of Naval Research Contracts N00014-86-C-0450 and N00014-88-C-0618, NSF Grant CCR-8707155, and a grant from the System Development Foundation. The first author is supported by a Research Fellowship of the Spanish Ministry for Education and Science.

that morphisms need not be functions in their advocated correspondence between category theory and logic.

A very important breakthrough has occurred with the recent introduction of linear logic by Girard [6,8]. One of the most promising prospects of this new logic for computer science applications is that it is a logic of *actions* that seems well suited for concurrent computation [7]. It is clear, and Girard stresses this point in his papers, that the old wineskins of the Curry-Howard correspondence cannot contain the *Beaujolais nouveau* of linear logic. The modes of thinking of functional computation are just inadequate. The Girard correspondence can be expressed as follows:

$$\begin{array}{lcl} \textit{Formulas} & \longleftrightarrow & \textit{States} \\ \textit{Proofs} & \longleftrightarrow & \textit{Transitions} \end{array}$$

The details of this correspondence for the case of Petri nets were first pointed out in a note by Asperti [1]. More recently, they have been further developed by Gunter and Gehlot [9].

The second author, in joint work with Ugo Montanari [22,23], has found a very fruitful isomorphism between concurrent systems and categories that can be summarized thus:

$$\begin{array}{lcl} \textit{States} & \longleftrightarrow & \textit{Objects} \\ \textit{Transitions} & \longleftrightarrow & \textit{Morphisms} \end{array}$$

Although this correspondence has been fully developed for the case of Petri nets, the paper [23] emphasizes its much greater scope by explaining how many other types of concurrent systems, such as term rewriting systems, or systems whose transitions have associated probabilities, can also be understood as categories with algebraic structure.

The main purpose of this paper is to put the last two isomorphisms together to get:

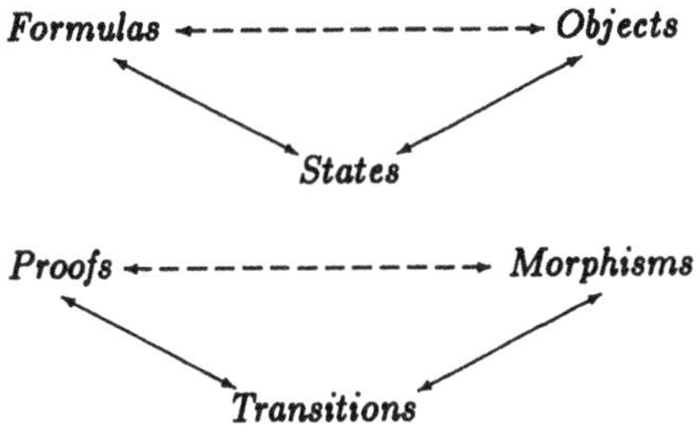

In this way, we can obtain very detailed information of what the Lambek-Lawvere correspondence for linear logic means in terms of concurrent systems. We view this as a potentially very fruitful *triangular correspondence*, between linear logic theories, concurrent systems, and categories, that provides powerful transfer principles between the three fields of logic, concurrency, and category theory. That a Lambek-Lawvere correspondence existed for linear logic had been already pointed out in the recent papers by Seely [28], who explains the relationship between linear logic and Barr's $*$-autonomous categories [2], and by de Paiva [25], who introduces a somewhat different categorical framework. However, none of these two papers developed any connections with the field of concurrent systems.

1.1 What is a proof?

This is a burning question for proof theorists, and Girard points out that his study of proof normalization and equivalence of proofs in linear logic aims at studying a "geometry of interaction" [8] that will liberate proof theory from the artificial and arbitrary syntactic distinctions that ordinary syntax creates. One of the fruits of the correspondence between linear logic, concurrent systems and categories developed in this paper is the suggestion[1] that the categorical theory of *coherence* [18,20] is the right mathematical framework to study this question and, given our correspondence, that coherence is also the right framework for answering the intimately related question,

What is a concurrent process?

which, for Petri nets, has been studied using coherence techniques in a joint paper of the second author with Pierpaolo Degano and Ugo Montanari [4]. A full development of the "triangular" implications of this point of view will have to wait for a later occasion.

1.2 Models for linear logic

One of the important benefits that the Lambek-Lawvere correspondence provides is a very general and flexible notion of *model* for a linear theory T. We follow the approach of Seely [28] and identify those models with interpretations of T in a *linear category*. Since the category $\underline{Cohl}$ of coherent spaces and linear maps [6] is a linear category, this general categorical semantics contains Girard's denotational semantics as a particular instance. This is entirely analogous to the way in which interpreting a typed lambda calculus theory in the category of Scott domains is a particular instance of taking general interpretations in cartesian closed categories. As in the typed lambda calculus case, completeness of the logic[2] now becomes an almost trivial property associated with the free linear category $\mathcal{L}[T]$ generated by the theory T. In this model, which is the initial model of T, proof theory and semantics are unified, just as they are in the case of algebraic data types. Therefore, the problem of equivalence of proofs is transformed into a study of initial models.

1.3 Using linear logic to specify concurrency

A Petri net N has an associated monoidal category $\mathcal{T}[N]$ whose objects are states, and whose morphisms are complex transitions, or processes, consisting of parallel and sequential compositions of atomic transitions [22]. On the other hand, a Petri net N can be viewed as a linear theory involving only the connective $\otimes$. The free linear category $\mathcal{L}[N]$ generated by the net N when regarded as a theory, essentially contains $\mathcal{T}[N]$ as a subcategory and therefore can also be viewed as a category whose objects are states and whose morphisms are processes. However, $\mathcal{L}[N]$ is so much more expressive than $\mathcal{T}[N]$ that we should think of its states and processes as, idealized or *gedanken*, states and processes. For example, ending in a state $A\&B$ means the possibility of choosing to end up in either state A or state B (external nondeterminism), and beginning in a state $A \oplus B$ means the possibility of beginning in either state A or state B (internal nondeterminism). We can use linear logic axioms to specify properties of the net N. Satisfaction of such specifications by N then means satisfaction in the model $\mathcal{L}[N]$ of its idealized computations and can in turn be reduced to provability of those specifications from the axioms provided by N when regarded as a theory. This is a nice correspondence between concurrent systems, models, and provability. The task ahead is to broaden this correspondence by allowing more expressive axioms and including a wider range of concurrent systems.

2 Petri nets

This section introduces Petri nets, and explains how they can be regarded as monoidal categories, and as linear logic theories involving only the conjunctive connective $\otimes$.

[1]This suggestion is not original with us. It belongs to the entire categorical tradition since the Lambek-Lawvere correspondence was discovered. The paper [21] by MacLane is an excellent exponent of that tradition.

[2]Originally established by Girard in [6] using the notion of phases, but not connected there to the coherent semantics, where other additional properties are satisfied.

2.1 Multisets and free commutative monoids

A multiset, also known as bag in the computer science literature, is a "set" in which the number of times that an element appears is taken into account, *i.e.*, each element has associated with it its multiplicity. One possible way of presenting a multiset over a set S is therefore as a function from S into $\mathbb{N}$, the set of natural numbers, that gives the multiplicity of each element. Of course, we can also define the union and other usual set-theoretic operations on multisets. We will be interested only in *finite* multisets, that is, multisets where the multiplicity is zero for all except a finite number of elements.

Definition 1 Given a set S, a *finite multiset* over S is a function $A : S \to \mathbb{N}$ such that the set $\{s \in S \mid A(s) \neq 0\}$ is finite.

Let A, B be two multisets over S.

The *union* of A and B, denoted $A \otimes B$, is the multiset given by $(A \otimes B)(s) = A(s) + B(s)$ for all $s \in S$. The operation $\otimes$ is associative and commutative because natural number addition is associative and commutative.

We say that A *is included in* B and write $A \subseteq B$ if $A(s) \leq B(s)$ for all $s \in S$.

If $A \subseteq B$, the *difference* $B - A$ is defined as the multiset given by $(B - A)(s) = B(s) - A(s)$ for all $s \in S$.

If $p \in \mathbb{N}$, the *p-th power* of the multiset A, denoted A^p, is the multiset defined by $A^p(s) = pA(s)$ for all $s \in S$.

If $s \in S$, its *singleton* multiset, also denoted s, is given by $s(s) = 1$ and $s(s') = 0$ for all $s' \in S$ such that $s' \neq s$. □

If A is a finite multiset over S, A can be expressed as a union of powers of singleton multisets $A = \bigotimes_{s \in S} s^{A(s)}$ in a unique way, with the order of the factors being immaterial. When all the exponents are zero, we obtain the "empty" multiset, that we will denote I.

The set of all finite multisets over S is denoted $S^{\otimes}$.

Given a K-indexed family $\{A_k\}_{k \in K}$ of finite multisets over S and a finite multiset $P = \bigotimes_{j=1}^{l} k_j^{n_j}$ over K, we write $\bigotimes_{k \in K} A_k^{P(k)}$ to denote the multiset $A_{k_1}^{n_1} \otimes \cdots \otimes A_{k_l}^{n_l}$.

In the same way that the set of (finite) lists over a set S, together with the empty list and concatenation is a free monoid over S, the set of (finite) bags or multisets over S together with the empty multiset and the union operation is a free commutative monoid over S.

Proposition 2 If S is a set, the set $S^{\otimes}$ of (finite) multisets over S, together with the union operation $\otimes$ and the element I, is a free commutative monoid over S. □

Remark: In the rest of this paper, a *multiset* will always mean a *finite* multiset.

2.2 Classical approach to Petri nets

In the usual approach to Petri net theory, a place/transition Petri net consists of a set of places and a disjoint set of transitions, together with a relation between them. To every transition, two multisets of places called preset and postset are associated. Global states consist of multisets of places called markings. More formally, we have the following definition.

Definition 3 A *place/transition Petri net* or, for short, a *Petri net* is a triple (S, T, F), where:

1. S is a set of *places*,
2. T is a disjoint set of *transitions*, and
3. F is a multiset over $(S \times T) + (T \times S)$ called the *causal dependency relation* (here the symbol $+$ denotes the disjoint union of sets).

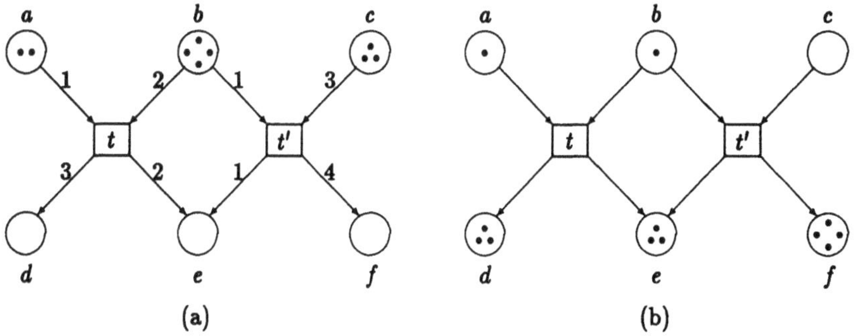

Figure 1: A Petri net before and after the concurrent firings of t and t'.

Given a transition $t \in T$, the associated *preset* is the multiset ${}^\bullet t$ over S defined by ${}^\bullet t(s) = F(s,t)$ for all $s \in S$. Analogously, the associated *postset* is the multiset $t^\bullet$ such that $t^\bullet(s) = F(t,s)$ for all $s \in S$.

Finally, a *marking* is a multiset over S. □

Example 4 Let us consider the Petri net in Figure 1(a). It has a set of places $S = \{a,b,c,d,e,f\}$ and a set of transitions $T = \{t,t'\}$. Incoming and outgoing arrows of a transition and the associated numbers specify presets and postsets, and of course also the causal dependency relation. We have ${}^\bullet t = a \otimes b^2$, $t^\bullet = d^3 \otimes e^2$, ${}^\bullet t' = b \otimes c^3$, and $t'^\bullet = e \otimes f^4$. □

From the above definition of preset and postset, it should be clear that the relation F of causal dependency and the presets and postsets for all $t \in T$ determine each other uniquely. Therefore, a Petri net is characterized by the sets S,T and the multisets ${}^\bullet t, t^\bullet$ for all $t \in T$.

Given a Petri net and a marking M, the number of *tokens* stored in a place $s \in S$ by the marking M is $M(s)$. The global state given by this marking can be changed by the *firing* of a transition $t \in T$. This firing decreases the number of tokens in a place s by ${}^\bullet t(s)$, the number of tokens in s consumed by t, and increases it by $t^\bullet(s)$, the number of tokens in s created by t. Therefore, for the firing of t to take place under a marking M, the number of tokens stored in each place s by M must be smaller than or equal to the number of tokens consumed by t in s. In such a case, we say that M *enables* the firing of t.

Example 5 Figure 1(a) shows a marking $M = a^2 \otimes b^4 \otimes c^3$ on the Petri net of Example 4, namely a marking where there are two tokens on place a, four on b and three on c.

The marking M enables the firing of both t and t' because ${}^\bullet t \subseteq M$ and also ${}^\bullet t' \subseteq M$. □

The firing of a single transition can be considered as the simplest step of computation in a Petri net. More generally, we may consider computations consisting of a sequence of firings with each firing involving a multiset of transitions.

Definition 6 Given a Petri net (S,T,F), a marking M, and a multiset U over T, the *parallel or concurrent firing* of U is enabled if $\bigotimes_{t\in T} {}^\bullet t^{U(t)} \subseteq M$ and in this case a step $M \stackrel{U}{\Longrightarrow} M'$ may take place, with M' the new marking or global state given by $M' = (M - \bigotimes_{t\in T} {}^\bullet t^{U(t)}) \otimes \bigotimes_{t\in T} t^{\bullet U(t)}$.

We write $M \stackrel{U_1;\ldots;U_k}{\Longrightarrow^*} M'$ to mean that there exist markings $M_1,\ldots,M_{k-1}$ such that $M \stackrel{U_1}{\Longrightarrow} M_1$, $\ldots, M_{k-1} \stackrel{U_k}{\Longrightarrow} M'$, *i.e.*, to denote the sequential firing of $U_1,\ldots,U_k$.

The notation $M \Longrightarrow M'$ means that there exists a multiset U of transitions such that $M \stackrel{U}{\Longrightarrow} M'$, and $\Longrightarrow^*$ denotes the reflexive and transitive closure of the relation $\Longrightarrow$ between markings. □

If we are interested only in the effect of a computation, *i.e.*, the resulting final global state, and not in the structure of the computations, it is possible to consider only sequences of firings of *single* transitions (see [4] for further discussion on this subject). However, the above definition is more general and connects nicely with the categorical approach given in Section 2.3.

Example 7 Let us consider again the net of Example 4 with the marking $M = a^2 \otimes b^4 \otimes c^3$ represented in Figure 1(a). Now Figure 1(b) describes the state $M' = a \otimes b \otimes d^3 \otimes e^3 \otimes f^4$ reached after the concurrent firings of t and t', *i.e.*, after the step $M \stackrel{t \otimes t'}{\Longrightarrow} M'$. □

Although in this section we have presented Petri nets from a classical point of view, we have already introduced some special notation for multisets that will be very useful in the more algebraic and categorical presentation given in the next section.

2.3 Petri nets as monoidal categories

In this section, we recall the categorical approach to Petri nets proposed in [22,23], and in particular the construction of the Petri category $\mathcal{T}[N]$ associated to a net N.

We have seen that a Petri net is given by two disjoint sets S, T and the presets and postsets associated with each transition $t \in T$. If we think of a transition t as an arrow $t : {}^\bullet t \to t^\bullet$, and remember that the set $S^\otimes$ of multisets over S is a free commutative monoid over S, it is natural to define a Petri net as a *graph* whose set of nodes is a free commutative monoid. This is equivalent to, yet simpler than, Definition 3.

Definition 8 A *Petri net* is a quadruple $N = (S^\otimes, T, \partial_0, \partial_1)$, where $S^\otimes$ is a free commutative monoid of *nodes* over a set of *places* S, T is a set of *transitions* and $\partial_0, \partial_1 : T \to S^\otimes$ are functions associating to each transition its *source* and *target* nodes respectively. □

As usual, if $t \in T$ with $\partial_0(t) = A$ and $\partial_1(t) = B$, we write $t : A \to B$.

Example 9 From this point of view, the net of Figure 1(a) is represented as a graph whose set of nodes is $\{a, b, c, d, e, f\}^\otimes$ and whose transitions are $t : a \otimes b^2 \to d^3 \otimes e^2$ and $t' : b \otimes c^3 \to e \otimes f^4$. □

A Petri net *morphism* h from $N = (S^\otimes, T, \partial_0, \partial_1)$ to $N' = (S'^\otimes, T', \partial'_0, \partial'_1)$ is a pair $\langle f, g \rangle$ with $f : T \to T'$ a function and $g : S^\otimes \to S'^\otimes$ a monoid homomorphism[3] and such that for any $t \in T$, $i = 0, 1$, $g(\partial_i(t)) = \partial'_i(f(t))$.

This defines a category <u>*Petri*</u> equipped with products and coproducts (see [23]).

When considering the computation consisting of the parallel firing of t and t', we encountered the multiset $U = t \otimes t'$. Can we extend the view of a Petri net as a graph to include these computations? Which is then the source and the target of $t \otimes t'$? If the source of t is ${}^\bullet t$ and its target is $t^\bullet$, which we can interpret as the global number of tokens produced and consumed by t, it is natural to view $t \otimes t'$ as an arrow ${}^\bullet t \otimes {}^\bullet t' \to t^\bullet \otimes t'^\bullet$.

Example 10 In Examples 4 and 9, we have $t \otimes t' : a \otimes b^3 \otimes c^3 \to d^3 \otimes e^3 \otimes f^4$ and $t^2 : a^2 \otimes b^4 \to d^6 \otimes e^4$. □

In this way, we can generate new arrows as multisets of the original set of arrows T, in such a way that if $U : A \to B$, $V : A' \to B'$ are such arrows with $U, V \in T^\otimes$, then $U \otimes V$ is an arrow $U \otimes V : A \otimes A' \to B \otimes B'$. In particular, we can introduce the arrow $I : I \to I$. Note that ∂_0 and ∂_1 then extend uniquely to monoid homomorphisms $\partial_0, \partial_1 : T^\otimes \to S^\otimes$, so that our Petri net extended in this way can now be viewed as a graph in the category of commutative monoids,

[3] *i.e.*, $g(I) = I$ and for any $A, B \in S^\otimes$, $g(A \otimes B) = g(A) \otimes g(B)$.

or (what amounts to the same) as a commutative monoid structure on a graph, such that its monoids of nodes and arrows are free.

Going one step further, we can also consider *idle* transitions, represented as identity arrows. For example, the identity $id_{a\otimes b} : a \otimes b \rightarrow a \otimes b$ is interpreted as the idleness of a token in place a and a token in place b.

These operations allow us to interpret algebraically the relation $M \stackrel{U}{\Longrightarrow} M'$ between markings defined in Section 2.2. If U is a multiset of transitions, its source is $\partial_0(U) = \bigotimes_{t\in T} {}^\bullet t^{U(t)}$ and its target is $\partial_1(U) = \bigotimes_{t\in T} t^{\bullet U(t)}$. Then $M \stackrel{U}{\Longrightarrow} M'$ iff $\partial_0(U) \subseteq M$ and $M' = M - \partial_0(U) \otimes \partial_1(U)$. If $A = M - \partial_0(U)$, then $M = \partial_0(U) \otimes A$ and $M' = \partial_1(U) \otimes A$, *i.e.*, writing $id_A : A \rightarrow A$ for the associate identity or idle transition, we have $M \stackrel{U}{\Longrightarrow} M'$ iff $U \otimes id_A : M \rightarrow M'$.

In addition, we wish to have a notion of *sequential composition* of computations[4] denoted by semicolon. Specifically, if $\alpha : A \rightarrow B$ and $\beta : B \rightarrow C$ are computations, then $\alpha;\beta : A \rightarrow C$ is also a computation. Notice that the target of α and the source of β must coincide. It is natural to assume that semicolon is associative. Also, given a computation $\alpha : A \rightarrow B$, the idle transitions id_A and id_B are its respective left and right identities.

If we are not interested in the internal structure of a computation, it is also natural to impose an equation that makes $\otimes$ a functorial operation: given $\alpha : A \rightarrow B$, $\alpha' : A' \rightarrow B'$, $\beta : B \rightarrow C$, $\beta' : B' \rightarrow C'$, we have $(\alpha;\beta) \otimes (\alpha';\beta') = (\alpha \otimes \alpha');(\beta \otimes \beta')$. The intuitive meaning of this equation is that the parallel or concurrent composition of two given independent computations has the same effect as a computation whose steps are the parallel compositions of the steps of the given computations. (We refer again to [4] for further discussion of this point, including finer distinctions among computations.)

The resulting structure obtained by combining the above operations of parallel ($\otimes$) and sequential (;) composition of computations is a category with a commutative monoid structure, that we will call a Petri category.

Definition 11 [22,23] A *Petri category* is a small category $\mathcal{C} = (S^\otimes, R, ;, id)$ whose set of objects is a free commutative monoid, and whose set of arrows has a commutative monoid structure $(R, \otimes, id_I)$, that is *not* necessarily free, and is compatible with the categorical structure in the sense that the source and target functions $\partial_0, \partial_1 : T \rightarrow S^\otimes$ are monoid homomorphisms and that $\otimes$ respects identities and (sequential) composition: given $\alpha : A \rightarrow B$, $\alpha' : A' \rightarrow B'$, $\beta : B \rightarrow C$, $\beta' : B' \rightarrow C'$, we have $(\alpha;\beta) \otimes (\alpha';\beta') = (\alpha \otimes \alpha');(\beta \otimes \beta')$ and $id_{A\otimes B} = id_A \otimes id_B$. □

Notice that in the above definition we are using the same symbol $\otimes$ for two different monoid structures: on objects and on arrows. This should not be confusing and in fact, what $\_\otimes\_$ denotes is a functor from $\mathcal{C} \times \mathcal{C}$ to $\mathcal{C}$; indeed, this is the meaning of the two equations.

Note also that the neutral element for $(R, \otimes)$ is id_I. The reason for this is that, if I' denotes such a neutral element, we can deduce from the above equations that $I' = id_I$.

Given two Petri categories $\mathcal{C}$ and $\mathcal{D}$, a Petri category *morphism* is a functor that is a monoid homomorphism when restricted to both the objects and the morphisms. This determines a category <u>*CatPetri*</u>.

There is an obvious forgetful functor $\mathcal{U}$: <u>*CatPetri*</u> $\longrightarrow$ <u>*Petri*</u> that forgets the categorical structure and the monoid structure on the arrows. This functor has a left adjoint $\mathcal{T}[\_]$: <u>*Petri*</u> $\longrightarrow$ <u>*CatPetri*</u>, that we describe in detail.

Given a Petri net $N = (\partial_0, \partial_1 : T \rightarrow S^\otimes)$, the Petri category $\mathcal{T}[N]$ is defined by the following generating rules:

$$(id)\ \frac{A \in S^\otimes}{id_A : A \rightarrow A \ in\ \mathcal{T}[N]} \qquad \frac{t : A \rightarrow B \ in\ N}{t : A \rightarrow B\ in\ \mathcal{T}[N]}$$

[4] Where by a *computation* we intuitively mean a possibly complex combination of parallel and sequential compositions of basic transitions.

$$(;)\ \frac{\alpha : A \to B,\ \beta : B \to C \textit{ in } \mathcal{T}[N]}{\alpha;\beta : A \to C \textit{ in } \mathcal{T}[N]}$$

$$(\otimes)\ \frac{\alpha : A \to B,\ \beta : C \to D \textit{ in } \mathcal{T}[N]}{\alpha\otimes\beta : A\otimes C \to B\otimes D \textit{ in } \mathcal{T}[N]}$$

subject to the following equational rules:

$$\frac{\alpha : A \to B,\ \beta : B \to C,\ \gamma : C \to D \textit{ in } \mathcal{T}[N]}{\alpha;(\beta;\gamma) = (\alpha;\beta);\gamma}$$

$$\frac{\alpha : A \to B \textit{ in } \mathcal{T}[N]}{id_A;\alpha = \alpha} \qquad \frac{\alpha : A \to B \textit{ in } \mathcal{T}[N]}{\alpha; id_B = \alpha}$$

$$\frac{\alpha : A \to B,\ \beta : B \to C,\ \alpha' : A' \to B',\ \beta' : B' \to C' \textit{ in } \mathcal{T}[N]}{(\alpha;\beta)\otimes(\alpha';\beta') = (\alpha\otimes\alpha');(\beta\otimes\beta')}$$

$$\frac{A, B \in S^{\otimes}}{id_A \otimes id_B = id_{A\otimes B}}$$

$$\frac{\alpha : A \to B,\ \alpha' : A' \to B',\ \alpha'' : A'' \to B'' \textit{ in } \mathcal{T}[N]}{\alpha\otimes(\alpha'\otimes\alpha'') = (\alpha\otimes\alpha')\otimes\alpha''}$$

$$\frac{\alpha : A \to B,\ \alpha' : A' \to B' \textit{ in } \mathcal{T}[N]}{\alpha\otimes\alpha' = \alpha'\otimes\alpha} \qquad \frac{\alpha : A \to B \textit{ in } \mathcal{T}[N]}{\alpha\otimes id_I = \alpha}$$

The objects of $\mathcal{T}[N]$ are the nodes of N, *i.e.*, the free commutative monoid $S^{\otimes}$, and the morphisms of $\mathcal{T}[N]$ are obtained from the transitions T of N by adding for every object A an identity morphism id_A and by closing freely with respect to the operations of *parallel composition* $\_\otimes\_$ and of *sequential composition* $\_;\_$, and then imposing the corresponding equations (associativity and identities for both operators, functoriality and commutativity for $\_\otimes\_$). The properties of the source and target functions are implicit in the notation used to write down the rules.

Theorem 12 [23] The forgetful functor $\mathcal{U}$: <u>CatPetri</u> $\longrightarrow$ <u>Petri</u> that views each Petri category as a Petri net has a left adjoint $\mathcal{T}[\_]$: <u>Petri</u> $\longrightarrow$ <u>CatPetri</u>, that is, if N is a Petri net, $\mathcal{T}[N]$ is the *free Petri category* generated by N. □

As we have already discussed, the intuitive meaning of the monoidal operation $\_\otimes\_$ is parallel composition, and of course the meaning of the categorical operation $\_;\_$ is sequential composition. Since the generators of the morphisms are the transitions of N, closing with respect to these operations gives origin to the generalized notion of a *net computation*.

Theorem 13 Given a Petri net N with set of places S, markings M and M' on N, and multisets $U, U_1, \ldots, U_k$ on S,

1. $M \stackrel{U}{\Longrightarrow} M'$ iff there exists $A \in S^{\otimes}$ such that $U \otimes id_A : M \to M'$ is a morphism in $\mathcal{T}[N]$ (remember the way in which we write multisets of transitions).
2. $M \stackrel{U_1;\ldots;U_k}{\Longrightarrow^*} M'$ iff there exist $A_1, \ldots, A_k \in S^{\otimes}$ such that $(U_1\otimes id_{A_1});\cdots;(U_k\otimes id_{A_k}) : M \to M'$ is a morphism in $\mathcal{T}[N]$.
3. $M \Longrightarrow^* M'$ iff there exists a morphism $M \to M'$ in $\mathcal{T}[N]$.

□

In the category $\mathcal{T}[N]$ the internal structure of the computations is deemphasized, so that different syntactic presentations of what semantically is intuitively the same computation are identified. However, we can consider other categories without so many identifications among computations (see [4]).

For the reader acquainted with category theory it should be clear that a Petri category is nothing but a strictly symmetric strict symmetric monoidal category[5] in which the monoid of objects is free.

2.4 Petri nets as theories

Girard's linear logic [6,7,8] presents itself explicitly as a logic of concurrent interaction in which resources are limited, and are consumed in such interactions. This is of course very similar to Petri nets, where resources are represented as tokens that are then consumed by transitions. At the proof-theoretic level, limitation of resources is expressed by forbidding the structural rules of weakening (which allows new resources to be obtained for free) and of contraction (which arbitrarily eliminates duplicated resources). Thus, Girard's conjunction $\otimes$ is not idempotent, *i.e.*, it does not satisfy $A \otimes A = A$. Indeed, if we regard a multiset over S as a "resource-conscious proposition," such a conjunction exactly corresponds to the union operation $A \otimes B$ of two multisets, and the logical rules for conjunction will reflect the properties of concurrent composition of computations. This correspondence of Girard's conjunction with concurrent computation in Petri nets was first pointed out in a note by Asperti [1], who showed that a Petri net can be regarded as a theory so that, identifying propositions with markings, there is an exact correspondence between provable sequents $A \vdash B$ and the relation $A \Longrightarrow^* B$ of Definition 6. Gunter and Gehlot have further developed this idea in their recent paper [9].

One of the essential questions in proof theory is: "When are two proofs the same?". For intuitionistic logic, the work of Prawitz [26] has made fundamental contributions to this problem. For Girard, this is also a key question in the context of linear logic, and he talks about a "geometry of interaction" [8] to overcome unnecessary syntactic distinctions and arrive at the right notion of proof. In the limited context of the fragment of linear logic involving only the conjunction operator $\otimes$, the categorical approach to Petri nets sketched in Section 2.3 provides an algebraic framework in which the question of equivalence of proofs can be naturally discussed. This further clarifies the relationship between Petri nets and linear logic already pointed out by Asperti [1] and by Gunter and Gehlot [9] in terms of the provability relation $\vdash$. Indeed, as the paper [4] shows, several different notions of equivalence between computations (and therefore between proofs) are possible. Here we shall limit ourselves to discussing the equivalence provided by the equational rules given for $\mathcal{T}[N]$. A more detailed proof-theoretic interpretation of other equivalences and of their relationship to the work of Girard on proof nets [6] will be given elsewhere.

An example may help clarify the key ideas. Consider the two actions of buying a pen p by paying one dollar \$, and of buying a candy bar c also by paying one dollar \$. In linear logic, we may express these two actions as axioms

$$(buy - p):\ \ \$ \vdash p \qquad\qquad (buy - c):\ \ \$ \vdash c$$

and then, the rules for conjunction $\otimes$ allow us to derive the sequent

$$\$ \otimes \$ \vdash p \otimes c$$

but we cannot derive the sequent $\$ \vdash p \otimes c$.

The corresponding Petri net describing this situation is given in Figure 2, where we need to provide *two* tokens in \$ in order to obtain one pen and one candy bar by means of the concurrent

[5] A strictly symmetric strict monoidal structure on a category $\mathcal{C}$ is just a —somewhat confusing— way of saying that $\mathcal{C}$ has the structure of a *commutative monoid* in the category Cat of small categories. In particular, $\mathcal{T}[N]$ is the commutative monoid structure on a category that is freely generated by N.

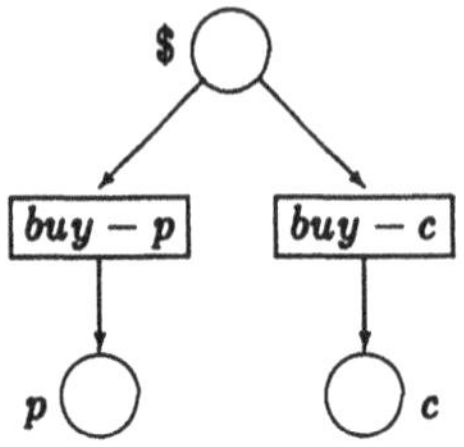

Figure 2: A Petri net for buying pens and candy bars.

action

$$buy - p \otimes buy - c : \ \$ \otimes \$ \to p \otimes c.$$

Of course, by the equational rules for $\mathcal{T}[N]$, we consider this action equivalent to that of first buying the pen and then the candy bar, or vice versa, *i.e.*, we have:

$$(buy - p \otimes idle_{\$}); (idle_p \otimes buy - c) = (idle_{\$} \otimes buy - c); (buy - p \otimes idle_c) = buy - p \otimes buy - c.$$

Note that, considering a set S of *constant propositions*, the propositional *formulas* generated by S with the sole use of the conjunction operator $\otimes$, which is assumed associative and commutative but not idempotent, are exactly the elements of the free commutative monoid $S^{\otimes}$ of multisets on S. We are now ready for the following

Definition 14 A *tensor theory* or *$\otimes$-theory* T is given by a set of constants S and a set Ax of sequents[6] of the form $\alpha : A \vdash B$ with $A, B \in S^{\otimes}$ which are called the *axioms* of T.

Given two $\otimes$-theories $T = (S, Ax)$ and $T' = (S', Ax')$ a $\otimes$-morphism $L : T \to T'$ maps formulas over S to formulas over S' preserving the operation $\otimes$, *i.e.*, it is a monoid homomorphism on formulas, and a sequent $\alpha : A \vdash B \ \in \ Ax$ to a sequent $L(\alpha) : L(A) \vdash L(B) \ \in \ Ax'$.

This defines a category $\underline{\otimes\text{-}Th}$ with $\otimes$-theories as objects and $\otimes$-morphisms as morphisms. □

The correspondence between Petri nets and $\otimes$-theories, such as for example that between the Petri net in Figure 2 and the $\otimes$-theory with constants \$, p, c, and with the two axioms $buy - p$ and $buy - c$, can now be made entirely precise.

Proposition 15 If N is a Petri net with set of places S and set of transitions T, the $\otimes$-theory associated to N is given by the set of constants S and for each $t \in T$, an axiom $t : {}^{\bullet}t \vdash t^{\bullet}$.

Conversely, if $T = (S, Ax)$ is a $\otimes$-theory, its associated Petri net has S as set of places and for every axiom $\alpha : A \vdash B \ \in \ Ax$ a transition α with ${}^{\bullet}\alpha = A$ and $\alpha^{\bullet} = B$.

Then $\otimes$-morphisms between theories correspond exactly to Petri net morphisms, and we have an isomorphism between the categories $\underline{Petri}$ and $\underline{\otimes\text{-}Th}$. □

Due to this isomorphism, we will represent also by N the $\otimes$-theory associated to the Petri net N.

Definition 16 Given a $\otimes$-theory $T = (S, Ax)$, a *proof expression* provable from T is a sequent $\alpha : A \vdash B$ with $A, B \in S^{\otimes}$ and α inductively generated from the axioms Ax and the axiom scheme

$$(id) \ \ id_A : A \vdash A$$

by the rules:

$$(cut) \ \ \frac{\alpha : A \vdash B, \ \beta : B \vdash C}{\alpha ; \beta : A \vdash C}$$

[6]Note that, for us, sequents come with an associated label α. This will be important in our formalization of proofs below, and makes the correspondence with Petri nets particularly clear.

$$(\otimes)\quad \frac{\alpha : A \vdash B,\ \beta : C \vdash D}{\alpha \otimes \beta : A \otimes C \vdash B \otimes D}$$

The *closure* $T^\diamond$ of T is the $\otimes$-theory with constants S and with axioms the set of all proof expressions provable from T, so that we have an obvious inclusion morphism $T \hookrightarrow T^\diamond$ in $\underline{\otimes\text{-}Th}$.
□

Note that, except for the slight change of notation from $\alpha : A \rightarrow B$ to $\alpha : A \vdash B$, the generating rules for the category $\mathcal{T}[T]$ associated to T (when viewed as a Petri net) are exactly the same as these for $T^\diamond$, *i.e.*, we begin with the basic axioms (transitions) and identities, and we generate all proof expressions by the rules (*cut*), corresponding to the rule (;) in $\mathcal{T}[T]$, and ($\otimes$), denoted the same in both cases. Note that, by construction, proof expressions in $T^\diamond$ and morphism expressions in $\mathcal{T}[T]$ are syntactically identical. Since in $\mathcal{T}[T]$ such expressions are identified by the equational rules of its definition, we can define a *proof* as the equivalence class of all proof expressions provably equal by those equational rules[7].

Theorem 17 If N is a Petri net with set of places S and set of transitions T, and M, M' are markings on N, there exists a morphism $[\alpha] : M \rightarrow M'$ in $\mathcal{T}[N]$ iff the proof expression $\alpha : M \vdash M'$ belongs to $N^\diamond$, the closure of the $\otimes$-theory N. □

We can, finally, make the connection with the more classical notation in Definition 6 by stating the following

Corollary 18 Given a Petri net N with set of places S, markings M and M' on N, and multisets $U, U_1, \ldots, U_k$ on S,

1. $M \stackrel{U}{\Longrightarrow} M'$ iff there exists $A \in S^\otimes$ such that the sequent $U \otimes id_A : M \vdash M'$ belongs to $N^\diamond$.
2. $M \stackrel{U_1;\ldots;U_k}{\Longrightarrow^*} M'$ iff there exist $A_1, \ldots, A_k \in S^\otimes$ such that $(U_1 \otimes id_{A_1}); \cdots ; (U_k \otimes id_{A_k}) : M \vdash M'$ is in $N^\diamond$.
3. $M \Longrightarrow^* M'$ iff there is a sequent $\alpha : M \vdash M'$ in $N^\diamond$.

□

This concludes our discussion of Petri nets. In the rest of this paper we further develop the intimate Lambek-Lawvere correspondence between proof theory and category theory, not just for Petri nets and its associated $\otimes$-fragment of linear logic, but also for the remaining linear logic connectives except the exponentials.

3 Linear logic and $*$-autonomous categories

In this section, we give a precise account of the correspondence between Barr's $*$-autonomous categories [2] and Girard's linear logic [6] that Seely motivates and outlines in [28]. The details turn out to be interesting. In particular, we are led to strengthen Barr's original definition of a $*$-autonomous category in order to express the connective $\invamp$ (dual in linear logic to $\otimes$) as a different symmetric monoidal product in the category. We also give precise definitions of the categories of linear theories, and of models of a linear theory, make explicit the details of the adjunction between linear theories and linear categories, define satisfaction of a sequent in a model, and show the expected soundness and completeness of linear logic for linear category models.

We keep as much as possible Girard's original notation. However, we replace Girard's **1** by I. As before, the composition of two morphisms $f : A \rightarrow B$ and $g : B \rightarrow C$ will be written in diagrammatic order and denoted $f;g$.

[7] As mentioned before, other equivalence relations, making finer distinctions among proof expressions are also possible [4].

3.1 Closed and $*$-autonomous categories

Essentially, a $*$-autonomous category is a closed symmetric monoidal category C with an involution $(\_)^{\perp} : C^{op} \longrightarrow C$. For the definition of a symmetric monoidal category, see [19].

Definition 19 A *closed symmetric monoidal category* is a symmetric monoidal category $(C, \otimes, I, a, c, e)$ such that, in addition, for each object A of C, the functor $\_ \otimes A : C \longrightarrow C$ has a (specific) right adjoint $A \multimap \_ : C \longrightarrow C$, that is, for all objects A, B, C of C, we have a natural (in B and C) isomorphism

$$\varphi^{A}_{B,C} : Hom_C(B \otimes A, C) \longrightarrow Hom_C(B, A \multimap C).$$

□

A *cartesian closed category* is a closed symmetric monoidal category such that the tensor product is the categorical product and the unit object is a final object.

The paper [23] studies the closed symmetric monoidal category structure of the categories Petri, CatPetri and other related categories.

Given a closed symmetric monoidal category $(C, \otimes, I, a, c, e, \multimap)$, we have the following (natural) isomorphism:

$$Hom_C(A, B) \overset{Hom(e,id)}{\longrightarrow} Hom_C(I \otimes A, B) \overset{\varphi}{\to} Hom_C(I, A \multimap B)$$

denoted $(\_)^{\sharp}$ (and its inverse $(\_)^{\flat}$), that gives an "internal" representation of the morphisms in C, following the idea that the object $A \multimap B$ is an "internal" representation of the set $Hom_C(A, B)$.

In particular, we have internal representations of identities:

$$j_A : I \longrightarrow (A \multimap A)$$

and composition:

$$m_{A,B,C} : (B \multimap C) \otimes (A \multimap B) \longrightarrow (A \multimap C).$$

Having seen how other categorical concepts are internalized in a closed symmetric monoidal category, it is natural to consider internalizing a functor. A functor $F : C \longrightarrow C$ is *strong* if its morphism part has an internal counterpart given by a family of arrows $s_{A,B} : (A \multimap B) \to (F(A) \multimap F(B))$ satisfying the expected functoriality properties with respect to the internal representation of morphisms, identities and composition explained in detail below: for example, if $f : A \to B$, $F(f)^{\sharp} = f^{\sharp}; s_{A,B}$.

Since we are specially interested in the case of a contravariant functor, we give the details for this case.

Definition 20 Let C be a closed symmetric monoidal category.

A *(contravariant) strong functor* from C into itself consists of a function F on the objects of C and of a family of morphisms $s_{A,B} : A \multimap B \to F(B) \multimap F(A)$ satisfying the commutative diagrams shown in Figure 3, that express internally the relationship with identities and composition. □

A strong functor can be externalized to yield an ordinary functor in the following way:

Proposition 21 Let C be a closed symmetric monoidal category and let $(F, \{s_{A,B}\}_{A,B \in Ob(C)})$ be a (contravariant) strong functor from C into itself.

If given a morphism $f : A \to B$ in C, we define $F(f) = (f^{\sharp}; s_{A,B})^{\flat}$, we get a (contravariant) functor $F : C^{op} \longrightarrow C$.

Moreover, the family of morphisms $s_{A,B} : A \multimap B \to F(B) \multimap F(A)$ is natural in A and B. □

Given a closed symmetric monoidal category C, an *involution* will be a strong (contravariant) functor $(\_)^{\perp} : C^{op} \longrightarrow C$ with a natural isomorphism $d_A : A \to A^{\perp\perp}$ subject to some conditions.

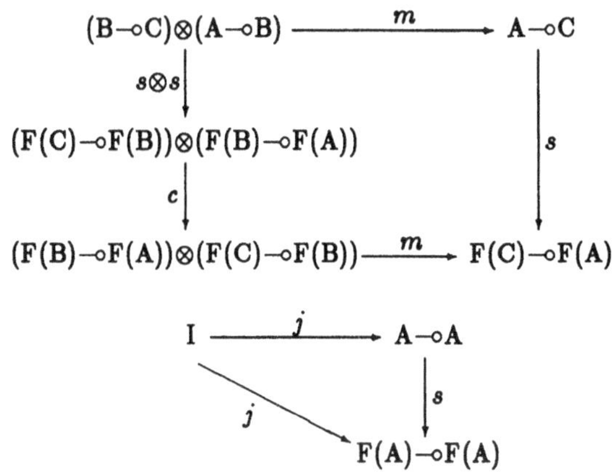

Figure 3: Commutative diagrams for a strong functor.

Suppose now given a (contravariant) strong functor $((\_)^{\perp}, \{s_{A,B}\})$ and a family of *isomorphisms* $d_A : A \to A^{\perp\perp}$ such that for each $A, B \in Ob(\mathcal{C})$ the following diagram commutes:

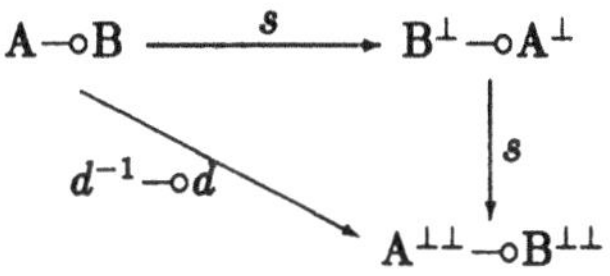

Then, if $f : A \to B$, we have

$$(f^{\perp\perp})^{\dagger} = f^{\dagger}; (d_A^{-1} \multimap d_B) = (d_A^{-1}; f; d_B)^{\dagger}$$

hence $d_A; f^{\perp\perp} = f; d_B$ and d is natural.

Moreover, from this diagram we also deduce that $s_{A,B}$ is an isomorphism for all $A, B \in Ob(\mathcal{C})$. And then, using the isomorphism $n_A : I \multimap A \to A$ that exists in any closed symmetric monoidal category, there is the (natural) isomorphism

$$A \multimap I^{\perp} \xrightarrow{d^{-1} \multimap id} A^{\perp\perp} \multimap I^{\perp} \xrightarrow{s^{-1}} I \multimap A^{\perp} \xrightarrow{n} A^{\perp}$$

denoted u_A, that allows us to call $I^{\perp}$ the *dualizing object*.

Using this isomorphism, we have the following (natural) isomorphism:

$$Hom_{\mathcal{C}}(A, B^{\perp}) \xrightarrow{Hom(id,u^{-1})} Hom_{\mathcal{C}}(A, B \multimap I^{\perp}) \xrightarrow{\varphi^{-1}} Hom_{\mathcal{C}}(A \otimes B, I^{\perp}) \longrightarrow$$

$$\xrightarrow{Hom(c,id)} Hom_{\mathcal{C}}(B \otimes A, I^{\perp}) \xrightarrow{\varphi} Hom_{\mathcal{C}}(B, A \multimap I^{\perp}) \xrightarrow{Hom(id,u)} Hom_{\mathcal{C}}(B, A^{\perp})$$

i.e., we have an adjunction

$$\psi_{B,A} : Hom_{\mathcal{C}^{op}}(B^{\perp}, A) \longrightarrow Hom_{\mathcal{C}}(B, A^{\perp})$$

"between the functor $(\_)^{\perp}$ and itself".

What is the aim of this discussion? Motivated by the correspondence with linear logic, we are interested in the connective $\parr$, which is the dual of $\otimes$. We can define a functor $\_\parr\_ : \mathcal{C} \times \mathcal{C} \longrightarrow \mathcal{C}$ by $A \parr B = (A^{\perp} \otimes B^{\perp})^{\perp}$, and by a similar expression for morphisms, an object $\perp = I^{\perp}$ and corresponding natural isomorphisms a', c', e', and we want $(\mathcal{C}, \parr, \perp, a', c', e')$ to also be a *symmetric* *monoidal* category. For this to be true, we need the equality $d_{A^\perp}; d_A^\perp = id_{A^\perp}$, that in general does not hold, as Barr has pointed out to us [3]. Moreover, if d is the unit (and the counit due to symmetry reasons, given that $\psi_{B,A}^{-1} = \psi_{A,B}$) of the adjunction given by ψ, then this equality is one of the "triangular equations" that hold in any adjunction. This condition is very natural, and we include it in our definition of $*$-autonomous category. Indeed, in Barr's own words [3],

> "The definition of $*$-autonomous [category] requires that *this* d be an isomorphism, not merely that there be a natural isomorphism. [...] that is —or should be— the definition."

Definition 22 A *$*$-autonomous category* consists of

1. A closed symmetric monoidal category $(C, \otimes, I, a, c, e, \multimap)$,
2. A (contravariant) strong functor from C into itself, called *involution*, given by a function $(\_)^\perp$ on objects and a family of morphisms $s_{A,B} : A \multimap B \to B^\perp \multimap A^\perp$,
3. A family of isomorphisms $d_A : A \to A^{\perp\perp}$, called *involution isomorphism*, such that the following diagram commutes:

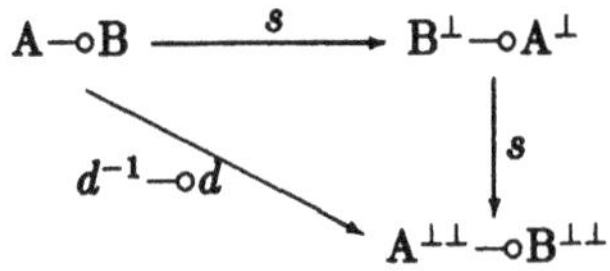

and such that d is the unit (and counit) of the adjunction

$$\psi_{B,A} : Hom_{C^{op}}(B^\perp, A) \longrightarrow Hom_C(B, A^\perp)$$

"between the functor $(\_)^\perp$ and itself" that we have just described above.

□

We will see some examples of $*$-autonomous categories in Example 26.
With this new definition, we can prove the following result:

Theorem 23 Let $(C, \otimes, I, a, c, e, \multimap, (\_)^\perp, s, d)$ be a $*$-autonomous category. If we define:

1. $A ⅋ B = (A^\perp \otimes B^\perp)^\perp$ for objects A, B
2. $f ⅋ g = (f^\perp \otimes g^\perp)^\perp$ for morphisms f, g
3. $\perp = I^\perp$
4. $a'_{A,B,C} = (id_{A^\perp} \otimes d_{B^\perp \otimes C^\perp})^\perp; (a^{-1}_{A^\perp,B^\perp,C^\perp})^\perp; (d^{-1}_{A^\perp \otimes B^\perp} \otimes id_{C^\perp})^\perp$
5. $c'_{A,B} = (c_{B^\perp,A^\perp})^\perp$
6. $e'_A = (d_I \otimes id_{A^\perp})^\perp; (e^{-1}_{A^\perp})^\perp; d^{-1}_A$

then $(C, ⅋, \perp, a', c', e')$ is a symmetric monoidal category. □

Notice also that in a $*$-autonomous category, the involution $(\_)^\perp : C^{op} \longrightarrow C$ is an *equivalence* of categories, so it preserves limits and colimits. In particular, if $A\&B$ is a (cartesian) product in C, and $\top$ is a final object in C, and we define $A \oplus B = (A^\perp \& B^\perp)^\perp$ and $0 = \top^\perp$, $A \oplus B$ is a coproduct in C and 0 is an initial object in C. This is a categorical version of de Morgan's laws.

The following proposition gives a list of (iso)morphisms that always exist in a $*$-autonomous category, and that will be very useful in the next section; some of them exist in any closed symmetric monoidal category.

Proposition 24 Let $(\mathcal{C}, \otimes, I, a, c, e, \multimap, (\_)^{\perp}, s, d)$ be a $*$-autonomous category. Let 0 (respectively $\top$) denote an initial (respectively final) object, and $\oplus$ (respectively &) denote binary coproduct (respectively product) whenever they exist. Then we have,

1. $A \multimap (B \multimap C) \;\cong\; (A \otimes B) \multimap C$
2. $I \multimap A \;\cong\; A$
3. $A \multimap I^{\perp} \;\cong\; A^{\perp}$
4. $A \multimap B \;\cong\; B^{\perp} \multimap A^{\perp}$
5. $A \;\cong\; A^{\perp\perp}$
6. $A \parr B \;\cong\; A^{\perp} \multimap B \;\cong\; (A^{\perp} \otimes B^{\perp})^{\perp}$
7. $A \otimes B \;\cong\; (A \multimap B^{\perp})^{\perp} \;\cong\; (A^{\perp} \parr B^{\perp})^{\perp}$
8. $A \multimap (B \parr C) \;\cong\; (A \multimap B) \parr C$
9. $A \otimes (B \oplus C) \;\cong\; (A \otimes B) \oplus (A \otimes C)$ and $A \otimes 0 \;\cong\; 0$
10. $A \parr (B \& C) \;\cong\; (A \parr B) \& (A \parr C)$ and $A \parr \top \;\cong\; \top$
11. $A \multimap (B \& C) \;\cong\; (A \multimap B) \& (A \multimap C)$ and $A \multimap \top \;\cong\; \top$
12. $(A \oplus B) \multimap C \;\cong\; (A \multimap C) \& (B \multimap C)$ and $0 \multimap A \;\cong\; \top$
13. $A \otimes (B \parr C) \to (A \otimes B) \parr C$ and dually, $(A \parr B) \otimes C \to A \parr (B \otimes C)$
14. $(A \otimes A') \otimes (B \parr C) \to (A \otimes B) \parr (A' \otimes C)$ and dually,
    $(A \parr B) \otimes (A' \parr C) \to (A \parr A') \parr (B \otimes C)$

□

3.2 Categorical interpretation of linear logic

In [28], Seely motivates and studies the appropriate categorical setting for the interpretation of linear logic. In this paper, we restrict ourselves to *propositional linear logic, without exponentials*[8], *but including negation.* Henceforth, we call this fragment *linear logic.* Instead of adopting Girard's original presentation [6], we follow Seely's [28], because it is better adapted to a categorical treatment of the subject. We give several examples of linear categories, and prove the soundness and completeness of linear logic for linear category models.

A *linear formula* is generated by the binary connectives $\otimes, \parr, \&, \oplus$ and $\multimap$ and by the unary operation $(\_)^{\perp}$ from a collection of propositional constants, including the logical constants $I, \perp, \top, 0$.

A *linear sequent* consists of ordered pairs of finite sequences of linear formulas, although the order of the formulas in both sides will be immaterial due to the rule (*perm*) below, together with a name to distinguish different derivations of the same sequent; in particular, this allows us to distinguish axioms with the same formulas, although we will not develop a complete language of proofs. Thus, a linear sequent has the form,

$$f : A_1, \ldots, A_n \vdash B_1, \ldots, B_m$$

where, in view of rules $(\otimes L)$ and $(\parr R)$ below, the commas on the left should be thought of as $(\otimes)$ conjunction and those on the right as $(\parr)$ disjunction.

In general, we will not write the name for the derivation, and only will make it explicit in some occasions, mainly when it becomes important to distinguish between two axioms involving the same formulas.

[8]These are the connectives *of course* (!) and *why not* (?), also known as *modalities.* They should not be confused with the linear implication $\multimap$, interpreted by an "internal hom" or "exponential" functor $\_\multimap\_ : \mathcal{C}^{op} \times \mathcal{C} \longrightarrow \mathcal{C}$ in a closed category $\mathcal{C}$.

Given a collection of (non-logical) propositional constants S, an *S-formula* is a linear formula constructed from the constants in S and the logical constants, and an *S-sequent* is a linear sequent whose formulas are S-formulas.

A *linear theory* T is given by a collection of (non-logical) propositional constants S and a collection Ax of S-sequents, called *axioms*.

Given a linear theory $T = (S, Ax)$, an S-sequent $\Gamma \vdash \Delta$ belongs to the *closure of* T, denoted T^*, if it can be derived (in the usual finite tree-like fashion) from the axioms Ax, and from S-sequents that are instances of the axiom schemes:

$$(id)\ \ A \vdash A$$

$$(IR)\ \ \vdash I \qquad\qquad (\bot L)\ \ \bot \vdash$$

$$(\top R)\ \ \Gamma \vdash \top, \Delta \qquad\qquad (0L)\ \ \Gamma, 0 \vdash \Delta$$

$$(neg1)\ \ A \vdash A^{\bot\bot} \qquad\qquad (neg2)\ \ A^{\bot\bot} \vdash A$$

using the following rules of inference:

Structural rules

$$(perm)\ \frac{\Gamma \vdash \Delta}{\sigma\Gamma \vdash \tau\Delta}\ \text{for any permutations } \sigma \text{ and } \tau$$

$$(cut)\ \frac{\Gamma \vdash A, \Delta \qquad A, \Gamma' \vdash \Delta'}{\Gamma, \Gamma' \vdash \Delta', \Delta}$$

Logical rules

(Negation)

$$(\bot var)\ \frac{\Gamma, A \vdash B, \Delta}{\Gamma, B^{\bot} \vdash A^{\bot}, \Delta}$$

(Multiplicatives)

$$(IL)\ \frac{\Gamma \vdash \Delta}{\Gamma, I \vdash \Delta} \qquad\qquad (\bot R)\ \frac{\Gamma \vdash \Delta}{\Gamma \vdash \bot, \Delta}$$

$$(\otimes L)\ \frac{\Gamma, A, B \vdash \Delta}{\Gamma, A \otimes B \vdash \Delta} \qquad\qquad (\otimes R)\ \frac{\Gamma \vdash A, \Delta \qquad \Gamma' \vdash B, \Delta'}{\Gamma, \Gamma' \vdash A \otimes B, \Delta, \Delta'}$$

$$(⅋ L)\ \frac{\Gamma, A \vdash \Delta \qquad \Gamma', B \vdash \Delta'}{\Gamma, \Gamma', A ⅋ B \vdash \Delta, \Delta'} \qquad\qquad (⅋ R)\ \frac{\Gamma \vdash A, B, \Delta}{\Gamma \vdash A ⅋ B, \Delta}$$

$$(\multimap L)\ \frac{\Gamma \vdash A, \Delta \qquad \Gamma', B \vdash \Delta'}{\Gamma, \Gamma', A \multimap B \vdash \Delta', \Delta} \qquad\qquad (\multimap R)\ \frac{\Gamma, A \vdash B, \Delta}{\Gamma \vdash A \multimap B, \Delta}$$

(Additives)

$$(\& L1)\ \frac{\Gamma, A \vdash \Delta}{\Gamma, A \& B \vdash \Delta} \qquad\qquad (\& L2)\ \frac{\Gamma, B \vdash \Delta}{\Gamma, A \& B \vdash \Delta}$$

$$(\& R)\ \frac{\Gamma \vdash A, \Delta \qquad \Gamma \vdash B, \Delta}{\Gamma \vdash A \& B, \Delta}$$

$$(\oplus L)\ \frac{\Gamma, A \vdash \Delta \qquad \Gamma, B \vdash \Delta}{\Gamma, A \oplus B \vdash \Delta}$$

$$(\oplus R1)\ \frac{\Gamma \vdash A, \Delta}{\Gamma \vdash A \oplus B, \Delta} \qquad (\oplus R2)\ \frac{\Gamma \vdash B, \Delta}{\Gamma \vdash A \oplus B, \Delta}$$

As we have remarked before, we are interested not only in knowing if a sequent belongs to the closure of a theory, but also in knowing its derivation.

The categorical notation that we have been using is intended to fit the logical notation introduced by Girard; in this presentation of formulas and sequents, the only departure from Girard's notation is the use of I instead of $\mathbf{1}$ for the unit of the tensor product.

Following [28], we have:

Definition 25 A *linear category* is a $*$-autonomous category $(\mathcal{C}, \otimes, I, a, c, e, \multimap, (\_)^{\perp}, s, d)$ with specific finite products, *i.e.*, a final object $\top$ and for any objects A, B, a binary product denoted $A \& B$.

A *linear functor* between two linear categories $(\mathcal{C}, \otimes, I, a, c, e, \multimap, (\_)^{\perp}, s, d, \top, \&)$ and $(\mathcal{C}', \otimes', I', a', c', e', \multimap', (\_)^{\perp'}, s', d', \top', \&')$ is a functor $F : \mathcal{C} \longrightarrow \mathcal{C}'$ that preserves all the additional structure in the category, *i.e.*, $F(A \otimes B) = F(A) \otimes' F(B)$, $F(a) = a'$, $F(A^{\perp}) = F(A)^{\perp'}$, $F(A \& B) = F(A) \&' F(B)$, etc.

The category <u>*LinCat*</u> has linear categories as objects and linear functors as morphisms. □

Example 26 Let R be a semiring. The category <u>$FSmod_R$</u>, whose objects are free R-semimodules over finite sets and whose morphisms are linear maps, is a linear category. Since a linear map is completely determined by its action over the elements of a basis, a linear map from the free R-semimodule over X, denoted $R(X)$ to the free R-semimodule over Y, $R(Y)$, is the same as a function $X \rightarrow R(Y)$, that we can view as an R-matrix of dimension $|X| \times |Y|$, where $|X|$ denotes the cardinality of the set X. Therefore, <u>$FSmod_R$</u> is equivalent to the category with finite sets as objects and "R-matrices" (functions $X \rightarrow R(Y)$) as morphisms.

Particularizing the semiring R, we obtain the following examples of linear categories:

1. For R the two element boolean algebra $\{0,1\}$, the category of finite sets and relations ("$\{0,1\}$-matrices"),
2. For R the set $\mathbb{N}$ of natural numbers, the category of finite sets and multirelations ("$\mathbb{N}$-matrices"), equivalent to the category of free commutative monoids (*i.e.*, sets $S^{\otimes}$ of multisets) over finite sets and homomorphisms,
3. For R a field K, the category of finite sets and "K-matrices," equivalent to the category of finite-dimensional vector spaces over K and linear maps.

For all these categories, the involution $(\_)^{\perp}$ is the well known duality of linear algebra. At the matrix level, the dual of an $|X| \times |Y|$-matrix is its transposed $|Y| \times |X|$-matrix.

A very important example of linear category is the category <u>*Cohl*</u> of coherent spaces and linear maps [6], that provided the first semantics for linear logic. <u>*Cohl*</u> also provides semantic interpretations for the exponentials and for second order linear logic. For a discussion of <u>*Cohl*</u> as a linear category see [28], and for a nice introduction to coherent spaces see [12].

All the above categories, including <u>*Cohl*</u>, have in common the property that their morphisms can be understood as *matrices* (called *traces* in <u>*Cohl*</u>). However, the involution $(\_)^{\perp}$ in <u>*Cohl*</u> is of a different nature than the usual linear algebra duality and has more the flavor of a complementation.

An example of linear category that tries to generalize the previous ones is the category <u>$Games_K$</u> defined by Lafont in [13]. Given a set K, the objects of <u>$Games_K$</u> are triples $\langle X, Y, e : Y \times X \rightarrow K \rangle$ where X and Y are sets and e is a function, and the morphisms from $\langle X, Y, e : Y \times X \rightarrow K \rangle$ to $\langle X', Y', e' : Y' \times X' \rightarrow K \rangle$ are pairs of functions $\langle f : X \rightarrow X', g : Y' \rightarrow Y \rangle$ such that $e(g(y), x) = e'(y, f(x))$ for each $x \in X$ and $y \in Y'$. The category <u>$Vect_K$</u> of vector spaces over a field K and linear maps is (isomorphic to) a full subcategory of <u>$Games_K$</u>; and <u>*Top*</u>, the category of topological spaces and continuous maps, and <u>*Cohl*</u> are (isomorphic to) full subcategories of <u>$Games_{\{0,1\}}$</u>. □

We have already pointed out that in a linear category we also have finite coproducts: an initial object 0 and binary coproducts denoted $A \oplus B$.

Given a linear category $\mathcal{C}$, a linear theory $T = (S, Ax)$, and an assignment of an object in $\mathcal{C}$ to each constant in S, it is clear that we can interpret any S-formula A as an object $|A|$ in $\mathcal{C}$. Then, we will associate to each derivation of an S-sequent $A_1, \ldots, A_n \vdash B_1, \ldots, B_m$ a corresponding morphism $|A_1| \otimes \cdots \otimes |A_n| \longrightarrow |B_1| \parr \cdots \parr |B_m|$ in $\mathcal{C}$. If $n = 0$, $|A_1| \otimes \cdots \otimes |A_n|$ reduces to I, and if $m = 0$, $|B_1| \parr \cdots \parr |B_m|$ reduces to $\perp$.

Then, an interpretation of a linear theory T in a linear category $\mathcal{C}$ will be given by the assignment of an object $|s| \in Ob(\mathcal{C})$ to each constant $s \in S$, extended freely to S-formulas, and of a morphism $|f| : |A_1| \otimes \cdots \otimes |A_n| \longrightarrow |B_1| \parr \cdots \parr |B_m|$ in $\mathcal{C}$ to each S-sequent $f : A_1, \ldots, A_n \vdash B_1, \ldots, B_m$ in Ax. However, we can define this concept in a more categorical and elegant way by using morphisms between linear theories.

Definition 27 Given two linear theories $T = (S, Ax)$ and $T' = (S', Ax')$, a *linear morphism* $L : T \to T'$ maps S-formulas to S'-formulas, preserving all the operations, *i.e.*, $L(I) = I$, $L(A \otimes B) = L(A) \otimes L(B)$, $L(A^{\perp}) = L(A)^{\perp}$, etc., and maps an S-sequent $f : A_1, \ldots, A_n \vdash B_1, \ldots, B_m \in Ax$, to an S'-sequent

$$L(f) : L(A_1), \ldots, L(A_n) \vdash L(B_1), \ldots, L(B_m) \in Ax'$$

This defines a category $\underline{LinTh}$ with linear theories as objects and linear morphisms as morphisms. □

It is clear that a $\otimes$-theory is a special case of linear theory[9] and that a $\otimes$-morphism $L : T \to T'$ between $\otimes$-theories is also a linear morphism when we consider T and T' as linear theories. Therefore we have an inclusion functor from $\underline{\otimes\text{-}Th}$ into $\underline{LinTh}$. Moreover, since the rule $(\otimes)$ of Definition 16 is a derived rule in linear logic, given a $\otimes$-theory T we have $T^{\diamond} \subseteq T^{\star}$.

Definition 28 Given a linear category $\mathcal{C}$, we have a linear theory $\mathcal{C}^{\circ}$ whose constants are the objects of $\mathcal{C}$ and such that

$$f : A_1, \ldots, A_n \vdash B_1, \ldots, B_m$$

is an axiom of $\mathcal{C}^{\circ}$ if and only if $f : A_1 \otimes \cdots \otimes A_n \longrightarrow B_1 \parr \cdots \parr B_m$ is a morphism in $\mathcal{C}$. □

Obviously, a linear functor $F : \mathcal{C} \longrightarrow \mathcal{C}'$ induces a linear morphism $F^{\circ} : \mathcal{C}^{\circ} \to \mathcal{C}'^{\circ}$, and therefore there is a "forgetful" functor $(\_)^{\circ} : \underline{LinCat} \longrightarrow \underline{LinTh}$.

Now, an *interpretation* of a theory T in a linear category $\mathcal{C}$ is just a linear morphism $T \to \mathcal{C}^{\circ}$ in $\underline{LinTh}$, and a *model* of T will consist of a linear category $\mathcal{C}$ together with an interpretation of T in $\mathcal{C}$.

Definition 29 Given a linear theory T, a *model* of T consists of a linear category $\mathcal{C}$ and a linear morphism $\mathcal{I} : T \to \mathcal{C}^{\circ}$.

A morphism[10] between two models of T, $(\mathcal{C}, \mathcal{I})$ and $(\mathcal{C}', \mathcal{I}')$ is a linear functor $F : \mathcal{C} \to \mathcal{C}'$ such that $\mathcal{I}; F^{\circ} = \mathcal{I}'$ in $\underline{LinTh}$.

This defines the category $\underline{Mod(T)}$ of models of T. □

The proof of the following theorem is a generalization of Proposition 13 in Chapter III of de Paiva's thesis [25]. de Paiva [25] considers interpretations of linear logic in a somewhat different categorical setting, not entirely compatible with $*$-autonomous categories.

[9]Strictly speaking, this is not entirely correct, since for $\otimes$-theories we have assumed that $\otimes$ is associative and commutative, whereas for linear theories we are not making any syntactic identifications. However, this is a very minor technical problem whose correction can safely be left as an exercise.

[10]This is not the most general definition of morphism that can be given. A more general definition would involve a natural transformation, when interpretations are viewed as linear functors (see Theorem 33).

Theorem 30 Given a linear theory $T = (S, Ax)$ and a model $(\mathcal{C}, \mathcal{I})$ of T, for each derivation of an S-sequent

$$A_1, \ldots, A_n \vdash B_1, \ldots, B_m \ \in \ T^*$$

there exists in $\mathcal{C}$ a morphism

$$\mathcal{I}(A_1) \otimes \cdots \otimes \mathcal{I}(A_n) \longrightarrow \mathcal{I}(B_1) ⅋ \cdots ⅋ \mathcal{I}(B_m).$$

Proof: By induction on the rules, using the morphisms of Proposition 24. □

We have already seen that linear categories can be regarded as linear theories. Conversely, given a linear theory we can associate to it a "free" linear category.

Definition 31 Given a linear theory $T = (S, Ax)$, there is a linear category $\mathcal{L}[T]$ whose objects are the S-formulas, and whose morphisms are equivalence classes of derivations of S-sequents $\Gamma \vdash \Delta \ \in \ T^*$ with respect to the congruence generated by the collection of equations that a category needs to satisfy in order to be a linear category. This collection can be obtained by selecting all the equational rules from Appendix A. □

Example 32 Given formulas A, B, C, D and morphisms $f : A \to B$, $g : C \to D$ corresponding to derivations of the sequents $A \vdash B$ and $C \vdash D$ respectively, the morphism $f \otimes g : A \otimes C \to B \otimes D$ is the equivalence class of the derivation

$$\cfrac{\cfrac{\begin{matrix} \vdots & \vdots \\ A \vdash B & C \vdash D \end{matrix}}{A, C \vdash B \otimes D}}{A \otimes C \vdash B \otimes D}$$

obtained using the rules $(\otimes R)$ and $(\otimes L)$. (By the way, this example proves our claim above, that the rule $(\otimes)$ of Definition 16 is a derived rule in linear logic).

Other examples can be found in the paper by Seely [28]. □

By now, it should be clear that every interpretation $\mathcal{I} : T \to \mathcal{C}^\circ$ of T in $\mathcal{C}$ can be extended uniquely to a linear functor $\mathcal{L}[T] \longrightarrow \mathcal{C}$.

Theorem 33 Given a linear category $\mathcal{C}$ and a linear theory T there exists a bijection between linear functors from $\mathcal{L}[T]$ into $\mathcal{C}$ and linear morphisms $T \to \mathcal{C}^\circ$. In other words, the functor $\mathcal{L}[\_] : \underline{LinTh} \longrightarrow \underline{LinCat}$ is left adjoint to the functor $(\_)^\circ : \underline{LinCat} \longrightarrow \underline{LinTh}$.

Moreover, there is an isomorphism of categories

$$\underline{Mod(T)} \simeq \mathcal{L}[T]/\underline{LinCat},$$

between the category of models of T and the *slice* category $\mathcal{L}[T]/\underline{LinCat}$ whose objects are linear functors $\mathcal{I} : \mathcal{L}[T] \longrightarrow \mathcal{C}$ and whose morphisms $F : \mathcal{I} \to \mathcal{I}'$ are linear functors $F : \mathcal{C} \longrightarrow \mathcal{C}'$ such that $\mathcal{I}; F = \mathcal{I}'$. □

Corollary 34 The model $(\mathcal{L}[T], T \hookrightarrow \mathcal{L}[T]^\circ)$ is initial in the category $\underline{Mod(T)}$, where $T \hookrightarrow \mathcal{L}[T]^\circ$ is the obvious inclusion of theories providing the unit of the adjunction $\mathcal{L}[\_] \dashv (\_)^\circ$. □

Corollary 35 If $T = (\emptyset, \emptyset)$ is the *pure* linear theory, the linear category $\mathcal{L}[T]$ is an initial object in the category $\underline{LinCat}$. □

These results motivate the idea that a linear category gives the right concept of "model" for a linear theory.

Definition 36 Given a linear theory $T = (S, Ax)$, a model $(\mathcal{C}, \mathcal{I})$ of T and an S-sequent $A_1, \ldots, A_n \vdash B_1, \ldots, B_m$, we say that $(\mathcal{C}, \mathcal{I})$ *satisfies* this sequent and write

$$(\mathcal{C}, \mathcal{I}) \models A_1, \ldots, A_n \vdash B_1, \ldots, B_m$$

if there is in $\mathcal{C}$ a morphism $\mathcal{I}(A_1) \otimes \cdots \otimes \mathcal{I}(A_n) \longrightarrow \mathcal{I}(B_1) ⅋ \cdots ⅋ \mathcal{I}(B_m)$.

Similarly, $(\mathcal{C}, \mathcal{I})$ satisfies a collection Q of S-sequents, denoted $(\mathcal{C}, \mathcal{I}) \models Q$, if $(\mathcal{C}, \mathcal{I})$ satisfies all of the sequents in Q. □

From this point of view, Theorem 30 asserts the soundness of the collection of inference rules with respect to this notion of model. We also have an easy completeness result:

Theorem 37 (Completeness) Given a linear theory $T = (S, Ax)$ and an S-sequent $\Gamma \vdash \Delta$,

$$\Gamma \vdash \Delta \in T^* \iff (\mathcal{L}[T], T \hookrightarrow \mathcal{L}[T]^\circ) \models \Gamma \vdash \Delta.$$

□

Corollary 38 Given a linear theory $T = (S, Ax)$ and a collection Q of S-sequents,

$$Q \subseteq T^* \iff (\mathcal{L}[T], T \hookrightarrow \mathcal{L}[T]^\circ) \models Q.$$

□

4 From Petri nets to linear logic

At the end of Section 2.3, we have noted that a Petri category is a symmetric monoidal category. Therefore we have an inclusion functor from the category <u>*PetriCat*</u> into the category <u>*MonCat*</u> whose objects are symmetric monoidal categories and whose morphisms are functors that preserve the symmetric monoidal structure. Since any linear category is a symmetric monoidal category, we also have a forgetful functor $\mathcal{V}$: <u>*LinCat*</u> $\longrightarrow$ <u>*MonCat*</u>. Taking into account the categories <u>*Petri*</u>, <u>$\otimes$ *-Th*</u> and <u>*LinTh*</u>, we have the situation described in the following diagram of functors:

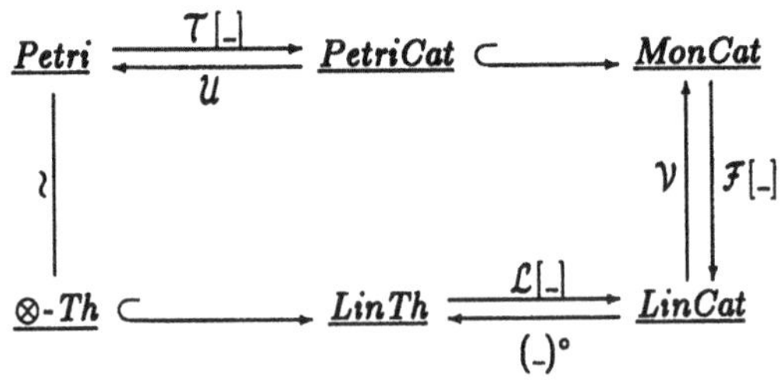

Here the arrows $\hookrightarrow$ denote inclusion functors, $\simeq$ denotes an isomorphism of categories, $\mathcal{U}$ and $\mathcal{V}$ are forgetful functors, $\mathcal{T}[\_]$ is left adjoint to $\mathcal{U}$, $\mathcal{L}[\_]$ is left adjoint to $(\_)^\circ$, and the functor $\mathcal{F}[\_]$: <u>*MonCat*</u> $\longrightarrow$ <u>*LinCat*</u> is a left adjoint to the forgetful functor $\mathcal{V}$: <u>*LinCat*</u> $\longrightarrow$ <u>*MonCat*</u> defined below. The spirit of the above diagram is that, after discarding the forgetful functors $\mathcal{U}$ and $\mathcal{V}$, the diagram should commute up to isomorphism: *i.e.*, for N a Petri net, we should have $\mathcal{F}[\mathcal{T}[N]] \simeq \mathcal{L}[N]$. However, this is not entirely correct, due to a mismatch between the strictness of $\otimes$ in $\mathcal{T}[N]$ and the nonstrictness of $\otimes$ in $\mathcal{L}[N]$. This is not at all a serious obstacle, and the desired isomorphism can be obtained by adopting a nonstrict treatment throughout.

Given a symmetric monoidal category $(\mathcal{C}, \otimes, I, a, c, e)$ we construct a *free linear category* $(\mathcal{F}[\mathcal{C}], \otimes, I, a, c, e, \multimap, (\_)^\perp, s, d, \top, \&)$ generated by $\mathcal{C}$. Notice the overloading of $\otimes, I, a, c, e$. This allows us to simplify the notation and also the number of rules needed to define $\mathcal{F}[\mathcal{C}]$, because we get rid of rules saying that the new operations coincide with the old ones when restricted to $\mathcal{C}$.

The objects of $\mathcal{F}[\mathcal{C}]$ are freely generated from those of $\mathcal{C}$ and a new object $\top$ by closing with respect to the operations $\otimes, -\!\circ, \&$ and $(\_)^\perp$. The morphisms of $\mathcal{F}[\mathcal{C}]$ are obtained from those of $\mathcal{C}$ and (families of) morphisms $id, a, a^{-1}, c, e, e^{-1}, \varepsilon, \pi, \pi', \langle\rangle, s, d, d^{-1}$, by closing freely with respect to the operations $\_;\_$, $\_\otimes\_$, $(\_)^\dagger$, $\langle\_,\_\rangle$ and by imposing the collection of equations that a category needs to satisfy in order to be a linear category.

The complete set of inference rules that define $\mathcal{F}[\mathcal{C}]$ is given in Appendix A, classified according to the structure that each subset of rules gives to $\mathcal{F}[\mathcal{C}]$:

- $\mathcal{C}$ is a subcategory of $\mathcal{F}[\mathcal{C}]$: every object and morphism of $\mathcal{C}$ is also in $\mathcal{F}[\mathcal{C}]$.
- $\mathcal{F}[\mathcal{C}]$ is a category: composition $\_;\_$, identities id and the expected equations for associativity and identities.
- $\mathcal{F}[\mathcal{C}]$ is symmetric monoidal: the functor $\_\otimes\_$ and the natural isomorphisms a, c, e satisfy the MacLane-Kelly coherence conditions (the object I is in $\mathcal{F}[\mathcal{C}]$ because it is already in $\mathcal{C}$).
- $\mathcal{F}[\mathcal{C}]$ is closed monoidal: $\_-\!\circ\_$ on objects, morphism ε and map $f \mapsto f^\dagger$ that give the adjunction
$$Hom_{\mathcal{F}[\mathcal{C}]}(B \otimes A, C) \longrightarrow Hom_{\mathcal{F}[\mathcal{C}]}(B, A -\!\circ C).$$
- $\mathcal{F}[\mathcal{C}]$ has finite products: a final object $\top$ and for every pair of objects A, B an object $A\&B$ and projections π, π', subject to the corresponding universal condition.
- $\mathcal{F}[\mathcal{C}]$ has an involution: a strong functor $(\_)^\perp, s$, and a family of isomorphisms d, satisfying the corresponding four equations expressed in terms of the notation introduced in Section 3.1: $j, m, \_-\!\circ\_$ for morphisms and the adjunction
$$\psi_{B,A} : Hom_{\mathcal{F}[\mathcal{C}]^{op}}(B^\perp, A) \longrightarrow Hom_{\mathcal{F}[\mathcal{C}]}(B, A^\perp)$$
"between the functor $(\_)^\perp$ and itself."

Theorem 39 The forgetful functor <u>*LinCat*</u> $\longrightarrow$ <u>*MonCat*</u> has a left adjoint $\mathcal{F}[\_]$: <u>*MonCat*</u> $\longrightarrow$ <u>*LinCat*</u>, that is, if $\mathcal{C}$ is a symmetric monoidal category, $\mathcal{F}[\mathcal{C}]$ is the *free linear category* generated by $\mathcal{C}$. □

5 Using linear logic to specify concurrency

This is of course the explicit intent of linear logic. However, much remains to be done to fully exploit the promising connections between linear logic and concurrency. This paper has developed a bridge between Petri nets and linear logic at the model-theoretic level of their categorical semantics. One immediate payoff of such a bridge is a precise definition of *satisfaction* of a linear logic formula by a Petri net, as described below.

In general, a good way of considering possible uses of (different variants of) linear logic for concurrent systems is by means of the recent notion of a concurrency calculus introduced by Milner in [24]. Roughly speaking, such a calculus consists of a logic $\mathcal{L}$ used for specifications and a class of concurrent systems $\mathcal{C}$, together with a relation $Q \models \varphi$ stating when a concurrent system Q in $\mathcal{C}$ satisfies a specification φ in $\mathcal{L}$. Milner [24] gives several examples, including the well-known one where $\mathcal{C}$ is CCS and $\mathcal{L}$ is the Hennessy-Milner logic. The case where $\mathcal{L}$ is propositional linear logic and $\mathcal{C}$ is the class of Petri nets is made explicit in the definition below.

Definition 40 Given a Petri net $N = (S, T)$, and an S-sequent $\Gamma \vdash \Delta$, we say that N *satisfies* $\Gamma \vdash \Delta$ and write $N \models \Gamma \vdash \Delta$ iff $(\mathcal{L}[N], N \hookrightarrow \mathcal{L}[N]^\circ) \models \Gamma \vdash \Delta$. □

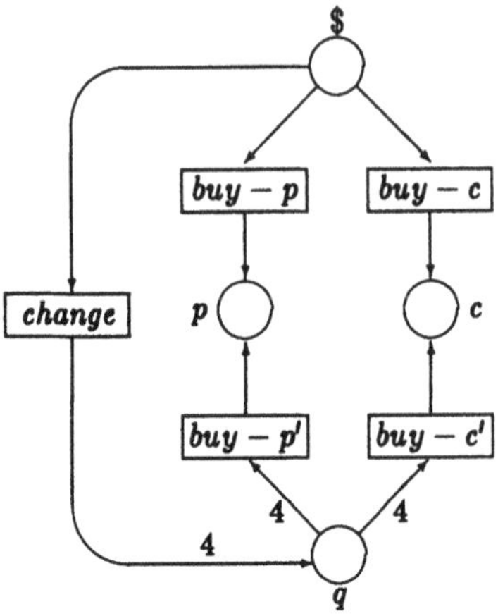

Figure 4: Another Petri net for buying pens and candy bars, and for getting change.

Due to the completeness Theorem 37, we know that $N \models \Gamma \vdash \Delta$ if and only if $\Gamma \vdash \Delta$ belongs to the closure N^*, *i.e.*, iff it is derivable from the axioms $\{t : {}^\bullet t \vdash t^\bullet \mid t \in T\}$ by using the axioms and rules of linear logic listed in Section 2.3. Therefore, we can consider N^* as the collection of all the specifications (written as S-sequents) in linear logic satisfied by the Petri net N and $\mathcal{L}[N]$ as the appropriate categorical model[11] in which to interpret such specifications.

The intuitive interpretation of the connectives & and $\oplus$ is external and internal choice, respectively. A formula written using only the connective $\otimes$ corresponds to the availability of resources, with $\otimes$ meaning resource accumulation; negating such a formula by $(\_)^\perp$ corresponds to a "debt" of resources, and $⅋$, being the dual of $\otimes$, corresponds to the accumulation of debts; this intuitive interpretation matches the duality $(A \otimes B)^\perp \cong (A^\perp ⅋ B^\perp)$. The connective of linear implication $\multimap$ expresses the possibilities of transition between two given states.

Example 41 The Petri net in Figure 4 extends that in Figure 2 by the new possibilities of buying a pen or a candy bar with four quarters instead of with a dollar, and of changing a dollar into four quarters. This net satisfies, among others, the following specifications:

1. $\$ \vdash p \,\&\, c \,\&\, q^4$
2. $q^4 \vdash p \,\&\, c$
3. $\$^2 \vdash p \otimes c$
4. $\$ \otimes q^4 \vdash q^8$
5. $\$ \oplus q^4 \vdash p$
6. $\$ \oplus q^4 \vdash q^4$
7. $\$ \oplus q^4 \vdash p \oplus c$
8. $\$^3 \vdash (p \otimes c \otimes q^4) \,\&\, q^{12} \,\&\, c^3 \,\&\, (c^2 \otimes \$)$
9. $\$ \vdash p \oplus \$^{1000}$
10. $p^\perp \vdash \$^\perp$
11. $p^\perp ⅋ c^\perp \vdash \$^\perp ⅋ \$^\perp$

[11] $\mathcal{L}[N]$ contains all the computations of the Petri net N and also additional "idealised computations" corresponding for example to possibilities of external choice (&), to reversals of causality ($(\_)^\perp$), etc.

12. $p^{\perp} \oplus c^{\perp} \vdash \$^{\perp}$

13. $\$ \vdash \$ \multimap (p \otimes c)$

The intuitive meaning of the above specifications is:

1. With one dollar, we can *choose* between buying a pen *and* (&) buying a candy bar *and* (&) changing it into four quarters.
2. With four quarters, we can *choose* between buying a pen *and* (&) buying a candy bar.
3. With *two* dollars, we can buy a pen *and* ($\otimes$) a candy bar.
4. With one dollar *and* ($\otimes$) four quarters we can get eight quarters.
5. With either a dollar *or* ($\oplus$) four quarters, we can buy a pen.
6. With either a dollar *or* ($\oplus$) four quarters, we can get four quarters.
7. With either a dollar *or* ($\oplus$) four quarters, we can buy either a pen *or* ($\oplus$) a candy bar.
8. With *three* dollars, we can *choose* between: (i) buying a pen *and* ($\otimes$) a candy bar *and* ($\otimes$) getting four quarters, *and* (&) (ii) changing them into twelve quarters, *and* (&) (iii) buying three candy bars, *and* (&) (iv) buying only two candy bars *and* ($\otimes$) saving one dollar.
9. With a dollar, we can get a pen *or* ($\oplus$) one thousand dollars. This is one example of *internal* choice, that in the above Petri net is always realized by choosing the first possibility.
10. We can replace a *debt* $((\_)^{\perp})$ of a pen by a *debt* $((\_)^{\perp})$ of a dollar.
11. We can replace a *debt* $((\_)^{\perp})$ of a pen *and* ($\parr$) a candy bar by a *debt* $((_)^{\perp})$of *two* dollars.
12. We can replace a *debt* $((\_)^{\perp})$ of either a pen *or* ($\oplus$) a candy bar by a *debt* $((\_)^{\perp})$ of a dollar.
13. Having a dollar, *if* we had another dollar, *then* ($\multimap$) we *could* buy a pen *and* ($\otimes$) also a candy bar.

□

The goal of using linear logic axioms to specify concurrency properties is also explicit in Asperti [1] and Gunter and Gehlot [9]. In particular, Asperti [1] suggests that axioms at the first order predicate logic level would be quite useful and would roughly correspond to descriptions at the level of predicate/events nets [27,5].

It seems to us that axioms in second-order propositional linear logic could be used to specify global properties of a concurrent system, since second order quantification intuitively corresponds to quantifying over the states of the system. However, the precise form and meaning of such specifications will require further research.

6 Concluding remarks

We have established a systematic triangular correspondence between linear logic theories, Petri nets and linear categories that permits the transfer of techniques and results between the fields of logic, concurrency and category theory. The use of linear logic axioms to specify concurrency properties of a net takes the form of a concurrency calculus in the sense of Milner [24].

Much remains to be done to further develop and exploit this correspondence. Three areas that, among others, should receive attention are:

1. The use of linear logic axioms as specifications of concurrent systems; specifically, the introduction of more powerful axioms, the comparison of such specifications with more traditional specification techniques, and the inclusion of other types of concurrent systems within the framework.

2. The study of equivalence and normalization for both proofs and processes using coherence techniques, and of the relationship of such techniques to the proof-theoretic methods already developed by Girard.

3. The uses of linear logic for concurrent programming; for functional computation, we already have the interesting work by Lafont [11].

Acknowledgements

We wish to thank Michael Barr for helping us in understanding $*$-autonomous categories. We also thank Carl Gunter and Andre Scedrov for helpful discussions.

References

[1] Andrea Asperti, *A Logic for Concurrency*, manuscript, November 1987.

[2] Michael Barr, *$*$-Autonomous Categories*, Volume 752 of *Lecture Notes in Mathematics*, Springer-Verlag, Berlin, 1979.

[3] Michael Barr, Personal communication, December 1988.

[4] Pierpaolo Degano, José Meseguer and Ugo Montanari, Axiomatizing Net Computations and Processes, in: *Proc. Logic in Computer Science*, Asilomar, 1989, pages 175–185.

[5] H. J. Genrich and K. Lautenbach, System Modelling with High-Level Petri Nets, *Theoretical Computer Science*, **13**, 1981, pages 109–136.

[6] Jean-Yves Girard, Linear Logic, *Theoretical Computer Science*, **50**, 1987, pages 1–102.

[7] Jean-Yves Girard, Linear Logic and Parallelism, in: *Proc. School on Semantics of Parallelism, IAC*, CNR, Roma, 1986.

[8] Jean-Yves Girard, Towards a Geometry of Interaction, in: J. W. Gray and A. Scedrov (eds.), *Proc. A. M. S. Conference on Categories in Computer Science and Logic*, Boulder, 1987.

[9] Carl Gunter and Vijay Gehlot, *A Proof-theoretic Operational Semantics for True Concurrency*, Preliminary Report, 1989.

[10] W. A. Howard, The Formulae-as-Types Notion of Construction, in: J. P. Seldin and J. R. Hindley (eds.), *To H. B. Curry: Essays on Combinatory Logic, Lambda Calculus and Formalism*, Academic Press, London, 1980, pages 479–490.

[11] Yves Lafont, The Linear Abstract Machine, *Theoretical Computer Science*, **59**, 1988, pages 157–180.

[12] Yves Lafont, Introduction to Linear Logic, *Lecture notes for the Summer School on Constructive Logic and Category Theory*, Isle of Thorns, August 1988.

[13] Yves Lafont, *From Linear Algebra to Linear Logic*, Preliminary Draft, November 1988.

[14] Joachim Lambek, Deductive Systems and Categories I, *Mathematical Systems Theory*, **2**, 1968, pages 287–318.

[15] Joachim Lambek, Deductive Systems and Categories II, in: *Category Theory, Homology Theory and their Applications I*, Volume 86 of *Lecture Notes in Mathematics*, Springer-Verlag, Berlin, 1969, pages 76–122.

[16] Joachim Lambek, Deductive Systems and Categories III, in: F. W. Lawvere (ed.), *Toposes, Algebraic Geometry and Logic*, Volume 274 of *Lecture Notes in Mathematics*, Springer-Verlag, Berlin, 1972, pages 57–82.

[17] F. W. Lawvere, Adjointness in Foundations, *Dialectica*, **23**, 1969, pages 281-296.

[18] Saunders MacLane, Natural Associativity and Commutativity, *Rice University Studies*, **49**, 1963, pages 28–46.

[19] Saunders MacLane, *Categories for the Working Mathematician*, Volume 5 of *Graduate Texts in Mathematics*, Springer-Verlag, Berlin, 1971.

[20] Saunders MacLane (ed.), *Coherence in Categories*, Volume 281 of *Lecture Notes in Mathematics*, Springer-Verlag, Berlin, 1972.

[21] Saunders MacLane, Why Commutative Diagrams Coincide with Equivalent Proofs, in: S. A. Amitsur, D. J. Saltman and G. B. Seligman (eds.), *Algebraists' Homage: Papers in Ring Theory and Related Topics*, Volume 13 of *Contemporary Mathematics*, American Mathematical Society, Providence, 1982, pages 387–401.

[22] José Meseguer and Ugo Montanari, Petri Nets Are Monoids: A New Algebraic Foundation for Net Theory, in: *Proc. Logic in Computer Science*, Edinburgh, 1988, pages 155–164.

[23] José Meseguer and Ugo Montanari, *Petri Nets Are Monoids*, Technical Report SRI-CSL-88-3, C.S.Lab., SRI International, January 1988, submitted for publication.

[24] Robin Milner, Interpreting One Concurrent Calculus in Another, in: *Proc. Int. Conf. on Fifth Generation Computer Systems*, Tokyo, 1988, pages 321–326.

[25] Valeria C. V. de Paiva, *The Dialectica Categories*, Ph. D. thesis, University of Cambridge, 1988.

[26] Dag Prawitz, *Natural Deduction: A Proof-Theoretical Study*, Almqvist and Wiksell, Stockholm, 1965.

[27] Wolfgang Reisig, *Petri Nets: An Introduction*, Springer-Verlag, Berlin, 1985.

[28] R. A. G. Seely, Linear Logic, *-Autonomous Categories and Cofree Coalgebras, in: J. W. Gray and A. Scedrov (eds.), *Proc. A. M. S. Conference on Categories in Computer Science and Logic*, Boulder, 1987.

A Inference rules defining $\mathcal{F}[\mathcal{C}]$

$\mathcal{C}$ is a subcategory of $\mathcal{F}[\mathcal{C}]$

$$\frac{A \in Ob(\mathcal{C})}{A \in Ob(\mathcal{F}[\mathcal{C}])} \qquad \frac{f : A \to B \text{ in } \mathcal{C}}{f : A \to B \text{ in } \mathcal{F}[\mathcal{C}]}$$

$\mathcal{F}[\mathcal{C}]$ is a category

$$\frac{f : A \to B,\ g : B \to C \text{ in } \mathcal{F}[\mathcal{C}]}{f;g : A \to C \text{ in } \mathcal{F}[\mathcal{C}]}\ (composition)$$

$$\frac{A \in Ob(\mathcal{F}[\mathcal{C}])}{id_A : A \to A \text{ in } \mathcal{F}[\mathcal{C}]}\ (identities)$$

$$\frac{f : A \to B,\ g : B \to C,\ h : C \to D \text{ in } \mathcal{F}[\mathcal{C}]}{f;(g;h) = (f;g);h}\ (associativity)$$

$$\frac{f : A \to B \text{ in } \mathcal{F}[\mathcal{C}]}{id_A; f = f}\ (id\ left) \qquad \frac{f : A \to B \text{ in } \mathcal{F}[\mathcal{C}]}{f; id_B = f}\ (id\ right)$$

$\mathcal{F}[\mathcal{C}]$ is symmetric monoidal

$$\frac{A, B \in Ob(\mathcal{F}[\mathcal{C}])}{A \otimes B \in Ob(\mathcal{F}[\mathcal{C}])}$$

$$\frac{f : A \to B,\ g : C \to D \text{ in } \mathcal{F}[\mathcal{C}]}{f \otimes g : A \otimes C \to B \otimes D \text{ in } \mathcal{F}[\mathcal{C}]}$$

$$\frac{f : A \to B,\ g : B \to C,\ f' : A' \to B',\ g' : C' \to D' \text{ in } \mathcal{F}[\mathcal{C}]}{(f;g) \otimes (f';g') = (f \otimes f');(g \otimes g')}\ (\otimes functor1)$$

$$\frac{A, B \in Ob(\mathcal{F}[\mathcal{C}])}{id_A \otimes id_B = id_{A \otimes B}}\ (\otimes functor2)$$

$$\frac{A, B, C \in Ob(\mathcal{F}[\mathcal{C}])}{a_{A,B,C} : A \otimes (B \otimes C) \to (A \otimes B) \otimes C \text{ in } \mathcal{F}[\mathcal{C}]}\ (associativity)$$

$$\frac{A, B, C \in Ob(\mathcal{F}[\mathcal{C}])}{a^{-1}_{A,B,C} : (A \otimes B) \otimes C \to A \otimes (B \otimes C) \text{ in } \mathcal{F}[\mathcal{C}]}\ (inverse)$$

$$\frac{A, B, C \in Ob(\mathcal{F}[\mathcal{C}])}{a_{A,B,C}; a^{-1}_{A,B,C} = id_{A \otimes (B \otimes C)}} \qquad \frac{A, B, C \in Ob(\mathcal{F}[\mathcal{C}])}{a^{-1}_{A,B,C}; a_{A,B,C} = id_{(A \otimes B) \otimes C}}$$

$$\frac{f : A \to A',\ g : B \to B',\ h : C \to C' \text{ in } \mathcal{F}[\mathcal{C}]}{a_{A,B,C}; (f \otimes g) \otimes h = f \otimes (g \otimes h); a_{A',B',C'}}\ (naturality)$$

$$\frac{A, B \in Ob(\mathcal{F}[\mathcal{C}])}{c_{A,B} : A \otimes B \to B \otimes A \text{ in } \mathcal{F}[\mathcal{C}]}\ (commutativity)$$

$$\frac{f : A \to A',\ g : B \to B' \text{ in } \mathcal{F}[\mathcal{C}]}{c_{A,B}; g \otimes f = f \otimes g; c_{A',B'}}\ (naturality)$$

$$\frac{A \in Ob(\mathcal{F}[\mathcal{C}])}{e_A : I \otimes A \to A \text{ in } \mathcal{F}[\mathcal{C}]}\ (unit)$$

$$\frac{A \in Ob(\mathcal{F}[\mathcal{C}])}{e_A^{-1} : A \to I \otimes A \text{ in } \mathcal{F}[\mathcal{C}]}\ (inverse)$$

$$\frac{A \in Ob(\mathcal{F}[\mathcal{C}])}{e_A; e_A^{-1} = id_{I\otimes A}} \qquad \frac{A \in Ob(\mathcal{F}[\mathcal{C}])}{e_A^{-1}; e_A = id_A}$$

$$\frac{f : A \to B \text{ in } \mathcal{F}[\mathcal{C}]}{e_A; f = (id_I \otimes f); e_B}\ (naturality)$$

$$\frac{A, B, C, D \in Ob(\mathcal{F}[\mathcal{C}])}{a_{A,B,C\otimes D}; a_{A\otimes B,C,D} = (id_A \otimes a_{B,C,D}); a_{A,B\otimes C,D}; (a_{A,B,C} \otimes id_D)}\ (coherence1)$$

$$\frac{A, B \in Ob(\mathcal{F}[\mathcal{C}])}{c_{A,B}; c_{B,A} = id_{A\otimes B}}\ (coherence2)$$

$$\frac{A, B \in Ob(\mathcal{F}[\mathcal{C}])}{a_{I,A,B}; (e_A \otimes id_B) = e_{A\otimes B}}\ (coherence3)$$

$$\frac{A, B, D \in Ob(\mathcal{F}[\mathcal{C}])}{a_{A,B,D}; c_{A\otimes B,D}; a_{D,A,B} = (id_A \otimes c_{B,D}); a_{A,D,B}; (c_{A,D} \otimes id_B)}\ (coherence4)$$

$\mathcal{F}[\mathcal{C}]$ is closed monoidal

$$\frac{A, B \in Ob(\mathcal{F}[\mathcal{C}])}{A \multimap B \in Ob(\mathcal{F}[\mathcal{C}])}$$

$$\frac{A, B \in Ob(\mathcal{F}[\mathcal{C}])}{\varepsilon_{A,B} : (A \multimap B) \otimes A \to B \text{ in } \mathcal{F}[\mathcal{C}]}\ (counit)$$

$$\frac{f : C \otimes A \to B \text{ in } \mathcal{F}[\mathcal{C}]}{f^{\dagger} : C \to (A \multimap B) \text{ in } \mathcal{F}[\mathcal{C}]}\ (bijection)$$

$$\frac{f : C \otimes A \to B \text{ in } \mathcal{F}[\mathcal{C}]}{(f^{\dagger} \otimes id_A); \varepsilon_{A,B} = f}\ (adjunction1)$$

$$\frac{g : C \to (A \multimap B) \text{ in } \mathcal{F}[\mathcal{C}]}{((g \otimes id_A); \varepsilon_{A,B})^{\dagger} = g}\ (adjunction2)$$

$\mathcal{F}[\mathcal{C}]$ has finite products

$$\frac{A, B \in Ob(\mathcal{F}[\mathcal{C}])}{A \& B \in Ob(\mathcal{F}[\mathcal{C}])}\ (product\ object)$$

$$\frac{f : C \to A,\ g : C \to B \text{ in } \mathcal{F}[\mathcal{C}]}{\langle f, g\rangle : C \to A \& B \text{ in } \mathcal{F}[\mathcal{C}]}\ (mediating\ morphism)$$

$$\frac{A, B \in Ob(\mathcal{F}[\mathcal{C}])}{\pi_{A,B} : A \& B \to A \text{ in } \mathcal{F}[\mathcal{C}]}\ (projection1)$$

$$\frac{A, B \in Ob(\mathcal{F}[\mathcal{C}])}{\pi'_{A,B} : A \& B \to B \text{ in } \mathcal{F}[\mathcal{C}]}\ (projection2)$$

$$\frac{f : C \to A,\ g : C \to B \text{ in } \mathcal{F}[\mathcal{C}]}{\langle f, g\rangle; \pi_{A,B} = f}\ (prod\ eq1)$$

$$\frac{f : C \to A,\ g : C \to B \text{ in } \mathcal{F}[\mathcal{C}]}{\langle f, g\rangle; \pi'_{A,B} = g}\ (prod\ eq2)$$

$$\frac{h : C \to A\&B \ in \ \mathcal{F}[C]}{\langle h;\pi_{A,B}\ ,\ h;\pi'_{A,B}\rangle = h}\ (prod\ eq3)$$

$$\top \in Ob(\mathcal{F}[C])\ (final\ object)$$

$$\frac{A \in Ob(\mathcal{F}[C])}{\langle\rangle_A : A \to \top \ in \ \mathcal{F}[C]}\ (morphism\ existence)$$

$$\frac{f : A \to \top \ in \ \mathcal{F}[C]}{f = \langle\rangle_A}\ (morphism\ unicity)$$

$\mathcal{F}[C]$ has an involution

$$\frac{A \in Ob(\mathcal{F}[C])}{A^{\perp} \in Ob(\mathcal{F}[C])}\ (involution1)$$

$$\frac{A, B \in Ob(\mathcal{F}[C])}{s_{A,B} : A \multimap B \to B^{\perp} \multimap A^{\perp} \ in \ \mathcal{F}[C]}\ (involution2)$$

$$\frac{A \in Ob(\mathcal{F}[C])}{d_A : A \to A^{\perp\perp} \ in \ \mathcal{F}[C]}\ (involution\ isomorphism)$$

$$\frac{A \in Ob(\mathcal{F}[C])}{d_A^{-1} : A^{\perp\perp} \to A \ in \ \mathcal{F}[C]}\ (inverse)$$

$$\frac{A \in Ob(\mathcal{F}[C])}{d_A; d_A^{-1} = id_A} \qquad \frac{A \in Ob(\mathcal{F}[C])}{d_A^{-1}; d_A = id_{A^{\perp\perp}}}$$

$$\frac{A \in Ob(\mathcal{F}[C])}{j_A; s_{A,A} = j_{A^{\perp}}}\ (strong\ functor1)$$

$$\frac{A, B, C \in Ob(\mathcal{F}[C])}{m_{A,B,C}; s_{A,C} = (s_{B,C} \otimes s_{A,B}); c_{C^{\perp}\multimap B^{\perp}, B^{\perp}\multimap A^{\perp}}; m_{C^{\perp},B^{\perp},A^{\perp}}}\ (strong\ functor2)$$

$$\frac{A, B \in Ob(\mathcal{F}[C])}{s_{A,B}; s_{B^{\perp},A^{\perp}} = d_A^{-1} \multimap d_B}\ (*-autonomous\ cat1)$$

$$\frac{A \in Ob(\mathcal{F}[C])}{\psi_{B,B^{\perp}}(id_{B^{\perp}}) = d_B}\ (*-autonomous\ cat2)$$

where j, m, $\_\multimap\_$ for morphisms and ψ are defined by the following equations:

$$j_A = e_A^{\dagger}$$

$$m_{A,B,C} = (a^{-1}_{B\multimap C, A\multimap B, A}; (id_{B\multimap C} \otimes \varepsilon_{A,B}); \varepsilon_{B,C})^{\dagger}$$

$$g \multimap f = ((id_{D\multimap A} \otimes g); \varepsilon_{D,A}; f)^{\dagger} \quad \text{if } f : A \to B \text{ and } g : C \to D$$

$$\psi_{B,A} = Hom_{\mathcal{F}[C]}(id_A, u_B^{-1}); (\varphi^{B}_{A,I^{\perp}})^{-1}; Hom_{\mathcal{F}[C]}(c_{B,A}, id_{I^{\perp}}); \varphi^{A}_{B,I^{\perp}}; Hom_{\mathcal{F}[C]}(id_B, u_A)$$

i.e.,

$$\psi_{B,B^{\perp}}(id_{B^{\perp}}) = (c_{B,B^{\perp}}; (u_B^{-1} \otimes id_B); \varepsilon_{B,I^{\perp}})^{\dagger}; u_{B^{\perp}}$$

with u_A the isomorphism $(A \multimap I^{\perp}) \to A^{\perp}$.

A Dialectica-like Model of Linear Logic

Valeria C. V. de Paiva
Dept. Informatica PUC-RJ
R.M.S.Vicente, 225 - Rio de Janeiro 22453
BRAZIL

The aim of this work is to define the categories **GC**, describe their categorical structure and show they are a model of *Linear Logic*. The second goal is to relate those categories to the Dialectica categories **DC**, cf.[DC], using different functors for the exponential "of course". It is hoped that this categorical model of Linear Logic should help us to get a better understanding of the logic, which is, perhaps, the first non-intuitionistic constructive logic.

This work is divided in two parts, each one with 3 sections. The first section shows that **GC** is a monoidal closed category and describes bifunctors for tensor "$\oslash$", internal hom "$[-,-]$", par "$\sharp$", cartesian products "&" and coproducts "$\oplus$". The second section defines linear negation as a contravariant functor obtained evaluating the internal hom bifunctor at a "dualising object". The third section makes explicit the connections with Linear Logic, while the fourth introduces the comonads used to model the connective "of course". Section 5 discusses some properties of these comonads and finally section 6 makes the logical connections once more.

This work grew out of suggestions of J.Y. Girard at the AMS-Conference on Categories, Logic and Computer Science in Boulder 1987, where I presented my earlier work on the Dialectica categories, hence the title. Still on the lines of given credit where it is due, I would like to say that Martin Hyland, under whose supervision this work was written, has been a continuous source of ideas and inspiration. Many heartfelt thanks to him.

1. The main definitions

We start with a finitely complete category **C**. Then to describe **GC** say that its objects are relations on objects of **C**, that is monics $A \overset{\alpha}{\rightarrowtail} U \times X$, which we usually write as $(U \overset{\alpha}{\leftrightarrow} X)$.

Given two such objects, $(U \overset{\alpha}{\leftrightarrow} X)$ and $(V \overset{\beta}{\leftrightarrow} Y)$, which we call simply A and B, a morphism from A to B consists of a pair of maps in **C**, $f: U \to V$ and $F: Y \to X$, such that a pullback condition is satisfied, namely that

$$(U \times F)^{-1}(\alpha) \leq (f \times Y)^{-1}(\beta), \tag{1}$$

where $(-)^{-1}$ represents pullbacks.

Using diagrams, we say (f, F) is a morphism in **GC** if there is a (unique) map in **C**, $k\colon A' \to B'$ making the triangle commute:

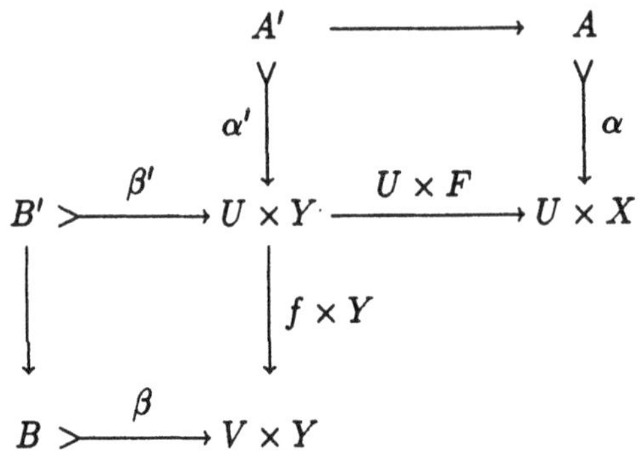

where A' is the pullback of α along $U \times F$ and B' the pullback of β along $f \times Y$. Note that we refer to the object $(U \overset{\alpha}{\nleftrightarrow} X)$ as "α", meaning the (equivalence class of the) monic, as well as A.

The intuition here is that, if we consider α and β set-theoretic relations, there is a morphism from α to β

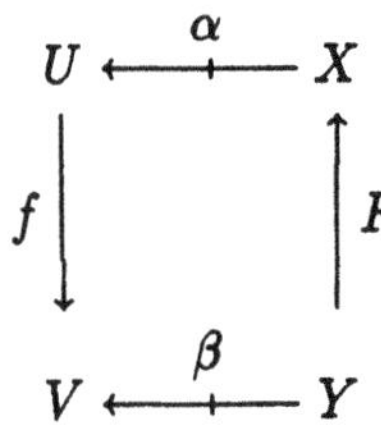

iff whenever $u \, \alpha \, F(y)$ then $f(u) \, \beta \, y$.

It is easy to see **GC** is a category, since composition is just composition in each 'coordinate'.

If the base category **C** is cartesian closed, as well as finitely complete, the category **GC** has a symmetric monoidal structure (tensor product) denoted by "$\oslash$" that can be made closed. This tensor bifunctor $\oslash$ seems somewhat involved and not very intuitive, but it is exactly what is needed to show that **GC** is symmetric monoidal closed, given the internal hom, $[-,-]_{\mathbf{GC}}$ defined below.

Definition 1. *There is an internal hom bifunctor in* **GC**, $[-,-]_{\mathbf{GC}}\colon \mathbf{GC}^{\mathrm{op}} \times \mathbf{GC} \to \mathbf{GC}$ *given by*

$$[A, B]_{GC} = (V^U \times X^Y \overset{\beta^\alpha}{\nleftrightarrow} U \times Y) \tag{2}$$

where intuitively the relation β^α reads as $(f,F)\beta^\alpha(u,y)$ iff whenever $u\alpha F(y)$ then $f(u)\beta y$.

Formally, we define β^α as the greatest subobject E of $V^U \times X^Y \times U \times Y$ such that $E \wedge A' \leq B'$, where A' is the pullback of A along the map

$$V^U \times X^Y \times U \times Y \xrightarrow{(\pi_3,\text{“}ev\text{”})} U \times X,$$

B' is the pullback of B along $V^U \times X^Y \times U \times Y \xrightarrow{(\text{“}ev\text{”},\pi_4)} V \times Y$ and "$\wedge$" means pullback again.

To guarantee the existence of such greatest subobject, we insist on $\mathbf{C}$ being *locally cartesian closed.* By that we mean that for any object A of $\mathbf{C}$, the slice category $\mathbf{C}/A$ is cartesian closed, cf. [See] 1984.

Definition 2. *Assuming $\mathbf{C}$ locally cartesian closed, consider the tensor product in $\mathbf{GC}$ given by the operation $\oslash\colon \mathbf{GC} \times \mathbf{GC} \to \mathbf{GC}$ which takes the pair of objects (A,B) to*

$$A \oslash B = (U \times V \overset{\alpha\oslash\beta}{\nleftrightarrow} X^V \times Y^U). \tag{3}$$

Intuitively, $(u,v)\alpha \oslash \beta(f,g)$ iff $u\alpha f(v)$ and $v\beta g(u)$.

The functor "$\oslash$" is not a categorical product, for example, projections do not exist necessarily, but it is associative and symmetric. The object $I = (1 \overset{1}{\nleftrightarrow} 1)$ is the unit for this tensor product.

Another tensor product, similar to the tensor bifunctor in $\mathbf{DC}$ can be defined, but it is not left-adjoint to the internal hom.

Definition 3. *The bifunctor $\otimes\colon \mathbf{GC} \to \mathbf{GC}$, which takes (A,B) to*

$$A \otimes B = (U \times V \overset{\alpha\otimes\beta}{\nleftrightarrow} X \times Y) \tag{4}$$

is associative and symmetric. It has the same unit $I = (1 \overset{1}{\nleftrightarrow} 1)$ as the bifunctor "$\oslash$" and intuitively the relation $(u,v)\alpha \otimes \beta(x,y)$ holds iff $u\alpha x$ and $v\beta y$.

Proposition 1. *The category $\mathbf{GC}$ is a symmetric monoidal closed category.*

Proof: It's enough to see the natural isomorphism

$$\mathrm{Hom}_{\mathbf{GC}}(A \oslash B, C) \cong \mathrm{Hom}_{\mathbf{GC}}(A, [B,C]_{\mathbf{GC}}). \tag{5}$$

Using diagrams,

$$\begin{array}{ccc} U \times V & \xleftarrow{\ \alpha\oslash\beta\ } & X^V \times Y^U \\ {\scriptstyle f}\downarrow & & \uparrow{\scriptstyle (F_1,F_2)} \\ W & \xleftarrow{\ \gamma\ } & Z \end{array} \qquad \begin{array}{ccc} U & \xleftarrow{\ \alpha\ } & X \\ {\scriptstyle (\overline{f},\overline{F}_2)}\downarrow & & \uparrow{\scriptstyle \overline{F}_1} \\ W^V \times Y^Z & \xleftarrow{\ \gamma^\beta\ } & V \times Z \end{array}$$

For symmetry reasons that will be apparent later, we want to introduce yet another bifunctor, to be called "$\sharp$", which is, in some sense, dual to the tensor product "$\oslash$" bifunctor. To define the bifunctor "$\sharp$" categorically we need **C** with *stable* (under pullbacks) and *disjoint coproducts* cf. [M/R] 1977.

Definition 4. *Consider the bifunctor "$\sharp$" that takes (A, B) to*

$$A \sharp B = (U^Y \times V^X \overset{\alpha\sharp\beta}{\leftrightarrow} X \times Y). \tag{6}$$

The relation defining $A \sharp B$ *says that* $(f,g)\alpha\sharp\beta(x,y)$ *iff* $f(y)\alpha x$ *or* $g(x)\beta y$.

Notice that the object $\perp = (1 \overset{0}{\leftrightarrow} 1)$, where 0 is the empty relation on 1×1, is the unit for the operation "$\sharp$" and that there is a natural map $\perp \to I$, but not conversely.

Notice as well that tensor "$\oslash$" and its dual "$\sharp$" have very similar "carriers", but duality here is transforming the metalanguage "and" into "or".

Now we define cartesian products and coproducts in **GC**.

Proposition 2. *If* **C** *is a finitely complete category with disjoint and stable coproducts then* **GC** *has categorical products and coproducts.*

Proof: Categorical products are given by the bifunctor $\&$: $\mathbf{GC} \times \mathbf{GC} \to \mathbf{GC}$, which takes the pair of objects (A, B) to

$$A\&B = (U \times V \overset{\alpha\&\beta}{\leftrightarrow} X + Y) \tag{7}$$

and the relation "$\alpha\&\beta$" is given, intuitively, by $(u,v)\ \alpha\&\beta\ \binom{x,0}{y,1}$ iff either $u\alpha x$ or $v\beta v$. Categorically, we take the coproduct map induced by the morphisms $A \times V \overset{\alpha\times V}{\rightarrowtail} U \times X \times V$ and $B \times U \overset{\beta\times U}{\rightarrowtail} V \times Y \times U$. The object $A\&B$ is a cartesian product, as can easily be checked and the object $\mathbf{1} = (1 \overset{e}{\leftrightarrow} 0)$ is the unit for the cartesian product and so a terminal object in **GC**.

The construction above can be dualised. Thus, if we take the coproduct map of $A \times Y \overset{\alpha\times Y}{\rightarrowtail} U \times X \times Y$ and $B \times X \overset{\beta\times X}{\rightarrowtail} V \times Y \times X$ that gives us

$$A \oplus B = (U + V \overset{\alpha\oplus\beta}{\leftrightarrow} X \times Y) \tag{8}$$

where the natural relation reads as $\binom{u,0}{v,1}\ \alpha \oplus \beta\,(x,y)$ iff either $u\alpha x$ or $v\beta y$. Clearly the endofunctor $\oplus$ defines coproducts in **GC**. The object $\mathbf{0} = (0 \overset{e}{\leftrightarrow} 1)$ is the unit for this construction.

A remark is that the intuitive "or" in the definitions of $\&$ and $\oplus$ is given by taking coproducts, while the one in the definition of $\sharp$ is a real "or".

2. Linear negation in GC

We define in **GC** a strong contravariant functor, which induces an involution on a subcategory of **GC**.

Recall that, given a symmetric monoidal closed category **C**, a *contravariant strong functor* $T: \mathbf{C} \to \mathbf{C}$ is a functor such that, for Ωevery pair of objects (A, B) in **C**, there is a family of maps $st_{(A,B)}: [A, B]_{\mathbf{C}} \to [TB, TA]_{\mathbf{C}}$ making the following diagrams commute.

$$\begin{array}{ccc} I & = & I \\ \downarrow & & \downarrow \\ [X, X] & \xrightarrow{st} & [TX, TX] \end{array} \qquad \begin{array}{ccc} [X, Y] \otimes [Y, Z] & \xrightarrow{M} & [X, Z] \\ \downarrow{\scriptstyle st \otimes st} & & \downarrow{\scriptstyle st} \\ [TY, TX] \otimes [TZ, TY] & \xrightarrow[M]{} & [TZ, TX] \end{array}$$

Definition 5. *Consider the internal hom bifunctor evaluated at* $\perp = (1 \overset{0}{\nleftrightarrow} 1)$ *in the second coordinate, that is consider* $[-, \perp]_{\mathbf{GC}}$*. This obviously defines a contravariant functor* $(-)^{\perp}: \mathbf{GC}^{op} \to \mathbf{GC}$.

More precisely to each object $(U \overset{\alpha}{\nleftrightarrow} X)$, the functor $(-)^{\perp}$ associates the object $(X \overset{\perp^{\alpha}}{\nleftrightarrow} U)$ where the relation "$\perp^{\alpha}$" intuitively says $x \perp^{\alpha} u$ iff whenever $u\alpha x$ then $\perp$. As "$\perp$" is the empty relation, it is never the case, so if we are dealing with decidable relations in **Sets**, $x \perp^{\alpha} u$ iff it is not the case that $u\alpha x$. Hence the name *linear negation.*

Proposition 3. *The functor* $(-)^{\perp}: \mathbf{GC}^{op} \to \mathbf{GC}$ *is a strong contravariant functor.*

Now we want to consider the subcategory "Dec GC", whose objects are the decidable objects in **GC**, that is decidable relations on **C**.

Definition 6. *By a decidable object on* **GC** *we mean that* $(U \overset{\alpha}{\nleftrightarrow} X)$ *is such that the canonical map from* $(U \overset{\alpha}{\nleftrightarrow} X)$ *to* $(U \overset{\perp\perp^{\alpha}}{\nleftrightarrow} X)$ *is an isomorphism.*

Our next proposition is to give names to structures. Following Barr, cf. [Bar] page 13, we say that a $*$-autonomous category comprises:

1. A symmetric monoidal closed category **C**.
2. A strong (contravariant) functor $*: \mathbf{C}^{op} \to \mathbf{C}$, thus the functor $*$ and a family of maps $st^*: [A, B]_{\mathbf{GC}} \to [B^*, A^*]_{\mathbf{GC}}$.
3. An equivalence $d = dA: A \to A^{**}$ such that the following diagram commutes

$$\begin{array}{ccc} [A, B]_{\mathbf{C}} & \xrightarrow{st} & [B^*, A^*]_{\mathbf{C}} \\ \| & & \downarrow{\scriptstyle st} \\ [A, B]_{\mathbf{C}} & \xrightarrow{[d^{-1}, d]} & [A^{**}, B^{**}]_{\mathbf{C}} \end{array}$$

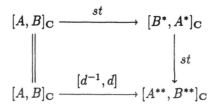

Proposition 4. *The subcategory* Dec **GC** *is an* $*$*-autonomous category, for* $* = (-)^\perp$.

□

It is easy to verify that the tensor product "$\oslash$" of **GC** distributes over the coproduct $\oplus$ and dually, that the bifunctor par "$\sharp$" distributes over the cartesian product "&". The following isomorphisms hold in **GC**

$$A \oslash (B \oplus C) \cong (A \oslash B) \oplus (A \oslash C) \quad \textit{and} \quad A\sharp(B\&C) \cong (A\sharp B)\&(A\sharp C).$$

Notice that 'multiplicatives' distribute over 'additives'. But there are also natural morphisms of the form $(A \oslash A') \oslash (B\sharp C) \xrightarrow{k} (A \oslash B)\sharp(A' \oslash C)$ and symmetrically $(A\sharp B)\oslash(C \oslash C') \xrightarrow{k'} (A \oslash C)\sharp(B \oslash C')$, which reduce to the morphisms $A \oslash (B\sharp C) \xrightarrow{i} (A \oslash B)\sharp C$ and $(A\sharp B) \oslash C' \xrightarrow{i'} A\sharp(B \oslash C')$.

3. Classical Linear Logic and GC

The category **GC** came into existence aiming to be a categorical model of Classical Linear Logic. It stems from a suggestion of Girard in Boulder 87, to whom I am very grateful, and to a great extent it fulfils its promise. In particular, the category **GC** is a very interesting model of Classical Linear Logic, since it does not collapse the units of "tensor" and "par" into a single object.

There are at least two equivalent presentations of Classical Linear Logic with slight variations in notation.

The original one, cf. [Gir] 1986 page 22, is very sleek and elegant, but it is hard to read off a categorical model from it.

Seely in [See] 1987, on the other hand, gives a presentation, which is geared towards the symmetries and thus more helpful. In his presentation a sequent has the form

$$G_1, G_2, \ldots, G_n \vdash D_1, D_2, \ldots, D_m,$$

where the commas on the left should be thought as some kind of conjunction and those on the right, some kind of disjunction.

A (propositional) Classical Linear Logic consists of formulae and sequents. Formulae are generated by the binary connectives $\otimes$, $\sharp$, &, $\oplus$ and $\multimap$ and by the unary operation $(-)^\perp$, from a set of constants including I, $\perp$, **1** and **0** and from variables.

The sequents are generated by the following rules, from initial sequents or axioms.

Axioms:

$$A \vdash A \quad \text{(identity)}$$
$$\vdash I \qquad \perp \vdash$$
$$\Gamma \vdash 1, \Delta \qquad \Gamma, \Delta \vdash \Delta$$
$$A \vdash A^{\perp\perp} \qquad A^{\perp\perp} \vdash A \quad \text{(negation)}$$

Structural Rules:

$$\frac{\Gamma \vdash \Delta}{\sigma\Gamma \vdash \tau\Delta} \ (\textit{permutation}) \qquad \frac{\Gamma \vdash A, \Delta \qquad A, \Gamma' \vdash \Delta'}{\Gamma, \Gamma' \vdash \Delta', \Delta} \ (\textit{cut})$$

Logical Rules:

$$\frac{\Gamma, A \vdash B, \Delta}{\Gamma, B^{\perp} \vdash A^{\perp}, \Delta} \quad (var)$$

Multiplicatives:

$$(unit_l)\frac{\Gamma \vdash \Delta}{\Gamma, I \vdash \Delta} \qquad (unit_r)\frac{\Gamma \vdash \Delta}{\Gamma \vdash \bot, \Delta}$$

$$(\otimes_l)\frac{\Gamma, A, B \vdash \Delta}{\Gamma, A \otimes B \vdash \Delta} \qquad (\otimes_r)\frac{\Gamma \vdash A, \Delta \qquad \Gamma' \vdash B, \Delta'}{\Gamma, \Gamma' \vdash A \otimes B, \Delta, \Delta'}$$

$$(\sharp_l)\frac{\Gamma, A \vdash \Delta \qquad \Gamma', B \vdash \Delta'}{\Gamma, \Gamma', A \sharp B \vdash \Delta, \Delta'} \qquad (\sharp_r)\frac{\Gamma \vdash A, B, \Delta}{\Gamma \vdash A \sharp B, \Delta}$$

$$(\multimap_l)\frac{\Gamma \vdash A, \Delta \qquad \Gamma', B \vdash \Delta'}{\Gamma, \Gamma', A \multimap B \vdash \Delta', \Delta} \qquad (\multimap_r)\frac{\Gamma, A \vdash B, \Delta}{\Gamma \vdash A \multimap B, \Delta}$$

Additives:

$$(\&_r)\frac{\Gamma \vdash A, \Delta \qquad \Gamma \vdash B, \Delta}{\Gamma \vdash A \& B, \Delta} \qquad (\&_l)\frac{\Gamma, A \vdash \Delta}{\Gamma, A \& B \vdash \Delta} \qquad \frac{\Gamma, B \vdash \Delta}{\Gamma, A \& B \vdash \Delta}$$

$$(\oplus_l)\frac{\Gamma, A \vdash \Delta \qquad \Gamma, B \vdash \Delta}{\Gamma, A \oplus B \vdash \Delta} \qquad (\oplus_r)\frac{\Gamma \vdash A, \Delta}{\Gamma \vdash A \oplus B, \Delta} \qquad \frac{\Gamma \vdash B, \Delta}{\Gamma \vdash A \oplus B, \Delta}$$

A remark on notation. Seely writes in his paper "$\times$" for "&", "+" for "$\oplus$", $\odot$ for $\sharp$ and $\neg$ for $(-)^{\perp}$, but we want to keep, as much as possible, the original notation from [Gir].

We would like **GC** with all the structure defined before, to be a categorical model of Classical Linear Logic. But it is clear that we do not have morphisms of the form $A^{\perp\perp} \to A$ for all objects A in **GC**. So, not all the objects are equivalent to their double linear negations, $A \cong A^{\perp\perp}$.

Thus, we omit from the system just presented the negation axiom $A^{\perp\perp} \vdash A$. It is interesting to note that, apart from the axiom, we only have to change the negation rule (var). Actually we transform the rule (var) into two rules, the rules (var_l) and (var_r) as below,

$$(var_r)\frac{\Gamma, A \vdash \Delta}{\Gamma \vdash A^{\perp}, \Delta} \qquad (var_l)\frac{\Gamma \vdash B, \Delta}{\Gamma, B^{\perp} \vdash \Delta}$$

Then only the rule (var_l) is satisfied in **GC**. That happens because the logic we are dealing with is really intuitionistic, at the bottom level. Thus, for example, in the model, the objects $(A \oslash B)^{\perp}$ and $(A^{\perp} \sharp B^{\perp})$ "look" exactly the same; they are both of the form $(X^V \times Y^U \leftrightarrow U \times V)$. But taking in consideration the relations, we only have a morphism in one direction: $(A^{\perp} \sharp B^{\perp}) \vdash (A \oslash B)^{\perp}$. This is just as it is the case in Intuitionistic Logic, thinking of "$\sharp$" as "*or*" and "$\oslash$" as "*and*". Looking at the rule (var_r) we see that it is not satisfied in general, since if we have a morphism $G \oslash A \to D$, that is equivalent to a morphism $G \to (A \multimap D)$. But from $A^{\perp} \sharp D$ we can prove intuitionistically, $A \multimap D$, but not conversely. Thus we have to leave out (var_r).

Let the new logical system, without $A^{\perp\perp} \vdash A$ and (var_r), be called $L.L_{\star}$ or sometimes just $L.L$.

Theorem. *The symmetric monoidal closed category* **GC**, *with bifunctors tensor product* $\oslash$; *"par"* $\sharp$; *internal hom* $[-,-]_{\mathbf{GC}}$; *cartesian product* &; *coproduct* $\oplus$ *and contravariant functor* $(-)^{\perp}$ *for linear negation, is a model of* $L.L_{\star}$. *Thus to each entailment* $\Gamma \vdash_{L.L_{\star}} A$ *corresponds the existence of a morphism in* **GC**,$(f,F)\colon |\Gamma| \to |A|$ *or* $|\Gamma| \vdash_{\mathbf{GC}} |A|$.

The proof is merely to check each of the axioms and rules.

Notice that rules $\oslash_l$ and $\sharp_r$ are fundamental, since they indicate how we should interpret the sequents in the category **GC**. They show that

$$G_1, G_2, \ldots, G_n \vdash D_1, D_2, \ldots, D_m$$

should be read as there exists a morphism in **GC**,

$$|G_n| \oslash \ldots \oslash |G_1| \to |D_k| \sharp \ldots \sharp |D_1|$$

Corollary. *The subcategory "Dec* **GC**" *is a model of Classical Linear Logic.*

4. Modalities in GC

Interpretations of the modal, or exponential, operators "!" and "?" of Linear Logic, in a categorical set-up, should correspond to a comonad and a monad, respectively, satisfying certain conditions. We discuss endofunctors on **GC**, which could play the role of the connective "!" in Classical Linear Logic. The first idea was, following the model of **DC**, to look at free monoids in **C** and see whether they would induce appropriate comonoids in **GC**.

Some definitions

Consider the monad given by the construction of free-monoids in **C**. Thus, suppose we are given an adjunction $F \dashv U\colon \mathbf{C} \to \mathbf{Mon\,C}$ and call S_0, or alternatively $*$, the composition $FU\colon \mathbf{C} \to \mathbf{C}$.

Intuitively X^* stands for "finite sequences of elements of X" and f^* for "f applied to each element of the sequence". Also S_0 has clearly a monad structure and it does not preserve products. Despite that S_0 induces an endofunctor $S\colon \mathbf{GC} \to \mathbf{GC}$ which has a natural structure as a comonad.

Definition 1. *The endofunctor S is given by $S(U \overset{\alpha}{\nleftrightarrow} X) = (U \overset{S\alpha}{\nleftrightarrow} X^*)$ on objects. The relation "$S\alpha$ " is given by the pullback below*

$$\begin{array}{ccc}
SA & \longrightarrow & A^* \\
\big\downarrow & & \big\downarrow{\scriptstyle \alpha^*} \\
U \times X^* & \xrightarrow[C_{U,X}]{} & (U \times X)^*
\end{array}$$

where the auxiliary maps $C_{(-,-)}$ are given by the serie of equivalences:

$$\begin{array}{c}
V \times Y \xrightarrow{\eta(V\times Y)} (V \times Y)^* \\ \hline
Y \xrightarrow{\overline{\eta}} (V \times Y)^{*V} \\ \hline
Y^* \xrightarrow{\overline{C}} (V \times Y)^{*V} \\ \hline
V \times Y^* \xrightarrow{C(V,Y)} (V \times Y)^*
\end{array}$$

The relation "$S\alpha$" reads intuitively as

"$u(S\alpha)(x_1, \ldots, x_k)$ iff $u\alpha x_1$ and ... and $u\alpha x_k$"

and S applied to a morphism (f, F) in **GC** is (f, F^*).

The functor $S\colon \mathbf{GC} \to \mathbf{GC}$ has a natural comonad structure, induced by the monad structure of S_0. Alas, this comonad has not the nice categorical properties it had with respect to the categories **DC**, due to the fact that the tensor product in **GC** is much more complicated than the one in **DC**. There are other very natural monads to consider in **C**, if **C** is cartesian closed.

Definition 2. *For each U in $\mathbf{C}$, let $T_U: \mathbf{C} \to \mathbf{C}$ be the endofunctor which takes $X \mapsto X^U$, $Y \mapsto Y^U$ and $f \in X^U$ to $fg \in Y^U$.*

That is clearly a monad in C with unit $(\eta_{T_U})_X: X \to X^U$ given by the "constant map", and multiplication $(\mu_{T_U})_X: X^{U \times U} \to X^U$, simply "precomposition with diagonal". We now turn our attention to defining a comonad in **GC** "induced by the monads T_U".

Definition 3. *Consider the functor T: **GC** $\to$ **GC** which takes $(U \overset{\alpha}{\nleftrightarrow} X)$ to the object $(U \overset{T\alpha}{\nleftrightarrow} X^U)$ and the object $(V \overset{\beta}{\nleftrightarrow} Y)$ to $(V \overset{T\beta}{\nleftrightarrow} Y^V)$. The relation "$T\alpha$" is defined by the pullback below*

$$\begin{array}{ccc} TA & \longrightarrow & A \\ \big\downarrow & & \big\downarrow \alpha \\ U \times X^U & \xrightarrow[(\pi_1, ev)]{} & U \times X \end{array}$$

Intuitively, it says that "$u(T\alpha)g$ iff $u\alpha g(u)$", where $g \in X^U$. To complete the definition say that T applied to a map $(f, F): A \to B$ is $(f, F(-)f): TA \to TB$.

It is easy to show that T has a comonad structure induced by the monad structures of the functors T_U.

Moreover, the monads T_U relate to S_0 in a very special way, described by Beck [69] as a "distributive law". More interesting is the fact that λ above, induces a new distributivity law Λ, this time between the comonads T and S in **GC**.

Proposition 1. *For each U and X in $\mathbf{C}$, we have a natural transformation*

$$\lambda_X: S_0 T_U X \to T_U S_0 X$$

corresponding to $(X^U)^* \to (X^*)^U$, satisfying the appropriate diagrams for a distributive law. There is also a natural transformation $\Lambda: TS \to ST$, at each object A, $\Lambda_A: TSA \to STA$ is given by $(1_U, (\lambda)_X)$ where $(\lambda)_X: (X^U)^* \to (X^*)^U$ is the distributive law in $\mathbf{C}$. This natural transformation Λ satisfies the conditions for a "distributive law of comonads".

Using Distributive Laws

It is widely known that the composition of monads is not always a monad, but given a distributive law λ, we can define the *composite monad* defined by λ, cf. [Beck]. We can also define the "lifting" of one of the monads and several relationships among the categories of algebras and Kleisli categories involved.

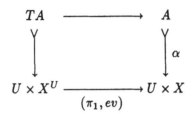

Definition 4. *The composite monad* $(T_U S_0)_\lambda$ *in* **C**, *takes* $X \mapsto (X^*)^U$. *Similarly, we have the composite comonad, induced by* Λ *and given by* (TS): **GC** $\to$ **GC**, *which takes* $(U \overset{\alpha}{\leftrightarrow} X)$ *to* $(U \overset{TS\alpha}{\leftrightarrow} (X^*)^U)$.

Besides the "composite monad", a distributive law provides a "lifting" of one of the monads to the category of algebras for the other monad.

Proposition 2. *The monad* $T_U: \mathbf{C} \to \mathbf{C}$ *lifts to the category of* S_0*-algebras. Dually, the comonad* T *in* **GC** *lifts to the category of* S*-coalgebras.*

We can describe another monad $\widetilde{T_U}: \mathbf{C}^{S_0} \to \mathbf{C}^{S_0}$.The endofunctor $\widetilde{T_U}$ applied to an S_0-algebra $(X, j: X^* \to X)$, gives ths S_0-algebra (X^U, h). The new structural map $h: (X^U)^* \to X^U$ is given by the composition $(X^U)^* \overset{\lambda}{\to} (X^*)^U \overset{(jx)^U}{\to} X^U$. The endofunctor $\widetilde{T}: (\mathbf{GC})^S \to (\mathbf{GC})^S$ has a comonad structure given by the monad structure of $\widetilde{T_U}$.

Clearly composing the functor T_U with the natural transformation $\eta: I \to S_0$ we have a monad morphism $\alpha: T_U \to T_U S_0$ which takes $X^U \mapsto (X^*)^U$. We also write β for $S_0 \to T_U S_0$, which is (η_{T_U}) applied to the functor S_0, thus taking $X^* \mapsto (X^*)^U$.

Similarly, there are comonad morphisms $\delta: TS \to T$ and $\kappa: TS \to S$, where for each A in **GC**, $\delta_A: TSA \to TA$ is given by $\delta_A = (1_U, \alpha_X)$ and $\kappa_A: TSA \to SA$ by $\kappa_A = (1_U, \beta_X)$.

Proposition 3. *The monad and comonad morphisms above induce:*

. maps in the categories of algebras, $\overline{\alpha}: \mathbf{C}^{T_U S_0} \to \mathbf{C}^{T_U}$ and $\overline{\beta}: \mathbf{C}^{T_U S_0} \to \mathbf{C}^{S_0}$;
. maps in the categories of coalgebras $\overline{\delta}: \mathbf{GC}^T \to \mathbf{GC}^{TS}$ and $\overline{\kappa}: \mathbf{GC}^S \to \mathbf{GC}^{TS}$.

Our next aim is to relate the categories $(\mathbf{C}^{S_0})^{\widetilde{T_U}}$ and $\mathbf{C}^{S_0 T_U}$ - dually $(\mathbf{GC}^S)^{\widetilde{T}}$ and $\mathbf{GC}^{ST}$.

Proposition 4. *There is an equivalence of categories of algebras,*

$$\Phi_0: (\mathbf{C}^{S_0})^{\widetilde{T_U}} \to \mathbf{C}^{S_0 T_U}$$

and respectively, of categories of coalgebras $\Phi: (\mathbf{GC}^S)^{\widetilde{T}} \to \mathbf{GC}^{ST}$.

The proofs of Propositions 2, 3 and 4 given in Beck's paper for algebras translate exactly to the coalgebras case, thus we omit them.

Clearly the monad S_0 does not lift to the category of T_U-algebras, since we cannot define the T_U-structural map for $S_0 X$ using λ, but it seems to lift to the T_U-Kleisli category, $\mathbf{C}_{T_U}$. Clearly we are talking about duality once more, but that is a more subtle case.

Proposition 5. *The monad S_0 "lifts" to the Kleisli category C_{T_U}. Dually, the comonad S lifts to the Kleisli category $\mathbf{GC_T}$.*

The endofunctor $\widetilde{S}_0 : C_{T_U} \to C_{T_U}$ takes X to X^* and a morphism in $\mathbf{C}_{T_U}$, $X \to Y$ - corresponding to a map $f: X \to T_U Y$ in $\mathbf{C}$ - to the composition $S_0(X) \stackrel{S_0(f_0)}{\to} S_0 T_U(Y) \stackrel{\lambda}{\to} T_U S_0(Y)$ which corresponds to $\widetilde{S_0}(f): S_0 X \to S_0 Y$ in $\mathbf{C_{T_U}}$. The functor $\widetilde{S}_0$ has a natural comonad structure.

This is a general consequence of the existence of the distributive law. The point here is that all the propositions above could be read off from Street's paper *"The formal theory of monads"*, by a clever 2-categorically minded reader. But we will not go into the 2-categorical aspects of the theory here.

Using the propositions above we can sum up the results of this section in the four "squares" below. Each square has three sides consisting of adjoint-pairs and the last side given by a natural morphism. In $\mathbf{C}$, relating algebras and Kleisli categories and in $\mathbf{GC}$ relating coalgebras and Kleisli categories.

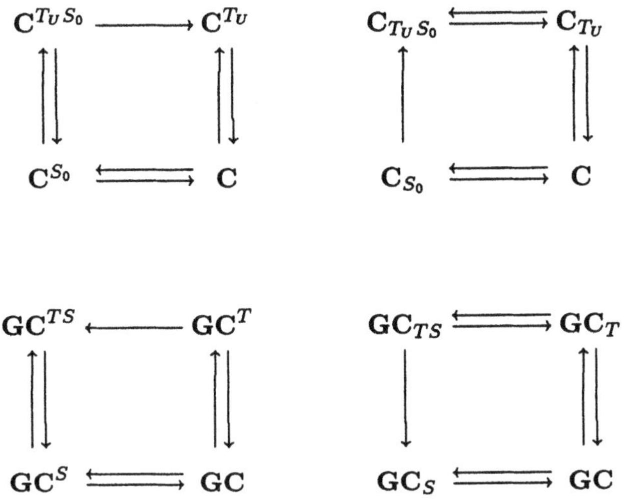

Note that if $\mathbf{C}$ has equalisers then, the two top squares are totally composed of adjoint-pairs, but we do not pursue it here, since it is not clear that equalisers in $\mathbf{C}$ would imply equalisers in $\mathbf{GC}$.

5. Properties of the comonads T and !

In the last section the endofunctor $T: \mathbf{GC} \to \mathbf{GC}$ was defined and shown to have a natural comonad structure. This endofunctor seems a reasonable candidate to represent the connective "!". For a start it has a "dual" endofunctor, to be denoted R, described in the next paragraph.

Definition 5. *The endofunctor R takes the object $(U \overset{\alpha}{\nleftrightarrow} X)$ to $(U^X \overset{R\alpha}{\nleftrightarrow} X)$, the object $(V \overset{\beta}{\nleftrightarrow} Y)$ to $(V^Y \overset{R\beta}{\nleftrightarrow} Y)$ and the map $(f, F): A \to B$ to the map $(f(-)F, F): RA \to RB$. The relation "$R\alpha$" is defined using the pullback of $A \overset{\alpha}{\rightarrowtail} U \times X$ along the evaluation morphism $U^X \times X \overset{(ev, \pi_1)}{\to} U \times X$ and intuitively it says "$g(R\alpha)x$ iff $g(x)\alpha x$".*

The functor R is the functor part of a monad, with unit $\eta_A: A \to RA$ given by the constant map $\eta_U: U \to U^X$ in the first coordinate and identity on X. Multiplication $\mu: R^2A \to RA$ is given by "restriction to the diagonal" $\mu_U: U^{X \times X} \to U^X$ in the first coordinate and identity on X.

We would like to have in **GC** results for "T" analogous to the ones for "!" in **DC**. For example, we would like the isomorphism relating categorical products to tensor products $!(A\&B) \cong !A \otimes !B$. But there is no obvious relationship between $T(A\&B) = (U \times V \nleftrightarrow (X+Y)^{U \times V})$ and $T(A) \oslash T(B) = (U \times V \nleftrightarrow X^{U \times V} \times Y^{U \times V})$. What we do have is a relation between the tensor products in **GC**.

Proposition 6. *There is a natural isomorphism in* **GC**, $T(A \otimes B) \cong TA \oslash TB$.

For a far more interesting result, ϐrecall that the T-Kleisli category $\mathbf{GC}_T$ has as objects the objects of **GC** but as maps from A to B, maps in **GC** from TA to B.

Proposition 7. *The maps from A to B in the T-Kleisli category $\mathbf{GC}_T$, are in 1-1 correspondence with maps from A to B in the category* **DC**.

Proof: We want to check

$$\mathrm{Hom}_{\mathbf{GC}_T}(A, B) = \mathrm{Hom}_{\mathbf{GC}}(TA, B) \approx \mathrm{Hom}_{\mathbf{DC}}(A, B).$$

The second equivalence holds, since a map $(f, F): TA \to B$ in **GC**, corresponds to $f: U \to V$ and $F: Y \to X^U$, satisfying the condition

$$(U \times F)^{-1}(\alpha) \leq (f \times Y)^{-1}(\beta). \tag{1}$$

That corresponds to $f: U \to V$ and, by exponential transpose, to $\overline{F}: U \times Y \to X$, satisfying a corresponding condition $(*)$ that is a map $(f, \overline{F}): A \to B$ in **DC**.

Since objects are the same in both categories **GC** and **DC**, Proposition 7 implies that there is an equivalence between categories $\mathbf{GC}_T$ and **DC**.

The comonad "!"

Consider now the composite comonad TS defined in the last section, with the difference that now S_0 denotes free *commutative* monoids in **C**. Thus $S(U \overset{\alpha}{\nleftrightarrow} X) = U \overset{S\alpha}{\nleftrightarrow} X^*)$ and S takes a morphism $(f, F): A \to B$ to $(f, F^*): SA \to SB$. Let the comonad TS be called "!". The functor part of "!" acts on objects as $!(U \overset{\alpha}{\nleftrightarrow} X) = (U \overset{!\alpha}{\nleftrightarrow} (X^*)^U)$ and on maps $!(f, F) = (f, F^*(-)f)$.

As we have shown "!" is the functor part of the composite comonad

$$(!, \varepsilon_!: !A \to A, \delta_!: !A \to !!A)$$

and we can consider the categories $\mathbf{GC}^!$ of !-coalgebras and $\mathbf{GC}_!$, the !-Kleisli category. Moreover, the objects "$!A$" have a natural comonoid-like structure, with respect to "$\oslash$".

Proposition 8. *There are natural morphisms in* **GC** *as follows*

1. *From the object $!A$ to I, given by the terminal map on U and the natural map $1 \to (X^*)^U$.*
2. *From $!A$ to $!A \oslash !A$, which is given by the diagonal map in* **C**, *$\Delta: U \to U \times U$ and the map $\theta: (X^*)^{U\times U} \times (X^*)^{U\times U} \to (X^*)^U$.*

The map θ is given, intuitively, by taking a pair of functions (ϕ, σ), each of them of the form $U \times U \to X^*$, to the product map $\phi \times \sigma$ precomposing it with the diagonal in U and post-composing it with the multiplication on X^*, as follows,

$$U \xrightarrow{\Delta} U \times U \xrightarrow{<\phi,\sigma>} X^* \times X^* \xrightarrow{\mu_*} X^*.$$

Proposition 9. *We have the following natural isomorphisms for each A and B in* **GC**, $!(A\&B) \cong !A \oslash !B$.

Proof: Look at the following series of equivalences :

$$!(A\&B) = TS(A\&B) \cong T(SA \otimes SB) \cong TSA \oslash TSB = !A \oslash !B.$$

Proposition 10. *The Kleisli category* $\mathbf{GC}_!$ *is cartesian closed.*

That is an easy corollary of the above, since

$$\mathrm{Hom}_{\mathbf{GC}_!}(A \oslash B, C) \cong \mathrm{Hom}_{\mathbf{GC}}(TS(A \oslash B), C) \cong \mathrm{Hom}_{\mathbf{GC}}(!A \oslash !B, C) \cong$$

$$\cong \mathrm{Hom}_{\mathbf{GC}}(!A, [!B, C]_{\mathbf{GC}}) \cong \mathrm{Hom}_{\mathbf{GC}_!}(A, [B, C]_{\mathbf{GC}_!}).$$

Corollary. *The morphisms from A to B in the category* $\mathbf{GC}_!$, *correspond naturally to morphisms in the category* $\mathbf{DC_S}$ *from A to B.*

6. Linear Logic with modalities

The composite comonad "!" defined in the last section satisfies the rules for the modality "!", but we would like also a monad "?" satisfying the rules for the dual connective, called by Girard "why not ?".

We recall the rules for the modality "!". These are:

$$\text{I.}\ \frac{\Gamma, A \vdash B}{\Gamma, !A \vdash B}\ (\textit{dereliction}) \qquad \text{II.}\ \frac{\Gamma \vdash B}{\Gamma, !A \vdash B}\ (\textit{weakening})$$

$$\text{III.}\ \frac{\Gamma, !A, !A \vdash B}{\Gamma, !A \vdash B}\ (\textit{contraction}) \qquad \text{IV.}\ \frac{!\Gamma \vdash A}{!\Gamma \vdash !A}\ (!)$$

Theorem. *The category* **GC** *with the composite comonad "!" defined in Section 5 is a model for Linear Logic enriched with modality "!".*

The proof is again to check the rules and it is straightforward. Moreover, using R the dual endofunctor to T we can get a monad to model ?. We just have to compose R with the monad in **GC** induced by the monad $U \mapsto U^*$ in **C**. The composite monad satisfies all necessary conditions.

Concluding remarks

To conclude it is perhaps worth mentioning some of the several questions that the work on the categories **DC** and **GC** prompts, apart from the ones already mentioned in the introduction.

1. Is there an interesting connection between the categorical models **DC** and **GC** and Girard's new work on the Geometry of Interactions ?
2. Since we think of maps in **DC** and **GC** as "linear morphisms", in opposition to the more usual morphisms in the Kleisli categories, can we characterize bilinear maps in this context ? There is some interesting work of Kock, but the obvious approach does not work, due to the fact that the comonads "!", or rather, their functor parts, are not strong functors.
3. We have shunned away from the 2-categorical aspects of everything discussed previously, but that is not, probably, the best policy, as was indicated by the need of distributive laws. More to the point, there is a very interesting question of using "spans" instead of relations in the construction of **DC** and **GC**, which was suggested by Aurelio Carboni.
4. We have worked only with commutative versions of the connectives, that is with symmetric tensor products, "par" bifunctors etc. There is a interesting case to look at, if this commutativity condition is dropped. On those lines there seems to be some connection with Joyal and Street's work on braided monoidal categories. In particular there is also a preprint by D. N. Yetter on "Quantales and (Non-commutative) Linear Logic".
5. Finally, there is the very promising, but as yet very vague idea of connecting Linear Logic with Concurrency and Parallelism. The idea being that Linear Logic may provide an *integrated logic*, where one would hope to model computational processes

in a less *ad hoc* fashion than it has been up to now. In particular, Petri Nets have been proposed as a model for Linear Logic, cf. [Gir] 1987.

References

[Bar] 1979 M. BARR
$*$-Autonomous Categories, LNM 752, Springer-Verlag, 1979.

[Bec] 1969 J. BECK
Distributive Laws, in Seminar on Triples and Categorical Homology Theory, LNM 80, Springer-Verlag, 119-140.

[Fox] 1976 T. FOX
Coalgebras and Cartesian Categories, Comm. Alg. **(7)** 4, 665-667.

[Gir] J-Y. GIRARD
1986 *Linear Logic*, Theoretical Computer Science **46**, 1-102.
1987 *Towards a Geometry of Interactions*, to appear in Proc. A.M.S Conference on Categories in Computer Science and Logic, Boulder, 1987

[G/L] 1986 J-Y. GIRARD and Y. LAFONT
Linear Logic and Lazy Computation, Proc. of TAPSOFT'87, Pisa.

[Göd] K. GÖDEL
1958 *Über eine bisher noch nicht benützte Erweiterung des finiten Standpunktes*, Dialectica **12**, 280-287.
1980 *On a Hitherto Unexploited Extension of the Finitary Standpoint*, Journal of Philosophical Logic **9**, 133-142.

[Koc] A. KOCK
1970 *Monads on Symmetric Monoidal Closed Categories*, Arch. Math **(21)**.
1971 *Bilinearity and Cartesian Closed Monads*, Math. Scand. **(29)**.

[M/R] 1977 M. MAKKAI and G. REYES
First-order Categorical Logic, LNM 611, Springer-Verlag.

[DC] 1987 V. C. V. de PAIVA
The Dialectica Categories, to appear in Proc. A.M.S Conference on Categories in Computer Science and Logic, Boulder, 1987.

[See] R. A. G. SEELY
1984 *Locally Cartesian Closed Categories and Type Theory*, Math. Proc. Cambridge Philosophical Society **95**, 33-48.
1987 *Linear Logic, *-autonomous categories and cofree coalgebras* to appear in Proc. A.M.S Conference on Categories in Computer Science and Logic, 1987.

[Str] 1972 R. STREET
The formal theory of monads, Journ. of Pure & Applied Alg. **2**, 149-1[illegible]

A Final Coalgebra Theorem

Peter Aczel and Nax Mendler
Manchester University

Abstract

We prove that every set-based functor on the category of classes has a final coalgebra. This result strengthens the final coalgebra theorem announced in the book "Non-well-founded Sets", by the first author.

1 Introduction

The theorem of this note is an improvement of a result that was first stated, but not proved in its full generality, in [NWFS]. The original inspiration for the result came from [SCCS]. There a language of infinitary expressions called, *agents*, is made into a labelled transition system via an operational semantics. Different agents can exhibit the same operational behaviour. A notion of *bisimulation relation* between agents is used to define the maximal bisimulation equivalence relation which captures formally the intuitive behavioural equivalence relation. The quotient of the class of agents with respect to this equivalence relation can itself be made into a labelled transition system which can be viewed as a mathematical interpretation of the language of agents. If the labels for a labelled transition system are taken from a set *Act* then the system may be viewed as a coalgebra for the functor $pow(Act \times -)$. Here *pow* is the natural covariant power class functor which associates with each class X the class $powX$ of its subsets. It turns out that the quotient of the coalgebra of agents by the maximal bisimulation relation can be characterised up to isomorphism as a final coalgebra for the functor. So this basic construction of Milner's is providing a proof of the existence of a final coalgebra for the functor.

The final coalgebra theorem is a general result about the existence of final coalgebras for a very general collection of functors on the category of classes. The functor *pow* itself is essentially a special case of $pow(Act \times -)$ where *Act* is a singleton set. The final coalgebras for *pow* give the full models for $ZFC^- + AFA$, a version of axiomatic set theory in which the Foundation Axiom, FA, is replaced by a dual Anti-Foundation Axiom, AFA, which expresses the existence of many non-well-founded sets.

The notion of a final coalgebra for a functor can be used to characterise up to isomorphism many other interesting mathematical structures. In this respect it can play the same kind of role as the dual notion of an initial algebra for a functor. For example associated with each single sorted signature, Ω, is a functor, F_Ω, on the category of classes whose initial algebras are the algebras isomorphic to the algebra of terms of the signature. Terms can be viewed as well-founded trees and if the well-foundedness restriction is dropped from the trees then the resulting algebra of possibly non-well-founded trees is essentially a final coalgebra for F_Ω. In this note we are not concerned to explore such instances of the final coalgebra theorem but rather give a proof of the result itself.

The reader may wonder why the category of classes is being used here rather than the more familiar category of sets. The reader may even be concerned about the foundational correctness of the use of such a superlarge category. The point is that the powerset functor on the category of sets cannot have a final coalgebra for cardinality reasons. And it is functors involving *pow* that provide the most interesting instances of final coalgebras, such as in connection with SCCS and set theory. The possible foundational problems concerning the category of classes can be overcome in a reasonably standard way, but it seems inappropriate to focus on those problems in this note. Use of the category of classes could be avoided by introducing cardinality restrictions at suitable points. But such a presentation would surely obscure the ideas.

A key issue in generalising Milner's quotient construction was the problem of formulating a suitable generalisation, to a coalgebra for a functor, of the notions of bisimulation and bisimulation equivalence on a labelled transition system. The notions we use in this note are that of pre-congruence and congruence, as defined in section 4. At the time [NWFS] was written the author of that work had in mind the slightly different notions of bisimulation and bisimulation equivalence on a coalgebra as defined in section 6 of this note. Although every bisimulation is a pre-congruence, a suitable converse cannot be proved without making an additional assumption about the functor. In [NWFS] the additional assumption used was that the functor preserves weak pullbacks. The relationship between bisimulations and pre-congruences is worked out in section 6.

The Final Coalgebra Theorem is formulated in the next section along with the necessary definitions and some key lemmas, including the small subcoalgebra lemma and the Main Lemma. The small subcoalgebra lemma is proved in section 3, while section 4 is concerned with the general construction of a maximal congruence. The proof of the final Coalgebra Theorem is completed in section 5 with a proof of the Main Lemma.

The final section of this note is concerned with a generalisation of the final coalgebra theorem obtained by replacing the category of classes by a superlarge category satisfying some fairly weak assumptions true for the category of classes.

2 The Theorem

A *coalgebra* for an endo-functor Φ on a category consists of (A, α) where A is an object of the category and $\alpha : A \rightarrow \Phi A$. A *homomorphism* $\pi : (A, \alpha) \rightarrow (B, \beta)$ between two coalgebras (A, α) and (B, β) is a map $\pi : A \rightarrow B$ such that $\beta \circ \pi = (\Phi\pi) \circ \alpha$. The coalgebras and homomorphisms form a category. So we can formulate the notion of a *final coalgebra*. This is a coalgebra (A, α) such that for any coalgebra (B, β) there is a unique homomorphism $(B, \beta) \rightarrow (A, \alpha)$.

The superlarge category of classes is the category whose objects are classes and whose maps are all class functions between classes. An endo-functor Φ on this category is called *set-based* if for each class A and each $a \in \Phi A$ there is a set $A_0 \subseteq A$ and $a_0 \in \Phi A_0$ such that $a = \Phi\iota_{A_0,A} a_0$, where $\iota_{A_0,A}$ is the inclusion map $A_0 \hookrightarrow A$. Our aim is to give a proof of the following result.

Theorem 2.1 (The Final Coalgebra Theorem.) *Every set-based functor has a final coalgebra.*

This is an improvement on the corresponding result in [NWFS]. There the functor was required to be standard and preserve weak pullbacks. A functor Φ is *standard* if and only if it is set-based and preserves inclusion maps; i.e. if $A \subseteq B$ then $\Phi A \subseteq \Phi B$ and $\Phi\iota_{A,B} = \iota_{\Phi A, \Phi B}$.

So the present theorem no longer requires the assumptions that the functor preserve inclusion maps and weak pullbacks. The assumption that the functor is set-based is not significantly more general than the stronger assumption that the functor is standard. Nevertheless it has the advantage of being category theoretic. In fact, it is not hard to show that a functor is set-based if and only if it is almost naturally isomorphic (i.e., naturally isomorphic on non-empty classes) to a standard functor. The remaining assumption that the functor is set-based will only play an explicit role in the proof of the following key lemma.

A coalgebra (A_0, α_0) is a *subcoalgebra* of (A, α) if $A_0 \subseteq A$ and the inclusion map $\iota_{A_0,A}$ is a homomorphism $\iota_{A_0,A} : (A_0, \alpha) \hookrightarrow (A, \alpha)$. The subcoalgebra is *small* if A_0 is a set.

Lemma 2.2 (The Small Subcoalgebra Lemma.) *If (A, α) is a coalgebra and X is a subset of A then $X \subseteq A_0$ for some small subcoalgebra (A_0, α_0) of (A, α).*

As in [NWFS] we call a coalgebra (A,α) *complete* [*weakly complete; strongly extensional*] if for each small coalgebra (A_0,α_0) there is a unique [at least one; at most one] homomorphism $(A_0,\alpha_0) \to (A,\alpha)$. We can use the same argument as sketched in [NWFS] to show that:

Lemma 2.3 *Every complete coalgebra is final.*

In brief the small subcoalgebra lemma guarantees that any coalgebra (A,α) is a colimit of its small subcoalgebras, so that by universality we get the existence of a unique homomorphism from (A,α) to a complete coalgebra. As in [NWFS] we can prove the result:

Lemma 2.4 *There is a weakly complete coalgebra.*

It is the coproduct of all the small coalgebras. The main task of this report is to prove:

Lemma 2.5 (Main Lemma) *For each coalgebra (A,α) there is a strongly extensional coalgebra $(\overline{A},\overline{\alpha})$ and a surjective homomorphism $(A,\alpha) \to (\overline{A},\overline{\alpha})$.*

This is essentially lemma 7.7 of [NWFS] except that in the book the functor was assumed to be a standard functor that preserved weak pullbacks, while here we only assume that the functor is set-based.

If this lemma is applied to a weakly complete coalgebra (A,α) then the strongly extensional $(\overline{A},\overline{\alpha})$ will still be weakly complete and hence will be complete and therefore final, concluding the proof of the final coalgebra theorem.

3 Proof of the small subcoalgebra lemma.

Let (A,α) be a coalgebra for the set based functor Φ. Let $\Theta : pow A \to pow(\Phi A)$ be given by

$$\Theta Y = \{\Phi\iota_{Y,A}a \mid a \in \Phi Y\}$$

for all $Y \in pow A$, where $pow A$ is the class of all subsets of A.

Lemma 3.1 $Y_1 \subseteq Y_2 \in pow A \implies \Theta Y_1 \subseteq \Theta Y_2$.

Proof: If $Y_1 \subseteq Y_2 \in pow A$ then

$$\iota_{Y_1,A} = \iota_{Y_2,A} \circ \iota_{Y_1,Y_2}$$

so that

$$\Phi\iota_{Y_1,A} = \Phi\iota_{Y_2,A} \circ \Phi\iota_{Y_1,Y_2}$$

and hence $\Theta Y_1 \subseteq \Theta Y_2$. □

Lemma 3.2 $a \in \Phi A \implies a \in \Theta Y$ *for some* $Y \in pow A$.

Proof: As Φ is set based, if $a \in \Phi A$ then $a = \Phi\iota_{Y,A}a_0$, for some $Y \in pow A$ and some $a_0 \in \Phi Y$, so that $a \in \Theta Y$. □

Lemma 3.3 $Y \in pow A \implies Y \subseteq \alpha^{-1}\Theta Y'$ *for some* $Y' \in pow A$.

Proof: For each $a \in A$, $\alpha a \in \Phi A$ and hence by lemma 3.2 there is $Y_a \in pow A$ such that $a \in \alpha^{-1}\Theta Y_a$. Now if $Y \in pow A$ let $Y' = \bigcup\{Y_a \mid a \in Y\}$. Then by lemma 3.1 $Y \subseteq \alpha^{-1}\Theta Y'$. □

Lemma 3.4 $X \in pow A \implies A_0 \subseteq \alpha^{-1}\Theta A_0$ *for some* $A_0 \in pow A$ *such that* $X \subseteq A_0$.

Proof: Given $X \in powA$ define $X_n \in powA$ for $n \in \mathrm{N}$ by

$$\begin{cases} X_0 &= X, \\ X_{n+1} &= X'_n \quad \text{for } n \in \mathrm{N}. \end{cases}$$

Let $A_0 = \bigcup\{X_n \mid n \in \mathrm{N}\}$. Then $A_0 \in powA$ and by lemmas 3.3 and 3.1, for each $n \in \mathrm{N}$

$$X_n \subseteq \alpha^{-1}\Theta X_{n+1} \subseteq \alpha^{-1}\Theta A_0.$$

Hence $A_0 \subseteq \alpha^{-1}\Theta A_0$. □

To prove the small subcoalgebra lemma it only remains to use the previous lemma and define a suitable map $\alpha_0 : A_0 \to \Phi A_0$. If $A_0 = \emptyset$ let $\alpha_0 : A_0 \to \Phi A_0$ be the empty map. If $A_0 \neq \emptyset$ then $\Phi\iota_{A_0,A}$ is injective. This is because $\iota_{A_0,A}$ is injective and hence there is a map $f : A \to A_0$ such that $f \circ \iota_{A_0,A}$ is the identity map on A_0 so that $(\Phi f) \circ (\Phi\iota_{A_0,A})$ is the identity map on ΦA_0. So we can define $\alpha_0 : A_0 \to \Phi A_0$ by

$$\alpha_0 a = (\Phi\iota_{A_0,A})^{-1}(\alpha a)$$

for $a \in A_0$. In either case (A_0, α_0) is a small subcoalgebra of (A, α), with $X \subseteq A_0$.

4 The Maximal Congruence on a Coalgebra.

In order to prove the main lemma we will need to form a quotient of a coalgebra (A, α) with respect to a suitable equivalence relation on A. If $q : A \to B$ then let

$$E(q) = \{(a_1, a_2) \in A \times A \mid qa_1 = qa_2\}.$$

Then $E(q)$ is an equivalence relation on A. If q is surjective and $R = E(q)$ then we say that $q : A \to B$ is a *quotient of A with respect to R*. By assuming a global form of the axiom of choice we can be sure that for each equivalence relation R on a class A there is a quotient of A with respect to R. In this section we will formulate the notion of a congruence on a coalgebra and prove the existence of a maximal congruence. If $R \subseteq A \times A$ and $A_0 \subseteq A$ let $R|A_0 = R \cap (A_0 \times A_0)$.

Lemma 4.1 *Let (A, α) be a coalgebra and let $q_i : A \to B_i$ for $i = 1, 2$. Then*

1. *If $A \neq \emptyset$, $E(q_1) \subseteq E(q_2) \iff \exists q : B_1 \to B_2\, [\, q \circ q_1 = q_2 \,]$*
2. *$E(q_1) \subseteq E(q_2) \implies E((\Phi q_1) \circ \alpha) \subseteq E((\Phi q_2) \circ \alpha)$*
3. *$E(q_1)|A_0 \subseteq E(q_2) \implies E((\Phi q_1) \circ \alpha)|A_0 \subseteq E((\Phi q_2) \circ \alpha)$, if (A_0, α_0) is a subcoalgebra of (A, α).*

Proof: Let (A, α) be a coalgebra and let $q_i : A \to B_i$ for $i = 1, 2$.

1. This is easy.
2. Trivial, if $A = \emptyset$, Otherwise, if $E(q_1) \subseteq E(q_2)$ then by 1 there is a map $q : B_1 \to B_2$ such that $q_2 = q \circ q_1$. Hence $\Phi q_2 = (\Phi q) \circ (\Phi q_1)$ so that $(\Phi q_2) \circ \alpha = (\Phi q) \circ ((\Phi q_1) \circ \alpha)$. By 1 again $E((\Phi q_1) \circ \alpha) \subseteq E((\Phi q_2) \circ \alpha)$.
3. Trivial, if $A = \emptyset$. Otherwise, if (A_0, α_0) is a subcoalgebra of (A, α) then $\alpha \circ \iota_{A_0,A} = (\Phi\iota_{A_0,A}) \circ \alpha_0$, so that

$$\begin{aligned} E((\Phi q_i) \circ \alpha)|A_0 &= E((\Phi q_i) \circ \alpha \circ \iota_{A_0,A}) \\ &= E((\Phi q_i) \circ (\Phi\iota_{A_0,A}) \circ \alpha_0) \\ &= E(\Phi(q_i|A_0) \circ \alpha_0). \end{aligned}$$

Hence we may apply 2 with (A_0, α_0) and $q_i|A_0$ replacing (A, α) and q_i. □

If (A, α) is a coalgebra and R is a relation on A then we may define

$$R^{\Phi} = E((\Phi q) \circ \alpha)$$

where $q : A \to B$ is a quotient of A with respect to R^e, the equivalence relation on A generated by R. By lemma 4.1.(2) R^{Φ} is independent of the particular choice of the quotient q. For if q_1, q_2 are both quotients of A with respect to R^e then $E(q_1) = R^e = E(q_2)$ so that by 4.1.(2) $E((\Phi q_1) \circ \alpha) = E((\Phi q_2) \circ \alpha)$. Note that R^{Φ} is itself an equivalence relation on A. Also note that by 4.1.(2), for all relations R, S on A

$$R \subseteq S \implies R^{\Phi} \subseteq S^{\Phi}.$$

Call a relation R on A a *pre-congruence* on a coalgebra (A, α) if $R \subseteq R^{\Phi}$ and a *congruence* on (A, α) if also it is an equivalence relation on A. Note that R is a pre-congruence if and only if R^e is a congruence.

Lemma 4.2 *Let (A_0, α_0) be a subcoalgebra of a coalgebra (A, α). If R is a precongruence on (A, α) then so is $R^e|A_0$.*

Proof: By 4.1.(3) if R is a relation on A then

$$R^{\Phi}|A_0 \subseteq (R^e|A_0)^{\Phi},$$

so that if also $R \subseteq R^{\Phi}$ then

$$R^e|A_0 \subseteq R^{\Phi}|A_0 \subseteq (R^e|A_0)^{\Phi},$$

so that $R^e|A_0$ is a precongruence. □

Lemma 4.3 *Any coalgebra has a (necessarily unique) maximal congruence.*

Proof: Let S be the union of all the small pre-congruences on a coalgebra (A, α). Then S is itself a precongruence. For if $a_1 S a_2$ then $a_1 R a_2$ for some small pre-congruence R and hence $a_1 R^{\Phi} a_2$ so that $a_1 S^{\Phi} a_2$, as $R \subseteq S$. S is a maximal pre-congruence on (A, α). For if R is a pre-congruence on (A, α) and $a_1 R a_2$ then, by the small subcoalgebra lemma, $a_1, a_2 \in A_0$ for some small subcoalgebra (A_0, α_0) of (A, α). Then $a_1(R^e|A_0)a_2$. But by lemma 4.2. $R^e|A_0$ is a pre-congruence on (A, α) so that it is a subset of S. So $a_1 S a_2$. Finally S is a congruence and hence a maximal congruence on (A, α) because, as S^e is a pre-congruence, $S^e \subseteq S$, so that S is an equivalence relation on A. □

5 Proof of the Main Lemma

The main lemma will be an immediate consequence of the three lemmas of this section.

Lemma 5.1 *Let R be a congruence on the coalgebra (A, α). Let $q_R : A \to A_R$ be a quotient of A with respect to R. Then there is a unique map $\alpha_R : A_R \to \Phi A_R$ such that q_R is a surjective homomorphism $q_R : (A, \alpha) \to (A_R, \alpha_R)$.*

Proof: For q_R to be a homomorphism α_R must be chosen so that $\alpha_R(q_R a) = (\Phi q_R)(\alpha a)$ for $a \in A$. The unique existence of α_R follows from the assumptions that q_R is surjective and that R is a congruence on (A, α). □

Call a coalgebra (A, α) *s-extensional* if for every congruence R on (A, α) if $a_1 R a_2$ then $a_1 = a_2$.

Lemma 5.2 *Every s-extensional coalgebra is strongly extensional.*

Proof:. Let (A, α) be s-extensional and let $\tau_1, \tau_2 : (B, \beta) \to (A, \alpha)$. We must show that $\tau_1 = \tau_2$. By the assumption that (A, α) is s-extensional it suffices to show that the equivalence relation R on A generated by $\{(\tau_1 b, \tau_2 b) \mid b \in B\}$ is a congruence on (A, α), as then if $b \in B$ then $(\tau_1 b)R(\tau_2 b)$ and hence $\tau_1 b = \tau_2 b$.

As $\alpha \circ \tau_i = (\Phi \tau_i) \circ \beta$ we get that $(\Phi q_R) \circ \alpha \circ \tau_i = \Phi(q_R \circ \tau_i) \circ \beta$ for $i = 1, 2$. Hence, as $q_R \circ \tau_1 = q_R \circ \tau_2$,

$$(\Phi q_R) \circ \alpha \circ \tau_1 = (\Phi q_R) \circ \alpha \circ \tau_2$$

so that $R \subseteq R^{\Phi}$. □

Lemma 5.3 *If R is the maximal congruence on (A, α) then (A_R, α_R) is s-extensional.*

Proof: Let S be a congruence on (A_R, α_R) and let

$$S' = \{(a_1, a_2) \in A \times A \mid (q_R a_1) S (q_R a_2)\}.$$

S' is an equivalence relation on A and $q_S \circ q_R : A \to (A_R)_S$ is a quotient of A with respect to S'. Also

$$\begin{aligned}(\Phi q_S) \circ \alpha_R \circ q_R &= (\Phi q_S) \circ (\Phi q_R) \circ \alpha \\ &= \Phi(q_S \circ q_R) \circ \alpha.\end{aligned}$$

Hence, as $S \subseteq E((\Phi q_S) \circ \alpha_R)$,

$$\begin{aligned}S' &\subseteq E((\Phi q_S) \circ \alpha_R \circ q_R) \\ &= E(\Phi(q_S \circ q_R) \circ \alpha) \\ &= S'^{\Phi};\end{aligned}$$

i.e. S' is a congruence on (A, α). Hence, as R is the maximal congruence on (A, α), $S' \subseteq R$. So if $x_1 = q_R a_1$ and $x_2 = q_R a_2$ then

$$\begin{aligned}x_1 S x_2 &\Longrightarrow a_1 S' a_2 \\ &\Longrightarrow a_1 R a_2 \\ &\Longrightarrow q_R a_1 = q_R a_2 \\ &\Longrightarrow x_1 = x_2.\end{aligned}$$

Thus (A_R, α_R) is s-extensional. □

To give an example, let's return to the initial example of the functor arising from a labelled transition system. For a fixed set *Act*, let the class functor Φ be defined by:

$$\begin{aligned}\Phi A &= pow(Act \times A) \\ (\Phi f)(A_0) &= \{(l, fa) | (l, a) \in A_0\}, \quad \text{for } f : A \to B,\ A_0 \in \Phi A.\end{aligned}$$

Φ is set-based. Define a weakly complete coalgebra (A, α) by

$$\begin{aligned}A &= \{(A_0, \alpha_0, a) | A_0 \text{ is a set, } \alpha_0 : A_0 \to \Phi A_0 \text{ and } a \in A_0\} \\ \alpha(A_0, \alpha_0, a) &= \{(l, (A_0, \alpha_0, a')) | (l, a') \in \alpha_0(a)\}.\end{aligned}$$

So for a small (A_0, α_0), the injection map $a \mapsto (A_0, \alpha_0, a)$ is a homomorphism into (A, α). An equivalence relation $R \subseteq A \times A$ is a congruence iff $R \subseteq R'$, where R' is defined as:

$$\begin{aligned}aR'b \Leftrightarrow\ & \forall (l, a') \in \alpha(a).\ \exists (l', b') \in \alpha(b).\ l = l' \wedge a'Rb' \\ & \wedge \\ & \forall (l, b') \in \alpha(b).\ \exists (l', a') \in \alpha(a).\ l = l' \wedge a'Rb'\end{aligned}$$

The maximal congruence S is the greatest fixedpoint of this. Let $q : A \to A_q$ be a quotient of A with respect to S. Define $\alpha_S : A_S \to \Phi A_S$ by

$$\alpha_S(q\, a) = \{(l, q\, a') | (l, a') \in \alpha\, a\}\ (= (\Phi\, q)(\alpha\, a)).$$

Then (A_S, α_S) is final.

6 Bisimulations

Call a relation R on A a *bisimulation* on the coalgebra (A, α) if there is a map $\beta : R \to \Phi R$ such that the projections ρ_1, ρ_2 of R on A are homomorphisms $(R, \beta) \to (A, \alpha)$.

Proposition 6.1 *Any bisimulation is a pre-congruence.*

Proof: Let R be a bisimulation on (A, α). So there is a map $\beta : R \to \Phi R$ such that ρ_1, ρ_2 are homomorphisms $(R, \beta) \to (A, \alpha)$. Let $q : A \to A_R$ be a quotient of A with respect to R^e. Then $q \circ \rho_1 = q \circ \rho_2$ so that

$$(\Phi q) \circ (\Phi \rho_1) = (\Phi q) \circ (\Phi \rho_2).$$

Hence if xRy then

$$\begin{aligned}(\Phi q)(\alpha x) &= (\Phi q)(\alpha(\rho_1(x,y)))\\ &= (\Phi q)((\Phi\rho_1)(\beta(x,y)))\\ &= (\Phi q)((\Phi\rho_2)(\beta(x,y)))\\ &= (\Phi q)(\alpha(\rho_2(x,y)))\\ &= (\Phi q)(\alpha y),\end{aligned}$$

so that $xR^{\Phi}y$. Thus R is a pre-congruence. □

A commutative square

$$\begin{array}{ccc} X_0 & \xrightarrow{p_1} & X_1\\ p_2\downarrow & & \downarrow q_1\\ X_2 & \xrightarrow{q_2} & Y\end{array}$$

is a *weak pullback* if for all $x_1 \in X_1, x_2 \in X_2$ such that $q_1x_1 = q_2x_2$ there is $x \in X_0$ such that

$$x_1 = p_1x \text{ and } x_2 = p_2x.$$

Proposition 6.2 *If Φ preserves weak pullbacks then any congruence on a coalgebra is a bisimulation.*

Proof: Let R be a congruence on the coalgebra (A, α). Let $q : A \to A_R$ be a quotient of A with respect to R. Then

$$\begin{array}{ccc} R & \xrightarrow{\rho_1} & A\\ \rho_2\downarrow & & \downarrow q\\ A & \xrightarrow{q} & A_R\end{array}$$

is a weak pullback. So if Φ preserves weak pullbacks then

$$\begin{array}{ccc} \Phi R & \xrightarrow{\Phi\rho_1} & \Phi A\\ \Phi\rho_2\downarrow & & \downarrow \Phi q\\ \Phi A & \xrightarrow{\Phi q} & \Phi A_R\end{array}$$

is a weak pullback. If xRy then $xR^{\Phi}y$ so that $\alpha x, \alpha y \in \Phi A$ are such that $\Phi q(\alpha x) = \Phi q(\alpha y)$. By the weak pullback property there is $u \in \Phi R$ such that $\alpha x = \Phi\rho_1 u$ and $\alpha y = \Phi\rho_2 u$. By the global axiom of choice we may define $\beta : R \to \Phi R$ so that if xRy then $\beta(x, y)$ is such a $u \in \Phi R$. It follows that ρ_1, ρ_2 are homomorphisms $(R, \beta) \to (A, \alpha)$ and R is a bisimulation. □

The naturally occurring functors on the category of classes are generally standard. Moreover they usually also preserve weak pullbacks. But not always. Here is a simple example of a standard functor that does not preserve weak pullbacks. For each class X let

$$\Phi X = \{(x, y, z) \in X^3 \mid card(\{x, y, z\}) < 3\}.$$

If $f : X \to Y$ then let $\Phi f : \Phi X \to \Phi Y$ be given by

$$\Phi f(x, y, z) = (fx, fy, fz)$$

for all $(x, y, z) \in \Phi X$. This functor Φ is easily seen to be standard. But consider the diagram

$$\begin{array}{ccc} A \times A & \xrightarrow{p_1} & A \\ p_2 \downarrow & & \downarrow q \\ A & \xrightarrow{q} & 1 \end{array}$$

where $A = \{0,1\}, 1 = \{0\}$ and p_1, p_2 are the two projections on A. This is a pullback and hence a weak pullback. Suppose that the diagram

$$\begin{array}{ccc} \Phi(A \times A) & \xrightarrow{\Phi p_1} & \Phi A \\ \Phi p_2 \downarrow & & \downarrow \Phi q \\ \Phi A & \xrightarrow{\Phi q} & \Phi 1 \end{array}$$

were a weak pullback. As $(0,0,1), (0,1,1) \in \Phi A$ with $\Phi q(0,0,1) = (0,0,0) = \Phi q(0,1,1)$ there should be $u \in \Phi(A \times A)$ such that $\Phi p_1 u = (0,0,1)$ and $\Phi p_2 u = (0,1,1)$. But this means that $u = ((0,0),(0,1),(1,1))$ which is not in $\Phi(A \times A)$ after all. Hence Φ cannot preserve weak pullbacks.

We can also use this example to see that congruences on a coalgebra are not always bisimulations. This shows that the weak pullback assumption cannot be left out of 6.2. Let $\alpha : A \to \Phi A$ be given by

$$\alpha 0 = (0,0,1),$$
$$\alpha 1 = (0,1,1).$$

Then (A, α) is a coalgebra. Note that the complete equivalence relation on A, $R = A \times A$, is the maximal congruence on (A, α), as $R^{\Phi} = R$. But R is not a bisimulation. For suppose that $\beta : R \to \Phi R$ such that ρ_1, ρ_2 are homomorphisms $(R, \beta) \to (A, \alpha)$. Then if $\beta(0,1) = ((x_1, x_2), (y_1, y_2), (z_1, z_2)) \in \Phi R$ then

$$(x_1, y_1, z_1) = (\Phi \rho_1)(\beta(0,1)) = \alpha(\rho_1(0,1)) = \alpha 0 = (0,0,1)$$

and

$$(x_2, y_2, z_2) = (\Phi \rho_2)(\beta(0,1)) = \alpha(\rho_2(0,1)) = \alpha 1 = (0,1,1).$$

So $\beta(0,1) = ((0,0),(0,1),(1,1))$ which cannot be in ΦR after all.

7 A General Final Coalgebra Theorem

In this section we give definitions and state results leading to a general final coalgebra theorem. The proofs of these results will appear elsewhere.

Given a full subcategory $\mathcal{S}$ of a category $\mathcal{C}$, we will define the notion of an *$\mathcal{S}$-based functor* $\Phi : \mathcal{C} \to \mathcal{C}$ and give sufficient conditions on $\mathcal{S}$ and $\mathcal{C}$ for every $\mathcal{S}$-based functor to have a final coalgebra. When $\mathcal{C}$ is the category of classes and $\mathcal{S}$ is the category of sets then the sufficient conditions hold and the notion of an $\mathcal{S}$-based functor coincides with the notion of a set-based functor.

So fix a full subcategory $\mathcal{S}$ of a category $\mathcal{C}$. For each object A of $\mathcal{C}$ let $\mathcal{S}(A)$ ($\mathcal{S}_m(A)$) be the full subcategory of the slice category $\mathcal{C}/A$ whose objects are the maps (monic maps) $f : B \to A$ of $\mathcal{C}$ such that B is in $\mathcal{S}$. We call $\mathcal{S}$ a *special subcategory* of $\mathcal{C}$ if for each object A of $\mathcal{C}$,

1. Every $f \in \mathcal{S}(A)$ has a factorization $f = gk$ with $g \in \mathcal{S}_m(A)$.

2. A is a "colimit of $\mathcal{S}_m(A)$;" i.e., (A, η^A) is a colimit cocone of the functor $H^A : \mathcal{S}_m(A) \to \mathcal{C}$.

Here if $f : s \to A$ is an object of $\mathcal{S}_m(A)$ then $H^A(f) = s$ and if $k : f \to g$ in $\mathcal{S}_m(A)$, where $f : s \to A$, $g : t \to A$ in $\mathcal{C}$, then $H^A(k) = k : s \to t$. Also (A, η^A) is the cocone of H^A, where for each object f of $\mathcal{S}_m(A)$,

$$\eta^A_f = f : H^A(f) \to A.$$

Lemma 7.1 *If $\mathcal{S}$ is a special subcategory of $\mathcal{C}$ then a colimit of the inclusion functor $\mathcal{S} \hookrightarrow \mathcal{C}$ is a final object of $\mathcal{C}$.*

Now fix a functor $\Phi : \mathcal{C} \to \mathcal{C}$, let $\mathcal{C}'$ be the category of coalgebras of Φ and let $\mathcal{S}'$ be the full subcategory of $\mathcal{C}'$ consisting of those coalgebras (A, α) of Φ with A in $\mathcal{S}$. Our aim is to apply lemma 7.1 to the inclusion functor $\mathcal{S}' \hookrightarrow \mathcal{C}'$. To that end we need to impose conditions on $\mathcal{S}$, $\mathcal{C}$ and Φ to insure that $\mathcal{S}'$ is a special subcategory of $\mathcal{C}'$ and that a colimit of $\mathcal{S}' \hookrightarrow \mathcal{C}'$ exists.

The next lemma will give conditions for $\mathcal{S}'$ to be special. Call a monic map *trivial* if its domain A is an initial object such that every map $A' \to A$ is an isomorphism. Φ *almost preserves monics* if Φf is monic for every non-trivial monic f. Φ is *$\mathcal{S}$-based* if Φ almost preserves monics and for every object A of $\mathcal{C}$ and every f in $\mathcal{S}(\Phi A)$ there is a g in $\mathcal{S}_m(A)$ such that $f = (\Phi g)k$ for some k.

For each object A of $\mathcal{C}$ the collection of objects of $\mathcal{S}_m(A)$ is pre-ordered by the relation $\leq$, where for f, g in $\mathcal{S}_m(A)$,

$$f \leq g \Leftrightarrow f = gk \text{ for some } k.$$

A special subcategory $\mathcal{S}$ of $\mathcal{C}$ is *strongly special* if for each object A of $\mathcal{C}$:

1. $\mathcal{S}_m(A)$ is directed by $\leq$; i.e., for all g_1, g_2 in $\mathcal{S}_m(A)$ there is an h in $\mathcal{S}_m(A)$ such that $g_1, g_2 \leq h$.
2. $\mathcal{S}_m(A)$ is closed under countable unions; i.e., if f_n is in $\mathcal{S}_m(A)$ for $n \in \omega$, then there is an f in $\mathcal{S}_m(A)$ such that
   (a) $f_n \leq f$ for $n \in \omega$.
   (b) If $g : A \to B$, then for all monic $h : Y \to B$ if $gf_n \leq h$ for all $n \in \omega$, then $gf \leq h$.

Lemma 7.2 *If $\mathcal{S}$ is a strongly special subcategory of $\mathcal{C}$ and Φ is an $\mathcal{S}$-based functor on $\mathcal{C}$ then $\mathcal{S}'$ is a special subcategory of $\mathcal{C}'$.*

We now turn to the formulation of sufficient conditions for the inclusion functor $\mathcal{S}' \hookrightarrow \mathcal{C}'$ to have a colimit. This will involve some supercardinality conditions on superlarge categories. Call a collection *class-sized* if it can be put in the form $\{A_i | i \in I\}$ where I is a class. Call a category class-sized if its collection of maps is class-sized. A category is *class-sized cocomplete* if every functor from a class-sized category into the category has a colimit.

Lemma 7.3 *Suppose that*

1. $\mathcal{C}$ is a category that is class-sized cocomplete.

2. $\mathcal{S}$ is a full subcategory of $\mathcal{C}$ that is class-sized and such that $Hom_{\mathcal{C}}(s, A)$ is class-sized for all objects s of $\mathcal{S}$ and all objects A of $\mathcal{C}$.

Then for every functor $\Phi : \mathcal{C} \to \mathcal{C}$ the inclusion functor $\mathcal{S}' \hookrightarrow \mathcal{C}'$ has a colimit.

We can now state the result we have been aiming at.

Theorem 7.4 (The General Final Coalgebra Theorem) *Suppose that $\mathcal{C}$ and $\mathcal{S}$ satisfy 1. and 2. of lemma 7.3 and also $\mathcal{S}$ is a strongly special subcategory of $\mathcal{C}$. Then every $\mathcal{S}$-based functor on $\mathcal{C}$ has a final coalgebra.*

This is an immediate consequence of the three lemmas.

References

[NWFS] Aczel, P. 1988. Non-Well-Founded Sets, CSLI Lecture Notes, Number 14, Stanford University.

[SCCS] Milner, R. 1983. Calculi for Synchrony and Asynchrony. *Theoretical Computer Science* 25:267-310.

Vol. 352: J. Díaz, F. Orejas (Eds.), TAPSOFT '89. Volume 2. Proceedings, 1989. X, 389 pages. 1989.

Vol. 354: J.W. de Bakker, W.-P. de Roever, G. Rozenberg (Eds.), Linear Time, Branching Time and Partial Order in Logics and Models for Concurrency. VIII, 713 pages. 1989.

Vol. 355: N. Dershowitz (Ed.), Rewriting Techniques and Applications. Proceedings, 1989. VII, 579 pages. 1989.

Vol. 356: L. Huguet, A. Poli (Eds.), Applied Algebra, Algebraic Algorithms and Error-Correcting Codes. Proceedings, 1987. VI, 417 pages. 1989.

Vol. 357: T. Mora (Ed.), Applied Algebra, Algebraic Algorithms and Error-Correcting Codes. Proceedings, 1988. IX, 481 pages. 1989.

Vol. 358: P. Gianni (Ed.), Symbolic and Algebraic Computation. Proceedings, 1988. XI, 545 pages. 1989.

Vol. 359: D. Gawlick, M. Haynie, A. Reuter (Eds.), High Performance Transaction Systems. Proceedings, 1987. XII, 329 pages. 1989.

Vol. 360: H. Maurer (Ed.), Computer Assisted Learning – ICCAL '89. Proceedings, 1989. VII, 642 pages. 1989.

Vol. 361: S. Abiteboul, P.C. Fischer, H.-J. Schek (Eds.), Nested Relations and Complex Objects in Databases. VI, 323 pages. 1989.

Vol. 362: B. Lisper, Synthesizing Synchronous Systems by Static Scheduling in Space-Time. VI, 263 pages. 1989.

Vol. 363: A.R. Meyer, M.A. Taitslin (Eds.), Logic at Botik '89. Proceedings, 1989. X, 289 pages. 1989.

Vol. 364: J. Demetrovics, B. Thalheim (Eds.), MFDBS 89. Proceedings, 1989. VI, 428 pages. 1989.

Vol. 365: E. Odijk, M. Rem, J.-C. Syre (Eds.), PARLE '89. Parallel Architectures and Languages Europe. Volume I. Proceedings, 1989. XIII, 478 pages. 1989.

Vol. 366: E. Odijk, M. Rem, J.-C. Syre (Eds.), PARLE '89. Parallel Architectures and Languages Europe. Volume II. Proceedings, 1989. XIII, 442 pages. 1989.

Vol. 367: W. Litwin, H.-J. Schek (Eds.), Foundations of Data Organization and Algorithms. Proceedings, 1989. VIII, 531 pages. 1989.

Vol. 368: H. Boral, P. Faudemay (Eds.), IWDM '89, Database Machines. Proceedings, 1989. VI, 387 pages. 1989.

Vol. 369: D. Taubner, Finite Representations of CCS and TCSP Programs by Automata and Petri Nets. X, 168 pages. 1989.

Vol. 370: Ch. Meinel, Modified Branching Programs and Their Computational Power. VI, 132 pages. 1989.

Vol. 371: D. Hammer (Ed.), Compiler Compilers and High Speed Compilation. Proceedings, 1988. VI, 242 pages. 1989.

Vol. 372: G. Ausiello, M. Dezani-Ciancaglini, S. Ronchi Della Roccha (Eds.), Automata, Languages and Programming. Proceedings, 1989. XI, 788 pages. 1989.

Vol. 373: T. Theoharis, Algorithms for Parallel Polygon Rendering. VIII, 147 pages. 1989.

Vol. 374: K.A. Robbins, S. Robbins, The Cray X-MP/Model 24. VI, 165 pages. 1989.

Vol. 375: J.L.A. van de Snepscheut (Ed.), Mathematics of Program Construction. Proceedings, 1989. VI, 421 pages. 1989.

Vol. 376: N.E. Gibbs (Ed.), Software Engineering Education. Proceedings, 1989. VII, 312 pages. 1989.

Vol. 377: M. Gross, D. Perrin (Eds.), Electronic Dictionaries and Automata in Computational Linguistics. Proceedings, 1987. V, 110 pages. 1989.

Vol. 378: J.H. Davenport (Ed.), EUROCAL '87. Proceedings, 1987. VIII, 499 pages. 1989.

Vol. 379: A. Kreczmar, G. Mirkowska (Eds.), Mathematical Foundations of Computer Science 1989. Proceedings, 1989. VIII, 605 pages. 1989.

Vol. 380: J. Csirik, J. Demetrovics, F. Gécseg (Eds.), Fundamentals of Computation Theory. Proceedings, 1989. XI, 493 pages. 1989.

Vol. 381: J. Dassow, J. Kelemen (Eds.), Machines, Languages, and Complexity. Proceedings, 1988. VI, 244 pages. 1989.

Vol. 382: F. Dehne, J.-R. Sack, N. Santoro (Eds.), Algorithms and Data Structures. WADS '89. Proceedings, 1989. IX, 592 pages. 1989.

Vol. 383: K. Furukawa, H. Tanaka, T. Fujisaki (Eds.), Logic Programming '88. Proceedings, 1988. VII, 251 pages. 1989. (Subseries LNAI)

Vol. 384: G.A. van Zee, J.G.G. van de Vorst (Eds.), Parallel Computing 1988. Proceedings, 1988. V, 135 pages. 1989.

Vol. 385: E. Börger, H. Kleine Büning, M.M. Richter (Eds.), CSL '88. Proceedings, 1988. VI, 399 pages. 1989.

Vol. 387: C. Ghezzi, J.A. McDermid (Eds.), ESEC '89. 2nd European Software Engineering Conference. Proceedings, 1989. VI, 496 pages. 1989.

Vol. 388: G. Cohen, J. Wolfmann (Eds.), Coding Theory and Applications. Proceedings, 1988. IX, 329 pages. 1989.

Vol. 389: D.H. Pitt, D.E. Rydeheard, P. Dybjer, A.M. Pitts, A. Poigné (Eds.), Category Theory and Computer Science. Proceedings, 1989. VI, 365 pages. 1989.

This series reports new developments in computer science research and teaching – quickly, informally and at a high level. The type of material considered for publication includes preliminary drafts of original papers and monographs, technical reports of high quality and broad interest, advanced level lectures, reports of meetings, provided they are of exceptional interest and focused on a single topic. The timeliness of a manuscript is more important than its form which may be unfinished or tentative. If possible, a subject index should be included. Publication of Lecture Notes is intended as a service to the international computer science community, in that a commercial publisher, Springer-Verlag, can offer a wide distribution of documents which would otherwise have a restricted readership. Once published and copyrighted, they can be documented in the scientific literature.

Manuscripts

Manuscripts should be no less than 100 and preferably no more than 500 pages in length.
They are reproduced by a photographic process and therefore must be typed with extreme care. Symbols not on the typewriter should be inserted by hand in indelible black ink. Corrections to the typescript should be made by pasting in the new text or painting out errors with white correction fluid. Authors receive 75 free copies and are free to use the material in other publications. The typescript is reduced slightly in size during reproduction; best results will not be obtained unless the text on any one page is kept within the overall limit of 18 x 26.5 cm (7 x 10½ inches). On request, the publisher will supply special paper with the typing area outlined.
Manuscripts should be sent to Prof. G. Goos, GMD Forschungsstelle an der Universität Karlsruhe, Haid- und Neu-Str. 7, 7500 Karlsruhe 1, Germany, Prof. J. Hartmanis, Cornell University, Dept. of Computer Science, Ithaca NY/USA 14850, or directly to Springer-Verlag Heidelberg.

Springer-Verlag, Heidelberger Platz 3, D-1000 Berlin 33
Springer-Verlag, Tiergartenstraße 17, D-6900 Heidelberg 1
Springer-Verlag, 175 Fifth Avenue, New York, NY 10010/USA
Springer-Verlag, 37-3, Hongo 3-chome, Bunkyo-ku, Tokyo 113, Japan

ISBN 3-540-51662-X
ISBN 0-387-51662-X